0° 20° 40° 60° 80° 100° 120° 140° 160° 180° 80°

70°

Ft

E

60°

E F

Dcb BSk 50°

E F D

Dcb

Dcb D

BWk BSk Dca

BWk Dca 40°

H Cf C

Dca

Cs Cf

BW Cf 30°

BS BWh

BWh BSh Cw Cf

BWh Ar Aw Ar 20°

BWh Aw Ar A

Bn Aw 10°

BW

Ar

EQUATOR 0°

A

Ar

Aw 10°

A

TROPIC OF CAPRICORN

BShw Aw 20°

BWh

BWh

BShs Cf

BS Cs Cs 30°

Cf Do

C Do 40°

D 50°

F 60°

70°

(Ice) Fi

20° 40° 60° 80° 100° 120° 140° 160° 180°

FLAT POLAR QUARTIC
EQUAL-AREA PROJECTION
ADAPTED BY ARTHUR H. ROBINSON

ELEMENTS OF GEOGRAPHY by G. T. Trewartha, A. H. Robinson,
and E. H. Hammond. © McGraw-Hill Book Co., N.Y., 1967.

AN INTRODUCTION TO CLIMATE

McGRAW-HILL SERIES IN GEOGRAPHY

John C. Weaver, *Consulting Editor*

Vernor C. Finch was Consulting Editor of this series from its inception in 1934 to 1951.

FOURTH EDITION

AN INTRODUCTION TO CLIMATE

GLENN T. TREWARTHA
Emeritus Professor of Geography
University of Wisconsin

Cartography by Randall D. Sale

McGRAW-HILL BOOK COMPANY
New York   St. Louis   San Francisco   Toronto   London   Sydney

PREFACE

The present book has three forebears. It probably should have more, since fourteen years have elapsed since the last edition appeared. In that time the added years of experience in teaching the geography of climate to college students and a greatly expanded body of materials in atmospheric science have provided the author with new ideas relative to the selection and organization of the materials to be incorporated in this fourth edition. It is no wonder that the book has had to be largely rewritten. Significant modifications of earlier editions include the following:

1. Much greater emphasis has been placed upon dynamic processes as genetic factors in climatic origins. Consequently Part One on the elements and controls of climate has been generally expanded, with particular attention to the materials on solar radiation and the earth's energy budget.

2. As indicated by the map inside the front cover, the classification of climates is new in many important respects, although it still has some resemblance to the modified Köppen scheme of the third edition. Teachers who still wish to use Köppen's classification will find it illustrated on the map following page 393 and described in the Appendix.

3. In Part Two on the world pattern of climates, synoptic weather types have been given more emphasis as ingredients of climatic description and interpretation. Climate is inseparably linked to the dynamic element of weather.

4. The appendix providing supplementary data for selected stations has been eliminated, and additional station data have been put into the body of the text. The sections in Part Two dealing with resource potentials of individual climatic realms have also been eliminated as belonging more properly to other courses.

5. Part Two's descriptions of each climatic type have been rounded out with brief analyses of certain individual climatic regions within the type on the various continents. The concept of "problem," or anomalous, climates is introduced in connection with regions which show significant departures from their general type.

6. Just as the text has been revised and modernized, so most of the old illustrations have been modified or redrawn, and scores of new ones have been added.

However, the overall two-part organization of *Introduction to Climate* remains the same as in previous editions. Part One, which deals with the individual elements and controls of climate, focuses on distribution and interrelation, both spatial and temporal. It describes world distribution patterns of the main climatic elements—solar energy, temperature, precipitation, and atmospheric circulations—and shows how their arrangements are conditioned by the climatic controls.

This emphasis on distribution reflects the book's design for what is mainly a geographic audience. The author regards descriptive analysis alone as inadequate, so that enough background on atmospheric processes has been provided to make the distribution patterns intelligible. For example, material on jet streams, mid-troposphere long waves, and general circulation is introduced mainly because these topics contribute to an understanding of atmospheric disturbances and fronts, and hence of temperature and precipitation distribution patterns. It is the author's conviction that a climatology which omits or seriously slights genesis and explanation is not only dull but also inadequate for the needs of geographers.

Part Two opens with a brief analysis of climatic classification and of the classificatory scheme used in this book. After introducing the concept of climatic regions, types, and groups deriving from a synthesis of the climatic elements, it describes the world pattern of climates based on these groups and types. Climatic genesis is emphasized by focusing on the characteristic locations of types and groups of climate with respect to latitude, wind and pressure systems, and air masses. On the other hand, precise definitions and the exact locations of climatic boundaries between types are deemphasized. As the text points out, these should be thought of an approximations at best. Description, distribution, and explanation of core climates form the basic approach.

Sections devoted to applied climatology, or the influence of climate on life forms, including humans, do not appear in the book because the author believes that the place for them is elsewhere. Also omitted are such topics as ancient climates and microclimates (those on the smallest scales). It is the purpose here to concentrate on climates having large dimensions—usually those on a world or regional scale.

Since the text has been kept relatively short so that it will provide a concise introduction to the geography of climate, some instructors and students will want more material on various topics than is given. For their convenience, supplementary readings and references have been listed for each of the two parts and for each chapter. With this framework a more comprehensive or a more advanced treatment of selected topics can be provided.

There is no long list of acknowledgments here, because the author is indebted to so many people that an inclusive list would be tedious to the point of defeating the purpose of gratitude. Prominent in any list of benefactors would certainly be my numerous students, who have also been my teachers. Special recognition should also be given to Professors Lyle H. Horn (meteorology) and Michael E. Sabbagh (geography) at the University of Wisconsin, and to Professor Wesley Calef (geography) at the University of Chicago, who have read parts or all of the manuscript and made invaluable improvements in it.

GLENN T. TREWARTHA

CONTENTS

THE EARTH'S OCEAN OF AIR

The planet earth is composed of a solid-liquid inner sphere whose radius is about 4,000 miles. This core is surrounded by an air ocean which, at its extreme limits, is several times as deep. Since the atmosphere is just as much a part of the planet as the solid-liquid core, the actual surface of the earth is the outer limit of the atmosphere. So life—including human life—dwells deep within the planet earth, not on its surface. It is only because air is so transparent, and so nearly weightless when compared with land or water, that the land-water surface is ordinarily referred to as the earth's surface. As a bottom-crawling organism in the densest part of the air ocean, man is of course greatly influenced by his atmospheric environment.

It is the air ocean that maintains the planet's habitable range of temperature and moisture. Without its fluid energy-distribution properties, the high latitudes would be lifeless because of unbearable cold, the tropics because of inconceivable heat. Through certain of its chemical and physical properties, the air also regulates the amount and kinds of energy that reach the land-water surface. Ozone in the high atmosphere shields terrestrial life from the sun's lethal ultraviolet radiation. The whole atmosphere above about 7 miles protects the surface from bombardment by highly charged particle radiation from the sun. Water vapor and carbon dioxide, which occur mainly in the lower atmosphere, act to retard the escape of the converted solar energy and thereby ensure adequate warmth at the surface.

The science of climate.   To laymen and specialists alike, climate relates to the air ocean and in this respect is one branch of atmospheric science. But climate is also a part of physical geography, which studies the physical earth as the home of man.

Climate is a layman's as well as a scientist's term, although scientists have given it various shades of meaning and have subdivided it into a number of specialized fields, such as dynamic climatology, synoptic climatology, and regional climatology. These need not be defined here; nor will even a comprehensive definition of climate itself be given in this introductory book. Instead, the following paragraphs will simply present some of the attributes of climate.

Climate and weather have much in common, yet they are not identical. The weather of any place is the sum total of the atmospheric variables for a brief period of time. Thus we speak of today's weather, or of last week's. Climate, on the other hand, refers to a more enduring regime of the atmosphere. It represents a composite of the day-to-day weather conditions, and of the atmospheric elements, for a long period of time. It is more than "average weather," for no adequate concept of climate is possible without an appreciation of seasonal and diurnal change and of the constant weather variations which occur. While the emphasis in a study of climate may be given to the norm, departures, variations, and extremes are also important.

Climate always applies to atmospheric conditions for a given period of time—a month, season, year. But not for one January, or one summer, or a single year. A sufficient number of these calendar units must be involved to provide a true composite.

Furthermore, climate is not an abstraction; it applies to places and regions. We speak of the climate of Chicago, the climates of Japan and of the American Great Plains. This place attribute of climate makes it highly geographic. And since much

geography has an anthropocentric quality, a geography of climate tends to emphasize those features and elements which have resource potential for man. Moreover, just as the science of geography in general emphasizes distribution, arrangement, and pattern, so also does the science of climatology.

Unfortunately, or so this writer believes, too much climatology has been largely descriptive, with little attempt to explain the phenomena described. But like any other science, climatology is potentially both descriptive and explanatory, the only limitations being the writer's competence and the adequacy of the data.

For its basic information, climatology depends mainly upon hourly and daily observations made at official weather stations, which form a network varying in density over the earth. The weather records of these stations are organized by calendar units—days, months, seasons, and years—based upon periodic sun control. Consequently it is only natural that climatic description has also been organized around these calendar units, which are based on the sun's control of the weather and hence are periodic (cyclical) in character.

Nevertheless, there is another important weather unit in addition to the periodic one reflecting sun control. This is the nonperiodic weather episode which is associated with a particular temporary weather disturbance or atmospheric circulation pattern. Such episodes are of various scales of size and duration, but they all have the attribute that they do not fit the solar-controlled calendar units. They may cover part of a day, several days, or even several weeks, but their beginnings and endings have little or nothing to do with day or night, noon or midnight. Hence the observations and records at official weather stations provide no summaries of data in terms of these nonperiodic units, or weather types. In the statistics, their effects are blurred by daily, weekly, and monthly summaries that cut across the periods in which individual episodes were in control of the weather.

Yet the nonperiodic weather element cannot be neglected in climatic analysis. It operates over the whole earth in varying degrees of intensity; in some regions and in some seasons it is more dominant than sun control. Unfortunately science cannot at present provide a comprehensive quantitative climatology of weather types. Much remains to be discovered about the kinds and distribution of the atmospheric disturbances which affect weather in different parts of the world, and likewise about the quantitative distribution of the weather elements (temperature, rainfall, etc.) associated with these disturbances. Moreover, every classification of weather types is open to question, since there is no single parameter which delineates the totality of weather conditions within a disturbance. As a consequence, statistical averages of the individual climatic elements are still used in this book as an important method of climatic description. On the other hand, they are supplemented wherever possible by information about the various kinds of disturbance types or weather situations, the composites of which comprise a climate. *For climate cannot be divorced from the dynamics of weather.*

Elements of weather and climate. As noted earlier, weather and climate are not identical. Yet the nature of both is expressed by the same elements in combination. Primarily these elements are solar energy, temperature, humidity-precipitation, and to a lesser degree, winds. These are called the *elements* of weather and climate; they are the variables out of which the many weather and climatic types are compounded.

Controls of weather and climate. Weather varies from day to day, and climate differs from region to region, because of variations in the amount, intensity, and distribution over the earth of the weather and climatic elements—particularly solar energy, temperature, and precipitation. What causes these climatic elements to vary temporally and regionally, so that some places and some seasons are hot and others are cold, some are wet and others dry? The answer is found in the operation of the climatic *controls.* Each of the climatic elements named in the previous section—solar energy, temperature, precipitation-humidity, and winds—also functions as

a climatic control and influences each of the other elements. Other important climatic controls are altitude, the distribution of land and water, mountain barriers, the great semipermanent high- and low-pressure cells, air masses, atmospheric disturbances of various kinds, and ocean currents. It is the controls, acting in different combinations and with different intensities, that produce the variations in temperature and precipitation which in turn give rise to the changing patterns of weather and climate.

Although it is climates and their world distribution that are of chief interest to geographers, some knowledge of the individual climatic elements and controls is essential to an understanding of world climates. In this book Part One, dealing with solar energy, temperature, winds and pressure, moisture and precipitation, air masses, and atmospheric disturbances, provides essential background for Part Two's discussion of the origin of the various types of climate and their patterned arrangement on the earth.

The nature of the atmosphere. Although the earth's atmosphere is a mixture of a number of gases, nitrogen (78 percent) and oxygen (21 percent) make up 99 percent of the total volume of dry air. What little remains is composed of argon, carbon dioxide, neon, krypton, ozone, and a number of other gases. The relative percentages of these permanent gases remain almost constant horizontally over the earth and also at different altitudes. But ordinary surface air differs somewhat from this description, for in addition to the permanent gases, it contains variable amounts of water vapor and many organic and inorganic impurities, including dust, and condensed water and ice particles. Although the amount of water vapor in the air varies greatly from time to time and place to place, it rarely exceeds 3 percent by volume, even in the humid tropics. Yet as far as climate is concerned, it is the paramount ingredient of the lower atmosphere, as will be explained later.

The nongaseous substances, collectively known as dust, held in suspension chiefly in the lower atmosphere, are essentially pollutants. Some have hygroscopic properties and so provide the nuclei around which atmospheric condensation takes place. In many localities, particularly in the vicinity of large cities and industrial districts, air pollution has become a grave and rapidly worsening problem. Its recent increase is the result of man's burgeoning consumption of the fossil fuels—coal, petroleum, and gas. A major source of these atmospheric impurities is factory smokestacks. The chief offender, however, is the exhaust fumes from the swelling tide of motor vehicles. Every day in the United States, 350,000 tons of pollutants are exhausted into the atmosphere by 90 million cars—probably more than that released by all other pollutant sources combined. Poisoning of the air over cities has become a serious menace to the health and comfort of urban populations.

The main characteristics of any gas, including the atmosphere, are its extreme mobility and its unusual capacity for expansion and compression. Gas has far greater fluidity and mobility than water. Air may appear weightless and insubstantial, but everyone knows how much force is exerted by a strong wind, and how much resistance is offered even by quiet air to an object moving rapidly through it.

Within the *troposphere*, which is approximately the lowest 6 to 10 miles of the air ocean, both air temperature and air density decline with altitude. The density decrease continues indefinitely, but the decrease in temperature does not. Since the main heat source for the lower atmosphere is the sun-heated land-water surface, it is understandable why normally the air is colder with increasing distance from that surface. But this decline stops at about 30,000 to 50,000 ft, at what is called the *tropopause*. The tropopause represents a boundary separating the troposphere, or mixing layer, below, where vertical air movement is widespread, from the *stratosphere*, or stratified layer, above, where rapid up and down movement is largely absent. In the stratosphere at a height between about 15 miles (24 km) and 31 miles (50 km), where the atmosphere's small amount of ozone is concentrated, temperature rises with altitude. This is because a small fraction of the incoming solar radiation is almost completely absorbed at the top of the ozone layer. Thus at an altitude of about 31 miles, air temperature reaches a second maximum—one which is nearly as high as that at the surface. Above this high-level maximum, the temperature declines up to about 50 to 55 miles, then rises rapidly again.

It is the lower 6 to 10 miles of atmosphere, the troposphere, which is the turbulent, convective layer. In it are concentrated a very large percentage of the atmosphere's dust, water vapor, and cloud, as well as most of the weather phenomena. The troposphere has several subdivisions. There is the extremely shallow laminar boundary layer which is in immediate contact with the earth's land-sea surface. Since it is so shallow that it lies below the level of weather-station instruments, information concerning this layer must be obtained largely from non-official sources. Only in the lowest surface layer is conduction, or molecular activity, an important factor in the vertical energy transfer between the land-sea surface and the air. In the laminar surface layer, vertical changes in temperature and moisture are exceedingly rapid. Above it is the so-called friction layer, which is about 1,000 meters (3,000 ft) deep. Here the vertical heat transfer is chiefly accomplished by turbulence or eddy motion. Here also, changes in temperature within the daily period are large. Above the friction layer is the free atmosphere, where winds are stronger and vertical energy transfer is carried out mainly through the formation of clouds.

GENERAL REFERENCES FOR PART ONE

Alissow, B. P., O. A. Drosdow and E. S. Rubinstein, *Lehrbuch der Klimatologie,* VEB Deutscher Verlag der Wissenschaften, Berlin, 1956.

Aviation Weather, U.S. Federal Aviation Agency and U.S. Weather Bureau, Washington, D.C., 1965.

Blair, Thomas A., and Robert C. Fite, *Weather Elements,* 5th ed., Prentice-Hall, Inc., Englewood Cliffs, N.J., 1965.

Blumenstock, David I., *The Ocean of Air,* Rutgers University Press, New Brunswick, N.J., 1959.

Blüthgen, Joachim, *Allgemeine Klimageographie,* Walter de Gruyter and Co., Berlin, 1964; 2d ed., 1966.

Brooks, C. E. P., *Climate through the Ages,* revised ed., McGraw-Hill Book Company, New York, 1949.

Brooks, Charles Franklin, *Why the Weather?* Harcourt, Brace & World, Inc., New York, 1935.

Byers, Horace Robert, *General Meteorology,* 3d ed., McGraw-Hill Book Company, New York, 1959.

Climate and Man, Yearbook of Agriculture, 1941, U.S. Department of Agriculture, Washington, D.C. See particularly Parts 1 and 4.

Compendium of Meteorology, Thomas F. Malone (ed.), American Meteorological Society, 1951.

Conrad, V., *Fundamentals of Physical Climatology,* Harvard University, Blue Hill Meteorological Observatory, Milton, Mass., 1942.

————, and L. W. Pollak, *Methods in Climatology,* 2d ed., Harvard University Press, Cambridge, Mass., 1950.

Court, Arnold, "Climatology: Complex, Dynamic and Synoptic," *Ann. Assoc. Amer. Geographers,* Vol. 47, no. 2, pp. 125–136, 1957.

Critchfield, Howard J., *General Climatology,* 2d ed., Prentice-Hall, Inc., Englewood Cliffs, N.J., 1966.

Garbell, Maurice A., *Tropical and Equatorial Meteorology,* Pitman Publishing Corporation, New York, 1947.

Geiger, Rudolph, *The Climate near the Ground,* Harvard University Press, Cambridge, Mass., 1965.

Gentilli, J., *A Geography of Climate,* The University of Western Australia Press, Perth, 1958.

Griffiths, John F., *Applied Climatology: An Introduction,* Oxford University Press, Fair Lawn, N.J., 1966.

Hare, F. K., *The Restless Atmosphere,* Harper & Row, Publishers, Incorporated, New York, 1963.

Haurwitz, Bernard, and James M. Austin, *Climatology,* McGraw-Hill Book Company, New York, 1944.

Heyer, Ernst, *Witterung und Klima,* B. G. Teubner Verlagsgesellschaft, Leipzig, 1963.

Kendrew, W. G., *Climatology,* 2d ed., Oxford University Press, Fair Lawn, N.J., 1957.

Koeppe, Clarence E., and George C. De Long, *Weather and Climate,* McGraw-Hill Book Company, New York, 1958.

Landsberg, Helmut, *Physical Climatology,* 2d ed., Gray Printing Company, Inc., Du Bois, Pa., 1958.

Leighly, John, "Climatology since the Year 1800," *Trans. Amer. Geophys. Union,* Vol. 30, no. 5, pp. 658–672, 1949.

Miller, A. Austin, *Climatology,* 3d ed., E. P. Dutton & Co., Inc., New York, 1953.

Pédelaborde, Pierre, *The Monsoon* (translated by M. J. Clegg), Methuen & Co., Ltd., London, 1963.

Péguy, Charles P., *Précis de climatologie,* Masson et Cie, Paris, 1961.

Petterssen, Sverre, *Introduction to Meteorology,* 2d ed., McGraw-Hill Book Company, New York, 1958.

Reiter, Elmar R., *Jet-stream Meteorology,* The University of Chicago Press, Chicago, 1963.

Reiter, Elmar R., *Jet Streams,* Doubleday and Company, Inc., Garden City, New York, 1967.

Riehl, Herbert, *Introduction to the Atmosphere,* McGraw-Hill Book Company, New York, 1965.

————, *Tropical Meteorology,* McGraw-Hill Book Company, New York, 1954.

Schwarzbach, Martin, *Climates of the Past: An Introduction to Paleoclimatology,* D. Van Nostrand Company, Ltd., London, 1963.

Sellers, William D., *Physical Climatology,* The University of Chicago Press, Chicago, 1965.

Symposium on Tropical Meteorology, Rotorua, New Zealand, 1963, J. W. Hutchings (ed.), New Zealand Meteorological Service, Wellington, 1964.

Watts, I. E. M., *Equatorial Weather,* University of London Press, Ltd., London, 1955.

Willett, Hurd C., and Frederick Sanders, *Descriptive Meteorology,* 2d ed., Academic Press Inc., New York, 1959.

PART ONE

THE ELEMENTS AND
CONTROLS OF CLIMATE

CHAPTER ONE

SOLAR RADIATION

Energy of the atmosphere.   Radiant energy from the sun provides 99.97 percent of the total energy of the atmosphere. It is also the prime source of energy for terrestrial life, as well as for the natural processes which go on in the oceans and in the upper layers of the solid earth. Temporal and regional differences in solar radiation give rise to weather and climate. Solar radiation which is absorbed by the atmosphere takes the form of either thermal energy (heat) or kinetic energy denoted by air in motion (wind). These two forms of atmospheric energy are readily converted from one to the other.

A huge and continuous flow of radiant energy streams out into interplanetary space in all directions from the sun, whose surface has an estimated temperature of about 12,000°F (6000°C). The planet earth, about 93 million miles distant, intercepts less than one five-billionth of the sun's total energy output. Nevertheless, solar radiation energy is the great engine which drives the earth's atmospheric and oceanic circulations, generates the weather, and makes the planet a livable place for plants and animals, including humans. The coal, gas, electricity, and petroleum that heat our homes and power our engines are only transformed solar energy.

Nature of radiation.   All bodies or objects, irrespective of their temperatures, give off radiant energy in the form of electromagnetic waves. These waves, traveling at a speed of about 186,000 miles a second, transport energy just as waves on the sea do. Electromagnetic radiation exists in a variety of conditions—solar energy, heat from a stove, radio and television transmission, warmth from the earth, x-rays. These differ from each other mainly in their wavelengths, or the linear distance between wave crests. Wavelengths of radiant energy range from the exceedingly short waves of cosmic rays and x-rays (10^{-8} to 10^{-5} microns[1]) to the long radio waves of several hundred meters (10^8 microns). In the whole range of known radiation wavelengths, those given off by the sun and earth are intermediate in their dimensions (between 10^{-3} and 10^2 microns).

Wavelength depends on the temperature of the radiating body: it increases as the temperature of the body decreases. Thus high-temperature solar radiation is in the form of shorter waves than those given off by the much cooler earth. Earth radiation waves are roughly 20 times longer than those of solar radiation. Only about one-fifth of the range of the solar spectrum can be perceived as light; there are other wavelengths, some shorter (ultraviolet), and others longer (infrared), which cannot be seen. Still, in terms of energy output, the visible part of the solar spectrum comprises a great share of the total solar radiation output.

Radiation is energy in transit. When it meets an object or substance it may be *transmitted, reflected,* or *absorbed,* in proportions which depend on the nature of the medium and the wavelength of the radiation. Only that part of the radiation which is absorbed by a medium is effective in heating it. Clear air or glass transmits most of the sunlight falling upon it and so retains very little heat. Fresh snow reflects most of the sunlight, so that it too is not much heated. But a dark soil absorbs a great part of the incident sunlight, and as a consequence its temperature is markedly increased.

A *blackbody* is a hypothetical object which by definition is the most efficient possible absorber and emitter of radiation. Theoretically it absorbs all the radiation incident upon it, reflecting and transmitting none. At the same time, it also emits the highest possible radiation for a given temperature. A true blackbody does not exist in nature, but radiation from the sun and from the earth's land-sea surface approximates that of a blackbody. By contrast, the atmosphere does not perform like a blackbody, for it absorbs and emits radiation only selectively; that is, in certain wavelengths but not in others.

Solar energy as a climatic element.   Since solar radiation is the single important source of atmospheric energy, its distribution over the earth is a major determinant of weather and climatic phenomena. Solar energy is the ultimate cause of all changes and motions of the atmosphere, and the single most important control of climate. But solar energy is not just a *control* of climate; it is also a climatic *element* of the highest importance. Quite apart from its influence on air temperature, solar radiant energy directly affects the growth and character of plants and animals, including humans. It does this both through illumination, as controlled by the visible part of the solar spectrum which is perceived as light, and through the nonvisible radiation.

The natural illumination regime, or the varying lengths of day and night, determines the period of photosynthesis, which for some plants is a critical factor. Solar radiation falling upon plants has three immediate and direct effects: At certain critical periods in their growth, it may produce burning; in green plants it largely determines the vegetative rates of growth; and it directly influences the rate of transpiration, or water loss, and consequently the water requirements of plants.

Human health and comfort, and the nature and distribution of human diseases, are likewise greatly affected by the duration and intensity of sunlight. For most outdoor workers, such as farmers, sunlight is an important part of the work environment. For people in general it is a feature of the natural environment which importantly affects outdoor recreation

[1]A micron equals one-millionth (10^{-6}) of a meter.

and enjoyment. At least for the cooler parts of the earth outside the tropics, sunshine is a relished feature of weather and climate. Its psychological effect is to add zest to living, buoy up the spirits, and dispel gloom, so much so that a cheerful, genial person is said to have a sunny disposition.

PRIMARY FACTORS DETERMINING DISTRIBUTION

Obviously the climate near ground level, where life forms are concentrated, is geographically of the greatest importance. But it is simpler at first to consider the distribution of solar radiation as it exists at the top of the earth's atmosphere, or as it would exist near ground level if there were no atmosphere.

Because of the spherical shape of the earth, half its surface is always in darkness, and even much of the lighted half receives solar beams at a fairly oblique angle. Consequently a unit of the surface at the outer boundary of the atmosphere receives on the average only one-fourth of the total solar radiation flux that would be intercepted by a similar plane surface at the same distance from the sun and perpendicular to the solar beam. This one-fourth of the solar radiation flux received by the earth is distributed very unevenly over its outermost atmospheric surface. The amount received in any latitude at any time of year depends primarily upon two factors: (1) the intensity of solar radiation, which is chiefly a function of the angle at which the beam of sunlight reaches that portion of the curved earth, and (2) the duration of solar radiation, or length of day compared with night.

When the noon sun is directly overhead in a particular area, and the beam of solar energy is therefore vertical at the surface, that area is receiving a maximum intensity of solar energy. When the noon sun is farther away from the zenith, so that its beam is more oblique, the area will intercept less energy. At the latitude of New York City, where in winter the altitude of the noon sun is about 64° away from the vertical, a unit surface intercepts only 44 percent of the solar energy which would be delivered by a perpendicular ray. In summer, when the noon sun at New York is only about 17° away from the zenith, the comparable figure is 95 percent. Of course, the sun's rays themselves don't actually change in intensity. Moreover, they are always parallel. Their varying angles, and therefore varying intensities, are only the result of the curvature of the earth's surface. Because an oblique solar ray is spread out over a larger segment of the earth's curved surface than a vertical one, it delivers less energy per unit area (Fig. 1.1). Moreover, although for the moment the effects of an atmosphere are being omitted, it should be added that an oblique ray also passes through a thicker layer of scattering, absorbing, and reflecting air which likewise greatly reduces its intensity (see table, page 13, top). For these two reasons, therefore, winter sunlight is much weaker than that of summer. Also, early morning and late afternoon sunlight is much weaker than that of midday.

As regards the second factor, duration of solar radiation, obviously the longer the period of daylight and the shorter the night, the greater the amount of solar energy received, all other conditions being equal (see table, page 13, bottom). In the latitude of southern Wisconsin, say, the longest summer days (15+ hr) have 6+ hr more of daylight than the shortest winter days (9− hr) do (Figs. 1.2, 1.3, 1.4). It is quite understandable, then, why in the middle latitudes summer temperatures are so much higher than winter temperatures: in summer (*a*) the sun's rays are less oblique and (*b*) the days are much longer.

Because length of day and angle of the sun's rays are equal on all parts of the same parallel, it follows that all places on a parallel receive the same amount of solar energy (except for differences in cloudiness and in transparency of the atmosphere). By the same reasoning, different parallels or latitudes receive varying amounts of solar energy which de-

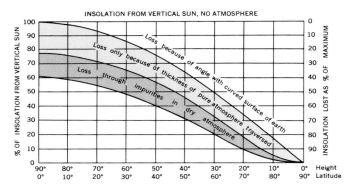

Figure 1.2 The effects of the angle of the incident ray on the intensity of solar radiation. Maximum solar radiation (100 percent on graph) is received from a vertical ray on a horizontal surface without an atmosphere. The top curve shows the intensity of the radiation received at varying angles from 90° to 0° (scale marked "height" of sun), disregarding the effects of an atmosphere. At 30° the intensity is half that at 90°. The middle curve on the graph shows the additional depleting effects of a pure, dry, cloudless atmosphere with a transparency of 78 percent. The third curve shows the further reduction resulting from impurities in the atmosphere. (*From S. Gentilli, A Geography of Climate.*)

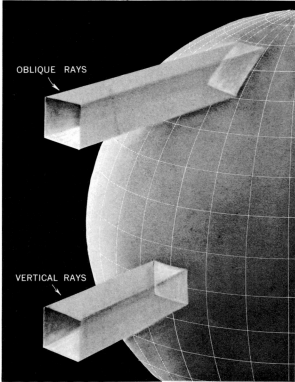

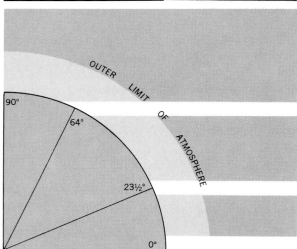

Figure 1.1 Oblique solar rays deliver less energy at the earth's surface than vertical rays, both because their energy is spread over a larger surface (*top*), and because they pass through a thicker layer of reflecting and absorbing atmosphere (*bottom*).

crease from equator to poles for the year as a whole.

If solar radiation were the only control of atmospheric phenomena, all places in the same latitude would have identical climates. And while they are certainly not identical, the strong climatic resemblances within latitude belts testify to the dominant, although not exclusive, rank of sun control.

A plane surface perpendicular to the solar beam intercepts the largest possible amount of radiation. At the outer limits of the atmosphere, such a surface receives solar energy amounting to 2 cal (calories) per sq cm per min. This is called the solar constant. Since the spherical earth's surface intercepts only one-fourth as much solar radiation as a perpendicular plane surface would, the solar constant for the earth is only 0.5 cal per sq cm of the earth's surface per min, averaged over the whole globe. There is an insignificant variation in the solar constant due to slight changes in the distance of the earth from the sun at different positions in the earth's orbit. The fact that the Northern Hemisphere is in the middle of its winter season at the time when the

Intensity of solar radiation (Transmission coefficient 78 percent)

Sun's altitude	Distances rays must travel through atmosphere*	Radiation intensity on a surface perpendicular to rays	Radiation intensity on a horizontal surface
90°	1.00	78	78
80°	1.02	77	76
70°	1.06	76	72
60°	1.15	75	65
50°	1.31	72	55
40°	1.56	68	44
30°	2.00	62	31
20°	2.92	51	17
10°	5.70	31	5
5°	10.80	15	1
0°	45.00	0	0

* Expressed in atmospheres.

earth is about 3 million miles closer to the sun only serves to emphasize that this item of distance is minor compared with the varying length of daylight and the angle of the sun's rays reaching the earth.

Earth and sun relations

The earth is held in space by the combined gravitational attraction of other heavenly bodies and has motions that are controlled by them. The two principal earth motions are rotation and revolution.

Rotation The earth rotates upon an imaginary axis, the ends of which are the North and South Poles. Owing to the earth's slight bulging at the equator and slight flattening at the poles, this axis is its shortest diameter. The time required for the earth to rotate once upon its axis determines the length of a day. During that time (approximately 24 hr), places on the sphere are turned alternately toward and away from the sun. They go through a period of light and a period of darkness, and they are swept over twice by the circle of illumination (i.e., boundary between light and dark): at dawn and again at twilight. The direction of earth rotation is toward the east. This not only determines the direction in which the sun, moon, and stars appear to rise and set, but is related to other earth phenomena of far-reaching consequence, such as the prevailing directions of winds and ocean currents.

Length of day in various northern latitudes (In hours and minutes on the 15th of each month)

Month	0°	10°	20°	30°	40°	50°	60°	70°	80°	90°
Jan.	12:07	11:35	11:02	10:24	9:37	8:30	6:38	0:00	0:00	0:00
Feb.	12:07	11:49	11:21	11:10	10:42	10:07	9:11	7:20	0:00	0:00
Mar.	12:07	12:04	12:00	11:57	11:53	11:48	11:41	11:28	10:52	0:00
Apr.	12:07	12:21	12:36	12:53	13:14	13:44	14:31	16:06	24:00	24:00
May	12:07	12:34	13:04	13:38	14:22	15:22	17:04	22:13	24:00	24:00
June	12:07	12:42	13:20	14:04	15:00	16:21	18:49	24:00	24:00	24:00
July	12:07	12:40	13:16	13:56	14:49	15:38	17:31	24:00	24:00	24:00
Aug.	12:07	12:28	12:50	13:16	13:48	14:33	15:46	18:26	24:00	24:00
Sept.	12:07	12:12	12:17	12:23	12:31	12:42	13:00	13:34	15:16	24:00
Oct.	12:07	11:55	11:42	11:28	11:10	10:47	10:11	9:03	5:10	0:00
Nov.	12:07	11:40	11:12	10:40	10:01	9:06	7:37	3:06	0:00	0:00
Dec.	12:07	11:32	10:56	10:14	9:20	8:05	5:54	0:00	0:00	0:00

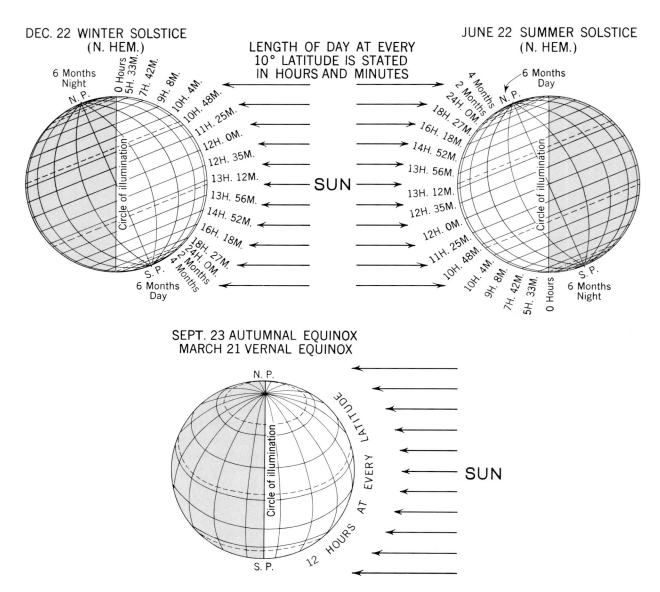

Figure 1.3 At the times of the two equinoxes, when the sun's noon rays are vertical at the equator, the circle of illumination cuts all parallels in half, so that days and nights are equal (12 hr) over the whole earth. At the times of the solstices, the sun's vertical noon rays have reached their greatest poleward displacement, 23½° north or south. The circle of illumination then cuts all parallels except the equator unequally, so that days and nights are unequal in length except at latitude 0°.

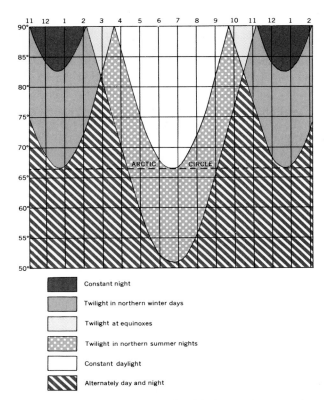

Constant night

Twilight in northern winter days

Twilight at equinoxes

Twilight in northern summer nights

Constant daylight

Alternately day and night

Figure 1.4 The annual march of light conditions poleward from 50°N. (*After W. Meinardus.*)

Revolution. The rotating earth revolves in a slightly elliptical orbit about the sun, keeping an average distance of about 93 million miles from it. The time required for the earth to pass once completely around its orbit fixes the length of the year. During the time of one revolution, the turning earth rotates on its axis approximately 365¼ times, thus determining the number of days in the year.

An imaginary plane passed through the sun and extended outward through all points in the earth's orbit is called the plane of the ecliptic. The axis of the earth's rotation has a position that is neither parallel with nor vertical to that plane. Instead, it is inclined about 66½° from the plane of the ecliptic (or 23½° from an imaginary line vertical to it). This position is constant, and therefore the axis at any time during the yearly revolution is parallel to the position that it occupies at any other time (Fig. 1.3). This is called the *parallelism* of the axis.

The degree of inclination of the earth's axis and the parallelism of that axis, together with the earth's shape, its rotation on its axis, and its revolution about the sun, produce a number of features in the earth's environment that are very important to man. One of the most fundamental is the changing distribution of solar energy over the earth and the accompanying changes of seasons.

The march of the seasons

Equinoxes: spring and fall. Twice during the yearly period of revolution, on about March 21 and September 23, the sun's noon rays are directly overhead, or vertical, at the equator (Fig. 1.3). At these times, therefore, the circle of illumination, marking the position of the tangent rays, passes through both poles and so cuts all the earth's parallels exactly in half. One-half of each parallel (180°) consequently is in light, and the other half is in darkness, so that days and nights are equal (12 hr each) over the entire earth. From this fact the two dates March 21 (spring equinox) and September 23 (autumn equinox) get their names, for *equinox* is derived from Latin words meaning equal night. At these seasons the maximum solar energy is received at the equator, from which it diminishes regularly toward either pole, where it becomes zero.

Solstices: summer and winter. On about June 22 the earth is midway in its orbit between the equinoctial positions, and the North Pole is inclined 23½° *toward* the sun (Fig. 1.3). As a result of the axial inclination, the sun's rays are shifted northward by that same amount (23½°), so that the noon rays are vertical at the Tropic of Cancer (23½°N), and the tangent rays in the Northern Hemisphere pass over the pole and reach the Arctic Circle (66½°N), 23½° on the opposite side of it. In the Southern Hemisphere the tangent rays do not reach the pole but terminate at the Antarctic Circle, 23½° short of it. Thus while all parts of the earth north of the Arctic Circle are in constant daylight, similar latitudes in the Southern Hemisphere (poleward from the Antarctic Circle) are entirely without sunlight. All parallels, except the equator, are cut unequally by the circle of illumination. Those in the Northern Hemisphere have the larger parts of their circumferences toward the sun, so that days are longer than nights. Longer days, plus more nearly vertical rays of the sun, result in a maximum receipt of solar energy in the Northern Hemisphere at this time. Summer, with its associated high temperatures, is the result, and north of the equator June 22 is known as the *summer solstice*. In the Southern Hemisphere, all these conditions are reversed: nights are longer than days, the sun's rays are relatively oblique, solar radiation is at a minimum, and winter conditions prevail (Fig. 1.4). The sun seems to stand briefly still for about 5 days at the time of the summer solstice before reluctantly starting southward, during which time there is almost no change in the time of sunrise or sunset. It appears almost as if the momentum gained during 6 months of northward travel must be first overcome before it can shift into reverse. The same happens at the winter solstice.

On about Dec. 22, when the earth is in the opposite position in its orbit, it is the South Pole that is inclined 23½° toward the sun (Fig. 1.3). The noon rays are then vertical over the Tropic of Capricorn (23½°S), and the tangent rays pass 23½° over the South Pole to the other side of the Antarctic Circle (66½°S). Consequently south of 66½°S there is constant light, while north of 66½°N there is none. All parallels of the earth, except the equator, are cut unequally by the circle of illumination, with days longer and the sun's rays more nearly vertical in the Southern Hemisphere. This, therefore, is summer south of the equator but winter in the Northern Hemisphere (*winter solstice*), where opposite conditions prevail.

EFFECTS OF AN ATMOSPHERE

In passing through the air ocean, the solar beam is depleted by an amount varying with latitude and with the season, and with local conditions. This weakening occurs because the sun's rays are scattered, reflected, and absorbed in varying degrees by the atmosphere's gases and by tiny particles suspended in the air (Fig. 1.5, left side).

Figure 1.5 To illustrate terrestrial heat balance. Note that the number of energy units absorbed by the earth's atmosphere and land-water surface just equals the units of energy radiated by the earth to space. Only about two-thirds of the total incoming solar radiation is utilized in heating the earth, one-third being reflected and scattered back to space. (*Adapted from Houghton.*)

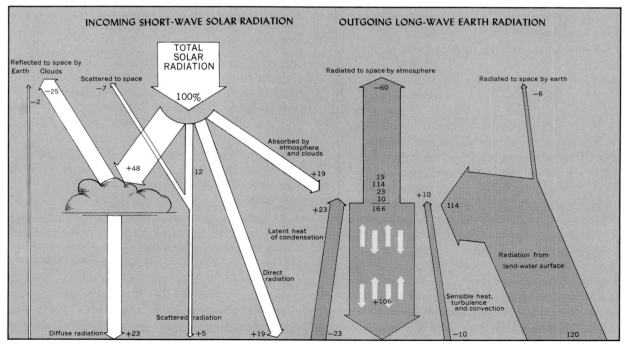

Scattering, reflection, and absorption. When a beam of solar energy passes through a relatively transparent medium such as air, some of the wavelengths are deflected from the direct beam by air molecules and very fine dust, including smoke haze, and sent in all directions. This phenomenon is known as *scattering*. True scattering occurs only when the diameters of the obscuring particles are smaller than the wavelengths of the radiation. Since the process is selective, with the shorter waves being affected most, various sky colors may result. A blue sky, for example, is the consequence of a more complete scattering by air molecules of the shorter wavelengths in the solar radiation beam. On the other hand, the sun appears red when seen through a smoke haze, because the blue light has been subtracted. Similarly, the ruddy cloud colors of sunset and sunrise result from cloud illumination by beams of light from which the blue portion of the spectrum has been removed by scattering. Because of scattering, there is a smaller amount of ultraviolet light in cities, where smoke is prevalent, than in the open country.

When the diameters of the obscuring particles are larger than the wavelengths of light, the result is *diffuse reflection,* which is caused by cloud droplets and larger dust and salt particles in the air. Being nonselective and therefore equally effective for all wavelengths, diffuse reflection does not affect sky color but rather light intensity. Thus the light reflected from a cloud when the sun is back of the observer is pure white. Also, the sun appears white when seen through a fog of water droplets.

The lengths of the solar waves do not change during the process of scattering and reflection, however; they are still fairly short waves and consequently the atmosphere does not absorb them readily.

The atmospheric gases are transparent to most of the wavelengths in the solar beam, and the air ocean transmits or reflects most sunlight. Those gases that do *absorb* are selective in their action, absorbing more of some wavelengths than of others. Water vapor is the controlling agent in atmospheric absorption; it absorbs six times as much solar radiation as all the other gases combined. Thus most absorption occurs in the lower atmosphere, where water vapor is found in greatest abundance.

The earth's albedo. A part of the solar energy which is scattered and reflected by clouds, the atmosphere, and the earth's land-sea surface is sent back to space and is therefore lost to the earth. This reflection from the earth (its *albedo*) causes the planet to shine brilliantly. It is earthshine that faintly illuminates the dark portion of the new moon, so that one sees the earth-lit part as "the old moon in the new moon's arms." The average earth albedo is estimated to be about 34 percent of the total solar radiation intercepted. The albedo varies with latitude, however. In the tropics and subtropics, probably about 30 percent of the solar radiation is reflected back to space. But in polar latitudes, the albedo increases to over 50 percent owing to the heavy cloud deck in high latitudes in summer, the long trajectory of oblique solar rays through the atmosphere, and the lasting winter snow cover which, like clouds, is an efficient reflecter. The albedo of different earth surfaces varies greatly, as these averages show:

Fresh snow	75–90 percent	Forests	3–10 percent
Old snow	50–70 percent	Grass	15–30 percent
Sand	15–25 percent	Bare ground	7–20 percent

Diffuse daylight. Some of the solar energy scattered and reflected in the atmosphere, however, is not lost to space but reaches the earth's land-sea surface in the form of diffuse blue light from the sky, called diffuse daylight. At the surface it is transformed into heat and other forms of energy. It is diffuse daylight which prevents absolute darkness on cloudy days, indoors, or in the shade, where direct sunlight is absent. The energy transmitted to the earth in this form may constitute 25 to 30 percent of the total incoming solar energy. The more oblique the ray of sunlight, the thicker the layer of reflecting and scattering atmosphere through which it must pass. When the sun is only 4° above the horizon, the solar rays have to penetrate an atmosphere more than twelve times thicker than those coming from the sun at altitude 90°. This explains why one can look at the sun at sunrise and sunset without being blinded. In the higher latitudes, where the sun is never very far above the horizon and direct sunlight is consequently weak, diffuse daylight is the source of a large part of the solar energy received at the earth's surface. At Stockholm, Sweden (59°23'N), for example, 25 percent of the total solar energy in May, and 80 to 90 percent in winter, is in the form of diffuse daylight. At the poles, probably not more than 18 percent of the total solar radiation reaches ground level.

Amount of depletion.   The amount of depletion of solar radiation by scattering, reflection, and absorption depends upon two factors: the length of the passage through the atmosphere (in other words, the angle of the sun's rays), and the transparency of the atmosphere (Fig. 1.2). The length of the passage can be computed mathematically, but the transparency varies according to time and place, for it is affected by the amount of cloudiness and the turbidity of the atmosphere, as Fig. 1.2 shows. Quantitative estimates of the depletions of solar radiation passing through the atmosphere are difficult to arrive at, vary considerably between authorities, and must be considered as only approximate. One such approximation (after Houghton and others), made in terms of 100 units (or 100 percent) of solar radiation is as follows:

Reflected to space		
By clouds	25	
By the earth's land-sea surface	2	
Scattered by molecules of air and fine dust	7	34 units
Absorbed in the atmosphere		
By the atmosphere	17	
By clouds	2	19 units
Transmitted through the atmosphere and absorbed at the land-sea surface		
As direct sunlight	19	
As diffused radiation through clouds	23	
As scattered radiation	5	47 units
		100 units

According to this estimate 34 percent of the total solar radiation arriving at the outer limits of the atmosphere is returned to space in its original short-wave form (Fig. 1.5, left part). It plays no part in heating either the land-sea surface or the atmosphere. Only 19 percent of solar radiation is absorbed in the atmosphere and thereby heats it directly. The remaining 47 percent reaches the land-water surface, is absorbed by it, and heats it.

Thus only about two-thirds of the total incoming solar radiation (19 percent absorbed by the atmosphere and its clouds directly, and 47 percent absorbed by the earth's land-sea surface) becomes effective for heating the atmosphere. Also, a much larger percentage (47 percent) is absorbed by the earth's surface than by the atmosphere directly (19 percent). The solar energy absorbed by the earth's land-sea surface, therefore, represents by far the largest part of the total solar energy absorbed by the whole earth environment—surface and atmosphere. Consequently the atmosphere gains most of its heat indirectly, i.e., by transfer from the earth's land-sea surface. For a shallow ground layer of air, this is almost entirely true. *It is of the utmost climatic significance that the atmosphere is fueled mainly from below.*

The amount of solar radiation absorbed at the earth's surface, varies greatly over short periods of time and from one region to another. This is mainly because of variations in the amount of cloudiness and contrasts in the albedos of different land surfaces, for to a large extent these factors determine the amount of solar energy reflected to space. Such temporal and regional variations naturally have marked climatic effects. For example, since the average surface-air temperature lags behind the maximum solar radiation by about 4 to 8 weeks, it is obvious that a region's excess or deficiency of solar radiation at the earth's surface will have much to do with determining the future air temperature over a period of time such as a week or a month. The land-water surface acts as a great heat reservoir. It is able to absorb vast amounts of solar radiation, and it is warmed to considerable depths by this energy. Subsequently the stored energy is given up gradually to the atmosphere.

DISTRIBUTION OF SOLAR RADIATION

As discussed earlier, the belt of maximum solar radiation swings back and forth across the equator during the course of a year, following the shifting rays of the sun. Two variables, the angle of the sun's rays and the length of day, largely determine the amount of solar energy received at any time or place (effects of the atmosphere omitted).

Variations from pole to pole. *For the year as a whole,* solar radiation reaches a maximum at the equator and diminishes gradually and regularly toward minima at either pole (see the table below and Fig. 1.6a). At the poles the total amount of solar radiation received for the entire year is about 40 percent of that received at the equator.

Total annual solar radiation for various latitudes (Effects of an atmosphere omitted)

Latitude	Thermal days*	Latitude	Thermal days*
0°	365.2	50°	249.7
10°	360.2	60°	207.8
20°	345.2	70°	173.0
30°	321.0	80°	156.6
40°	288.5	90°	151.6

* The thermal day is the average total daily solar energy at the equator.

At the time of the equinoxes (about March 21 and September 23), when the sun's noon rays are vertical at the equator and the tangent rays reach the poles, latitudinal distribution of solar radiation resembles that for the year as a whole, since the maximum is at the equator with minima at the poles (Fig. 1.6b). Although the symmetrical curves representing annual and equinoctial latitudinal distributions of solar radiation are relatively similar in form, they differ in that solar radiation declines to zero at the poles at the time of the equinoxes, while this is not the case with annual distribution.

Not only does the latitudinal distribution of solar energy in spring and fall resemble that for the year as a whole, but patterns of temperature, pressure, winds, and precipitation over the earth in these transition seasons most closely approximate yearly averages. This is because it is in spring and fall that the Northern and Southern Hemispheres are receiving approximately equal amounts of solar radiation. The result is that temperature conditions in the two hemispheres are most nearly similar then, and pressure, wind, and precipitation conditions are likewise more in balance to the north and south of the equator.

At the time of the two solstices (about June 22 and December 22), when the noon rays of the sun are vertical on the 23½° parallels north or south, and the length of day increases from 66½° in the winter hemisphere to the pole in the summer hemisphere,[2] solar radiation is very unequally distributed in the two hemispheres. The summer hemisphere actually receives two to three times the amount of solar radiation received by the winter hemisphere. Neglecting for the moment the effects of the atmosphere, on about June 22 the zonal solar energy curve, beginning at zero at the Antarctic Circle, continues to rise steadily up to about latitude 44°N, in spite of the fact that the sun's rays are increasingly more oblique north of 23½°N (Fig. 1.6c). North of this latitude, however, there is a slight decline in solar radiation, which continues to about latitude 62°N, because the more oblique rays of the sun offset the increased length of day. But the solar energy curve again rises north of 62°N and reaches an absolute maximum at the North Pole. The counterpart of this solar radiation curve, simply reversed as to hemisphere, occurs at the time of the winter solstice (Fig. 1.6d). The curve of solar energy distribution from pole to pole at the time of the solstices is complicated because it is a compromise between the effects of two important controls of solar radiation—angle of the sun's rays and length of day—which do not coincide in their latitudes of maximum effect. Thus on June 22, although

[2] By "summer" hemisphere is meant the hemisphere that has summer. Thus in July the Northern Hemisphere is the summer hemisphere and the Southern Hemisphere is the winter hemisphere. In January this situation is just reversed.

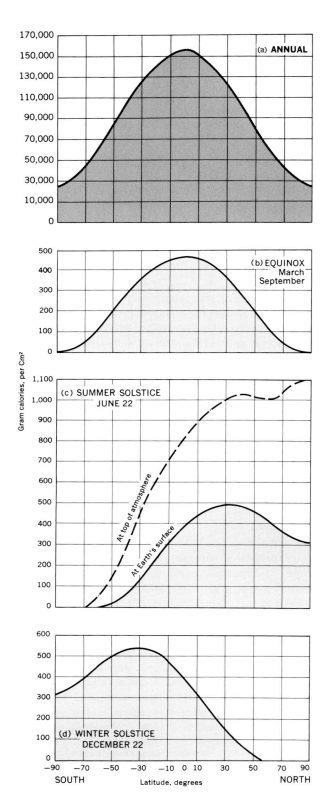

the sun's noon rays are vertical at parallel $23\frac{1}{2}°$N, days are longest beyond the Arctic Circle.

It can be seen from Fig. 1.6c that at the time of a solstice the latitudinal profiles of solar radiation are significantly different at the top of the atmosphere and at the earth's land-sea surface. Not only is there a general depletion of solar energy as a consequence of its passage through the atmosphere, but also there are some differences in the latitudinal locations of the peaks and hollows in the two profiles. For example, at the land-sea surface the peak of solar radiation does not occur at a pole; instead there is a broad maximum zone in the vicinity of latitude 40°, while the pole receives less radiation than the equator. In the summer hemisphere, solar energy is distributed fairly uniformly with latitude, so that there is no sharp maximum in any particular latitude, and the parallels in middle latitudes receive more than the equator does. Therefore the maximum surface-air temperatures in summer usually occur over land masses in the lower middle latitudes rather than at the equator. In the winter hemisphere, the latitudinal rate of change (gradient) in solar energy is greater.

These characteristics of solar radiation distribution at the seasonal extremes of summer and winter furnish the basic explanations for many features of global weather and climate. Some of these are (1) the striking latitudinal shifts of temperature, pressure, wind, and precipitation belts from summer to winter following a similar migration of solar radiation; (2) the high summer temperatures of the lower middle-latitude continents, where solar energy receipts are at a maximum; (3) the much greater latitudinal rate of change in temperatures in the winter hemisphere compared with the summer

Figure 1.6 Latitudinal distribution of solar energy at the earth's surface. For the year as a whole and at the two equinoxes, solar energy is symmetrically distributed between the Northern and Southern Hemispheres. There is a maximum in equatorial latitudes and minima at the North and South Poles. At the solstices, solar energy is very unequally distributed, with the summer hemisphere receiving two to three times the amount that the winter hemisphere does.

hemisphere—a situation which approximates the distribution of solar radiation; and (4) the greater storminess and variability of weather in the winter hemisphere compared with the summer hemisphere —a feature which is related to the steeper rate of temperature change (gradient) in the former.

Annual march by latitudes. Annual solar radiation curves fall naturally into three groups: low-, middle-, and high-latitude curves (Fig. 1.7).

1. The low-latitude, or tropical, curve, which is characteristic of the belt lying between the Tropic of Cancer and the Tropic of Capricorn, remains constantly high, with little seasonal variation. This accounts for the continuously high year-round temperatures of the tropics. During the course of a year, all places situated between the two tropics are passed over twice by the vertical rays of the sun, so that the solar radiation curve for the low latitudes shows two weak maxima and two weak minima.

2. The middle-latitude curve for the belts lying between $23\frac{1}{2}°$ and $66\frac{1}{2}°$ in each hemisphere shows a single strong maximum and a single minimum, both of which coincide with the solstices. Like the tropics, the middle latitudes have no period

when solar radiation is absent, so that the curve does not reach zero at any time. Nevertheless, it shows far greater seasonal extremes, which are reflected in greater seasonal extremes of temperature.

3. The polar solar radiation curve for latitudes poleward of the Arctic and Antarctic Circles resembles that of the middle latitudes in that it has one maximum and one minimum, coinciding with the summer and winter solstices. It differs from both the other curves in that it does reach zero; during a portion of the year, the high latitudes have no direct sunlight (Fig. 1.4).

Annual march at top and bottom of atmosphere. At stations in high latitudes, there is a close correspondence between the annual march of solar radiation upon a horizontal surface at ground level and that at the top of the atmosphere. In other words, for both levels the maximum and minimum occur close to the solstice dates.

In middle latitudes, while the correspondence between the march of solar radiation at the top and bottom of the atmosphere is not as close as in high latitudes, the two curves still show peaks and depths occurring at the same general times. The maximum ground-level radiation is most commonly received during May to July in the Northern Hemisphere, the minimum during November to January—periods which include the solstices.

But in the wet tropics of equatorial latitudes, there is little correspondence between the annual march of solar radiation at the top of the atmosphere and that at ground level. Indeed, they are characteristically out of phase, in part because of the small annual range of solar radiation in low latitudes, but mostly because the belts of cloud (and rainfall) commonly follow the sun in low latitudes, so that cloudiness is at a maximum during the period of high sun. This is less true outside the tropics. Consequently the equatorial latitudes have a wider diversity of low-level solar radiation regimes than the middle and high latitudes do.

At the equator, for example, the annual variation in the daily value of solar radiation at the top of the atmosphere is only 13 percent, and at latitude

Figure 1.7 Annual march of solar radiation received at the outer limits of the earth's atmosphere at different latitudes. In the very low latitudes close to the equator, the amount of solar energy received at the top of the atmosphere is large, and varies little throughout the year. In the middle and higher latitudes, there are large seasonal differences in the receipts of solar energy.

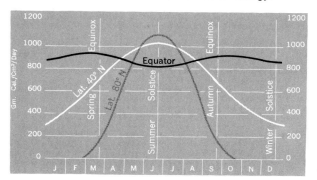

10° N and S, it is only 22 percent. But because of variations in cloudiness, atmospheric turbidity, and diffuse reflection in the atmosphere, radiation received at the ground in the tropics ranges anywhere from about 20 to 80 percent of that received at the outer atmosphere. With the solar radiation regime at the top of the atmosphere so nearly uniform within the broad belt of the wet tropics, radiation at ground level becomes chiefly a function of cloudiness. For instance, Rangoon, Burma, at about 17°N, receives nearly 75 percent more ground-level radiation in low-sun February than in high-sun July, even though in July the outer-atmosphere radiation is 20 percent greater. Greater cloudiness in July offsets the effects of higher sun. Thus in equatorial latitudes there is a close inverse relationship between cloudiness and ground-level solar radiation.

Distribution of observed solar energy. The distribution of observed annual solar radiation by latitude belts at the earth's land-sea surface is shown in Fig. 1.8. Note the slight decline in radiation in the cloudy equatorial latitudes. (Compare Figs. 1.8 and 1.6a.) On the map (Fig. 1.9) showing average *annual* distribution of solar radiation at ground level, the following facts can be observed:

1. The values range from a low of under 70 kilogram calories per square centimeter, found only in the high latitudes, to a high of over 220 in the eastern Sahara of North Africa. The highest value is thus more than three times the smallest.

Figure 1.8 Total annual solar radiation, direct and diffuse, received at the earth's land-sea surface by latitude belts (in kg cal per cm² per year). (*Data from Budyko.*)

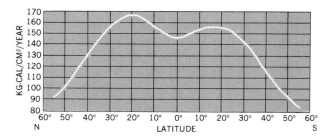

2. A general zonal pattern of distribution is conspicuous in the middle and higher latitudes, with the values declining poleward. Solar radiation in higher latitudes is primarily a seasonal phenomenon: little radiation is received in winter, and a very large part of the year's total is concentrated between the spring and autumn equinoxes.

3. In the high latitudes, the lowest annual radiation values are over oceans. In general these areas of least radiation coincide with the abundant cloud associated with the great semipermanent, oceanic, low-pressure cells, which will be described in Chapter 3.

4. A zonal pattern of distribution is not conspicuous in the tropics, where the amount of cloud is highly variable from region to region.

5. Even though equatorial latitudes have the greatest receipts of solar radiation at the outer limits of the atmosphere, they show a weak secondary minimum at the land-water surface (Figs. 1.8, 1.6a). This reflects the cloudiness of the inner tropics. Within the equatorial belt, centers of least solar radiation tend to coincide with the warm continents, where convection is strong and cumulus cloud abundant. Thus cloudy inner Amazonia in South America receives less than 120 kg cal per cm²—about on a par with central and southern Japan, the Great Lakes region of North America, and southern France and northern Italy. Equatorial West Africa also shows relatively low radiation values.

6. The zonal maxima of solar radiation are in the subtropics, where high-pressure cells, many large deserts, and relatively little cloud, especially over the land, are the rule. The largest continuous area of high solar radiation values generally coincides with the extensive dry lands of northern Africa and southwestern Asia. Other important centers of high radiation values, all of which are dry regions with little cloud, are located in the southwestern United States and northern Mexico, northern Chile and northwestern Argentina, southern (and especially southwestern) Africa, and Australia.

Seasonal maps. At the time of the solstices in June and December, when the sun's vertical rays

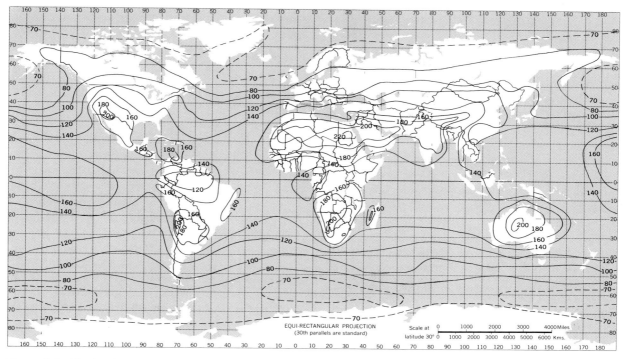

Figure 1.9 Total solar radiation received at the earth's land-sea surface during the entire *year* in kg cal per cm² per year. (*After Landsberg.*)

have shifted to their extreme northern and southern limits, naturally the distribution of solar radiation at the earth's land-sea surface is markedly different from the annual averages.

In *June,* as during the whole year, the zonal pattern of distribution still continues to be much more conspicuous in the high latitudes than in the low latitudes (Fig. 1.10). The highest values, of course, are in the Northern Hemisphere, although their centers (usually located between about latitudes 20 and 40°N over the land) are not very far north of the centers of highest value on the annual map. The most striking changes are those in radiation values for the high latitudes of the Northern Hemisphere and the middle latitudes of the Southern Hemisphere. For example, in June the latitudinal rate of change (gradient) in radiation over the Northern Hemisphere is weak, so that Arctic North America

receives almost as much radiation as the region of maximum values in the subtropics, and more than most equatorial regions. In the winter hemisphere south of the equator, radiation values are relatively low, as would be expected. Actually, northernmost North America receives up to thirteen times as much radiation as the cloudy southernmost extremity of South America.

In *December* there is a large-scale reversal of the June distribution, although a zonal pattern still prevails in the higher latitudes of both hemispheres (Fig. 1.11). In the Northern Hemisphere high latitudes, where days are short and the sun's rays oblique, daily mean values of radiation drop below 50 gm cal. The highest values are in the relatively cloudless continental deserts of the Southern Hemisphere, which receive much more radiation than the cloudier equatorial latitudes.

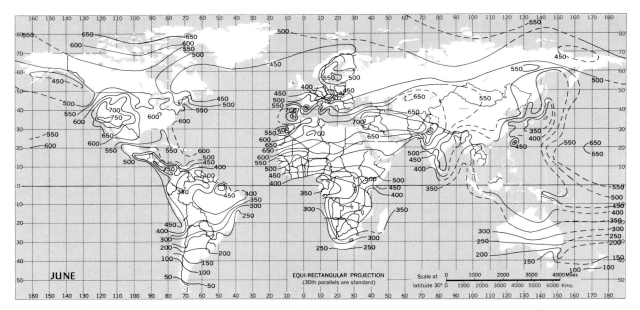

Figure 1.10 Total solar radiation received at the earth's land-sea surface in *June*, in gm cal per cm² per day. (*After Löf, Duffie, and Smith.*)

Figure 1.11 Total solar radiation received at the earth's land-sea surface in *December*, in gm cal per cm² per day. (*After Löf, Duffie and Smith.*)

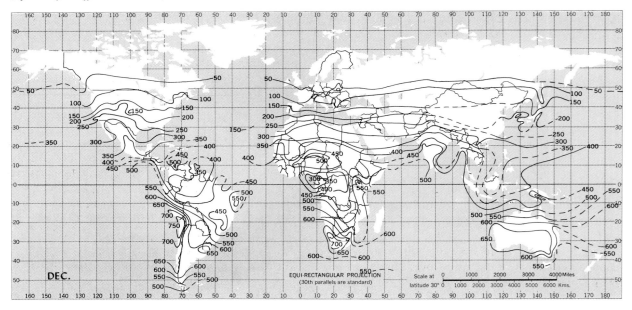

SELECTED REFERENCES FOR FURTHER STUDY OF TOPICS IN CHAPTER ONE

Note: References are inclusive. In order to avoid repeating long titles, reference many times is made to a particular book through the author's last name only. For the complete title of any book, its publisher, and date of publication, see the list of general references on pp. 4–5.

1. **Solar energy: its nature, transmission, and distribution.** Blair and Fite, pp. 61–78; Blüthgen, pp. 44–61; Byers, pp. 1–44; *Compendium of Meteorology*, pp. 13–33, pp. 379–389; Critchfield, pp. 14–22; Conrad, pp. 1–15; Haurwitz and Austin, pp. 1–21; Kendrew, pp. 1–13; Landsberg, pp. 120–134; Alissow, Drosdow, and Rubinstein, pp. 23–61; Petterssen, pp. 87–99; Riehl, *Introduction to the Atmosphere*, pp. 31–47.

2. **General references on solar radiation.** M. I. Budyko, *The Heat Balance of the Earth's Surface* (translation), U.S. Weather Bureau, Washington, D.C., 1958; M. I. Budyko et al., "The Heat Balance of the Surface of the Earth," *Soviet Geog.: Rev. Transl.* Vol. 3, pp. 3–16, May, 1962; Farrington Daniels, *Direct Use of the Sun's Energy*, Yale University Press, New Haven, Conn., 1964; H. E. Landsberg, "Solar Radiation at the Earth's Surface," *Solar Energy*, Vol. 5, no. 3, pp. 95–98, 1961; H. E. Landsberg, H. Lippmann, K. H. Paffen, and C. Troll, *World Maps of Climatology*, 2d ed., E. Rodenwaldt and H. J. Jusatz (eds.), Springer-Verlag OHG, Berlin, 1965: Map 1, Mean January Sunshine (hours), by Lippmann; Map 2, Mean July Sunshine (hours), by Lippmann; Map 3, Total Hours of Sunshine (annual), by Landsberg; Map 4, Generalized Isolines of Global Radiation (Kcal/cm^2/yr), by Landsberg; Map 5, Seasonal Climates of the Earth, by Troll and Paffen (each map 1 : 45,000,000); G. O. G. Löf, J. A. Duffie, and C. O. Smith, *World Distribution of Solar Radiation*, Engineering Exp. Sta. Rep. 21, The University of Wisconsin, Madison, 1966.

CHAPTER TWO

TEMPERATURE OF THE ATMOSPHERE

As discussed in Chapter 1, the ultimate source of the atmosphere's energy, thermal and otherwise, is solar energy. But because high-temperature solar energy is in the form of such short wavelengths, probably less than 20 percent of the solar rays entering the earth's atmosphere are absorbed directly by it. In order to be readily absorbed by the air, short-wave solar energy first must be converted into low-temperature, long-wave terrestrial energy.

This conversion takes place mainly at the earth's land-sea surface, which is able to absorb much more of the incident solar energy than the relatively transparent atmosphere can. Solar energy absorbed at the earth's land-sea surface is converted into heat and other forms of energy. The earth itself then becomes a radiating body, but at a much lower temperature (averaging about 60°F or 15°C)[1] than the sun, so that the wavelengths of radiation are far longer. Thus the atmosphere receives

---

[1] Unless stated otherwise, temperatures are given in °F. Conversion from °F to °C, and vice versa, can be made by means of the following formulas (see also Fig. 2.1):

°C = $\frac{5}{9}$ (°F + 40) −40 or °C = $\frac{5}{9}$ (°F − 32)

°F = $\frac{9}{5}$ (°C + 40) −40 or °F = $\frac{9}{5}$ °C + 32

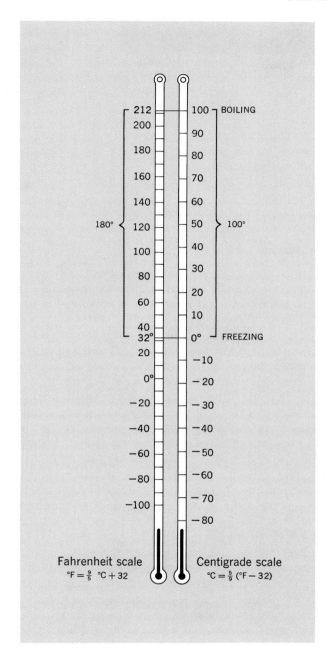

Fahrenheit scale
$°F = \frac{9}{5}°C + 32$

Centigrade scale
$°C = \frac{5}{9}(°F - 32)$

Figure 2.1 Temperatures can be converted from °F to °C, and vice versa, by a direct comparison of the two scales, or by using a conversion formula.

most of its heat indirectly from the sun and directly from the earth's surface, which in turn had previously absorbed, and itself been warmed by, solar energy. It has been estimated that the atmosphere absorbs nearly seven times more energy from the earth's land-sea surface than from solar radiation directly. On a sunny day, the striking decrease in temperature from the solar-heated hot sand or soil to the air above at a height of a few feet demonstrates that the main heat source for the lowest atmosphere is the ground. Thus a sandy beach may reach a midday temperature of 140°; in the air a quarter inch above the sand, the temperature can still be 135°; but at 3 ft, it may be down to 105°; and at the height of man, or about 6 ft, it may be only 98°—a difference of 42° within a 6-ft layer.

Differential heating of land and water surfaces. Various kinds of land and water surfaces react differently to incident solar energy in terms of reflection, absorption, and transmission, so that their potential for heating the atmosphere varies. Naturally, the surfaces which absorb relatively larger proportions of solar energy and retain it in a relatively thin surface layer are the ones which will themselves heat (and cool) most quickly, and so will have the promptest effect on the air above.

Considering land areas alone, there are marked differences in the heating characteristics of various types of solid surface. Dissimilar reflecting properties account for some of the differences, since solar energy does not heat a surface that reflects it. Snow, which reflects 70 to 90 percent of the solar radiation that falls on it, heats very slowly indeed. On the other hand, dry black soil reflects only 8 to 14 percent of the sun's rays, grass 14 to 37 percent, and coniferous forest less than 10 percent. Solar energy retained by the ground is disposed of mainly in three ways: by heat flux downward into the soil, heat flux upward to the air, and evapotranspiration (evaporation + transpiration) from moist surfaces, including vegetation. Surfaces from which evaporation is great heat more slowly. The same is true of those with high soil conductivity, which transmit a larger proportion of the heat they receive downward to lower layers.

Although the heating properties of the many kinds of land surfaces vary considerably, the greatest contrasts are those between land and water surfaces, which react so differently to solar radiation that they produce two unlike climates: oceanic and continental. The surfaces of relatively deep bodies of water heat and cool slowly compared with land surfaces. By far the most important reason for this slowness of temperature change is that in highly mobile matter such as water (and also air), redistribution of heat occurs mainly through turbulence. Heat in the solid earth is redistributed by molecular heat conduction, and so proceeds by moving from particle to particle; but ocean waves, drifts, currents, tides, and convectional overturning systems all act to disperse the absorbed solar energy throughout a large mass. Obviously no such mixing can occur on land. Therefore when land and water surfaces have identical amounts of solar energy falling on them, the land surface heats more rapidly and reaches a higher temperature.

Moreover, when a water surface begins to cool, vertical convectional movements are set up, and the cooler, heavier surface layers sink to be replaced by warmer water from underneath, Consequently the whole mass of water must be cooled before the surface layers can be brought to a low temperature. Thus a larger mass of water than land can impart its heat to the surrounding air, and for a longer period of time. As a consequence the water surface cools more slowly.

There are other contrasts between the properties of land and water which contribute to their differential heating and cooling, but these are much less important compared with mixing due to turbulence and convectional overturning. The effects on surface temperatures of land-water differences in reflection, specific heat, transmission of solar radiation, and evaporation all appear to be minor. For example, the average reflection (albedo) from the great variety of the earth's land surfaces is not very much different from that of the oceans. And volume for volume,

the specific heat of water is only 2.5 times greater than that of soil. As for transmission of solar radiation, it might seem that water's transparency would cause the rays of the sun to penetrate a much greater volume of it than of the opaque land, with the result that this would be one of the major reasons why a land surface heats more rapidly and intensely. Actually, however, the longer of the solar spectrum waves do not penetrate water very far, so that the differential transparency of land and water, while not insignificant, is less important than might be assumed.

Differential evaporation is also one of the minor factors creating temperature differences between water and land surfaces, since over water, evaporation is always at a maximum. In large areas of the continents, on the other hand, effective evaporation is less than potential evaporation, especially in the interiors of middle-latitude continents. But this differential does not always hold true, for in the wet tropics, evaporation from the continents is greater than from the oceans. Large-scale evaporation and its retarding effects on excessive heating and cooling is therefore not restricted to ocean surfaces.

Thus the turbulence and mixing which occur in water, together with the minor factors mentioned above, cause water to heat and cool to great depths compared with land. Observations show that daily temperature variations do not penetrate more than about 3 ft into the solid earth, and even annual temperature changes go to a depth of only about 47 ft. In quiet water, however, daily temperature changes are felt at least 20 ft below the surface. And it is believed that in oceans and deep lakes a layer 600 to 2,000 ft thick is subject to annual temperature variations, owing chiefly to turbulence. The result is that radiation in the summer or during a sunny day in any season will slowly heat a relatively thick layer of water to a moderate temperature, but on land a thin layer will usually be heated to a high temperature. In winter and at night, water draws upon heat reserves accumulated within a large volume and so cools slowly, but a land surface, being shallowly heated, loses heat rapidly.

Continental and oceanic climates. With the same amount of incident solar energy, a land surface reaches a higher temperature, and reaches it more quickly, than a water surface. Conversely, a land surface cools more rapidly and cools to a lower temperature than a water surface. Land-controlled, or continental, climates are therefore characterized by large daily and seasonal extremes of temperature, with the maximum and minimum air temperatures closely following the times of maximum and minimum solar radiation. By contrast, ocean-controlled, or marine, climates have moderate air temperatures, smaller daily and seasonal ranges of temperature,

and a greater time lag of the maximum and minimum seasonal temperatures after the maximum and minimum of solar radiation. In the middle and higher latitudes there are striking differences between the temperatures of marine and continental climates, principally because the solid earth does not store up so large an amount of heat. The ocean, on the other hand, has a vast potential for heat storage, mostly because it distributes energy absorbed at the surface throughout a great volume. At the ocean surface, temperatures probably do not change more than 1° between day and night, and seasonal changes also are very small.

HEATING AND COOLING THE ATMOSPHERE

The atmosphere is heated by solar energy, directly through absorption, and indirectly through processes involving conduction, radiation, and condensation. The heat acquired by these processes is then transferred from one part of the atmosphere to another by vertical and horizontal currents.

Atmospheric absorption of solar radiation

As Chapter 1 discussed, only some 19 percent of the incoming solar radiation is absorbed by the gases of the atmosphere, mostly by water vapor. About one-half of this absorption takes place in the lower 2 km of air, where water vapor is concentrated. But since a layer of air 2 km thick is a large mass through which to spread 9 to 10 percent of the original solar beam, the process of absorption is not a large factor in producing the normal daytime rise in surface-air temperature. For example, on a clear, cold winter day, when the land surface is blanketed by a reflecting snow cover, surface air temperatures may remain extremely low in spite of a bright sun. At the same time, the air close to the south side of an absorbing brick wall or building, where short-wave sun energy is being converted into long-wave terrestrial energy, may be comfortably warm.

Conduction

When two bodies of unequal temperature are in contact with one another, energy in the form of heat passes from the warmer to the colder until they both attain the same temperature. In the absence of a snow cover, the solid earth during daylight hours reaches a higher temperature than the air above. This is because land is a more efficient absorber of solar energy than air. Then, by conduction, a thin layer of air resting immediately upon the warmer earth becomes heated. But air is a poor conductor, so that heat from the warmed lower layer of air is transferred only very slowly to the layers above. Heating the atmosphere by molecular conduction, which is primarily a daytime and a summer phenomenon, is negligible in importance, except where turbulence constantly replaces the heated lowest surface layer of air. Even so, however, conduction is not a major process of atmospheric heating and heat transfer.

Just as a warm land surface on a summer day heats the air layer next to it by conduction, so a cold surface, chilled by terrestrial radiation on a winter night, cools the adjacent air. As a result of radiation, and to a lesser degree conduction, the lowest layer of air on clear, calm winter nights frequently becomes colder than the layers some hundreds of feet higher. This effect, called a temperature inversion, will be discussed later in the chapter.

Earth radiation

As described earlier, the 47 percent of the short-wave solar energy which is absorbed at the earth's land-sea surface is transformed into heat or consumed in evapotranspiration. Through this absorption and conversion of solar energy, the heated land-water surface becomes a radiating body whose average temperature is about 60°. As already discussed, the spectrum of earth radiation is composed of long-wave energy, so that unlike solar radiation, it is not visible to the human eye. Although the atmosphere is relatively transparent to short-wave solar energy, it is able to absorb and retain perhaps as much as 80 to 90 percent of the outgoing long-wave earth radiation (see Fig. 1.5).

Among the atmosphere's gases, the chief absorber and emitter of earth radiation is water vapor, and to a lesser degree, carbon dioxide. Water vapor absorbs selectively in the terrestrial radiation spectrum, being much more transparent to certain wavelengths than to others. The larger the amount of water vapor in the air, the greater the amount of absorption. In deserts, for example, where dry air and clear skies permit a swift escape of earth energy to space, rapid night cooling is the rule.

It can be seen that the effect of the atmosphere is that of a heat trap, which permits 70 to 75 percent (47 units) of the effective solar energy (66 units) to reach the ground before it is absorbed, but allows only some 5 percent (6 units) of the upward moving long-wave radiation from the ground (120 units) to escape directly to space (Fig. 1.5). It is because the atmosphere is able to absorb so much of the earth radiation, and to retard its eventual escape to space through a repeated and complex radiation exchange between ground and atmosphere, that night cooling is slowed. This process is sometimes called the *greenhouse effect* of the atmosphere, for the glass roof and sides of a greenhouse permit easy entrance of short-wave solar energy but largely absorb the

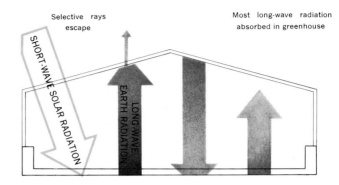

Figure 2.2 Illustrating the "greenhouse effect" of the earth's atmosphere. The glass in the roof and sides of the greenhouse, like the atmosphere, is relatively transparent to short-wave solar energy, but relatively opaque to long-wave earth radiation.

long-wave heat energy and so greatly delay its escape (Fig. 2.2, right side). A more universal illustration of the greenhouse effect is the superheating of the inside of an automobile that occurs on a sunny day when the car is parked with its windows closed.

If there were no atmosphere, probably about 80 units of the total solar radiation would be absorbed by the ground, and the average loss to space (albedo) would approximate 20 percent. This would produce a mean earth temperature of about 15° instead of the present 60°. Thus the greenhouse effect creates an average earth temperature which is markedly higher than it would be otherwise. A further consequence of the greenhouse effect is that it greatly retards night cooling and thereby moderates the diurnal range of surface temperature.

Radiation under clear skies.   Although, as the previous section said, an average of 6 units of the total earth radiation (5 percent) escapes directly to space, the rate of loss by regions varies greatly with differing atmospheric conditions. The maximum loss occurs under cloudless skies. But even the rate of loss with clear skies is variable, since the water vapor in the air is by far the most effective absorber of earth radiation among the atmosphere's gases, and the amount of water vapor is extremely variable both with regions and with time. Heat loss is least when the air's moisture content not only is high, but also is concentrated in the lower few thousand feet of air, where temperatures are highest.

Most of the long-wave earth radiation is absorbed by the water vapor, which in turn sends out long-wave radiation to the layers of atmosphere both above and below it, and also back down to the ground surface. Each successive absorbing layer of water vapor goes through the same process of absorption and radiation. The total result of this complex pattern of upward and downward streams of radiation is the greenhouse effect: retardation of the loss of heat from the earth to space. An atmosphere with high moisture content in its lower layers absorbs and reradiates downward more earth radiation than a dry atmosphere. This is a main reason why the daily range of temperature is greater in dry than in humid air. Since water vapor, the chief absorber of both solar and terrestrial radiation, diminishes rapidly with altitude (half of all the atmosphere's water vapor lies below 6,600 ft), both the incoming and the outgoing radiation increase with altitude. Thus high mountains may be called the "radiation windows" of the earth.

Radiation of terrestrial energy from the earth's surface upward toward space is a continuous process. During the daylight hours up to about midafternoon, however, more energy is received from the sun than is radiated from the earth, so that surface-air temperatures usually continue to rise until two to four o'clock in the afternoon. During the night, when receipts of solar energy cease, a continued loss of energy through earth radiation results in a cooling of the land surface and a drop in air temperature. Being a more efficient radiator than air, the ground during the night usually becomes cooler than the air above it. When this happens, the lower layers of atmosphere lose heat by radiation to the colder ground as well as upward toward space. During long winter nights when skies are clear and the air is dry and relatively calm, excessively rapid and long-continued out-radiation takes place.

Effects of a cloud cover. Radiation conditions under a cloud cover are very different from those under a clear sky, even when the air is equally humid in both cases. Water vapor, a gas, is selective in its absorption of earth radiation, letting some wavelengths escape and absorbing others. But a cloud sheet, composed of water droplets or ice particles, absorbs and radiates in all wavelengths, like a blackbody (the hypothetical object described in Chapter 1). That is, at the base of a cloud all upward-moving radiation in all wavelengths is completely absorbed. The returning stream of radiation from cloud to earth consequently includes all the wavelengths the cloud received, whereas water vapor, since it allows some wavelengths to escape through the atmosphere to space, sends less total radiation back to earth.

Because of the total absorption by clouds, the effect of a cloud cover is to considerably retard night cooling of the solid earth and the lower atmosphere. (In addition, the cloud deck may actually be warmer than the land surface, either because the clouds have moved in from warmer latitudes, or because of the extra heat which condensation releases into the air.) Under a night sky covered with low cloud, the net loss of heat from the ground is only about one-seventh the loss with clear skies. The higher the cloud cover, the less effective it is in retarding the heat loss from the ground. Thus the net loss under cirrostratus clouds located at a height of about 4 miles is nearly 80 percent of that for clear skies, while the loss is only 14 percent under nimbus and stratus clouds somewhat less than 1 mile high. In contrast to the effects of clouds at night, a daytime cloud cover reflects much of the incoming solar radiation and sends it back to space, in this way retarding surface heating and reducing the stored-up energy in the earth.

Since a cloud deck makes for warmer nights and cooler days, its net effect on *average* temperature for the 24 hours may be small. So while clear skies promote high daytime and low nighttime temperatures, and therefore large diurnal ranges, clouds result in a greater uniformity of temperature and small diurnal ranges. In the Sahara's cloudless skies and dry air, there are cases of daytime temperatures of 90° or above being followed by night temperatures slightly below freezing.

Effects of a snow surface.   A snow cover favors low temperatures. Daytime heating during short winter days is slight, for most of the solar radiation is reflected by the snow and thus does not heat the ground under it. At night the snow, which is an efficient radiator but a very poor conductor of heat, allows little heat to come up from the soil below to replenish that lost by radiation from the top of the snow surface. As a result the snow surface becomes excessively cold, and then so does the air layer resting upon it.

Effects of a water surface.   Water, like land, is a good radiator, but the cooled surface waters keep constantly sinking and are replaced by warmer water from below. Extremely low temperatures over water bodies are therefore impossible until they are frozen over, after which they act like a snow-covered land surface.

Latent heat of condensation

A large proportion of the solar energy which reaches the earth's surface is used in evaporating moisture from the land-water surface and from vegetation. The process of evaporation changes the moisture into water vapor, one of the atmosphere's gases. This transformed solar energy is thus contained in the air in latent or potential form. When condensation takes place, the latent energy is again released into the atmosphere and heats it. The air retains this heat when the condensed water falls to the ground as precipitation. (See Fig. 1.5.)

It appears that the single most important heat transfer process at the earth-atmosphere interface is evaporation, which provides about half the atmosphere's fuel, in the form of water vapor.[2] More than

[2] Michael Garstang. "Sensible and Latent Heat Exchange in Low Latitude Synoptic Scale Systems." *Proc. Sea-Air Interaction Conference.* Tallahassee, Fla., Feb. 23–25, 1965. U.S. Weather Bureau, Washington, D.C., Aug. 20, 1965.

half this fuel is supplied to the lower atmosphere by tropical oceans between 30°N and 30°S. So latent energy represented by water vapor derived from evaporation constitutes the most important single component of the atmosphere's heat budget. Above a shallow surface layer a few thousand feet thick (in which turbulence is the major factor), the latent heat contained in clouds is responsible for most of the upward heat transfer throughout the troposphere.

Transfer of heat by air currents

The energy acquired by the atmosphere through conduction, radiation, and condensation, as well as through direct absorption of solar energy, can be transferred from one part of the atmosphere to another by vertical and horizontal currents. Such a transfer, and the accompanying mixing processes, leads to changes in air temperature.

Vertical mixing by turbulence.   While for the atmosphere as a whole, the air remains suspended without rising or falling (see Chap. 3), individual masses of air do move upward and downward, and by such a process transfer heat. This is called turbulence. (See Fig. 1.5.)

Some turbulence is thermally induced. On a clear summer morning the ground heats up from solar radiation and, by conduction and radiation processes, in turn heats a shallow layer of air adjacent to it. The expanded warm air is buoyant, like a cork held under water, and tends to rise. The original small updrafts often combine into columns of rising air called *thermals.* The rising warm air mixes with that which it displaces, and the warm air particles give up some of their heat to the cooler air above. In the spaces between the thermals or rising air columns, there is a compensating downward motion. Such a vertical exchange, or convection system, creates a mixed layer which gradually becomes thicker as the day advances, shrinks as evening approaches, and disappears during the night as the ground air becomes chilled and convection halts. Over oceans, where daily variations in temperature are almost absent, vertical mixing by thermal turbulence continues both night and day as long as sea-surface temperature exceeds air temperature.

Another form of up-and-down motion is created by a process called mechanical turbulence. Internal eddying produces some mechanical turbulence, but most of it is the result of friction in fast-moving air as it passes over a rough land surface. The main roughnesses are associated with terrain features, such as mountains; but trees and buildings can also cause mechanical turbulence. Its effect on vertical temperature distribution is the same as that of thermal convection: heat carried upward from the land-sea surface creates a mixed lower layer of atmosphere, varying from a few hundred to several thousand feet in thickness. This is called the friction layer. As a method of heat transfer, turbulence of whatever kind is most important in that part of the atmosphere below the height of the cloud-base level.

Horizontal mixing by advection.   Advection refers to the transportation of any atmospheric property, such as temperature or humidity, by the horizontal movement of air (that is, by wind). In the Northern Hemisphere's middle latitudes, for example, a south wind is usually associated with unseasonably high temperatures. As an air mass of tropical origin advances northward over the central and eastern United States, it carries with it the heat and temperatures acquired in its source region, where high temperatures are normal.

Such an importation of southerly warmth in winter results in mild weather, with melting snow and sloppy streets. In summer, several days of south wind may produce a heat wave, with maximum temperatures higher than 90°. In the same way, advection of air from polar latitudes, or from the cold interior of a winter continent, causes a drop in temperature. Air from over an ocean brings a continent not only moderating temperatures, but also vast amounts of water vapor, creating an environment favorable to cloud and precipitation. An invasion of dry desert air, on the other hand, sharply diminishes the atmosphere's moisture content.

HEAT BALANCE OF THE EARTH

Referring again to the left-hand portion of Fig. 1.5, which shows the effects of the atmosphere upon incoming solar radiation, note that in all, some 66 percent of the solar radiation received at the outer limits of the earth's atmosphere is absorbed either by the atmosphere (19 percent) or by the earth's land-sea surface (47 percent). This 66 percent is called the *effective solar radiation,* for it is the portion which is available for heating the atmosphere. The remaining 34 percent is simply returned to space in the form of short-wave solar radiation. Since the yearly mean temperature of the earth as a whole is neither increasing nor decreasing, it follows that the 66 percent of solar radiation gained is balanced by an equal amount of energy radiated back to space in the form of long-wave earth radiation. This is spoken of as the *terrestrial heat balance.*

The right-hand side of Fig. 1.5 illustrates one estimate of the operation of the terrestrial heat balance. Of the 120 units of energy radiated upward from the ground, 6 units pass directly through the atmosphere to space, and consequently play no part in heating the air. The remaining 114 units are absorbed by the atmosphere. But 106 of these are reradiated back to the ground surface, resulting in a net gain of 8 units by the atmosphere. The atmosphere also gains 10 units by heat transport from the earth's surface through turbulence and convection currents. Another 23 units are transferred as evaporated moisture from the ground to the atmosphere, where subsequently they will be released as latent heat of condensation.

So far, we have 19 (solar energy units absorbed directly by the atmosphere) + 114 + 10 + 23 = 166 units, or the total earth radiation absorbed by the atmosphere. Of this, 106 units are reradiated

Incoming solar radiation and outgoing earth radiation (Northern Hemisphere)*

Latitude	0°	10°	20°	30°	40°	50°	60°	70°	80°	90°
Incoming radiation absorbed	0.339	0.334	0.320	0.297	0.267	0.232	0.193	0.160	0.144	0.140
Outgoing radiation lost	0.271	0.282	0.284	0.284	0.282	0.277	0.272	0.260	0.252	0.252
Difference	+0.068	+0.052	+0.036	+0.013	−0.015	−0.045	−0.079	−0.100	−0.108	−0.112

* Annual means are given for each latitude in gm cal per cm² per min.

Source: After Simpson.

back to the ground, while 60 are radiated by the atmosphere to space. These 60, plus the 6 which were radiated directly to space by the ground, total 66 earth radiation units—just equal to the 66 units of solar radiation absorbed by the ground and the atmosphere.

On first thought, it may seem odd that the earth's land-sea surface radiates more than it receives from the sun, but actually this is logical in view of the complex streams of up-and-down radiation involved in the atmosphere's greenhouse effect. Keep in mind, however, that the figures given above are largely intelligent estimates subject to revision as more accurate measurements or computations are made.

Latitudinal heat balance

While for the earth as a whole, gains in solar energy equal losses to space of earth energy, this is not the case for most latitudes (see Fig. 2.3 and the table above). In the low latitudes equatorward from about 35–40°, the earth and atmosphere together gain more heat from solar radiation than they lose to space by earth radiation; in the middle and high latitudes poleward from 40°, the opposite is true. Thus there is a continuing net gain of energy in low latitudes and a net loss in middle and higher latitudes. The tropics are not becoming progressively warmer and the polar regions colder, however;

Figure 2.3 In the lower latitudes equatorward of about 37°, the annual amount of incoming solar radiation exceeds the losses from outgoing earth radiation. The reverse is true for the middle and higher latitudes, where losses from outgoing earth radiation exceed the gains from incoming solar radiation.

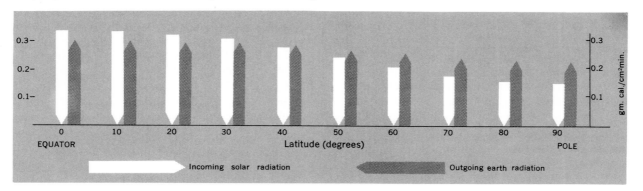

instead, a continuous large-scale transfer of heat from low to high latitudes is carried out by atmospheric and oceanic circulations.

In order to preserve the heat balance in the various latitudes of the earth, the following average quantities of heat must be carried poleward:

Lat, North	10^{19} cal per day
0°	0
10°	4.05
20°	7.68
30°	10.46
40°	11.12
50°	9.61
60°	6.68
70°	3.41
80°	0.94
90°	0.00

Heat transfer from the tropics poleward takes place throughout the year, but at a much slower rate in summer than in winter. The reason is that the temperature difference between low and high latitudes is considerably smaller in summer than in winter—only about half as large in the Northern Hemisphere. As would be expected, the winter hemisphere has a net energy loss and the summer hemisphere a net gain. Most of the summertime gain is stored in the surface layers of land and ocean, mainly in the ocean.

The unequal latitudinal distribution of solar and terrestrial radiation is the ultimate cause of the earth's atmospheric and oceanic circulations. Only through the transfer of heat by winds and ocean waters can latitudinal energy imbalances be equalized. Of the total heat transport in subtropical and middle latitudes, about 75 percent is carried out by atmospheric circulation and 25 percent by ocean water movements. Not only is the great system of planetary winds involved, but also most other phenomena of weather and climate, including atmospheric disturbances (storms). As the table above shows, the greatest transfer of heat is required in the middle latitudes. Naturally the greatest turbulence and storminess is found in regions where there is a maximum of advection and air-mass movement.

Components of the heat budget

As described above, the polar and middle latitudes consistently show a negative heat balance and the tropics a positive one. But both in the troposphere and at the land-sea surface, these net balances have different components of heat gain and loss in various latitudes.

Heat budget components in the troposphere.   In the tropical belt, 0–15°, there are three components of heat gain (solar radiation, 60 units; heat of condensation, 150; vertical transfer of sensible heat by

Figure 2.4 Heat budget constituents of the atmosphere's main layer of weather activity by 15° belts, in millicalories per cm² per min. The algebraic sum of the four gains or losses is shown as the centrally located numeral in each 15° belt. Note that the two belts 0°–15° and 15°–30° show an excess of energy; all the others a deficiency. (*After Heinz H. Lettau.*)

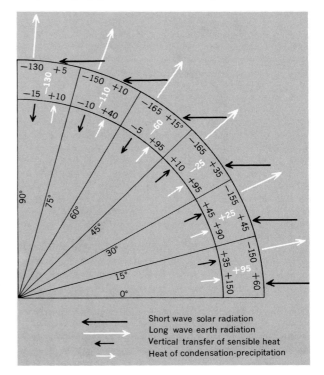

Short wave solar radiation
Long wave earth radiation
Vertical transfer of sensible heat
Heat of condensation-precipitation

Components of the heat balance at the earth's surface (Mean values in kcal per cm² per year)

Latitude	Oceans				Land			Earth			
	R	E	T	C	R	E	T	R	E	T	C
70–60°N	23	33	16	+26	20	14	6	21	20	9	+ 8
60–50°	29	39	16	+26	30	19	11	30	28	13	+11
50–40°	51	53	14	+16	45	24	21	48	38	17	+ 7
40–30°	83	86	13	+16	60	23	37	73	59	23	+ 9
30–20°	113	105	9	+ 1	69	20	49	96	73	24	+ 1
20–10°	119	99	6	−14	71	29	42	106	81	15	−10
10–0°	115	80	4	−31	72	48	24	105	72	9	−24
0–10°S	115	84	4	−27	72	50	22	105	76	8	−21
10–20°	113	104	5	− 4	73	41	32	104	90	11	− 3
20–30°	101	100	7	+ 6	70	28	42	94	83	15	+ 4
30–40°	82	80	8	+ 6	62	28	34	80	74	11	+ 5
40–50°	57	55	9	+ 7	41	21	20	56	53	9	+ 6
50–60°	28	31	10	+13	31	20	11	28	31	10	+13
Earth as a whole	82	74	8	0	49	25	24	72	59	13	0

R = Net radiation balance
E = Loss of heat by evaporation
T = Turbulent heat transfer

C = Redistribution of heat by ocean currents
Source: After Budyko et al.

turbulence, 35; total, 245 units) and one of heat loss (150 units lost through earth radiation)—resulting in a net troposphere gain of +95 units (Fig. 2.4). This is by far the largest net gain for any of the six 15° latitude belts. With increasing distance from the equator, the four components of heat exchange vary in different degrees. Obviously solar radiation will steadily decline poleward, but the other components are not so consistent (see Fig. 2.4). For example, in latitude belt 45–60° there are only two positive components: +15 units of solar radiation and +95 units of heat of condensation, or a total of +110 units. Negative components are a loss of −5 units by vertical transfer of sensible heat from the atmosphere to the ground, and −165 units lost by earth radiation—a total of −170. The result is a net loss of −60 units.

Heat balance of the earth's land-sea surface.[3] Distribution of the components of the surface heat bal-

[3] M. I. Budyko et al. "The Heat Balance of the Surface of the Earth." *Soviet Geog.: Rev. and Translation.* Vol. 3, pp. 3–16, May, 1962.

ance by latitude belts is shown in the above table. Only one component, the radiation balance (R), has a similar distribution on land and sea. Over both continents and oceans, the maximum net heat gain from radiation is in the tropics and the minimum values are in higher latitudes. In both cases too, the balance of radiation remains almost constant within the tropics. The greatest land-water contrasts in amounts of net gain from radiation are also in the tropics, where the values for land are only about two-thirds those for oceans. With increasing latitude, land and water balances become more alike (Fig. 2.5).

As the table shows, the values for heat loss by evaporation (E) over land and oceans change with latitude in different ways. On land there is a strong maximum in equatorial latitudes, where heavy rainfall and lush vegetation provide abundant evaporable moisture, and a rapid decline in subtropical latitudes, where settling air in anticyclones creates dry climates (Fig. 2.6). But over oceans the loss of heat by evaporation is at a maximum in the subtropics, where solar radiation is greater because

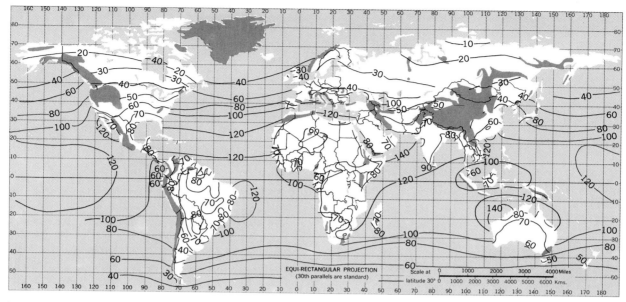

Figure 2.5   Radiation balance; annual values in kg cal per cm² per year. (*After Budyko et al.*)

Figure 2.6   Expenditures of heat by evaporation; annual values in kg cal per cm² per year. (*After Budyko et al.*)

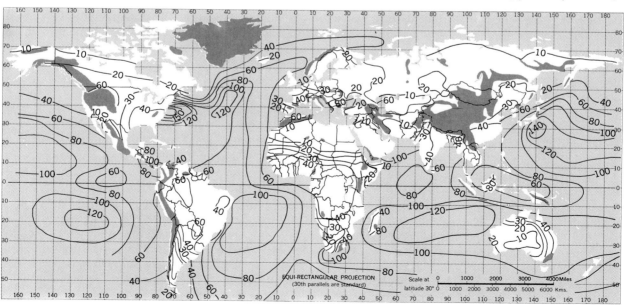

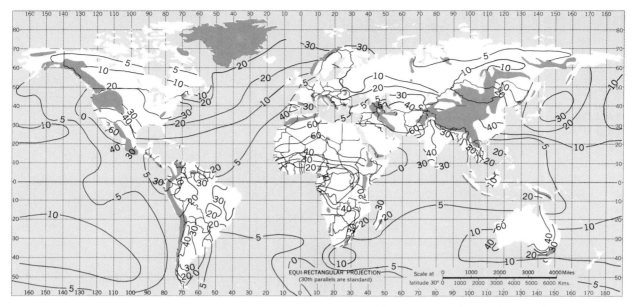

Figure 2.7   Turbulent heat exchange; annual values in kg cal per cm² per year. (*After Budyko et al.*)

there is less cloudiness. Closer to the equator, greater cloudiness reduces the solar energy received, and evaporation over oceans declines.

From the columns headed T in the table, it can be seen that the values of turbulent heat transfer from the ocean surface to the atmosphere are relatively small in all latitudes (Fig. 2.7). One reason, a point which was discussed earlier, is that sunlight does not heat a water surface intensely, as it does a land surface. On land the surface layer can become so hot that it causes strong convectional overturning in the air above it. Turbulent heat values over oceans increase with higher latitude, where warm currents warm the air in the cold season. On land, not only is turbulent heat transfer greater than it is over water, but the latitudinal distribution is also different. The highest values are in the subtropical latitudes where, because of the average minimal cloudiness and dry air, solar surface heating and convection are strong, while the expenditure of heat for evaporation is low.

As for the redistribution of heat by ocean currents (C), the table shows that heat is transported out of

a zone situated between 20°N and 20°S. The currents carry it poleward and release most of it to the atmosphere in latitudes 50° to 70° (see Fig. 3.42).

Many other facts about the earth's heat balance can be learned from this table of latitudinal heat balance components. The student of climate should also become familiar with world maps of radiation balance, heat loss by evaporation, and turbulent heat exchange (Figs. 2.5, 2.6, and 2.7).

Heat balance of individual continents.   As shown in the table on page 39, average heat balances vary widely among continents. Greatest net radiation gains (R) are in South America, Australia, and Africa, all of which lie in tropical-subtropical latitudes. The radiation balance is lowest in Europe, whose average latitude is relatively high. South America loses by far the most heat from evaporation (E), since it is located largely in the humid tropics. Africa, which is bisected by the equator, ranks next, but well below South America because of its extensive regions with dry climates. Drought is also the cause of Australia's low evaporation,

Heat balance of individual continents (In kcal per cm² per year)

	R	E	T
Europe	39	24	15
Asia	47	22	25
Africa	68	26	42
North America	40	23	17
South America	70	45	25
Australia	70	22	48

R = Net radiation balance
E = Loss of heat by evaporation
T = Turbulent heat transfer

Source: After Budyko et al.

while similarly low values in Asia and North America result both from their extensive regions with dry climates and from their high-latitude location. In Europe, on the other hand, the chief cause of low evaporation is high latitude alone. Turbulent heat transfer (T) is greatest in Australia and Africa, where hot deserts and steppes cover vast areas.

TEMPORAL DISTRIBUTION OF TEMPERATURE

All average temperatures for a month, a season, a year, or even a long period of years are built upon the *mean daily temperature* as the basic unit. The daily mean is thus the individual brick out of which the general temperature structure is formed. It is usually computed by the formula

$$\frac{\text{maximum temperature} + \text{minimum temperature}}{2}$$

or in other words, by taking an average of the highest and the lowest temperatures recorded during the 24-hr period.

Daily march, or cycle. The mean daily "march" of air temperature, or rhythm between night and day, closely follows the temperature of the land surface—which in turn reflects the balance between incoming solar radiation and outgoing earth radiation (Fig. 2.8). From about sunrise until 2 to 4 P.M., when energy is being supplied by incoming solar radiation faster than it is being lost by earth radiation, the air-temperature curve usually rises (Figs. 2.8, 2.9). From about 2 to 4 P.M. to sunrise, when loss by terrestrial radiation exceeds receipts of solar energy, the curve usually falls. It is noticeable, however, that the time of highest temperatures (2 to 4

P.M.) does not exactly coincide with that of maximum solar radiation (12 noon, sun time). This lag occurs because temperature continues to rise as long as the amount of incoming solar radiation exceeds the outgoing earth radiation. Thus although energy receipts begin to decline after noon, they continue to exceed the energy losses until ±3 P.M. Moreover, it requires some time for heat to pass up from the ground to the height where thermometers are usually located.

While daily air-temperature variations closely follow those of the land surface, the range is greatest

Figure 2.8 The march of incoming solar radiation and of outgoing earth radiation for the daily 24-hr period at about the time of an equinox, and their combined effects upon the times of daily maximum and minimum temperatures.

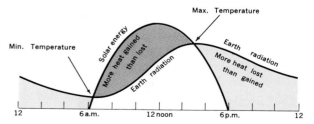

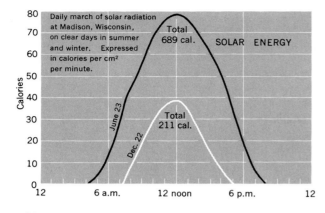

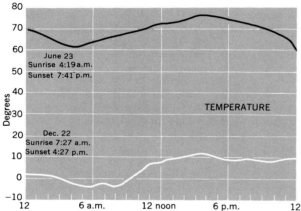

Figure 2.9 Daily march of solar radiation and temperature on clear days at the times of the summer and winter solstices at Madison, Wisconsin. The total solar energy recorded was 3.27 times as great on June 23 as on December 22. Note that temperature lags somewhat behind solar radiation. Advection by southerly winds prevented normal night cooling on Dec. 22.

Mean daily temperature range at different heights on the Eiffel Tower, Paris

Height, ft	Range, °F
6.6	13.7
403.0	9.5
646.0	8.5
991.0	7.0

is similarly retarded. The result is a flattened daily temperature curve (Fig. 2.10). Small diurnal ranges are also characteristic of oceans and their windward coasts, for since a water surface shows little temperature change during the 24-hr period, neither does the air above it (Fig. 2.11). Continental stations have larger diurnal ranges (Figs. 2.11, 2.12).

Often the daily air temperature curve does not have a periodic rise and fall corresponding to the rise and fall of solar radiation. Along coasts, land and sea breezes may modify this diurnal rhythm. And atmospheric disturbances—both because of their variable cloudiness which affects both incoming and outgoing radiation and because of their

Figure 2.10 Daily march of air temperature at Washington, D.C., on clear and cloudy days. The data represent deviations from the 24-hr mean. Each curve is the mean of 10 days at about the time of the autumn equinox, when days and nights are nearly equal. Days were selected when there was a minimum of advection. (*After Helmut Landsberg, Physical Climatology.*)

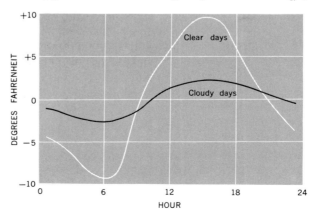

next to the ground and decreases upward. The daily range is also much larger on clear days than on cloudy ones. On a clear day solar radiation quickly warms the solid earth, and it in turn heats the air above it. On a clear night rapid out-radiation from the earth similarly results in quick cooling. But as an earlier section has described, an overcast sky greatly reduces incoming solar radiation by day and so decreases daytime heating, while night cooling

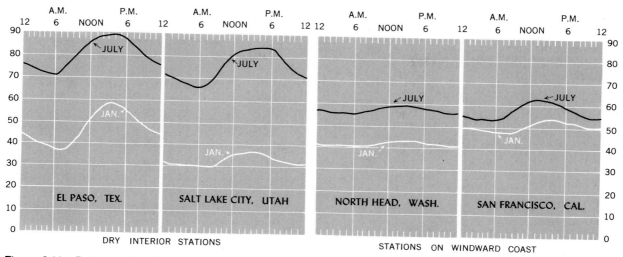

Figure 2.11 Daily march of temperature at marine and continental stations, calculated using average hourly temperatures. This is not the same as the daily range, which is based on the daily extremes regardless of hours of occurrence. Small daily variations of temperature are characteristic of marine climates (North Head and San Francisco), and large ones of the interior stations. (*After Kincer, in Atlas of American Agriculture, U.S. Department of Agriculture.*)

wind systems which advect air of contrasting temperatures from higher and lower latitudes—greatly alter the sun-imposed rhythm of night and day temperatures. In Fig. 2.9, for example, south winds and a cloud cover prevented a normal decline of temperature on the afternoon and night of December 22. In the stormy middle latitudes it is even possible for midnight to be warmer than noon.

If a station's daily maximum and minimum temperatures are plotted for a period of several weeks, two controls usually stand out. First of all, the periodic or diurnal control determined by the sun induces a regular rise of temperature by day and a fall at night. But superimposed upon this sun-controlled pattern is a second one, nonperiodic in character, that is controlled largely by the passage of atmospheric disturbances and their associated air masses. Such disturbances, which occur at irregular periods, drastically modify temperatures through their associated cloud decks and their wind systems. Periodic and nonperiodic controls of temperature are illustrated again and again in Part Two of this book, in the graphs showing maximum and minimum daily temperatures for warm and cold months at selected stations. For example, see Figs. 8.7, 8.19, 9.19, 10.17. In general these graphs demonstrate that while diurnal, or sun, control of temperature is dominant in the tropics and often in middle-latitude summers, nonperiodic control associated with

Figure 2.12 Change in the amplitude of daily temperature (°C) variation with distance from the ocean. Latitude about 40°; elevations not over 500 meters (1650 ft). Numerals indicate locations of stations. Station 1 is an island. Station 2 is at the coast. (*After H. Landsberg, Physical Climatology.*)

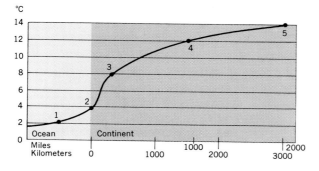

atmospheric disturbances and their air masses is very important in the middle and high latitudes, especially in winter.

In the middle latitudes particularly, the temperature differences between successive days, or interdiurnal temperature changes, are greater in winter than in summer. This reflects the steeper temperature gradients and stronger air-mass control usually found in winter. As the table below shows, interdiurnal temperature changes are also larger in continental than in marine locations.

Temperature differences between successive days in middle latitudes

	Winter	Summer	Year
Central North America and central Asia	8.5°	4.1°	6.3°
West coast of North America	3.6°	1.8°	2.7°

Annual march. The annual march or cycle of temperature reflects the daily increase in solar energy (and therefore in heat accumulated in the air and ground) from midwinter to midsummer, and the decrease from midsummer to midwinter (Fig. 2.13). This is the rhythm of the changing seasons. Usually air temperature lags 30 to 40 days behind the periods of maximum or minimum solar radiation, a delay which reflects the balance between incoming solar and outgoing terrestrial energy. The lag is even greater over the oceans in middle latitudes, where in the Northern Hemisphere, August is often the warmest month and February the coldest. Strongly ocean-dominated land climates—for instance, west coasts in middle latitudes—also may show this typical oceanic lag (Fig. 2.14). In the interior parts of middle-latitude continents, however, July in the Northern Hemisphere is almost universally the warmest month, and January the coldest. Amplitude or range is modest in marine climates and larger in the continental interiors (Fig. 2.14).

Annual temperature is the average of the 12 monthly means, and each of these means in turn

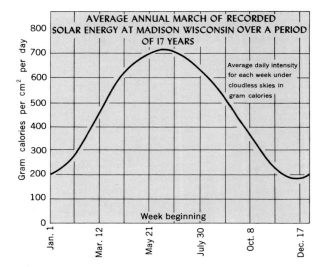

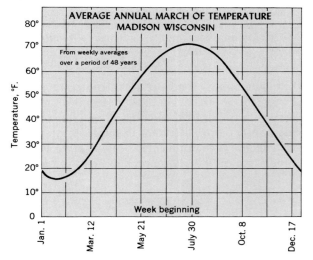

Figure 2.13 Note that the temperature maximum and minimum lag a month or so behind maximum and minimum solar radiation. The solar radiation curve has been smoothed slightly.

is the average of the 28 to 31 daily means. Both the average daily and annual marches, or variations, of temperature for any station can be represented by means of thermoisopleths on a single diagram. These show lines of equal temperature, both for hours of the day and for months of the year, and

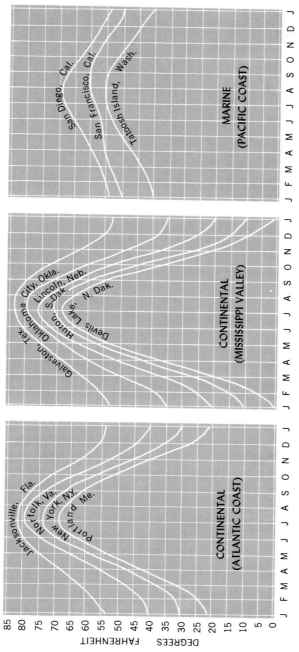

Figure 2.14 Annual march of temperature at marine and continental stations in the United States. Temperature range is large in the interior and along the leeward Atlantic coast, although continentality is somewhat modified in the latter location. Range is small along the windward Pacific coast. At the continental stations, temperature change from north to south is much greater in winter than in summer. (*After Kincer.*)

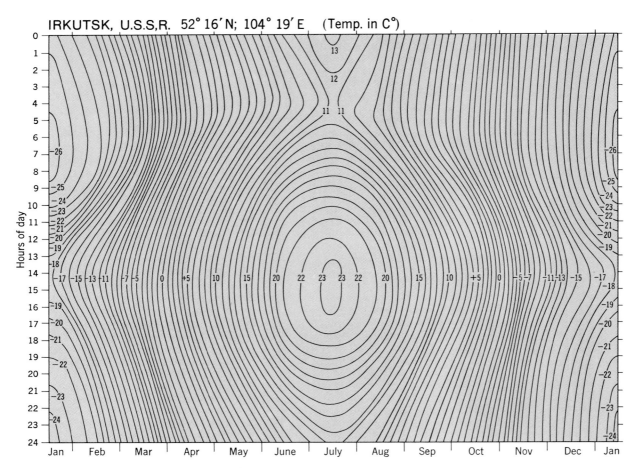

Figure 2.15 Thermoisopleth graphs for two contrasting stations. Irkutsk, U.S.S.R., represents a severe continental climate with large annual and daily variations in temperature. Macquarie Island typifies a strongly marine climate in

thereby make it easy to compare the annual and diurnal temperature regimes. Examples of thermo-istopleth diagrams, which appear throughout Part Two of this book, are given here in Fig. 2.15. One is a severe-climate continental station in the middle latitudes (Irkutsk, U.S.S.R.); the other represents a cool oceanic climate (Macquarie Island).

Temperature singularities. When plotted from monthly means, a station's annual variation of temperature is a fairly smooth curve. But if the annual variation is plotted from daily means obtained over

a long period of observation, or even from 5-day means, the result is a somewhat zigzag line whose general form follows the curve plotted from monthly means. In the annual profile obtained from daily means there appear to be sudden jumps and drops in temperature, signifying warm spells and cold spells which occur in many years at about the same calendar date.

These spells of weather, called *singularities,* do not match in all parts of the world; they are regional rather than universal. Among the better-known singularities are the following: (1) "January thaw"

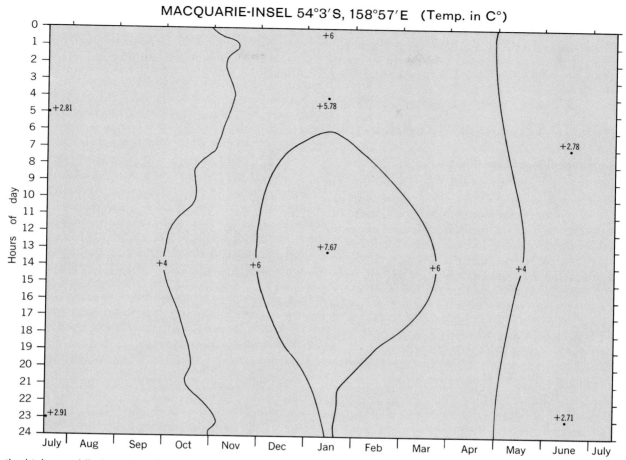

MACQUARIE-INSEL 54°3′S, 158°57′E   (Temp. in C°)

the higher middle latitudes of the Southern Hemisphere, where the annual and daily temperature variations are very small. (*After Troll.*)

of the northeastern coastal United States, a warm period centered around January 20 to 23; (2) "Ice Saints" of western and central Europe, a cold spell bringing frost in May; (3) "Indian Summer," a spell of sunny, warm autumn weather in late October and November in much of the central and eastern United States; (4) a cold spell around mid-October in the central United States, which is often the first genuine outbreak of cold polar air ushering in the winter season.

The reality of climatic singularities is still a controversial topic. Even the most pronounced singular-ities occur in only 50 to 60 percent of the years on record, and in any particular year a singularity may be absent or untimely. They are not a very useful tool in long-range weather forecasting.

The causes of singularities, while not wholly understood, may be of the following nature. The principal control of the annual course of weather is the march of solar radiation. This in turn generates the atmospheric circulation and its seasonal changes. Such fixed terrestrial controls as the distribution and arrangement of land and sea, and the position of great mountain ranges, also play a part in the sequence of weather. In the annual march of weather events, contrasts in such atmospheric elements as pressure, wind systems, and temperature gradually build up between different latitudes and between land-surface features in much the same fashion year after year. When these contrasts reach certain threshold values, sudden shifts in large-scale pressure and circulation patterns appear to occur and in turn to initiate temperature and humidity-precipitation shifts.

VERTICAL DISTRIBUTION OF TEMPERATURE

Normal decrease with altitude

As a general rule, throughout the troposphere or lower 5 to 7 miles of atmosphere, temperature decreases with increasing elevation. Exceptions to this rule occur mostly in the lowest several thousand feet of the air ocean, known as the friction layer, where diurnal temperature changes are greatest and nighttime inversions of temperature are characteristic. The rate of temperature decrease with altitude is not uniform; it varies with the time of day, season, and location. The average, however, is approximately 3.5°F for each 1,000-ft rise in elevation (0.65°C per 100 m), as shown in Fig. 2.16B. This is known as the *normal lapse rate.*

The lapse rate, or vertical temperature gradient, is approximately 1,000 times greater than the average horizontal rate of temperature change with latitude. The steeper the lapse rate, or the more rapid the vertical temperature decrease, the closer to a horizontal position is the temperature-altitude line on a graph chart. A vertical temperature-altitude line indicates an isothermal condition in which there is no change in temperature with altitude.

The fact that air temperature is normally highest at low elevations next to the earth clearly indicates that most of the atmosphere's energy is received directly from the earth's surface and only indirectly from the sun. But the lower air is warmer not only because it is closest to the direct source of heat, but also because it is denser and contains more water vapor, water particles, and dust. These make it a more efficient absorber of terrestrial radiation than the thinner, drier, clearer air aloft. Also remember that vertical temperature distribution throughout the lowest few thousand feet of atmosphere is maintained not only through radiation processes, but also through upward heat transfer by turbulent overturning. Higher in the troposphere, above the cloud-base level, release of heat of condensation in clouds is an additional important mechanism affecting the vertical temperature structure. Above the tropopause, which separates the troposphere from the stratosphere, temperature distribution is more complex, as was described in the introductory section to Part One.

Temperature inversions

Although normally the lower several miles of atmosphere show a decrease in temperature with increasing altitude, often this condition is reversed at certain levels, so that temperatures temporarily or locally increase with altitude. When the colder air lies underneath the warmer air, and closer to the earth's surface, the normal temperature lapse rate is reversed, a phenomenon which is appropriately named a *temperature inversion.*

Surface inversions. One of the commonest forms of temperature inversion occurs at low levels close to the land surface. Over land areas in middle and high latitudes, surface inversions probably develop on a majority of nights. Usually they are the result of radiation cooling of the lower air by the underlying cold land surface. Since a land surface is a more effective radiator than the atmosphere, nighttime cooling is more rapid at the ground than higher in the air, and consequently the coldest air may be found next to the land surface (Figs. 2.16A, 2.20).

Ideal conditions for surface inversions are (1)

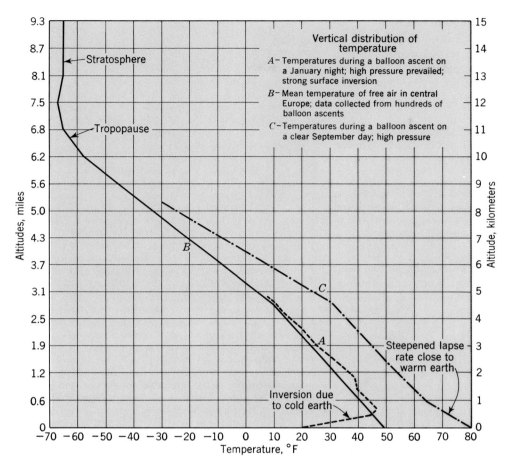

Figure 2.16 Contrasts in the vertical temperature distribution (lapse rates) *A,* on a winter night; *C,* on a summer day; and *B,* under average conditions. On the *B* lapse rate, plotting of temperature is continued up to 15 km (9+ miles) so that the tropopause and the stratosphere are represented.

long nights, as in winter, in which there is a relatively long period when outgoing earth radiation exceeds incoming solar radiation; (2) clear skies, or skies with only high cloud, so that loss of heat by terrestrial radiation is rapid and unretarded; (3) relatively dry air that absorbs little earth radiation; (4) calm air, so that little vertical mixing occurs and the surface stratum can rapidly acquire the temperature of the underlying ground surface; and (5) a snow-covered surface which, owing to reflection of solar energy, heats little by day, and being a poor conductor, re-

tards the upward flow of heat from the ground below.[4] A porous sandy soil with low heat conductivity further lowers the temperature of air next to the ground.

While night and winter provide the best conditions for them, surface inversions are common even in summer, although then they usually disappear

[4] At Milton, Massachusetts, on February 9, 1934, C. F. Brooks recorded a temperature of −27° on the surface of the snow and +24° 7 in. below the surface—a difference of 51°.

by day because of surface heating. Throughout the year at the Eiffel Tower in Paris, temperature increases upward from base to top between midnight and 4 A.M. Over polar areas, temperature inversions are the normal condition, in summer as well as winter. They are also normal for the night hours of the colder months over the snow-covered land masses of the higher middle latitudes (Fig. 2.16). The mean depth of the polar inversion layer is about 0.6 miles, while in the middle latitudes it is usually only a few hundred feet.

When cooler, denser air underlies warmer air, as it does in an inversion, the circumstances are opposed to all upward movement and hence to the formation of rain-bearing clouds. The air is strongly stable; convective overturning is stifled. When nighttime surface inversions form over cities, dust, smoke from chimneys, automobile exhaust fumes, and other atmospheric impurities collect underneath the inversion lid to form dense smoke fogs, or smog. Whether in city or country, surface inversions favor the development of lowland fog, since temperatures are relatively low and relative humidity is usually high within the stable layer (see Chap. 4). A close relationship also exists between surface temperature inversions and frost, for the same conditions are conducive to both (see page 64). While surface inversions are common on flattish land surfaces, they occur far more often in topographic depressions.

Besides being produced by radiation cooling, low-level temperature inversions may be caused by the advection of cold air at the surface, or of warmer air aloft. Such an advection inversion ordinarily covers a more extensive area than inversions of radiation origin. The widespread and intense surface inversions over the great lowland areas of northern Eurasia and North America in the colder months are at least partly caused by advected air masses.

Contrasts between continents and oceans. Since a land surface ordinarily has a wide range of temperatures between day and night, the layers of air next to the land also show large diurnal variations in vertical structure. During the day the surface air has a steep lapse rate, with the warmest air closest

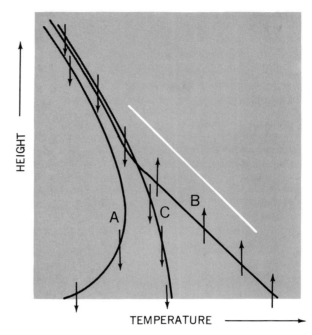

Figure 2.17 Diurnal variations in vertical distribution of temperature, and of stability conditions, over a land surface. *A,* night and early morning; *B,* midday; *C,* evening. White line is the dry adiabat. Arrows indicate direction of eddy transfer of heat. (*From Petterssen.*)

to the ground. At night there is usually a surface inversion (Fig. 2.17).

But over seas, where the daily water-temperature change is slight, the changes in surface air temperature are also slight. In fact, somewhat higher up, where temperature changes are controlled largely by radiation, diurnal variations are greater than they are close to the ocean surface. This results in a slightly steeper lapse rate by night than by day —just the reverse of the situation over land (Fig. 2.18).

Surface inversions and air drainage. In regions where the land surface is uneven, the density of the cold inversion stratum of air causes it to drain gradually off the uplands and slopes into the surrounding lowlands (Figs. 2.19, 2.33). The collecting of cold, dense surface air in valleys and lowlands, where as a result the inversion becomes deeper and more

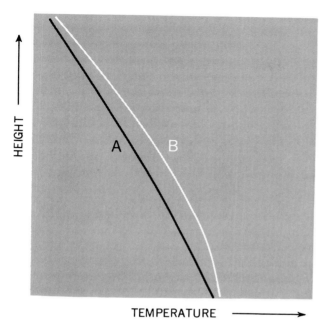

Figure 2.18 Diurnal variations in vertical distribution of temperature over oceans. *A,* night; *B,* midday. Note that the lapse rate in the lower layers is steeper at night than at midday—the reverse of conditions over land. (*After Petterssen.*)

intense, is a phenomenon referred to as *air drainage*. It is because of air drainage that the first freeze in autumn and the last in spring occur in bottomlands, and that the lowest minima on calm, clear winter nights are found in low spots. On one occasion

during a winter cold spell, a minimum night temperature of −8.9° was recorded on top of Mount Washington, New Hampshire, while the surrounding valleys registered temperatures as low as −23° and −31°.

In California, citrus orchards, which cannot tolerate frost, are planted on slopes rather than in the chillier valleys. Coffee plantations in Brazil are situated on the rolling uplands, not in the valleys. Many resort hotels in the Swiss Alps are built on the sunnier, milder slopes above the cold, foggy valleys. Some high hill lands and mountains in middle latitudes have a "thermal belt"—an intermediate zone along the slope. Here nighttime freezing temperatures are less likely to occur than at lower levels (where air drainage creates a greater frost hazard) or at higher levels (where low temperatures are related to the normal lapse rate).[5] (See Fig. 2.19.)

Above-surface inversions. Inversions may also develop at various atmospheric levels well above the earth's surface. A few of these are due to radiation at high levels, but this is a relatively unimportant cause. Others are frontal in origin—that is, they are associated with the convergence of air masses which have contrasting temperatures and densities. Still others are caused by mechanical processes such as subsidence and turbulence. Because of their wide-

[5] Gary S. Dunbar. "Thermal Belts in North Carolina." *Geog. Rev.,* Vol. 56, pp. 516–526, 1966.

Figure 2.19 Air drainage and thermal belts on the slopes of some middle-latitude valleys.

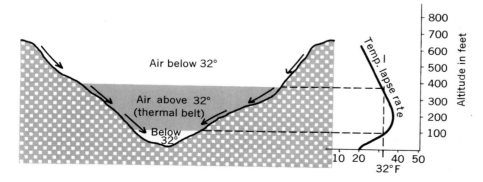

spread distribution and their inhibition of rainmaking processes, the inversions caused by subsidence are climatically the most significant.

Subsidence, or slow sinking of the air, has two major effects. First, the subsiding air is heated by compression, which reduces its relative humidity and hence increases its aridity. Second, it is made more stable (see Chap. 4), or nonbuoyant, so that it opposes upward movement, which is essential for cloudy condensation and precipitation.

Above-surface inversions associated with subsidence are produced in the following way. Turbulence, involving upward and downward movements of air currents, is almost always present near the ground, because of friction created by land-surface features. Since this prevents the lower air layers from participating in the general subsidence, they do not share in the general rise of temperature, and the inversion ordinarily does not reach down to ground level. Consequently a temperature inversion develops between the lower turbulent layer and the subsiding warmed layers aloft (Fig. 2.20). Within the lower turbulent air the lapse rate is normal, so that temperature decreases with altitude. But at the top of the turbulence layer, there is a sharp increase in temperature and a marked decrease in relative humidity. Throughout the inversion layer, which may reach a thickness of 5,000 to 10,000 ft, temperature continues to *increase* with elevation. Further aloft above the subsidence, however, the normal lapse rate is resumed. The inversion layer may extend down to within 1,500 ft of ground level, or even lower. The lower the inversion layer, the thicker it is, and the lower the humidity within it, the less favorable the conditions are for rainmaking processes. The higher the base of the inversion and the greater the turbulence and convection in the lower layers of surface air, the greater the opportunity for rainfall. The inversion layer acts as a lid or ceiling which limits the height of ascending air currents and thus stifles the precipitation mechanism. It is not uncommon, however, for a layer of stratus cloud to develop at the base of the inversion layer.

Subsidence is characteristic of high-pressure systems or anticyclones, especially those which are stationary or slow-moving. Hence the most exten-

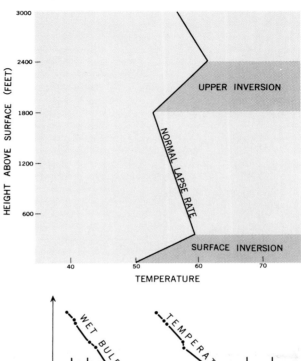

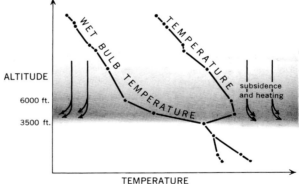

Figure 2.20 *Top:* A vertical temperature distribution showing a radiation surface inversion, and a subsidence inversion aloft.
Bottom: Vertical temperature observations taken in the heart of the maritime trade winds in the eastern South Atlantic Ocean, showing a well-developed subsidence inversion. Above the base of the inversion layer, temperature rises sharply, while the wet-bulb temperature, reflecting a corresponding fall in relative humidity, declines. (*After Kuhlbrodt and Reger.*)

sive and perfectly developed above-surface inversions are found in the warm anticyclones typical of subtropical latitudes and the cold anticyclones which are common in winter over large continents in the middle and high latitudes. Largely because of these inversions, annual rainfall is modest in the subtropical latitudes and winter precipitation is light over the interiors of cold continents. The earth's largest deserts are associated with the subtropical anticyclones and their trade winds. It is especially the poleward parts of the trade winds along the equatorial flanks of the high-pressure cells which are characterized by strong inversions, but these become weaker and are located at increasingly higher levels as the winds approach the equatorial, convergence zone (see Figs. 2.21 and 3.35). This helps to explain why the trade winds, or "trades," are dry near their source region in the subtropical highs, and increasingly showery as they approach the equatorial convergence. Within the subtropical anticyclonic cells over the oceans, the inversion is ordinarily much stronger and lower in the eastern portion of each of the cells than along its western margin. Greater stability and intensified drought therefore characterize the eastern parts of the oceanic subtropical highs and their trade winds.

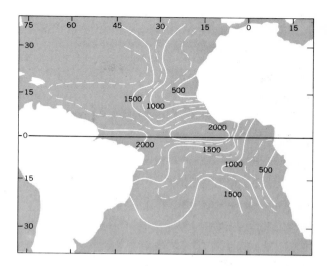

Figure 2.21   Height in meters of the base of the trade-wind inversion over the tropical Atlantic Ocean. Note that the inversion is lowest in altitude toward the eastern side of the oceanic trades. (*After H. Riehl, in Tropical Meteorology.*)

HORIZONTAL DISTRIBUTION OF TEMPERATURE

Isothermal maps.   Temperature distribution over the earth is shown in Figs. 2.22 and 2.23 by means of isotherms—lines connecting places which have the same air temperature. Thus all points on these maps through which any one isotherm passes have identical average temperatures for the period indicated. A world map covered with numerals representing the temperatures of hundreds of stations would be very cumbersome to use. But on a map which shows isotherms, many of the significant facts of thermal distribution can be seen at a glance.

In Figs. 2.22 and 2.23 all temperatures have been reduced to sea level to eliminate the effects of altitude. If this were not done, the complications caused by mountains and other relief forms would obscure the general worldwide isothermal patterns resulting from latitude and land-water distribution. For farmers, engineers, and others who need to put their data to practical use, these maps of sea-level isotherms

are not so useful as maps showing actual surface temperatures.

As Figs. 2.22 and 2.23 show, isotherms in general trend east-west, roughly following the parallels. This is logical, since all places along the same parallel receive identical amounts of solar energy, except for differences in the atmosphere's transparency associated chiefly with cloudiness. The east-west trend of isotherms is additional evidence that solar radiation, which is a function of latitude, is the single greatest cause of temperature contrasts. On no parallel of latitude at any season are temperature differences so great as those between poles and equator.

In some parts of the earth the isotherms are closely spaced, indicating that air temperature changes rapidly in a horizontal direction. In other parts they are widely spaced, showing that horizontal temperature differences are slight. The rate of

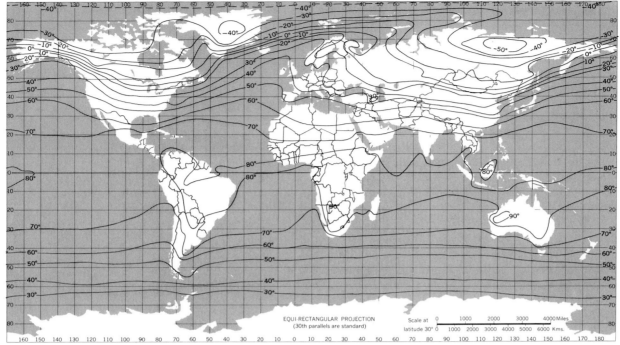

AVERAGE SEA-LEVEL TEMPERATURES
JANUARY

Figure 2.22

change of temperature is called the *temperature gradient*. Closely spaced isotherms signify a steep horizontal temperature gradient, widely spaced ones a weak temperature gradient.

Average annual temperature distribution.   In general the average annual temperature of a place is not a very significant figure, since for much of the earth it is the mean of large seasonal extremes. Only in equatorial latitudes is the air temperature difference between the extreme months small enough so that the average annual temperature becomes meaningful.

The highest average annual temperatures are in the tropics and subtropics, where the largest amounts of solar energy are received in the course of a year. Lowest average temperatures are found in the vicinity of the poles, the regions of least annual

solar energy (Fig. 1.8). In most parts of a low-latitude zone about 45 to 50° wide, there are only very slight temperature differences in a meridional direction. This near-absence of north-south temperature gradients within much of the tropics is partly a result of more cloud near the equator. In these rainier parts of the inner tropics, less solar energy is received at ground level than in the clearer subtropical areas farther from the equator.

The table on page 53 shows that mean annual temperatures for comparable latitudes are usually somewhat higher in the Northern Hemisphere than in the Southern. In part this reflects the gigantic cold source of Antarctica; for tropical and subtropical latitudes it is also evidence of the larger land masses north of the equator. The table also shows that the belt of highest mean annual temperature is not located at the equator but at about 10° N.

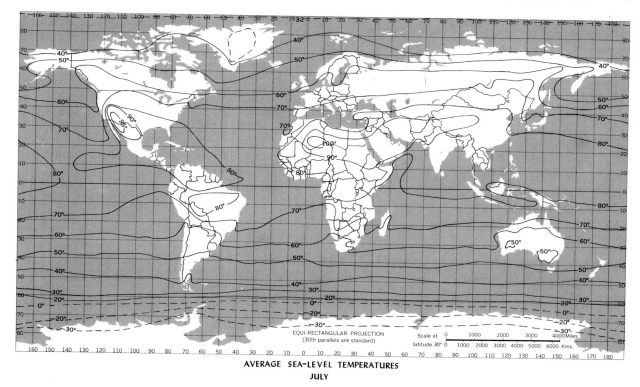

Figure 2.23

This belt, from which temperatures decrease toward either pole, is called the thermal equator. For the year as a whole the main cold pole of the earth is the Antarctic icecap, while the ice plateau of Greenland forms a secondary cold center.

January and July temperatures. For the earth in general, January and July temperatures ordinarily represent the seasonal extremes. As can be seen in Figs. 2.22 and 2.23, isotherms tend to be straighter, have fewer "kinks," and are also more widely spaced in the Southern Hemisphere, whose surface is more homogeneous, being largely water. The isotherms deviate most from an east-west course where they seem to buckle as they pass from continents to oceans, or vice versa. These kinks are caused by the thermal effects of ocean currents and by the unlike heating and cooling properties of land and water surfaces. Next to latitude or sun control, land and water contrasts are the second greatest control of temperature

Mean annual temperature for different latitudes (°F)

	0°	10°	20°	30°	40°	50°	60°	70°	80°
Northern Hemisphere	79.2	80.1	77.5	68.5	57.2	42.3	30.2	14.0	1.9
Southern Hemisphere	79.2	77.5	73.4	65.1	53.6	42.1	31.6	11.3	−3.6

Source: From Landsberg.

distribution. Cool ocean currents off the coasts of Peru and northern Chile, southern California, and southwestern Africa result in a strong equatorward bending of the isotherms. Similarly, warm currents in higher latitudes cause isotherms to bend poleward, as they do off the coast of northwestern Europe.

Following are some of the more significant features of temperature distribution as shown on the seasonal maps in Figs. 2.22 and 2.23:

1. Comparison of the two maps reveals a marked latitudinal shifting of the isotherms between July and January, following the latitudinal migration of solar radiation belts.

2. The migrations of isotherms are much greater over continents than over oceans, because land has greater seasonal extremes of temperature than water (Fig. 2.24).

3. The highest temperatures are found in the tropical-subtropical land areas of the summer hemisphere.

Figure 2.24 Comparative amplitude of isotherm seasonal migration over land and over water in middle latitudes. The difference results from the contrasting heating and cooling properties of land and water surfaces.

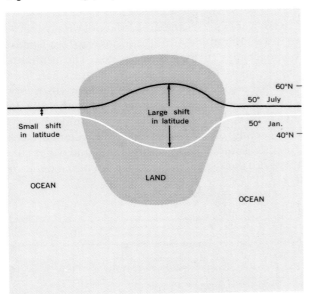

4. The lowest temperatures occur over land in the winter hemisphere. In January, the cold poles are northeastern Asia, northwestern Canada, and Greenland in the Northern Hemisphere, and ice-covered Antarctica in the Southern Hemisphere, even though it is summer there. In July, the earth's cold pole is Antarctica.

5. In the Northern Hemisphere the January isotherms bend abruptly equatorward over the colder continents and poleward over the warmer oceans. In July exactly the opposite happens.

6. No such seasonal contrasts between land and water are to be found in the Southern Hemisphere, for it has no large land masses in the higher middle latitudes.

7. On the January map the lowest temperature is over northeastern Asia—the leeward side of the largest land mass in higher middle latitudes. The next-lowest temperatures are over Greenland and North America.

8. Temperature gradients, like solar energy gradients, are steeper in winter than in summer, especially in the Northern Hemisphere with its large continents. Close spacing of isotherms is particularly conspicuous over the continents of the Northern Hemisphere in January. Gradients are less steep in the Southern Hemisphere, although between about 42 and 50°S over the open ocean, where frigid, heavy water from Antarctica sinks below the warmer surface water of the westerly wind belt, the rate of latitudinal temperature change is rapid.[6]

9. Great seasonal reversals of temperature are not characteristic of the Southern Hemisphere, again because its land area is negligible in the higher middle latitudes.

Annual range of temperature. The difference between the average temperatures of the warmest and coldest months, usually January and July, is the annual range of temperature for a given location. The largest annual ranges occur over the Northern Hemisphere continents, which become alternately hot in summer and cold in winter (Fig. 2.25). Ranges

[6] Harry van Loon. "On the Annual Temperature Range over the Southern Oceans." *Geog. Rev.,* Vol. 56, pp. 499–503, 1966.

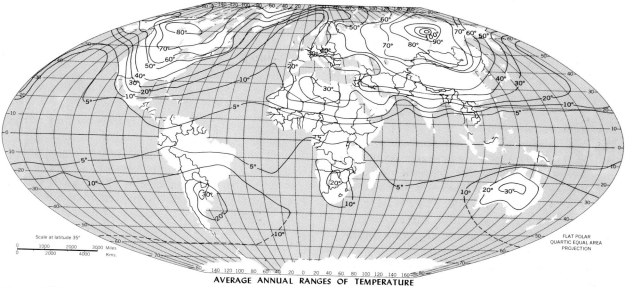

Figure 2.25

are never large near the equator, where solar radiation is fairly constant throughout the year. Nor are there great variations over large water bodies—which accounts for the small ranges found everywhere in the middle latitudes of the Southern Hemisphere. In general, annual ranges increase toward the higher latitudes, but much more markedly over the continents than over the oceans (Fig. 2.26).

Extreme temperatures.   The extremes of heat and cold on the earth, as measured 5 ft above the land-water surface, cover an unbelievably large range of 261°F (146°C)—from 136° above zero in Libya in northern Africa to 125° below zero in east-central Antarctica. At elevations closer to the ground than 5 ft, extremes of heat and cold are considerably larger.

Most of the world's highest temperatures have been recorded in and close to the Sahara in North Africa. For extremes of winter cold, next to Antarctica comes northeastern Siberia, where one station has recorded a minimum of about 110° below zero and another −92°. It is fortunate that over most of the earth the seasonal extremes do not ap-

proach these figures. About one-sixth of the earth's land area has temperatures which do not rise above 100° or fall below zero.

Figure 2.26   Mean surface-air temperature for January and July, and seasonal temperature range, around the earth at 50°N. Note the abrupt change at the North American west coast compared with the gradual change across Europe. (*After H. Riehl, Introduction to the Atmosphere.*)

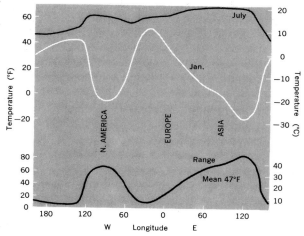

Mean annual range of temperature for different latitudes (°F)

	0°	10°	20°	30°	40°	50°	60°	70°	80°
Northern Hemisphere	1.4	2.0	11.5	22.9	34.4	45.2	53.6	59.4	63.5
Southern Hemisphere	1.4	4.3	9.7	13.0	11.9	9.7	19.4	38.5	45.0

Source: From Landsberg.

Isanomalous temperatures. The temperature distribution shown on an isothermal map of the world is caused by (1) the balance between incoming solar radiation and outgoing earth radiation on a homogeneous surface, upon which is superimposed (2) a field of disturbances produced by such terrestrial phenomena as unequal heating and cooling of land and water surfaces, ocean currents, prevailing streams of airflow, etc. The effects of (1) and (2) can be separated by computing the mean temperatures of the parallels as they would be if temperature were controlled only by radiation on a homogeneous earth, and subsequently determining the difference between the observed mean temperatures of places and the mean temperatures of their parallels. This difference, called a thermal anomaly, represents the amount of deviation from the latitude normal. Lines called *isanomals* can be drawn on a world map to join places which have equal thermal anomalies, producing an isanomalous map (Figs. 2.27, 2.28).

The greatest thermal anomalies are in the Northern Hemisphere, where not only oceans but also extensive land masses occupy middle-latitude locations. In the relatively homogeneous Southern Hemisphere, temperature deviations from the latitude normal are much smaller.

On the *January* map in the Northern Hemisphere,

Figure 2.27   Isanomalies of temperature, January. Temperatures in °C; °F equivalents in parentheses. (*After Hann-Süring.*)

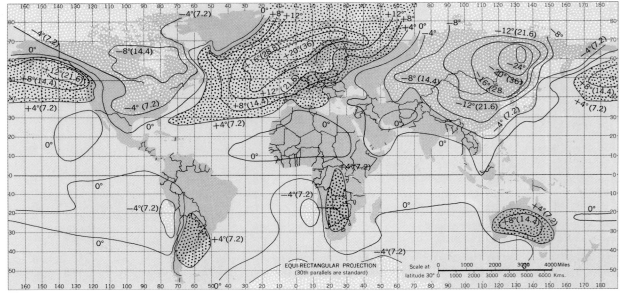

JANUARY

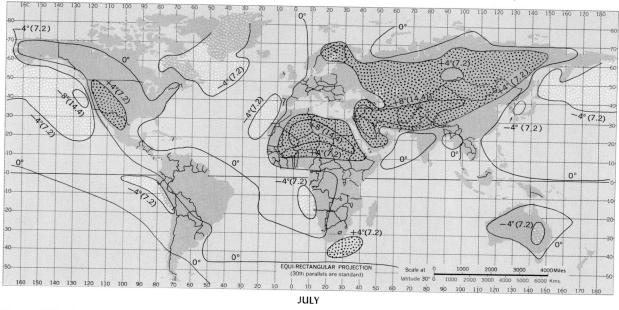

JULY

Figure 2.28 Isanomalies of temperature, July. (*After Hann-Süring.*)

positive anomalies are found over the oceans and negative ones over the continents. The greatest positive anomalies are along the eastern sides of middle-latitude oceans and the adjacent western margins of the continents—locations where warm ocean waters prevail. Nothwestern Europe has the extreme positive anomaly on the earth: a small town on the Norwegian coast at 63°N is 46° too warm for its latitude in January. The extreme negative anomaly is in northeastern Siberia, which is 46° colder than the average for its parallel. In the middle latitudes, where westerly winds prevail, both the positive anomalies over the oceans and the negative anomalies over the continents are shifted away from the centers of the land or water areas toward the eastern or leeward sides. In *July* the situation is reversed, and the Northern Hemisphere continents have positive anomalies while the oceans have negative ones. For the year as a whole, the continents show negative anomalies poleward from about latitude 40° and positive anomalies farther equatorward.

Air temperature and sensible temperature

Correct air temperature can be obtained only from an accurate thermometer properly exposed. One of the main principles of correct exposure is that the thermometer must not be in the sun; otherwise it receives energy not only from the surrounding air but from the absorption of solar radiation as well. It should also be protected from direct radiation from the ground and adjacent buildings and from the open sky.

Sensible temperature refers to the sensation of temperature that the human body feels, as distinguished from the actual air temperature recorded by a properly exposed thermometer. Unlike a thermometer, which has no temperature of its own, the human body, as well as that of every warm-blooded animal, is a heat engine, generating energy at a relatively fixed rate when at rest. Consequently anything that affects the rate of loss of heat from the body affects physical comfort.

A certain amount of heat produced by the body

must be got rid of in order to maintain the proper heat balance. If the climate is so hot that it prevents sufficient heat loss, the person feels uncomfortable, and may even have a heat stroke if the condition continues. On the other hand, if the weather is so cold and windy that heat losses from the body cannot be compensated for by internal heat generation, then chilling, and eventual death, may result.

The factors controlling the thermal comfort of the human body are numerous.[7] Air temperature is the major influence, but wind, relative humidity, and radiation also have important effects. Thus a humid hot day is more uncomfortable than a dry hot one, since loss of heat by evaporation is retarded when the air is humid. A windy cold day feels uncomfortable because the body is losing heat too rapidly; not only does the wind speed up evaporation, but it conducts heat more rapidly away from the body by continually replacing the warm air surrounding it with cold air. A sunny day in winter feels less cold than its air temperature may indicate, owing to the body's absorption of direct solar radiation. Cold air containing liquid moisture particles is especially penetrating, since the skin becomes

moist and evaporation results, while further loss of heat follows from contact with the cold water. Because of its sensitiveness to these factors, the human body is not a very accurate instrument for measuring air temperature.

Although no instrument can accurately measure the human body's feeling with respect to a temperature condition, a wet-bulb thermometer probably comes closest to it. This is just an ordinary thermometer whose bulb is covered with saturated gauze or some other freely evaporating surface. Evaporation from the saturated gauze reduces the temperature of the wet-bulb thermometer below that of the dry bulb by an amount varying inversely with the relative humidity of the atmosphere. For example, in southwestern Arizona the average depression of the wet-bulb thermometer in July during the hottest part of the day is over 30°. Such a depression indicates a very dry atmosphere with a high potential evaporation. Because of this distinctly lower sensible temperature, the extremely high daytime temperatures of that region are not so unbearable as they might appear.

OTHER TEMPERATURE DATA AND THEIR USES[8]

Means, frequencies, and deviations.   The arithmetic mean, which is the sum of all observations divided by the number of observations, is the factor most often used to represent climatic conditions. Mean annual temperature, mean monthly temperature of January, July, or some other month, mean of the daily maximum and minimum temperatures for July or January—such commonly used means have

been referred to again and again in this chapter. Their computation is simple. For example, the mean January temperature for a station is found by adding the average January temperatures for each of the years in which observations exist, and dividing this sum by the number of years represented.

Means alone, however, give an incomplete picture of actual climatic conditions. For example, the mean annual temperatures of Boston and Oxford (England) are both approximately 49°F.; yet the monthly means from which the annual means were computed are very different. At Oxford, the monthly means are grouped more closely about the annual mean, the coldest month being only 11° below and the warmest month 13° above the mean. At Boston, the warm month is 22° warmer than the yearly

[7] See Werner H. Terjung. "Physiologic Climates of Conterminous United States; A Bioclimatic Classification Based on Man." *Ann. Assoc. Amer. Geographers,* Vol. 56, pp. 141–179, 1966.

[8] For a brief treatment of the use of climatological data, see Helmut Landsberg, *Physical Climatology,* 2d ed., Gray Printing Co., Du Bois, Penn., pp. 66–106. For a more detailed treatment, see V. Conrad and L. W. Pollok, *Methods in Climatology,* 2d ed., Harvard University Press, Cambridge, Mass., 1950.

Mean monthly and annual temperatures (°F)

	J	F	M	A	M	J	J	A	S	O	N	D	Yr
Boston	27	28	35	45	57	66	71	69	63	52	41	32	49
Oxford	38	40	42	47	53	59	62	61	57	49	44	40	49

average, and the cold month 22° colder. It is obvious that there are striking temperature differences between Boston and Oxford even though they have similar mean annual temperatures. The same criticism applies to other temperature means, and to means of rainfall, winds, and the rest of the climatic elements.

A more comprehensive picture of climate is obtained by combining arithmetic means with data on the frequency of occurrence of certain climatic values. Frequency is defined as the number of times a value occurs within a specified interval; for example, the frequency of hours during which the temperature at a station was within a certain limit. A very important type of frequency is the number of occurrences of a specific deviation or departure from the mean. For example, the average temperature at Philadelphia for a period of 60 years was 54.4°. The departures of the individual yearly averages from 54.4° were computed and their mean determined, with the results shown in the following table. When the

Frequencies of annual temperature deviation from 60-yr mean at Philadelphia (°F)

Temperature deviation	−3	−2	−1	0	+1	+2	+3
Number of cases		4	5	10	17	18	4	2

Source: From Landsberg.

Mean departures of monthly means (°C)

	Winter	Summer	Mean
Interior of North America	2.54	1.20	1.95
Northern Russia	3.43	1.61	2.33
Northern Germany	2.02	0.93	1.28
Northern slope of Alps	2.28	1.06	1.56
England	1.41	0.95	1.24

Source: From Hann.

same technique is applied to monthly means, the deviations are ordinarily more striking. For example, the mean departure of the monthly mean temperature for winter months in northern European Russia is as much as 3.43°C, and in interior North America it is 2.54°C.

Annual temperature curve (march of temperature). The annual course or march of temperature, as represented by a line joining the 12 monthly means, provides information on a number of temperature characteristics. One is the *range of variation,* i.e., the difference between the temperature of the warmest and coldest month, or *annual range.*

A second characteristic is the *phase,* which refers to the month with the highest and the month with the lowest temperature. At most land stations in the middle latitudes, July is the warmest month and January the coldest, so that the lag of temperature behind solar radiation is about a month or a little less. But over the oceans and at various marine stations located on islands or on windward coasts, August and February may be the months of maximum and minimum (Fig. 2.14). Although January is the coldest month at San Francisco, for example, September is the warmest. In regions of strong monsoons, such as India, the highest temperature is frequently in May, before the summer solstice. This is because the monsoon, which is the time of maximum cloudiness and rainfall, comes in June, July, and August, and greatly diminishes the amount of solar radiation received at the surface during those months (Fig. 8.17).

A third characteristic of the annual temperature curve is *symmetry* or *asymmetry.* There are two axes of symmetry, and therefore symmetry can be analyzed in two directions. As shown in Fig. 2.29, one axis of symmetry is the straight line B-7, which is

SAN DIEGO, CALIFORNIA

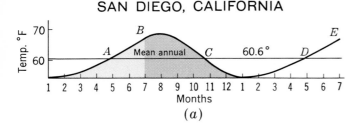

(a)

BARNAUL, SOVIET RUSSIA

(b)

Figure 2.29 Contrasting symmetry of the annual temperature curves for a marine and a continental weather station.

the ordinate of the month of July. In a symmetrical curve the area 1-A-B-7 should equal the area 7-B-C-1. In the annual curve for Barnaul, a typical continental station, this is approximately true, but it is not the case in the San Diego curve. At San Diego the latter half of the year is considerably warmer than the first half; that is, fall is warmer than spring. The importance of this fact for all forms of life is obvious.

The second axis of symmetry (Fig. 2.29) is the straight line ACD, which is the mean of the monthly mean temperatures. Symmetry with respect to this axis is determined by comparing the length of the upper part of the curve with that of the lower part. In the Barnaul curve, AC is somewhat longer than CD, while in the San Diego curve, CD is longer than AC. On the whole, curves that are above the mean a longer time than they are below indicate a longer period of vegetation growth.

Mean duration of certain temperatures. There are certain temperatures which are recognized as critical: for example, 32°, the freezing point of water; 40 or 42°, the approximate temperature at which seeds germinate and plants begin to grow; and perhaps 50°, often given as the lower limit of human comfort. The dates on which an annual temperature curve rises above such important thresholds and sinks below them, and the length of time between these dates, are of great significance climatically.

Although most plants are not killed by cool temperatures above 32°, many species do not start growing until the daily temperature rises to 40° or a little more. The effectiveness of temperature in promoting plant growth, or the *temperature efficiency,* may therefore be measured by the number of days with a temperature above 40°, and the amount of temperature rise above that point (Fig. 2.30). A high temperature efficiency is obviously an advantage, for more different kinds of crops can be grown. Each crop has its own basic temperature at which growth begins, and each crop also requires a certain number of heat units to bring it to maturity. Thus peas germinate at a temperature of 40°F, but sweet corn requires 50°F. As a rule the rate of maturing is approximately doubled for each 10°C, or 18°F, increase in temperature. For peas, each Fahrenheit degree above a mean daily temperature of 40° is a *growing degree-day,* or heat unit. For sweet corn, each unit above the mean daily temperature of 50° is a heat unit. In the case of corn, a series of 3 days whose mean temperatures are 55, 60, and 65° would be counted as 5, 10, and 15 degree-days, or a total of 30 in all. Data on growing degree-days serve as a guide in crop planting and in forecasting dates of crop maturity. If 1,500 heat units are required

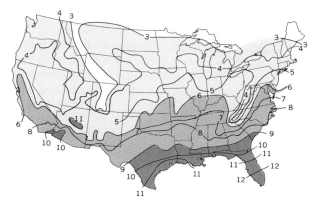

records compiled over a period of many years are used by the Weather Bureau in establishing monthly averages of degree-days for the entire country. These form the basis for estimating the fuel requirements. Shortly after the close of each month, the Weather Bureau computes daily degree-day totals for more than 300 locations and publishes them.

Period without freeze.   One of the most important limiting values in temperature is the freezing point of water—32°F or 0°C. The period between the last day in spring which has a minimum temperature below 32° and the first day in fall with the same minimum is often called the "frost-free season" or "growing season." In reality, however, there is no simple relationship between plant growth and

Figure 2.30 Temperature efficiency during the average frost-free season. In most plants active growth in spring does not begin until the mean daily temperature rises to, or somewhat exceeds, 40°F. The temperature efficiency of an area is the effectiveness of temperature in promoting plant growth. It varies with the number of days that have mean temperatures above 40°F, and with the amount of rise above this point. Numbers on the map represent efficiency units in hundreds.

Figure 2.31 Growing degrees increase rapidly in late spring and summer. (*U.S. Weather Bureau.*)

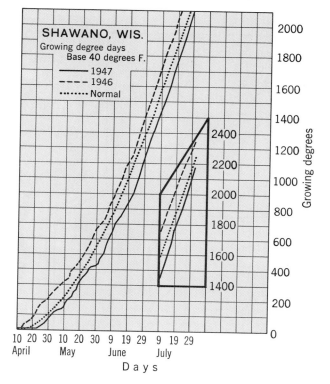

for maturity, corn will theoretically be harvested when this number has actually accumulated. However, the tendency in agriculture at present is toward a use of more complex indices for plant growth than the concept of degree-days provides (Fig. 2.31).

Another type of degree-day is used by heating engineers, who have discovered that the difference between 65°F and the mean outside temperature for a day, a week, or any other period, is an accurate index of the fuel required for heating a building. Thus the heating fuel consumed in a month which has an average daily mean temperature of 15°F, as compared with a month whose average daily mean is 40°F, would be $(65 - 15)/(65 - 40) = 50/25$, or twice as great. The unit used is the *heating degree-day,* which is defined as a departure of 1° per day from 65°. If the mean outside temperature for a day is 25°, the number of degree-days for that 24-hr period would be $65 - 25$, or 40. The degree-days for any longer period are computed by determining the number of degrees by which the mean temperature for each individual day falls below 65° and totaling them for the period. Degree-day a freezing temperature. Some hardy plants can withstand severe freezes, while other sensitive ones are injured by low temperatures well above freezing. The term frost is also rather ambiguous. "Killing frost" is too vague to be very helpful. In addition, the occurrence of frost is extremely localized, its distribution depending largely on the irregularities of the land surface. Thus the soundest concept to use is the 32° point marking the limit of freezing temperatures, and it is in relationship to this point that the terms frost and growing season are used here.

A short freeze-free period greatly limits the number and kind of crops that can be grown in a region. Where the freeze-free period is less than 90 days, the opportunities for agricultural development are extremely meager. The greater diversity of crops which is possible with a long freeze-free season increases the stability of agriculture and tends to make it more profitable. Figure 2.32 shows in a very

generalized way the duration of the freeze-free period for the land areas of the earth. As a rule, the extensive areas in the tropics where a freezing temperature has never occurred are not the regions of largest present-day agricultural development. These regions do have great potentialities, however.

Protection from freeze

Throughout almost all the middle latitudes, freezing temperatures are most serious as a menace to crops in autumn and spring. In subtropical latitudes such as California and Florida, however, midwinter freezes are the critical ones, because sensitive crops are grown during that season. In the southern parts of boreal or subarctic Canada and the U.S.S.R., summer freezes sometimes do serious damage to cereal crops. In tropical lowlands, freezing temperatures usually do not occur.

Figure 2.32 Average length of the frost-free period, or growing season, in days. (*From The Great Soviet World Atlas, Vol. 1.*)

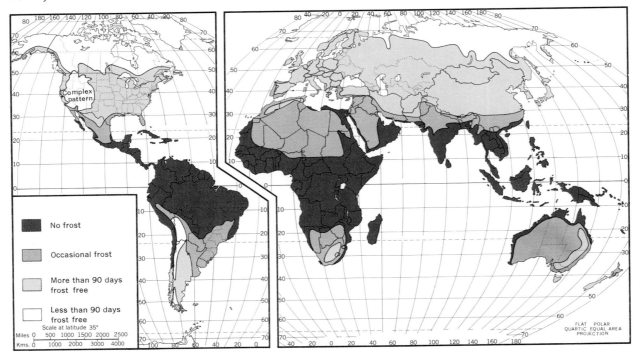

Conditions favorable to frost. Ideal conditions for a freeze are those which favor rapid and prolonged surface cooling: importation of a mass of chilly polar air, followed by a clear, dry, calm night during which the surface air, by radiation and conduction, is reduced below freezing. The original importation provides the necessary mass of air which is already cool, although still somewhat above freezing. Further rapid loss of heat by earth radiation during the following clear night is all that is necessary to reduce the temperature of the surface air below freezing. But even though these conditions may prevail over extensive areas, the destructive effects of the frost ordinarily are local and patchy, depending chiefly on surface configuration, air drainage, and soil conductivity.

Artificial protection. The problem of artificial protection from freeze is important only in regions of highly sensitive and valuable crops which occupy restricted areas. It is obviously impossible to protect such extensively grown field crops as corn or small grains, even when Weather Bureau warnings are issued 12 to 24 hr in advance. Most field crops in the middle latitudes are annuals planted late enough in spring to avoid frost. If fall-sown, they do not reach a stage in which they can be injured by frost until the danger from a spring freeze is past. Ordinarily these crops of annuals are harvested in summer or early fall before frosts are likely to occur, although in the latitude of southern Wisconsin, corn is sometimes damaged by early fall frosts.

Orchards such as the valuable citrus groves of California and Florida, and important truck and market gardens occupying small areas, are the crops which require special precautions to minimize the danger of frost. This may be done either by planting them on a site where there is least danger of a freeze, or by adopting special measures for preventing frost occurrence. Two types of sites which are freer from frost than the surrounding countryside are the lower slopes of highlands and coasts situated on the windward side of fairly large bodies of water (Fig. 2.33). Slopes are warmer than valleys owing to air drainage, which was discussed in an

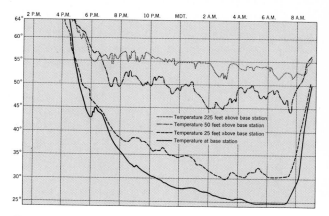

Figure 2.33 Actual temperature recordings from 4 P.M. to 9 A.M. at the base, and at different heights above the base, along a steep slope on a clear, still night in spring. In this instance, air drainage resulted in a minimum temperature of 25°F at the foot of hill, causing freeze damage to fruit. This was 26°F lower than the temperature 225 ft up the slope. (*U.S. Weather Bureau.*)

earlier section. Coasts are protected because an adjacent body of water has regulatory effect upon the temperatures of land, a feature which is conspicuous only along its windward side. The major fruit belts along the southeastern shore of Lake Michigan, and the south shores of Lakes Erie and Ontario, illustrate the protective influence of water bodies.

But dangerous freezes are occasionally inevitable even on the most favorable sites. Damage to sensitive crops which occupy restricted areas may be prevented or lessened in two ways: either the loss of heat from the earth can be retarded, or a temperature above freezing can be maintained immediately around the threatened crop by artificial heating. Heat loss from the land may be checked by spreading over the crop some nonmetallic covering such as paper, straw, or cloth, thereby intercepting the heat being radiated from the ground and plants. The purpose of the cover, quite obviously, is not to keep the cold out but to keep the heat in. This inexpensive type of frost protection is the one used by the housewife to save her garden plants from freezing. It is not so well suited to the protection of extensive orchard areas, however.

In California and Florida, the huge losses in the citrus areas resulting from an occasional freeze have inspired the most careful and sustained experimentation in frost-fighting methods. A single freeze in California can cause citrus losses amounting to tens of millions of dollars. Many protective devices, some of them of little or no practical value, have been constructed and tried out in the citrus groves. The two most effective and practical ones appear to be orchard heaters, consisting of sheet metal cylinders in which diesel fuel is burned, and huge fans, which agitate the lower few meters of air so that the surface inversion is dissipated. In the cranberry areas of Wisconsin and New England, the bogs are usually flooded if killing frost seems probable. The water surface created by flooding cools more slowly than a land surface and thereby lessens the freeze hazard. Moreover, a protective cover of light fog is likely to develop over the flooded fields.

SELECTED REFERENCES FOR FURTHER STUDY OF TOPICS IN CHAPTER TWO

1. **Heating and cooling the atmosphere.** Byers, pp. 11–44; Haurwitz and Austin, pp. 14–22; Landsberg, pp. 118–126, 141–146; Petterssen, pp. 77–99; Hare, pp. 9–16; Kendrew, *Climatology,* pp. 13–22; Miller and Parry, pp. 38–57.

2. **Distribution of temperature over the earth.** Blüthgen, pp. 66–97; Byers, pp. 45–60; Critchfield, pp. 28–36; Conrad, pp. 16–38; Haurwitz and Austin, pp. 23–43; Kendrew, *Climatology,* pp. 23–30; Koeppe and De Long, pp. 42–47; Landsberg, pp. 146–165; Miller and Parry, pp. 48–54; Willett and Saunders, pp. 38–81.

3. **General references on temperature.** M. I. Budyko, *The Heat Balance of the Earth's Surface* (translation), U.S. Weather Bureau, Washington, D.C., 1958; Arnold Court, "Duration of Very Hot Temperatures," *Bull. Amer. Meteorol. Soc.,* Vol. 33, no. 4, pp. 140–149, 1952; D. M. Houghton, "Heat Sources and Sinks at the Earth's Surface," *Meteorol. Magazine,* Vol. 87, pp. 132–142, 1958; Henry G. Houghton, "On the Annual Heat Balances of the Northern Hemisphere," *J. Meteorology,* Vol. 11, pp. 1–9, 1954; W. H. Ransom, "Solar Radiation and Temperature," *Weather,* Vol. 8, pp. 18–23, 1963; K. A. Salishchev, "The Cold Pole of the Earth," *Geor. Rev.,* pp. 684–685, 1935; R. Stone, "Solar Heating of Land and Sea," *Geography,* Vol. 40, p. 288, 1955; Carl Troll, "Thermische Klimatypen der Erde," *Petermanns Geograph. Mitt.,* Vol. 89, pp. 81–89, 1943.

CHAPTER THREE

WIND AND PRESSURE SYSTEMS

It is chiefly as *controls* of temperature and precipitation, rather than as *elements* of climate, that atmospheric pressure and winds are crucial to life on earth. Pressure as a climatic element does not seem to make a great difference to life forms with its slight variations and changes that occur at the earth's surface. Winds are more important than pressure as an element, for wind speed does have an effect on plant life, and also on man's comfort and safety. Moreover, wind influences the rate of evaporation and therefore affects sensible temperatures. But on the whole, the winds of a place do not contribute as much to its climatic environment as solar energy, temperature, and precipitation do.

As climatic controls, however, pressure and winds rank among the highest. Although very small pressure changes are imperceptible to human beings, they may induce remarkable variations in general weather conditions. Also, they are usually the cause of wind, and wind is a powerful determinant of the air temperature and moisture conditions of a place. In fact, it can be said that pressure and winds operate as climatic controls mainly through their effects upon temperature and precipitation—two climatic elements of critical importance as far as living things are concerned.

Winds serve two basic climatic functions.

1. By transporting heat, in either sensible or latent form, from the lower to the higher latitudes, winds are the principal agent in maintaining the latitudinal heat balance of the earth. They operate to correct the unbalanced receipt of solar radiation between low and high latitudes.

2. Winds also provide the land masses with much of the moisture necessary for precipitation, since they transport the water vapor which is evaporated over the oceans to the lands, where subsequently a large part is condensed and falls as rain. Because rainfall distribution over the earth is much affected by the atmospheric circulation, this chapter on pressure and winds precedes Chapter 4's discussion of moisture and precipitation.

Measurement of atmospheric pressure. Atmospheric pressure is a measure of the weight of the atmosphere above the height of observation. The total weight of the air ocean is over 5,600 trillion tons. If this weight of air were to be replaced by the same weight of water, the water would form a layer about 34 ft (10+ meters) deep over the whole earth. In spite of its great total weight, the atmosphere as a whole does not fall toward the solid-liquid earth. Rather, it remains suspended, neither rising nor sinking, except for individual air masses. This is because the downward pull of gravity on the air ocean is just balanced by another force urging it upward, a force created by the decrease in atmospheric pressure with height. Such equilibrium is called *hydrostatic balance.*

A column of air 1 sq in. in cross-sectional area extending from sea level to the top of the atmosphere weighs nearly 15 lb. This weight is balanced by a column of mercury 29.92 in., or 760 mm, tall having the same cross-sectional area. The value 29.92 in., or 760 mm, is accepted as the normal value of atmospheric pressure at sea level at latitude 45°. Thus it is customary to measure air pressure in terms of its equivalent weight expressed in inches or millimeters of a column of mercury—in other words, to measure it by a mercurial barometer.

Another measurement unit of atmospheric pressure, called a millibar, has been widely adopted by the national weather services of the world. The millibar is a force equal to 1,000 dynes per cm^2 and a dyne is a unit of force approximately equal to the weight of a milligram. Sea-level pressure (29.92 in. or 760 mm) under this system of measurement is 1,013.2 millibars (mb). One-tenth of an inch of mercury is approximately equal to 3.4 mb. On all United States weather maps published since January, 1940, pressure readings have been in millibars instead of inches. These maps have *isobars* (lines of equal surface pressure) drawn for every 4 mb of pressure, an arrangement that closely corresponds to the interval of $1/10$ in. previously used.

Barometric pressure equivalents

Millibars	Inches of mercury	Millimeters of mercury
940	27.76	705.1
950	28.05	712.5
960	28.35	720.1
970	28.65	727.7
980	28.94	735.1
990	29.24	742.7
1,000	29.53	750.1
1,010	29.83	757.7
1,013.2*	29.92*	760.0*
1,020	30.12	765.0
1,030	30.42	772.7
1,040	30.71	780.0
1,050	31.01	787.7

* Pressure at sea level at latitude 45°.

Types of pressure systems and their origins. Pressure systems vary greatly in both size and duration. At one extreme are the immense semipermanent cells of high and low pressure which can be seen on average monthly, seasonal, and annual pressure charts of the earth. But in addition there are many smaller, short-lived moving pressure systems which show up chiefly on the daily weather map. These are closely identified with daily weather changes.

The two main types of pressure are high-pressure systems and low-pressure systems. Centers of relatively low pressure are designated as *depressions,*

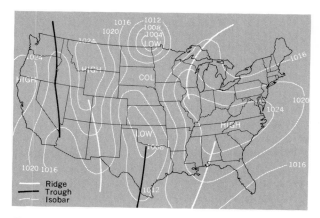

Figure 3.1 Pressure systems. (*From Aviation Weather.*)

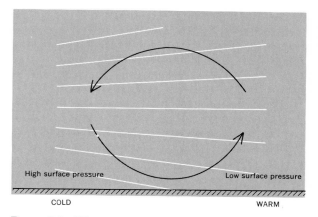

Figure 3.2 What may be called the expected relationship between air temperature on the one hand and surface pressure and atmospheric circulation on the other. Sloping white lines indicate surfaces of equal pressure.

cyclones, or *lows.* Elongated lows are called *troughs.* Centers of high pressure are called *anticyclones* or *highs.* An elongated high is a *ridge* (Fig. 3.1).

At present several features of average pressure distribution over the earth, and many of the pressure changes which occur from day to day, cannot be satisfactorily explained. Since a column of hot air is less dense and weighs less than a column of cold, heavy air of the same length at the same height, it seems likely that some surface lows are at least partly caused by high surface temperature, and some highs by low surface temperature (Fig. 3.2). Most pressure systems, however, appear to be related to nonthermal, or mechanical, factors affecting air in motion, such as frictional effects, centrifugal force, and the blocking effects of highlands. These mechanical forces operate in various ways to either remove or pile up air, thereby producing low- and high-pressure systems. Thus while pressure differences can induce air movement, the circulation of the air in turn may act to generate new pressure differences. So it is not always easy to say which comes first and is the primary cause.

DISTRIBUTION OF ATMOSPHERIC PRESSURE

Vertical distribution

Since air is very compressible, air weight, or pressure, decreases rapidly with height. The lower layers of the atmosphere are the densest, because the weight of all the layers above rests upon them. The change in pressure with height is proportional to density, so that it is possible to determine approximate elevation by the pressure of the atmosphere at that level.[1] For the first few thousand feet above sea level, the rate of pressure decrease is about 1 in., or 34 mb, of pressure for each 900 to 1,000 ft of elevation. At higher altitudes the air rapidly becomes much thinner and lighter. Although the air ocean extends upward for thousands of miles, at an elevation of about 18,000 ft one-half the atmosphere by weight is below the observer, and at 36,000 ft three-quarters lies below. Since the human body is not physiologically adjusted to the low pressures and low oxygen content of the air at high altitudes,

[1] Accurate determination would require an adjustment for temperature.

nausea, faintness, and nosebleed often result from a too rapid ascent. Oxygen tanks are a part of the normal equipment of aircraft operating at high altitudes.

Atmospheric environment

Altitude (ft)	Pressure		Temperature (° C)
	Inches of mercury	Millibars	
70,000	1.3	44.0	− 55.2
50,000	3.4	115.1	− 56.5
35,000	7.0	237.0	− 54.0
18,000	14.9	506.0	− 20.6
10,000	20.6	697.5	− 4.8
5,000	24.9	843.1	5.1
Mean sea level	29.92	1013.2	15.0 (59.0° F)

Horizontal distribution at sea level

As the accompanying table shows, pressure declines rapidly with increasing altitude. Consequently it is necessary to adopt a standard height of observation when comparing pressure readings for stations at different elevations. This standard is the sea-level surface.

Average annual conditions. Just as temperature distribution is represented on maps by isotherms, so atmospheric pressure distribution is shown by isobars. These are lines connecting places having the same atmospheric pressure at a given elevation. Pressure distribution charts are constructed for sea level and also for a number of different constant-pressure surfaces in the free atmosphere—commonly the 700-mb surface at about 10,000 ft and the 500-mb surface at about 18,000 ft. On the sea-level pressure charts (see Figs. 3.6 and 3.7) the effects of different elevations on the continents have been eliminated by reducing all pressure readings to sea level.

Where isobars are closely spaced, there is a rapid change of pressure in a direction at right angles to the isobars. The rate and direction of pressure change is called the *pressure gradient*. The gradient

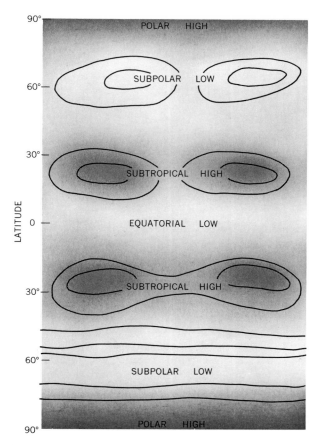

Figure 3.3 Idealized arrangement of zonal sea-level pressure. Except in the higher latitudes of the Southern Hemisphere, the zonal pressure "belts" are cells of low or high pressure arranged in belts which are concentrated in particular latitudes.

is weak where isobars are widely spaced and steep where they are closely spaced. The horizontal pressure gradient may therefore be defined as the decrease in pressure per unit distance in the direction in which the pressure decreases most rapidly (see Figs. 3.6, 3.7, and 3.9).

The earth's average sea-level pressure shows something of a zonal arrangement in latitudinal belts (Fig. 3.3). But while a belted pattern is fairly conspicuous, especially when pressure is averaged for all longitudes, some of the so-called belts are more accurately described as centers, or cells, of pressure whose long axes are aligned roughly east-west. The belted arrangement is much more conspicuous in the relatively homogeneous Southern Hemisphere—a fact which suggests that the great continents, with their seasonal temperature contrasts, their frictional effects on winds, and their highland barriers that obstruct the free flow of air currents, have much to do with the cellular pattern visible on the seasonal pressure maps.

Generalized sea-level pressure patterns as represented in Figs. 3.3 and 3.8 show these outstanding features:

1. The dominant element is the series of high-pressure centers or cells which form discontinuous belts of high pressure, one in either hemisphere, located at about 25–35° N and S. These are the subtropical highs. Although they represent what is probably the key factor in the earth's surface pressure, their origin is not fully understood—except that it is definitely mechanical or dynamic and not directly thermal.

2. Surface pressure declines equatorward from the two subtropical high-pressure belts. Thus there is a general trough of low pressure between them in the vicinity of the geographic equator.

3. Pressure likewise decreases poleward from the subtropical highs toward cells of low pressure located at about the latitudes of the Arctic and Antarctic circles. These are the subpolar cells or troughs of low pressure. They form individual oceanic cells of low pressure in the Northern Hemisphere, but in the Southern Hemisphere they consist of a deep and continuous circumpolar trough of low pressure. The causes of the subpolar lows are also not clear, but these pressure systems can scarcely be of direct thermal origin.

4. Poleward from about latitude 65°, aerological data are so scarce that the pressure distribution pattern cannot be described with certainty. It is generally assumed that anticyclones, partly of thermal origin, occupy the inner polar areas.

On a weather map of the world for any particular day and time, the arrangement of average zonal surface pressure shown in Figs. 3.3 and 3.8 is not so evident. This fact suggests that the generalized

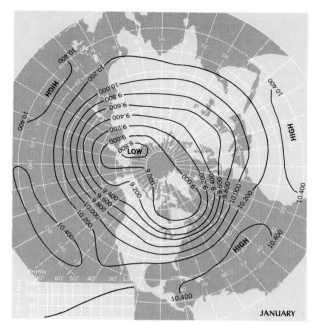

Figure 3.4 January: 700-mb pressure surface in the Northern Hemisphere. (*After Namias and Clapp.*)

Figure 3.5 July: 700-mb pressure surface in the Northern Hemisphere. (*After Namias and Clapp.*)

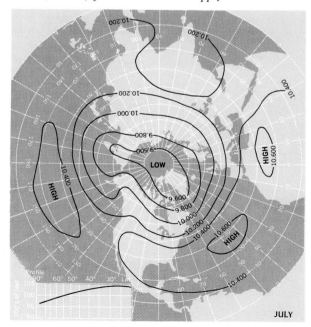

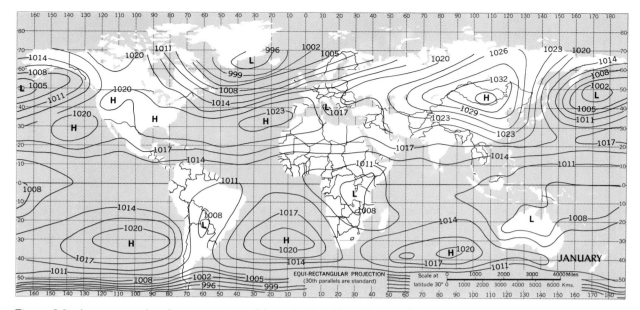

Figure 3.6   January sea-level pressure, in millibars (mb). (*After Mintz and Dean.*)

Figure 3.7   July sea-level pressure, mb. (*After Mintz and Dean.*)

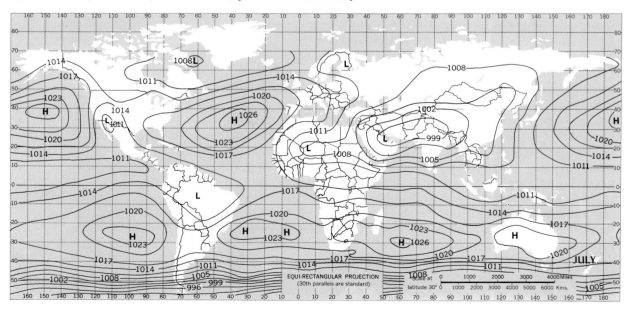

cells and belts of surface pressure are partly, at least, statistical averages of complicated day-to-day systems of smaller traveling highs and lows.

With increasing elevation above the earth's surface, the cellular pattern of pressure distribution so conspicuous at sea level gradually becomes less clear. It is only faintly discernible in the mid-troposphere at about 10,000 ft, or the 700-mb level. At this elevation there is a broad plateau of high pressure in tropical latitudes, from which pressure decreases to minima in the polar regions (Figs. 3.4 and 3.5).

Distribution in January and July. Average sea-level pressures for the seasonal extremes, January and July, are shown in Figs. 3.6, 3.7, and 3.8. Although pressure distribution at these times retains many of the features of average annual conditions, there are some noteworthy departures:

1. Pressure belts and cells, like belts of solar energy and temperature, shift northward in July and southward in January—a fact of crucial importance to climate. This latitudinal migration is most readily observed in Fig. 3.8 showing the seasonal profiles of pressure.

2. In general, pressure is higher and pressure gradients are steeper in the winter hemisphere. Thus between summer and winter there is an enormous mass transfer of atmosphere from the warmer summer hemisphere to the colder winter hemisphere. It has been calculated that the July-to-January net flow of air across the equator amounts to about two trillion tons, or 1/2,500 of the total mass of the atmosphere.

3. The subtropical belts of high pressure are more continuous in the winter hemisphere; in the summer hemisphere, they are weakened by the heated continents. As a rule these subtropical cells are best developed over the eastern sides of the oceans and tend to be weaker toward the western sides. In the Northern Hemisphere in July, the subtropical highs along the eastern sides of the oceans extend well poleward and into the middle latitudes, strongly affecting the climate of the western sides of the continents.

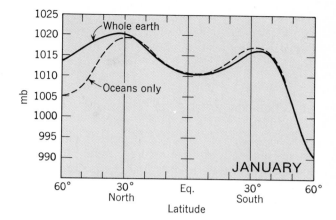

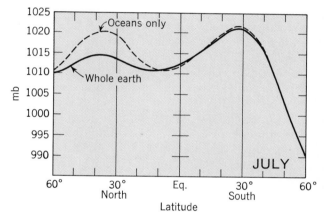

Figure 3.8 Profiles of sea-level pressure from 60°N to 60°S, averaged for all longitudes, at the time of the extreme seasons. Equatorial low, subtropical highs, and subpolar lows are conspicuous features of both profiles. Note the seasonal north-south displacements of the pressure belts following the sun.

4. In the Southern Hemisphere, the subpolar low is very deep and forms a continuous circumpolar trough in both winter and summer. But because of the great continents in the Northern Hemisphere, the subpolar low there is represented by individual oceanic cells which are strongly seasonal in character. In January a deep cell of low pressure occupies both the North Pacific (Aleutian Low) and the North Atlantic (Iceland Low), but in July these oceanic lows are only faintly discernible.

5. In January a strong cell of high pressure develops over the higher middle latitudes of the cold Eurasian continent; a weaker one appears over smaller North America. In July these same continents, now warm, develop weaker thermal lows in somewhat lower latitudes.

RELATION OF WINDS TO PRESSURE

Chapter 2's discussion of heat balance in the atmosphere pointed out that there is a general convective transport of heat from lower to higher altitudes, and an advective or horizontal transport of heat from lower to higher latitudes, in order to correct the earth's radiational imbalance. It is for this reason that the atmosphere is a restless medium in which circulations of all sizes are normal. While vertical motions (properly spoken of as currents) are very important climatically because they help produce cloud and precipitation, their total magnitude is small in comparison with the horizontal movements of wind. Wind is simply air moving in a direction which is essentially parallel with the earth's surface. The atmosphere is fixed to the solid-liquid earth in gravitational equilibrium and so moves with the earth in its west-to-east rotational movement. Wind, therefore, is air movement *in addition to* that associated with rotation.

In large-scale circulations covering several thousand miles, horizontal motion or wind greatly exceeds vertical motion. Thus, a wind which takes several days or even a week or more to cross an ocean may move up or down only a few miles. The vertical component of movement is much greater in small-scale circulations such as thunderstorms and tornadoes. In a thunderstorm, air may ascend to the top of the troposphere (± 6 miles) in a half hour to an hour.

Wind is complex in origin. Usually its direct cause lies in differences between atmospheric densities resulting in horizontal differences in air pressure. That is, it represents nature's attempt to correct pressure inequalities. When these horizontal pressure differences develop, so that there is a high in one location and a low in another, a slope or gradient of pressure exists between them, analogous to the slope or gradient of level which separates a hill from an adjacent valley. But in spite of the direct part played by pressure differences, the ultimate source of energy for generating and maintaining winds against the drag of friction is derived mainly from the differences in heating and cooling between high and low latitudes.

Forces governing winds

Four forces operate to determine the speed and direction of winds: (1) pressure gradient force, (2) the deflecting force of the earth's rotation (Coriolis force), (3) friction, and (4) centrifugal force.

Pressure gradient force. This force sets the air in motion and causes it to move with increasing speed along the gradient. If a mass of air is subjected to higher pressure on one side than on the other, there is a net pressure force moving it from the higher toward the lower pressure.

Pressure gradient force has both magnitude and direction. The *magnitude* is inversely proportional to the isobar spacing: the closer the spacing, and hence the steeper the gradient, the stronger the accelerating force and the greater the wind speed. Weak winds are associated with widely spaced isobars; calms with a near-absence of isobars (Figs. 3.9 and 3.10). The *direction* of the pressure gradient force is represented by a line drawn at right angles to the isobars. Since the gradient slopes downward, from high to low pressure, direction of airflow is from high to low pressure, or along the pressure gradient. But because of the operation of another force, that of earth rotation, the trajectory of an air particle moving from high to low pressure is very indirect, except close to the equator.

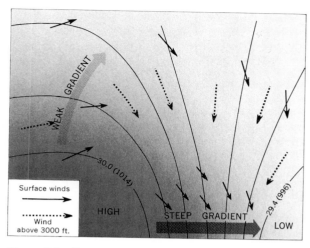

Figure 3.9 Pressure gradient is the rate and direction of pressure change. It is represented by a line drawn at right angles to the isobars. Gradients are steep where isobars are closely spaced and weak where they are far apart. Winds move from high to low pressure, but because of earth rotation they do so indirectly. At the surface, friction causes them to cross the isobars at an angle of 10 to 30°, but aloft they nearly parallel the isobars.

Figure 3.10 Relation of isobaric surfaces in the atmosphere to isobaric lines at the earth's sea-level surface (*xy*).

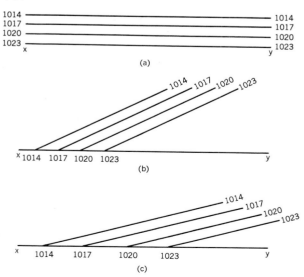

Coriolis force. The second force, which affects only the direction of winds, is *the deflecting force of the earth's rotation,* known as the *Coriolis force.* Except at the equator, winds and all other moving objects, no matter what their direction, are deflected to the right of the gradient in the Northern Hemisphere and to the left in the Southern Hemisphere (Fig 3.11). The force always acts at right angles to the direction of motion.

In reality, Coriolis force is not a real force but only an apparent one. It arises because a moving

Figure 3.11 The apparent deflection of the planetary winds on a rotating earth. White arrows indicate wind direction as it would develop from pressure gradient force alone. Black arrows indicate the direction of deflected winds resulting from earth rotation.

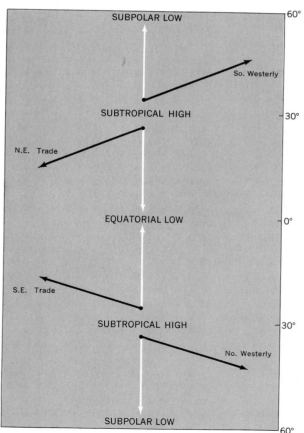

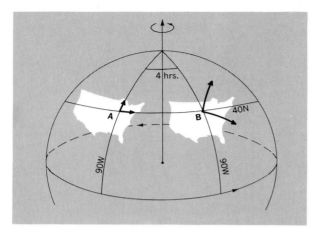

Figure 3.12 Showing the rotation of the United States (in white) around the earth's axis as the earth turns during about 4 hr, or one-sixth of a complete rotation. In its initial position *A*, one wind is shown blowing from the south and another from the west at 40°N and 90°W. But 4 hr later these winds are from the southwest and the northwest, as indicated in the second position *B*. They have undergone an apparent deflection to the right. (*After H. Riehl in Introduction to the Atmosphere.*)

object like the wind tends to keep its original direction fixed with respect to a point in space. By contrast, directions on the earth's surface are related to a coordinate system of meridians and parallels which themselves rotate and change absolute direction as the earth turns on its axis.

This is illustrated in Fig. 3.12, where the United States is shown as moving from west to east as the earth turns on its axis. Assume for the moment that there is no friction and no horizontal pressure force. The wind will therefore maintain its original absolute direction with respect to a point in space, while the earth underneath turns. In position *A* a wind from the south is shown along the 100°W meridian and a wind from the west along the 40°N parallel. Four hours later, the earth has turned 60° on its axis. Meridians and parallels have rotated counterclockwise and changed absolute directions, and the United States area has similarly rotated with respect to the axis. But these rotations are not imparted to the winds above. So to a person on the

ground, unaware that he and his directions have been turning, it will appear that the winds have been deflected to the right (or to the left, if he is in the Southern Hemisphere) of their original starting direction. And so they have, from his point of observation on the rotating earth where directions are changing. Thus four hours later, the original south wind appears to be from the southwest, and the west wind from the northwest. To an observer on the earth, it appears that the winds have been deflected to the right by some force exerted to the right of the wind. (The right-hand deflection of winds in the Northern Hemisphere and their left-hand deflection in the Southern Hemisphere will not be apparent, however, unless the observer's back is toward the wind so that he faces the direction in which the wind is blowing.)

This deflective force of earth rotation can be illustrated by an analogy. Figure 3.13 shows a circular disk rotating counterclockwise, as indicated by the arrows. If an object at point *C* is moved toward point *P* beyond the rotating disk, an outside observer would see the object move from *C* to *P* along the straight dashed line. An observer *T* riding *on* the disk, however, would be at point *T* when the object started from *C*, but because of rotation of the disk, would find himself at *T'* when the object arrived

Figure 3.13 The deflecting force of earth rotation.

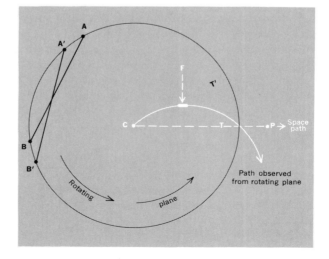

at *P*. It would appear to this observer on the disk that a force *F* had caused the moving object to be pushed to the right of its original path. This apparent force is the Coriolis force. Both observers would be right in their observations, for the one in space observed the object's absolute motion, while the one on the rotating disk observed its relative motion (see also Fig. 3.14).

Thus on a nonrotating earth, air set in motion by pressure gradient force will move in a straight line from high to low pressure. On a rotating earth, air starts toward the low pressure, but to an observer on the earth it is

Figure 3.14 Deflective effect of earth rotation at the North Pole. Illustrated is a flat disk tangent at the North Pole. The arrows paralleling the circumference indicate the west-to-east rotation of the earth. Any body that is set in motion in a given direction tends to keep that direction unless acted upon by some external force. Based upon the preceding principle, a south wind, illustrated by arrow 1*A*, continues to keep its direction constant as the earth rotates. When the earth has turned through 45°, the original south wind is still moving in the same direction with respect to space, for arrow 2*A* is parallel with 1*A*. But in terms of directions on the rotating earth, whose meridians and parallels are constantly changing positions, the 2*A* is from the southwest. Similarly, the west wind 1*B* becomes northwest wind 2*B*; north wind 1*C* becomes northeast wind 2*C*; and east wind 1*D* becomes southeast wind 2*D*. (*After Koeppe.*)

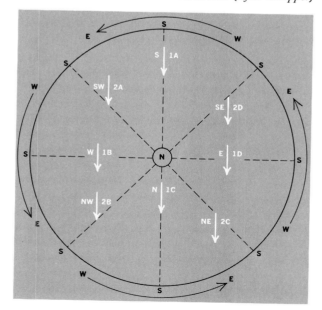

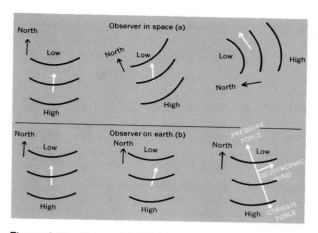

Figure 3.15 The solid black lines are isobars, drawn parallel with the latitude circles. The white arrows indicate the position of the air particle and its direction of motion as seen *a* by an observer in outer space, and *b* by an observer in a fixed position on the earth. N = local north. (*After H. Riehl in Introduction to the Atmosphere.*)

deflected more and more to the right of its initial path until, in the absence of friction, it moves parallel with the isobars (Fig. 3.15). To an observer in space the situation looks different. He sees the air start out toward the low pressure along a straight line in space. But it cannot get to the low directly, because the pressure system rotates as the earth turns on its axis. The air is obliged to continuously turn in response to the changing positions of low and high pressure. Eventually it moves parallel with the isobars and at right angles to the gradient.

The magnitude of the Coriolis force is proportional to (1) the speed of the wind, and (2) the degree of inclination of the earth's spherical surface with respect to the earth's axis (i.e., the latitude). The effect of the changing inclination of the earth's surface is illustrated in Fig. 3.16, which makes it necessary to distinguish clearly between two different rotations. There is, first, the turning of the entire earth on its axis once in 24 hr. But second, there is a rotation of meridians and parallels, and so of surface areas on the earth with respect to the axis. At the pole, where the earth's surface is inclined 90° to the axis, an area will rotate in a plane perpendicular to the axis and make a complete

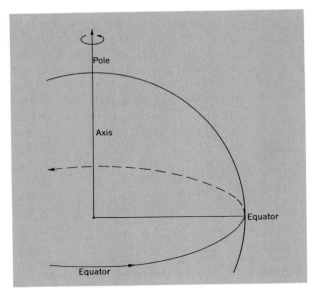

Figure 3.16 The surface of the earth is perpendicular to the axis at the pole and parallel to it at the equator. So during a complete earth rotation a person near the pole would turn through 360° and consequently experience a complete change in directions. But a person at the equator would experience no change in direction. (*After H. Riehl in Introduction to the Atmosphere.*)

360° rotation once in 24 hr. For this reason Coriolis force is strongest in high latitudes. With increasing distance from the pole, the earth's spherical surface is more and more turned away from the axis, so that the component of area rotation acting parallel to the earth's surface decreases. At latitude 30°, this component is only half what it is at the pole. At the equator, where the earth's surface is parallel with the axis, the whole rotation of surface areas takes place in a plane perpendicular to the earth's surface. Hence Coriolis force disappears there.

When a wind (in this case called a *geostrophic wind*) blows parallel with the isobars, the Coriolis force is just balanced by the pressure gradient force (Fig. 3.17). Outside the atmosphere's friction layer, which extends 1,500 to 3,000 ft above the earth's surface, winds actually do blow in a direction almost parallel with the isobars, with low pressure on the left and high pressure on the right (in the

Northern Hemisphere). The air may then be thought of as flowing between the isobars, in the same way that a river flows between its banks. Where the channel is narrow, the flow is faster; where it is wider, the flow is slower, If the spaces between the isobars are thought of as tubes of constant transport, the speed of the air above the friction layer is nearly inversely proportional to the width of the channel, or the space between the isobars (Fig. 3.18).

Friction.   Both wind speed and wind direction are affected by friction, the third of the forces governing winds. Friction between the moving air and the earth's land-sea surface tends to slow the air's movement—a factor which is important mainly in the lower few thousand feet of the air ocean. It likewise produces turbulence, or vertical motion. An air mass itself in which the wind changes with height also has a slight amount of internal friction.

Because of the frictional effects of the land-sea

Figure 3.17   A geostrophic wind results when pressure gradient force is balanced by the deflecting force. Straight (or nearly straight) isobars and a similar airflow are assumed. The geostrophic wind blows parallel with the isobars, with high pressure to the right (Northern Hemisphere). Winds are essentially geostrophic in character above about 1,000 m. (When isobars are strongly curved, and the path of a moving parcel of air is as well, the effects of a third force, centrifugal force, should be included. Then the flow of air which is necessary to balance pressure gradient, deflection, and centrifugal forces is called a *gradient wind*.)

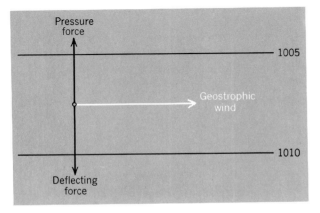

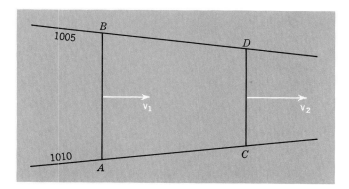

Figure 3.18 The spaces between the isobars may be thought of as channels of constant transport, so the speed of the air is inversely proportional to the width of the channel. Consequently the amount of air that flows through section CD equals that flowing through section AB where the channel diameter is greater. It follows that the air must have a higher velocity (V_2) at CD than at AB (V_1).

surface upon air flowing over it, surface air does not flow essentially parallel with the isobars as it does aloft, but instead crosses them at an oblique angle (Fig 3.19). The greater the friction, the wider

Figure 3.19 The effects of friction on the direction of surface winds. Over the sea surface, where friction is slight, winds cross the isobars at a small angle. The angle is much greater over a rough land surface.

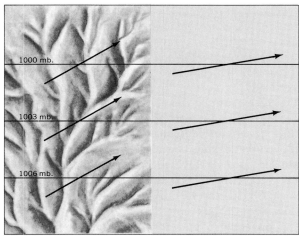

ROUGH LAND SURFACE   WATER SURFACE

is the angle the wind direction makes with the isobars; i.e., the closer the wind direction is to the direction of the pressure gradient. Winds over irregular land surfaces usually form angles varying from 20 to 45° with the isobars. But over oceans, where the frictional drag is less, the angle may be as little as 10°.

Centrifugal force. Fourth of the great influences on winds is centrifugal force,[2] which operates only when air moves in a curved path. It acts in a direction radially outward from the center of curvature of the air's path, and increases both with wind speed and with the shortening of the radius of curvature. Centrifugal force is a major factor only when the wind is strong and the radius of curvature small—as they are in tropical hurricanes, tornadoes, and the centers of a few unusually well-developed cyclonic storms.

Wind direction and speed

Winds are always named by the direction they come from. Thus a wind from the south, blowing toward the north, is called a south wind. The wind vane points into the wind. *Windward* refers to the direction a wind comes from, *leeward* to the direction it blows toward. Thus a windward coast is one along which the air is moving onshore; on a leeward coast, winds blow offshore. When a wind blows more frequently from one direction than from any other it is called a *prevailing* wind. Wind direction is expressed in terms of directions on a 32-point compass and is referred to by the compass point, by letter abbreviations of the directions, or by the number of degrees east of north.

Wind speed increases rapidly with height above the ground level, as frictional drag declines. Wind is commonly not a steady current but is made up of a succession of gusts, slightly variable in direction, separated by lulls. Close to the earth the gustiness is caused by irregularities of the surface, which

[2] This is scarcely a real force; it is rather a consequence of the unbalancing of the other forces when isobars are curved and wind flow is too.

create eddies. Larger irregularities in the wind are caused by convectional currents. All forms of turbulence play a part in the process of transporting heat, moisture, and dust into the air aloft.

Cyclonic and anticyclonic circulations

As discussed earlier in the chapter, sea-level pressure patterns are commonly cellular in form. On an isobaric chart they are represented as systems with closed isobars. When such a system has the lowest pressure at the center, it together with its circulation is designated as a cyclone. When a system of closed isobars has the highest pressure at the center, it together with its circulation is called an anticyclone (Fig. 3.20). The combined effects of pressure gradient force, Coriolis force, and friction cause airflow in a cyclone to be *convergent,* with surface winds flowing obliquely across the isobars toward the center. In an anticyclone the circulation is *divergent,* with surface winds moving obliquely across the isobars outward from the center.

Figure 3.20 A cyclonic circulation is a converging system of airflow around a low-pressure center, counterclockwise in the Northern Hemisphere and clockwise south of the equator. An anticyclonic circulation is a diverging system of airflow around a high-pressure center, clockwise in the Northern Hemisphere and counterclockwise in the Southern.

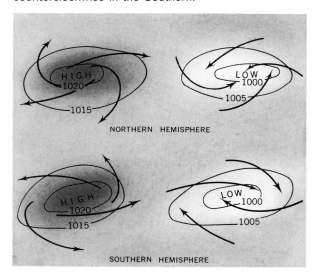

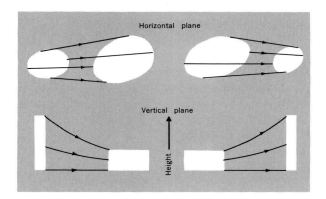

Figure 3.21 Mass convergence and divergence of the atmosphere, shown in horizontal and vertical planes. In the lower diagram the lowest streamlines of airflow are shown horizontal because the base of the air columns is the ground. (*After H. Riehl in Introduction to the Atmosphere.*)

Convergence and divergence

Although vertical air movement occurs on a much smaller scale than horizontal transfer, it is very important in the formation of cloud and precipitation. Vertical movement has various causes. It can be produced by thermal convection due to heating of air near the ground, or by mechanical turbulence caused by terrain irregularities. A major cause of vertical movement is lifting and subsidence associated with convergence and divergence in horizontal airstreams. Convergence occurs when the horizontal area occupied by a given mass of air decreases; divergence occurs when the area increases (Fig. 3.21). Air which converges laterally must spread vertically or stretch. In an airstream bounded by the earth's surface, this results in upward movement. Divergence and horizontal spreading in surface flow result in vertical shrinking and downward movement, or subsidence. In large-scale circulations, convergence or divergence may take place over areas covering 50,000 to 500,000 square miles. In such systems the vertical movement has a speed of only about a half mile a day.

Convergence and lifting in an airstream also occur when its speed decreases downstream. This results in a piling up of air to the rear and consequently in upward movement there. Conversely, when speed increases downstream, the air diverges and settles. In a curved airstream, convergence occurs when the air curves counterclockwise (forming a cyclonic circulation) in the Northern Hemisphere. When the curvature is clockwise (anticyclonic), divergence prevails. Air moving equatorward into latitudes of decreasing Coriolis force tends to diverge and subside; moving poleward, it slowly converges and lifts.

GENERAL CIRCULATION OF THE EARTH'S ATMOSPHERE

The force required to drive the gigantic circulation of the earth's atmosphere is provided by the energy contrasts between the warm energy source in tropical latitudes and the cold sources in polar latitudes. Obviously these latitudinal temperature contrasts originate because the low latitudes receive much more solar energy than the high latitudes. The atmospheric circulation (and the oceanic circulation as well) transports heat from a zone of excess energy in the tropics to a region of deficiency close to either pole. In other words, it acts like a heat engine. The poleward heat transport is mainly in the form of sensible heat (heat that is felt), and to a lesser degree in the form of latent heat traveling as water vapor acquired by evaporation. Although ultimately there must be an exchange of energy in a north-south direction (that is, there must be a meridional circulation of the atmosphere), the average circulation actually is overwhelmingly zonal, or west-east —a feature due to earth rotation and its resulting Coriolis force.

On a nonrotating earth, warm in the low latitudes and cold at the poles, the atmosphere would circulate in the form of two gigantic convectional systems, one in each hemisphere. The warm air would rise in the low latitudes and flow poleward at higher levels, where it would slowly subside to ground level and give up its heat. The cold polar air would flow equatorward as a surface current. Under these circumstances, surface pressure would be high in the polar areas and low in equatorial latitudes.

But on a rotating earth, both low-level currents moving equatorward and high-level ones moving poleward are acted upon by earth rotation. The result is that the earth's atmospheric pressure arrangement, as well as its atmospheric circulation, becomes unusually complicated. In fact, it is so complex that some elements of the circulation still remain uncertain, and there are no widely accepted explanations for some features that are known. For example, just how the required meridional heat exchange is accomplished in a circulation which is essentially zonal has always been something of a puzzle.

In the low latitudes between about 0 and 30°, where Coriolis force and the rotation component of meridians and parallels about the earth's axis are weak (dropping to zero at the equator), the atmospheric circulation resembles that which one would expect in a simple convectional system (Fig. 3.22). The air flows obliquely equatorward at the surface and poleward aloft, and though the high-tropospheric flow is variable, the net transfer is poleward.

This direct thermal type of circulation prevails within the tropics in spite of the fact that significant north-south contrasts in sensible temperature are lacking. The equatorial heat source is maintained chiefly by heat of condensation released in cumulus clouds in the relatively narrow belt of heavy equatorial rainfall. Significantly, the belt between 25°N and 25°S furnishes 60 percent of the earth's evaporated moisture. Tropical cumuli have effects far beyond the local region in which they originate. Indeed, they play a vital role in the fire-box and fuel-pump functions of the atmosphere's heat engine, which maintains the tropical circulation. The *mean* tropical atmosphere does not generate large-diameter cumulus towers. The trade-wind cumuli do act, however, to raise and weaken the trade inversion along the air trajectory and thereby create the pressure head which maintains the steady

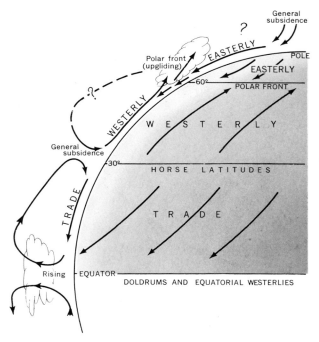

Figure 3.22 Important elements of the general circulation of the atmosphere. Some features of the upper circulation are not at present sufficiently well understood, or are too complicated, to be represented in a simple diagram.

easterly flow. But a few thousand simultaneously active giant cumulonimbus "hot towers" in the equatorial-trough zone, as well as occasional hurricanes, condense two-thirds of all the water evaporated into the atmosphere and transport aloft several hundred times as much energy as is consumed in all the global wind systems combined.[3] It is these giant hot towers, mainly equatorial, that perform the fire-box function in the tropical atmosphere's heat engine. They condense the water vapor carried equatorward by the trades, thereby releasing prodigious amounts of heat of condensation.

The tropical "cold source" is broader in latitudi-

[3] Joanne S. Malkus. "Tropical Convection: Progress and Outlook." In J. W. Hutchings (ed.). *Proc., Symposium on Tropical Meteorology, Rotorua, New Zealand,* 1963. New Zealand Meteorological Service, Wellington, 1964. Pp. 247–277.

nal spread, covering much of the whole volume of the atmosphere between about latitudes 5 or 10° and 30°. It is a consequence of radiational cooling at various levels and is accompanied by a broad area of slow settling and drying of the air in the subtropics.

In the middle latitudes, the circulation does not resemble that of a large convectional system, for in the westerly wind belts of these latitudes the surface flow is from a warm source toward a cold one. Since Coriolis force is strong, the rotation component of meridians and parallels about the earth's axis is large. Consequently the north-south heat transfer in the middle latitudes is accomplished less by steady flow than by strong, sporadic air-mass exchanges across the latitude belts. These take the form of intermittent thrusts of cold surface polar air equatorward, and of tropical air moving poleward. Outside the tropics, airflow is variable at all heights. Such individual air-mass exchanges accompany the succession of cyclones and anticyclones which move from west to east in the middle latitudes. They are also linked with the upper-tropospheric long waves (discussed in the next section), which occur in the same latitudes. It appears, therefore, that the contrasting types of general circulation inside and outside the tropics are strongly affected by variations in Coriolis force with latitude, in the same way as is the component of rotation of meridians and parallels about the earth's axis.

It is not so difficult to produce a vertical profile model of the atmospheric circulation in the low latitudes, where the airflow is reasonably steady. But constructing any such model of the mean circulation outside the tropics presents almost insuperable difficulties because of the sporadic air-mass exchange.

Least is known about the polar circulation. Although it is sporadic in nature, its overall direction of flow most likely resembles that of the tropics more than that of the middle latitudes. In other words, probably the surface-air movement is largely from higher to lower latitudes, or from a colder to a warmer source. Consequently the low-troposphere flow in middle latitudes must move in a direction

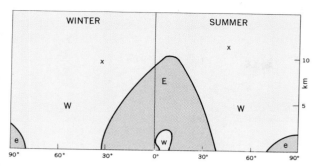

Figure 3.23 A pole-to-pole cross section of the planetary winds up to about 8 or 9 miles above the earth's surface. E = tropical easterlies or trades; W = westerlies; x = average location of the jet stream; w = the somewhat less certain belt of equatorial westerlies; e = polar easterlies. (*After Flohn.*)

opposite to those on both its high- and low-latitude margins. This cannot help but result in a clash of polar and tropical air masses within the middle-latitude westerlies. Such a clash necessarily has major effects on temperature, precipitation, and storminess. Another fact of great importance, particularly in relation to cloud and precipitation, is that the mechanism of the general circulation results in ascent of air both in equatorial and in middle latitudes, but in settling or subsidence in the subtropics and probably in the polar regions.

Figure 3.23 shows a generalized meridional cross section of the normal state of the circulation. Much the greatest volume of the atmosphere has a westerly flow. In middle latitudes these vast and deep circumpolar westerly circulations reach down to the earth's surface. The westerly direction is less obvious in the surface flow, however, because of turbulence and numerous disrupting atmospheric disturbances. Both steadiness of direction and speed of flow increase aloft in the westerlies. A sharp maximum in speed is reached in the vicinity of the jet stream (which is described in the following sections), at an altitude of about 10 or 12 km. The existence of the smaller isolated area of westerly flow in the vicinity of the equator has not been completely confirmed.

As seen in Fig. 3.23, there are two main systems of easterly circulation. First of these is the broad and deep belt of tropical easterlies, or trade winds, which extend out 30° and more on either side of the equator. The second, located in the high latitudes of each hemisphere, is the much shallower and more fickle polar easterlies, whose existence has not yet been decisively verified.

General circulation features of the upper troposphere

Airflow is predominantly in the form of wave motion. Within the troposphere there are air waves of all sizes and at all altitudes. It is the atmosphere's wave patterns, and their associated convergence and divergence, which largely govern weather occurrences on time scales ranging from periods as short as a day to those as long as a month or even more.

Upper-air long waves. The middle- and upper-troposphere westerlies contain two features—the long waves and the jet streams—which themselves are closely linked, and which are both intimately related to surface weather. Superimposed upon the circumpolar vortex of westerly winds is a system of huge, slow-moving long waves, several thousand miles in length, which radiate out from the polar cyclonic vortex (Fig. 3.24). These long waves in the upper westerlies form troughs and ridges in the constant pressure surfaces. Although they vary from three to six in number, the atmosphere appears to have a strong inclination to return to a system of four principal waves. The long waves either are stationary or move very slowly. Moreover, because the earth's surface is not uniform, the atmosphere's troughs and ridges seem to have certain preferred positions. (Fig. 3.25). Major troughs commonly are positioned as follows: (1) extending from Labrador toward Florida in eastern North America or the adjacent Atlantic, (2) extending from northern Scandinavia toward the Mediterranean, (3) extending from Alaska toward the central Pacific, and (4) possibly another from central Siberia toward Burma. However, the long waves and their troughs and

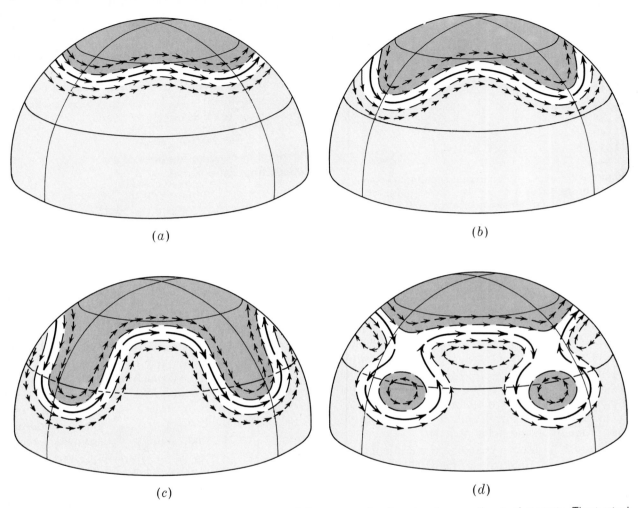

(a)

(b)

(c)

(d)

Figure 3.24 Upper-air waves on the jet stream, which bring periods of contrasting weather in their train. The typical sequence, called an index cycle, is shown in the four diagrams. The undulating jet goes into increasingly large oscillations. North of the jet lies cold polar air, and south of it warm tropical air. The great oscillations carry polar air into the middle and low latitudes and tropical air into the middle and high latitudes. Finally, the extended waves are cut off, leaving cells of cold air in the south and cells of warm air in the north. (*After Namias.*)

ridges often depart from their normal positions for variable periods. Such shifts in location cause the weather patterns of a hemisphere to shift also.

Large-scale wave motion is characteristic of the atmosphere in the middle and high latitudes, but is found much less often in the tropics. This fact is related in a cause-and-effect sense to two others

of more than ordinary importance: (1) Coriolis force increases with latitude, and (2) heat transport in the tropics is achieved by a fairly steady flow resembling a convectional system, but in the westerlies of the middle latitudes, where the long waves prevail, heat exchange takes place by sporadic thrusts of cold air equatorward and warm air poleward.

In experiments with water-filled rotating basins, it was discovered that slow rotation gave rise to a steady flow, with the highest speeds near the inner cold core. At faster rotations a sinuous wave pattern developed which resembled that of the westerlies, with high speeds concentrated in a narrow band similar to the jet-stream type of current described in the next section. The basin experiments suggest that it is the more rapid rotation of earth surfaces with respect to the axis (stronger Coriolis force) in middle and high latitudes that is responsible for the long waves there, and for the associated sporadic method of heat transport. Slower rotation and weaker Coriolis force fit the tropical situation, with its near-absence of wave motion and its steady process of heat transport.

Unquestionably, the strength, frequency, and positioning of the long-wave pressure troughs and ridges have much to do with the broad-scale regional patterns of surface weather. Thus there is a close genetic connection between the upper waves and the cyclones and anticyclones at the surface. Since the air on the front of a meridionally oriented ridge or rear of a trough is from higher latitudes, polar outbreaks of deep layers of cold air often accompany the passage of a trough. The equatorial margins of these polar thrusts become the hemisphere's polar front. Subsidence below a high-level ridge is conducive to dry weather at the surface. By contrast, a trough preceded by a poleward flow of tropical air favors high humidity and rain.

Over the United States, the average midwinter

Figure 3.25 Upper-air troughs are closely associated with the location of major frontal zones, which in turn are concentrated regions of cyclogenesis. Eastern North America–western Atlantic, and eastern Asia–western Pacific are two Northern Hemisphere regions with a noteworthy concentration of upper troughs, frontal zones, and cyclone development. (*After Flohn.*)

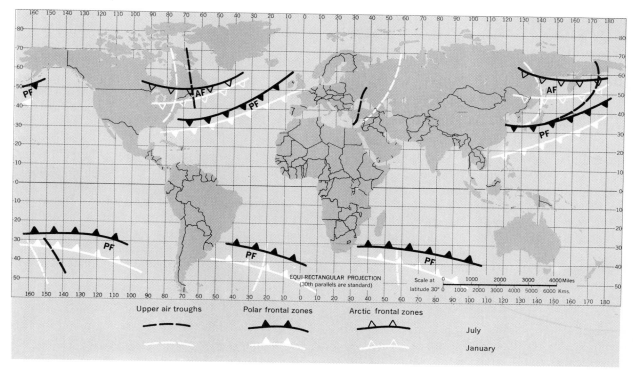

flow at 500 mb shows a meridional pressure ridge over the western half of the United States and a trough over the eastern half. In an instance cited by Riehl, the average flow in January, 1958, resembled the normal pattern except that the troughs and ridges were unusually strong. This arrangement drew warm air into the western states, while a cold northerly flow dominated the East. As a consequence the West was warm and the East abnormally cold. Precipitation was above normal along much of the Pacific Coast and also the Atlantic and Gulf Coasts. It was below normal over the interior. In January, 1949, the positioning of troughs and ridges was reversed. An upper trough with cold air dominated the West and a ridge with warm air the East. The result was cold surface temperatures in the West and warmth in the East—exactly opposite to the situation in January, 1958. Rainfall was below normal along the Pacific Coast and well above normal in the southern interior.

Troughs and ridges in the upper westerlies are definitely linked with frontal waves and cyclones and anticyclones at the earth's surface. A cold trough in the upper atmosphere usually prevails to the rear of a cyclone family at the surface, and an upper-atmosphere warm ridge of high pressure is usually positioned over, or to the east, of it. Thus the cyclone family chiefly exists in the warmer air, but it is commonly followed by a deep polar outbreak of cold air.

There is reason to believe that these dynamically induced long waves in the upper westerlies have a great deal to do with the cellular structure of pressure and atmospheric circulation at the earth's surface. The troughs and ridges in the high westerlies create convergences and divergences which probably lead to the development of the great semipermanent centers of high and low pressure at low levels. No doubt this dynamic arrangement is modified by the seasonal contrasts in surface heating over continents and oceans. Fundamentally, however, the subtropical high-pressure cells and the subpolar low-pressure cells appear to be the result of nonthermal controls which are associated with convergences and divergences caused by the high-altitude troughs and ridges. The shallow surface highs of the polar regions, on the other hand, are probably the result of radiational surface cooling. This idea of a relationship between upper-air waves and surface pressure cells is supported by the fact that the waves in the circumpolar vortex are much less pronounced (have a smaller amplitude) in the Southern Hemisphere than in the Northern Hemisphere, and the cellular structure of sea-level pressure is correspondingly less strongly marked in the Southern Hemisphere.

Jet streams. This second feature of the upper westerlies, which is intimately related both to upper long waves and to the surface weather, resembles a system of meandering torrential streams. Jet streams are relatively narrow bands of strong winds bounded by slower-moving air (Fig. 3.26). Their wind speeds vary considerably in different longitudes of their circumpolar course, but become faster in winter than summer. In winter jets, rates of 100 to 150 miles per hour are common, and instances of 300 miles per hour are known. The highest speeds occur at altitudes of about 30,000 to 40,000 ft.

On the average, the jet is positioned over the subtropical anticyclones at the surface (30 to 35°N), where horizontal temperature contrasts are great. Across the jet (or jets) horizontally, temperature changes rapidly, with cold air on its poleward side and warmer, often tropical, air on its equatorward side (Fig. 3.24). In reality, while it is true that the average "statistical jet" lies above the subtropics, there are usually several jets, with the more northerly one (or ones) showing higher speeds and also more inconstancy in position. It is because of this inconstancy of the northerly streams that the *mean* position falls in the subtropics, where the jet is fairly stationary in location. As might be expected, the jet or jets shift their average latitudinal positions with the seasons.

While the average direction of air movement in the jets is from west to east, an individual jet often shows a very sinuous or meandering pattern which is in harmony with that of the long waves described previously. It is especially the more northerly jets

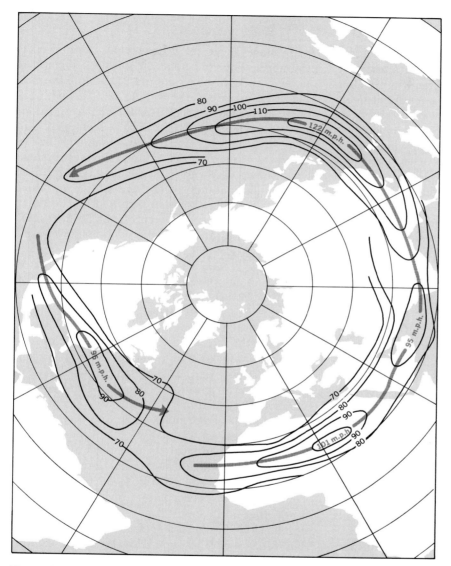

Figure 3.26 Jet-stream map for the Northern Hemisphere in January, showing average positions and speeds at elevations of 35,000 to 40,000 ft. (*After Namias.*)

which vary considerably in position and sinuosity, and it is these which are closely associated with middle-latitude weather disturbances. The meanders of the northerly jet aloft can be identified with the position of the surface polar front, where the strongest north-south temperature gradients prevail.

It is along this front that large numbers of cyclonic storms develop. It may be too much to say that the jet creates the cyclones, but doubtless it has much to do with the courses they follow over the earth's surface. By thus steering the cyclones, jets greatly influence the distribution of surface precipitation as well as temperature. This is because cyclones not only are important generators of precipitation, but also act to steepen surface temperature gradients. Rainfall appears to be concentrated in areas lying underneath the jet, where the effects of a traveling cyclone may be focused. When a jet is relatively stable in position, there may be a detectable zone of seasonal precipitation concentration underneath it, as there is in subtropical China in winter. Such a concentration is not obvious, however, when a jet and its related cyclonic storms frequently shift positions, as is the case in the central and eastern United States.

Jets also probably set in motion those deep and extensive air masses whose movements over the earth's surface bring sustained periods of remarkable heat and cold, drought and flood. The ultimate cause of the varying patterns of the jet, with their important repercussions upon world weather, is a question not completely answered.

Nonperiodic variations in the upper-air circulation. The normal state of the general circulation as represented on hemisphere or world maps is one which is never repeated in that precise form on any daily weather map, or on maps showing the mean for any week or month. For the "normal" condition is only a statistical average of the varying circulations which prevail from week to week or for other shorter or longer periods of time. The most striking of the variations are those in the circumpolar vortex west-wind circulation, which involve the interrelated phenomena of long waves, jet streams, polar fronts, and large-scale weather situations. Although they are nonperiodic, these variations are not purely random, for there is an observable sequence of episodes. The variations may be looked upon as contractions and expansions of the circumpolar westerly wind vortex toward and away from the pole. During these contractions and expansions the long waves change in length, the jet streams shift positions and alter velocity, and the planetary polar front and its associated storm tracks undergo wide oscillations. Such a period of change is called an

index cycle and usually takes from 4 to 6 weeks to complete (Fig. 3.24).

In stage 1, the jet stream is positioned well to the north. The cold polar air masses are confined to the higher latitudes, and most regions in the middle latitudes are covered by mild air. The high westerlies are strongly zonal in character, so that long waves and jet streams are weakly developed. Surface westerlies also are strongly zonal, and they lie poleward of their normal position. Pressure systems are oriented east-west, cyclonic activity is confined to the higher latitudes, and there is little latitudinal air-mass exchange between high and low latitudes. This stage is spoken of as representing a high zonal index, for the rate of change in pressure along a meridian is great.

In stage 2, the jet stream moves farther equatorward, and its waves increase in amplitude. This results in thrusts of polar air equatorward and of tropical air poleward. Stage 3 consists of a further increase in the amplitude of the jet waves. Associated with the increase are strong north-south movements of tropical and polar air masses, causing great east-west contrasts in temperature. At this stage the mean position of the jet stream is much closer to the equator.

In stage 4 of the index cycle, the waves in the jet have increased so greatly in amplitude that immense pools of tropical air are cut off and isolated in higher latitudes, while similar large masses of polar air are left stranded in lower latitudes. This condition is described as one of low zonal index, since the north-south contrasts in pressure are small. There has been a complete fragmentation of the zonal westerlies into closed cellular centers; air-mass and temperature contrasts are at a maximum in an east-west direction, and there is a north-south orientation of pressure centers and frontal systems. Such a period of strongly agitated circulation is the occasion of topsy-turvy surface weather, when Alaska may be warmer than Florida. Subsequently the circulation returns to the zonal pattern characteristic of stage 1.

As mentioned earlier, the great convolutions in

the circumpolar vortex and the jet stream are a part of the overall mechanism of latitudinal heat balance on the earth. Their mass exchanges of air between high and low latitudes represent nature's chief method of equalizing the radiation imbalance between polar and equatorial regions.

THE EARTH'S SURFACE-WIND SYSTEMS

A diagram of surface winds

It is possible to superimpose on the generalized sketch of sea-level pressure (Fig. 3.3) a very simple system of zonal surface winds (Fig. 3.27). These, of course, reflect the four forces previously described which determine the direction and speed of winds—especially pressure gradient force, Coriolis force, and friction.

Two main systems of surface winds are recognized in each hemisphere. From the subtropical highs, whose crests are located at about 25° to 35° N and S, surface winds flow along the pressure gradient toward the low-pressure trough near the equator. Coriolis force turns these into easterly winds, so that they are known as tropical easterlies or trade winds: northeast trades north of the equator and southeast trades to the south (Fig. 3.27). Poleward from the subtropical high-pressure ridges in each hemisphere, winds flow toward the subpolar troughs of low pressure. These winds are turned by Coriolis force so that their general direction of flow is from west to east. They are the middle-latitude westerlies: southwesterly in the Northern Hemisphere and northwesterly in the Southern Hemisphere. Above the shallow friction layer at the surface, both trades and westerlies are strongly zonal, flowing nearly parallel with the isobars.

Poleward from the westerlies, weather observations are scarce and the prevailing direction of surface winds is not known with any certainty. If it may be assumed that on the average there are shallow highs of thermal origin in the general vicinity of the poles, then outflowing surface easterly winds should characterize the very high latitudes.

Figure 3.27 A much idealized representation of the earth's surface winds. Average airflow is predominantly from an easterly direction in most of the low latitudes, and from a westerly direction in the middle latitudes.

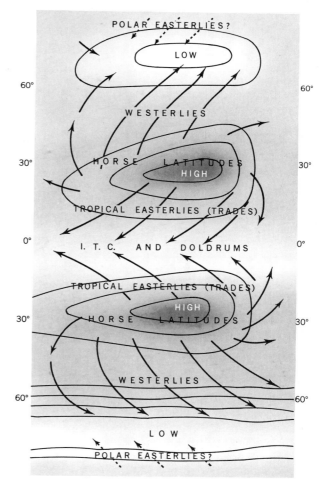

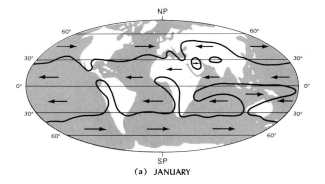

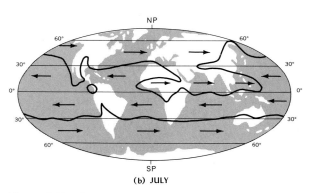

Figure 3.28 Geographical distribution of the west-east component of the mean surface wind. The heavy lines separate tropical easterlies from westerlies, both in equatorial and in middle latitudes. (*After Mintz and Dean.*)

And in fact there is considerable evidence to support this concept of polar easterlies.

To summarize surface winds, easterly winds predominate in the tropics, or low latitudes, and westerly winds prevail in the middle latitudes. Less information exists concerning average wind direction in polar latitudes (Figs. 3.28 and 3.29).

Data on average wind directions near the equator are also inadequate. It is known, however, that winds are usually variable and weak here between the converging trades in the vicinity of the equatorial pressure trough. This transition zone has been given various names—among them intertropical

convergence zone (ITC), doldrums, and equatorial belt of variable winds and calms. Along the axis of the equatorial trough there is a large-scale ascent of warm, moist air.

Figure 3.29 Zonal component (direction and speed) of the mean surface wind, averaged for all longitudes, over the oceans in meters per second. Negative speed represents easterly winds; positive speed, westerly. The zonal component drops to zero in equatorial latitudes in July. Note the latitudinal shifting of winds from January to July. (*After Mintz and Dean.*)

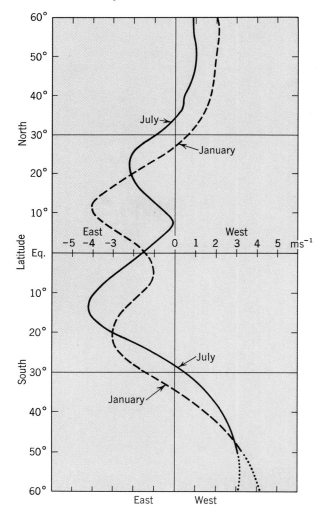

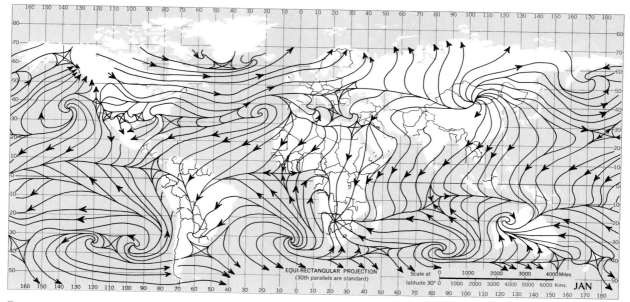

Figure 3.30 January: Direction of mean surface resultant wind. (*After Mintz and Dean.*)

Another belt of weak and variable winds occupies the intermediate area between trades and westerlies—a zone which coincides with the crests of subtropical highs. These are the *horse latitudes*, lo-

Figure 3.31 July: Direction of mean surface resultant wind. (*After Mintz and Dean.*)

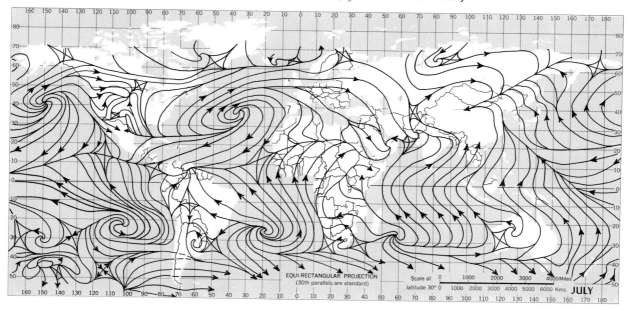

cated at about 30° in either hemisphere. In the horse latitudes there is a slow but widespread settling or subsidence of air, which is fed into this area aloft from both polar and tropical sources. Along the western flanks of the individual subtropical cells, genuinely tropical air invades the realm of the middle-latitude westerlies (see Fig. 3.27); along the eastern flanks of the cells, cool polar air intrudes into the tropics.

A third transition zone occurs where polar easterlies clash with the westerlies, which have been supplied with air from the subtropics and tropics. This zone lies in the subpolar trough along a line of discontinuity known as the *polar front* (Fig. 3.22). The fluctuating and wavy polar front is the breeding ground for many of the cyclonic disturbances which infest the westerlies. In fact, probably the subpolar trough of pressure is simply the statistical average of the annual or seasonal procession of cyclonic disturbances.

Surface winds in January and July

In Figs. 3.30 and 3.31 the direction of the mean resultant surface winds in January and July, the two extreme seasons, is shown by means of atmospheric streamlines.[4] These charts portray a much more complicated pattern of surface winds than those in

[4]Resultant winds are obtained by adding separately the north and south and the east and west components of the observed winds, averaging them, and then applying these average components to obtain a resultant. This procedure is illustrated in Fig. 3.32. Here the vector pointing east (E) represents the average of the east and west components. The vector pointing south (S) represents the average of the north and south components. *R* is the resultant of these two vectors.

Figure 3.32 Procedure for obtaining resultant surface winds.

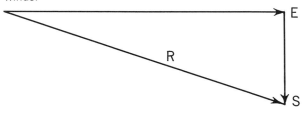

the previous section, which emphasized a simple zonal system of tropical easterlies and middle-latitude westerlies (Fig. 3.27). On the streamline maps there is, to be sure, ample evidence of the dominance of easterly winds in the low latitudes and westerlies in the middle latitudes. But at the same time these maps show modifications of, and departures from, the very elementary pattern outlined before.

Yet even the January and July streamline maps are only means of the highly variable and complicated short-time airflow patterns given on the daily weather map. Still, these mean seasonal maps are a very useful climatic tool, capable of explaining important climatic features.

Because of the meager data, no attempt is made in Figs. 3.30 and 3.31 to show wind directions poleward of latitude 65° N and S. In the other latitude belts, mean surface resultant winds exhibit cellular as well as zonal arrangements. Main circulations occur in the form of anticyclonic and cyclonic circulations around high-pressure and low-pressure centers, or cells. Some of these cellular circulations appear to persist throughout the year, for they can be seen on both the January and July charts. Others are definitely seasonal in character. As a general rule the cellular pattern of pressure and winds, and the seasonal changes in position and size of the centers of cyclonic and anticyclonic winds, are not nearly so pronounced in the more homogeneous Southern Hemisphere as they are in the Northern Hemisphere. North of the equator, the earth's wind-flow patterns are modified by thermal contrasts between large continents and oceans in winter and summer, as well as by the frictional effects of the continents and the blocking action of their high cordilleran systems. These modifications lead to cellular (as contrasted with zonal) circulations in the Northern Hemisphere.

The most prominent features of the mean surface-wind flow are the large systems of divergent anticyclonic circulation concentrated in subtropical latitudes. The equatorward flow of air from these anticyclonic circulations consists of the well-known tropical easterlies or trades, while the poleward flow is the middle-latitude westerlies. Thus the subtropical anticyclones are pivotal to both trades and westerlies, a fact which has earned them the designation "centers of action." Yet they are really centers of inaction as far as bad weather and storms are concerned, for such disturbances are steered away from their cores to circulate around their margins.

On both January and July maps the subtropical anticyclonic systems are dominant; there are no equally persistent systems of convergent cyclonic flow. The nearest approach is found in the smaller and weaker centers of converging cyclonic circulation over the oceans in the higher latitudes of the Northern Hemisphere, especially in winter. Their weakness may seem strange in view of the well-developed low-pressure centers which occupy the same areas, as shown on Fig. 3.6. In reality, however, the winter lows over the North Atlantic and North Pacific largely represent statistical averages of the numerous moving low-pressure centers whose paths concentrate on these areas. They are not durable features of the daily weather map.

January circulation. On the January map of surface winds (Fig. 3.30), divergent anticyclonic systems are relatively weak over the Northern Hemisphere subtropical oceans, and they are largely confined to the eastern parts of the North Atlantic and North Pacific. In the Southern Hemisphere, by contrast, the subtropical anticyclonic circulations over the oceans are well developed and relatively continuous in an east-west direction. In addition to the anticyclonic wind systems over the subtropical oceans in both hemispheres, there is also one over middle-latitude eastern Asia. This diverging circulation, the Asiatic winter monsoon, is associated with an average winter high-pressure cell whose origin is related to moving cold anticyclones and to the accumulation of frigid surface air over the cold continent.

In January not only are the subtropical anticyclonic systems over the Northern Hemisphere oceans weakened, but also there is a spreading and deepening of the oceanic low-pressure centers and associated cyclonic circulations in somewhat higher latitudes. Convergent cyclonic circulations around the Aleutian Low over the North Pacific and the Iceland Low over the North Atlantic are best developed at this season. In the Southern Hemisphere, where it is summer, a relatively strong low-pressure cell and associated cyclonic wind system develop over heated Australia, and somewhat weaker ones appear over South America and southern Africa.

July circulation. In July (Fig. 3.31), well-developed highs and strong anticyclonic circulations cover much of the North Atlantic and North Pacific Oceans with their centers at about 35° and 40°N. The pattern of the subtropical anticyclonic circulations in the Southern Hemisphere has not greatly changed from what it was in January, except that the cells are more continuous in an east-west direction as a result of the cooler continents.

In the Northern Hemisphere, the subpolar oceanic lows and their associated cyclonic circulations are weaker than in January. On the other hand, thermally induced lows with cyclonic circulations occur over the heated land areas of northwestern India, eastern Asia, and western interior North America. In India this is the well-known summer monsoon.

Lines of horizontal divergence and convergence. Prominent features of the pattern of surface-wind flow shown in Figs. 3.30 and 3.31 are the extended lines of divergence and convergence of wind direction. The most conspicuous of the great lines of divergence are those which pass through the centers of the anticyclonic systems and extend poleward and eastward, and equatorward and westward, from these centers. Along such lines of divergence in surface-wind flow there must be a compensating feeding in of dry air from aloft, with accompanying slow but extensive subsidence.

Somewhat less well developed are the extended lines of convergence of wind direction associated with the centers of the smaller, weaker cyclonic circulations. Such convergence zones are areas in which the atmosphere is undergoing slow lifting. The most prominent of all the lines of convergence ·lies between the two trades, and therefore in the vicinity of the equatorial low-pressure trough (also see Figs. 3.36 and 3.37).

SURFACE-WIND SYSTEMS AND THEIR WEATHER

Winds of the tropics

Scientific understanding of the atmospheric circulation within the tropics is still fragmentary. Even today the network of weather observatories is widely spaced in most tropical regions, their instrumental equipment is rudimentary, and competent observers are scarce. During the Second World War, the stepped-up tempo of military aviation greatly increased the number of observations both of surface and of upper winds in the low latitudes. One consequence was that many long-held concepts concerning tropical winds and weather were shattered. But the newly added observations are not always sufficiently numerous, well distributed, and long continued to permit the formulation of satisfactory substitute hypotheses and models. Thus at present our knowledge of the tropical circulation and associated weather patterns remains basically negative. Meanwhile the professional literature is full of hasty generalizations based on inadequate local observations, which in turn lead to a mass of contradictory explanations and resulting confusion.

Many of the features of weather, and the tools of weather analysis, which are accepted as standard in the middle latitudes must be discarded in the tropics. For instance, Coriolis force is weak (and absent at the equator), and the usual relation of pressure gradient to wind does not hold. Isobaric patterns are poorly defined, and recognizable pressure and wind systems seldom exist. Pressure gradients are so weak that many times they are obscured by instrumental errors. The scale of the synoptic patterns is unknown, and the same is largely true of synoptic models. Moreover, tropical disturbances often are phenomena of the middle and upper troposphere and may not appear on surface charts,

so that they are more difficult to detect and observe. Fronts have little importance in the tropics.

Because of the data collected during and since World War II, the previously held concept of zonal uniformity in winds and weather in low latitudes has had to be modified. It is now known that the tropics are affected by more atmospheric disturbances and by different kinds of them than was previously believed. Accordingly, students of the atmosphere are increasingly aware that tropical weather is not as exclusively sun-controlled and periodic as they used to think. They have also discovered that the really steady trades occupy only a fraction of the total oceanic area within the tropics, while the antitrades[5] aloft are not present everywhere.

Winds of the equatorial convergence zone. In the trough of low pressure situated between the converging trades from opposite hemispheres, wind conditions are complex and not well understood. In this zone of intertropical convergence or ITC, some longitudes have a prevalence of calms and light variable winds, a condition known as *doldrums*. In certain regions and seasons, light easterly winds are common; in others westerly winds predominate. The equatorial trough with its light and variable winds is not always continuous around the earth. In some places and on some occasions it may be greatly constricted or even wiped out by encroaching trades or monsoons; then again it may expand to twice its normal width.

It is the meridional component of the trade winds which results in their convergence in equa-

---

[5] High-altitude winds above the surface trades, flowing in a direction opposite to them.

torial latitudes. Also, it is the convergence of the trades, even more than thermal convection, which creates the widespread ascent of air in equatorial areas. Thus the intertropical convergence zone (ITC) is predominantly a region of rising air, and this in turn causes a lowering of pressure and an abundance of cumulus clouds and heavy showery precipitation. Vast amounts of latent heat of condensation are released into the air, adding to its instability. It is mainly this equatorial heat source that drives the tropical circulation. Pressure gradients are weak. The unstable and stagnant air masses are very sensitive to local variations in relief, vegetation cover, land and sea breezes, and mountain and valley breezes, so that weather is extremely localized. At four weather stations on Singapore, an island whose dimensions are only 24 by 13 miles, the same weather phenomena are rarely observed at the same time.

The concepts of air masses and density fronts,[6] which are crucial in middle-latitude weather analysis, are of negligible importance in tropical regions. Temperatures are so nearly uniform in the tropics that convergence of airstreams, including convergence of the two trade winds, rarely results in density fronts; or if it does, they appear in much weakened form.

But while genuine density fronts are rare in the tropics, horizontal convergences lacking in frontal structure are common. It is within such convergence zones that an important part of tropical weather develops. When homogeneous air masses converge and density contrasts are absent, there is no tendency for one air mass to be shoved obliquely upward above the other, separated by a sloping surface of discontinuity or front. (This is the pattern of middle-latitude weather disturbances, as will be described in Chapter 5.) Rather, the air along the axis of a pronounced field of convergence is forced vertically upward simply because of the squeeze to which it is subjected by the oppositely advancing air masses. Condensation occurs in these rising air currents

along a convergence zone, creating tall, multilayer cumulonimbus clouds, with associated convective showers and thunderstorms.

The most extended of the convergence lines or zones within the tropics is the intertropical convergence (ITC), situated between the trades arriving from the opposite hemispheres (Fig. 3.33). Some

Figure 3.33 Pressure and circulation patterns *a* when the ITC is located near the equator, and *b* when the ITC is displaced some distance north-south of the equator. (*After Sawyer.*)

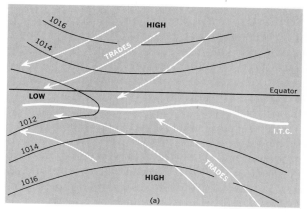

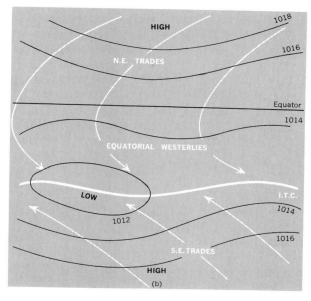

---

[6] Boundary surfaces separating air masses of contrasting temperatures and hence densities.

authorities consider the ITC a single zone which, on the average, is continuous around the earth, is shifted north and south with the march of the seasons, and is governed chiefly by the outward thrusting of air from the winter hemisphere. Others describe the ITC more realistically as discontinuous, for ordinarily it cannot be traced around the earth on the daily weather map. Still others describe the ITC as double in character, involving a northern intertropical convergence (NITC) and a southern one (SITC), with true equatorial air masses in the form of weak westerlies between the two (Fig. 3.34).

It has also been suggested that the NITC and SITC are not single lines of convergence but multiple ones. In this view, evaporation of the rain falling from the clouds which are developed at the convergence, and additional cooling of the air within the ITC by radiation from the cloud tops, may cause the ITC to become a local cool source with descending air. The result is the formation of new ITCs on either side of the original one. The total effect is the development of multiple ITCs and the local northward and southward displacements of the zone of convergence. From this brief description it can be seen that the literature on the intertropical convergence and its possible multiple convergences is very confusing and sometimes contradictory.

Equatorial westerlies. A somewhat controversial element of the tropical circulation concerns the existence of an equatorial west-wind zone in the intertropical convergence area between the broader belts of easterlies (Figs. 3.23 and 3.33a). Some weather specialists view equatorial westerlies as re-curved trades which have crossed the equator into the summer hemisphere (Fig. 3.33b). Others believe that they represent the normal situation over extensive longitudes within the inner tropics. The latter group thinks of the equatorial westerlies as being located between a northern intertropical convergence, or NITC, and a southern intertropical convergence, or SITC (Fig. 3.34). With the data now available it does not seem possible to determine whether the zonal wind, averaged for all longitudes over land as well as water, would show a mean westerly current near the equator at any season. Some assert that such an equatorial west-wind zone is reasonably well established from present observations for nearly 200° of longitude, extending from western and central Africa across the Indian Ocean and including the equatorial western Pacific, and probably also the eastern equatorial Pacific and parts of northern South America. They consider the southwest monsoons of southern Asia to be only one element of the equatorial westerlies. Other students of the atmosphere are less certain that the equatorial westerlies constitute a basic element of the planet's general circulation.

The explanations offered for the equatorial westerlies, if they actually do exist as a basic current, are too diverse to warrant their presentation here. But the significant features of these winds are worth summarizing: (1) They are most extensive in the summer hemisphere. (2) They are less well developed in the transition seasons. (3) They have a directional component toward the thermal low-pressure centers located over the lands. Hence they blow toward the poles, so that they are southwest in the northern summer and northwest in the southern summer. (4) They appear to be best developed over continental regions and adjacent oceanic areas which are more or less under continental control. (5) They are more variable in time and space than the trades. And because they are directionally less constant than trades, rainfall contrasts along their windward and leeward coasts are less marked.

The relationship between direction of winds in equatorial latitudes and the frequency and amount of precipitation has unusual climatic importance. Studies by H. Flohn show that winds with either westerly or poleward components (i.e., equatorial westerlies and tropical monsoons) are more unstable and have a greater abundance of rain than tropical easterlies (trades), which normally have an equatorward component. For a rectangular area in the equatorial Atlantic bounded by the parallels 10°N and 10°S, Flohn found that westerly winds are accompanied by rain in 25.1 percent of all cases, whereas with easterly winds the rain frequency is only 8.1 percent. Winds blowing toward the equator are accompanied by rainfall in 8.1 percent of the cases,

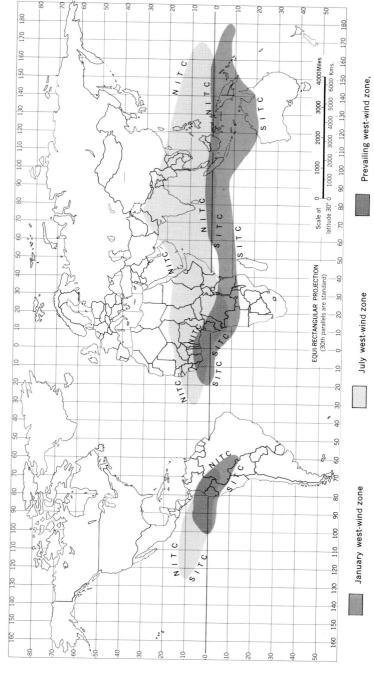

Figure 3.34 Displacement of the equatorial west-wind zone, situated between a northern and a southern ITC, in the extreme seasons. Greatest poleward displacement is over southern Asia in July. Displacement is also large over northern Africa in July and over Australia in January.

January west-wind zone

July west-wind zone

Prevailing west-wind zone,

EQUI-RECTANGULAR PROJECTION
(30th parallels are standard)

Scale at
latitude 30°

compared with 12 percent for winds blowing away from the equator. In part, the greater rainfall of equatorial westerlies and tropical monsoons possessing a poleward component could be a matter of air-mass character. Probably the decisive factor is the greater instability of such circulations, with their increased convergence and greater vertical movement. These in turn are related to latitudinal change in Coriolis force. In short, this suggests that winds with

$$\frac{\text{westerly}}{\text{easterly}} \quad \text{and/or} \quad \frac{\text{poleward}}{\text{equatorward}} \quad \text{components}$$

have a statistical tendency toward

$$\frac{\text{upward}}{\text{downward}} \text{ movement}$$

and thus to $\dfrac{\text{instability}}{\text{stability}}$ and $\dfrac{\text{raininess}}{\text{dryness}}$

as produced by Coriolis force on the rotating earth (Flohn).

Tropical easterlies or trade winds.   Without doubt the trades are more uniform in direction and speed, and are characterized by fewer atmospheric disturbances and spells of bad weather, than either the equatorial or middle-latitude westerlies. But as mentioned earlier, the classical picture of trade-wind uniformity in direction, speed, and weather is a fiction. Steady fair-weather trades with moderate-to-fresh breezes averaging 10 to 15 miles an hour are mainly characteristic of their central latitudes and their eastern oceanic sectors. Constancy declines toward their poleward and equatorward margins. Their western oceanic parts are also more variable because of frequent interruptions there by atmospheric disturbances. In the west also, the trades have a smaller north-south directional component, and in some places they may blow parallel with the equator, or even away from it.

Weather and climate also vary in different areas of the trades. Their poleward parts are closest to the centers of the subtropical anticyclones, where subsidence and horizontal divergence are strong, and the inversion is low. Consequently these loca-

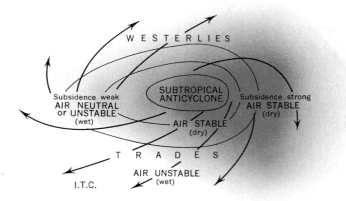

Figure 3.35   The circulation pattern around a subtropical anticyclone, with general areas of stability and instability shown. The eastern end of the oceanic cell and its trade winds are much more stable than the western end. The poleward parts of the trades are more stable than the equatorward parts. Degree of stability is suggested by the intensity of shading.

tions have few atmospheric disturbances and much dry weather and sunshine (Fig. 3.35). Here the air is stable and nonbuoyant, opposed to deep convectional overturning. As the trades move equatorward, two types of modifications gradually occur. First, as a result of traveling over great expanses of tropical ocean, they acquire large additions of moisture (hence latent energy) through evaporation. Second, the trade-wind air gradually leaves an area of strong subsidence in the subtropical anticyclones, where the temperature inversion is low, and approaches the equatorial trough of low pressure, where horizontal convergence and lifting are characteristic. Both modifications—the addition of moisture at the base, and the increasing convergence and lifting—make the trades more unstable, buoyant, and deeply humidified in their equatorward parts. In such an environment, atmospheric disturbances develop more readily, deep convectional overturning is facilitated by the absence of an inversion, and clouds and precipitation are more common. The trades evaporate great quantities of water and transport the water vapor, representing potential energy, toward the equatorial pressure trough. There it is released and returned to the land-sea surface in the form of heavy rain showers falling from towering cumulonimbus clouds, which are concentrated in synoptic disturbances.

Not only do the trades change character in a north-south direction; they also differ in their eastern and western oceanic parts and the adjacent margins of the continents. Normally the subtropical cells, on whose equatorward slopes the trades originate, are best developed toward the eastern sides of oceans. It is in these same eastern parts, where subsidence is pronounced and the inversion is strong and reaches down to low levels, that the trades are most stable and nonbuoyant, so that rain-producing disturbances are fewest (Fig. 3.35). Farther west, the anticyclone is weaker, subsidence is less, the inversion level is higher, and the air is more buoyant and inclined to rise. The result is that weather disturbances are more numerous and better developed, and clouds and precipitation are more frequent. Thus the trades are likely to be dry, fair-weather winds in their poleward and eastern oceanic parts, while their equatorward and western oceanic sectors have more weather disturbances and greater amounts of cloud and precipitation. (See Fig. 2.21.)

Winds of the subtropics

Latitudes from about 25° to 35° N and S are the centers of great anticyclonic circulations around a series of oceanic high-pressure cells elongated in an east-west direction. In these warm anticyclones there is deep and extensive subsidence of air. Their centers are associated with extended lines of mean surface-wind divergence, from which trades flow equatorward and westerlies poleward. As a consequence of subsidence and divergence the air is stable and nonbuoyant, so that these anticyclonic systems produce drought and fair weather. Pressure gradients are weak, average wind speeds are slow, constancy of wind direction is low, and resultant wind velocities are small. Thus the subtropical highs resemble the equatorial trough of low pressure in many features of their surface winds. However, the two regions are greatly unlike in general weather, because one is a region of convergence and ascent, while air subsides and diverges in the other.

Although in general, constancy of wind direction is likely to be low in the subtropics, or horse latitudes, it does vary considerably from one subtropical anticyclone to another. When a particular anticyclone has a quasi-stationary character, so that its size and shape remain much the same from one daily weather map to another, the constancy of wind direction may be relatively high. But where the subtropical cell is really a mean of a large number of anticyclones migrating across the area, the constancy of direction is much lower.

Actually, the zones of the subtropical anticyclones are not longitudinally as similar throughout in winds and weather as the above description might indicate. On the margins of the anticyclonic cells, wind velocity and constancy of direction are both higher than at the centers. And what is more important, the eastern flanks of each cell show stronger subsidence, lower inversion, and fewer atmospheric disturbances than the western margins (Fig. 3.35). This pattern is like that of the trades described in the previous section.

The strong subsidence in the eastern parts of an anticyclonic cell produces a marked inversion whose base reaches down to about 500 to 1,500 meters above sea level. The air above the inversion is dry, and that together with the temperature inversion is responsible for the meager rainfall. Some of the driest regions of the earth are located along the east sides of oceans and the adjacent west coastal margins of continents at about 20° to 30° N and S. These areas come under the influence of the strong subsidence in the eastern parts of the oceanic subtropical anticyclones. The intensified aridity of the coastal portions of the Sahara Desert in western North Africa, the Kalahari Desert in southwestern Africa, the Chilean-Peruvian Desert in South America, and the Sonoran Desert in northwestern Mexico testify to the drying influence of the eastern margins of the cells.

By contrast, the western side of a subtropical oceanic anticyclone is usually characterized by moist, unstable or neutral air and by fairly abundant rainfall. This condition is the exception in anticyclonic systems. Regions occupying positions coincident with the western ends of the oceanic anticyclones include the lands bordering the Caribbean Sea and Gulf of Mexico in North America, eastern China and southern Japan in Asia, southeastern Brazil in South America, southeastern Africa, and eastern Australia. All of these are rainy places and therefore stand in sharp climatic contrast to the arid regions along the western sides of the continents in similar latitudes, for the dry west-side areas are influenced by the subsident eastern margins of the oceanic anticyclones.

Various causes have been suggested for the marked air-mass and weather contrasts between the eastern and western ends of the oceanic subtropical anticyclones. A group of Norwegian meteorologists views the circulation in the subtropical cells as taking place in such a way that the planes of the ellipses which represent the trajectories of the air particles are tilted upward from east to west. Accordingly, the eastern end of the cell may be considered as tilted downward and the western end upward. So a mass of air in the cellular circulation (Northern Hemisphere) tends to subside as it moves eastward on the northern side of the cell and to be lifted as it moves westward on the southern side. The total vertical displacement of the air as it passes from one end of the cell to another may be ½ mile to a mile. It is to the tilted circulation in the subtropical oceanic cells, with subsidence prevailing at the eastern ends and lifting at the western ends, that these meteorologists would ascribe the greater stability in the east.

Other contributing factors which have been suggested are these: (1) Colder waters at the eastern end tend to chill and stabilize the air masses. (2) The air masses on the west have had a longer oceanic trajectory over warmer waters, so that they are more deeply humidified. (3) On the east the air masses are of more recent polar origin, while those in the west are old tropicalized air. (4) In the east

the atmospheric circulation shows a strong component directed toward the equator, so that decreasing Coriolis force acts to stabilize the air. In the west the circulation may even be directed poleward, with resulting increased instability. (5) In the west where the easterlies come onshore, they are frequently made unstable by orographic lifting (lifting caused by highlands).

When the opposite-moving air currents of two adjacent cells overlap, as they may do in the cols of slightly lower pressure between the cells, the current from the equatorial side of the eastern cell may be warmer than the other coming from the northern side of the western cell. As a consequence the warmer air may be forced to ascend over the cooler along an inclined surface of discontinuity whose slope is approximately 1:200. The cols therefore mark the boundaries between the cells and are characterized by converging air masses separated by discontinuity surfaces along which storms and rainfall may originate. These weak fronts formed in the cols between the cells are sometimes called meridional fronts, or trade fronts.

Middle-latitude westerlies

Moving with the gradient from the subtropical highs to the subpolar lows (roughly 35 or 40° to 60 or 65°) are the middle-latitude winds known as the stormy westerlies. However, the westerly circulation is not just composed of air supplied from the poleward flanks of the subtropical highs. Much genuinely tropical air from farther equatorward enters the westerlies through the extensive longitudinal gaps between the individual high-pressure cells. These invasions are especially conspicuous around the western ends of the oceanic cells and in the form of the summer monsoon in eastern Asia. The poleward boundary of the westerly wind belt is a particularly fluctuating one, shifting with the seasons and over shorter periods of time as well. Both in speed and direction the westerlies are highly variable. "Spells" of weather are one of their distinguishing characteristics. At times, especially in the winter, they blow with gale force; at other times they are

Mean wind directions at Scilly Islands, 49.38° N, 6.16° W (Frequency in percent)

N	5	E	7	S	6	W	11
NNE	5	ESE	5	SSW	5	WNW	10
NE	5	SE	4	SW	6	NW	9
ENE	4	SSE	4	WSW	7	NNW	4
				Calm 3			

Source: After Kendrew.

mild breezes. Most of their winds are moderate to strong, with stormy winds more prevalent than weak winds. Calms are uncommon. Although they are called westerlies, since the most frequent and strongest winds are from that direction, air actually blows from all points of the compass.

The variability of the westerlies in both direction and strength is largely the result of the procession of atmospheric disturbances (cyclones and anticyclones) and their fronts that travel from west to east in these latitudes. Such storms, with their local systems of converging and diverging winds, tend to modify or disrupt the general westerly flow. Moreover, on the eastern side of Asia, and to a lesser degree of North America, continental wind systems called monsoons tend to disturb the westerly flow, especially in summer.

It is in the Southern Hemisphere, which has few land masses in latitudes 40 to 65°, that the stormy westerlies can be observed in their least interrupted development. There over great expanses of ocean, winds of gale strength are common in summer as well as winter. Early mariners called these the "roaring forties," and when later navigators penetrated higher latitudes, they discovered the "furious fifties" and the "shrieking sixties." In the vicinity of Cape Horn, violent winds often make east-west traffic around the Cape not only difficult but dangerous. It is a wild region, where gale follows gale with brief intervening lulls; a place with raw chilly weather, cloudy skies, and mountainous seas.

As this chapter has frequently mentioned, the great Northern Hemisphere land masses with their seasonal pressure reversals cause all wind systems north of the equator to be much more complex. The

westerlies there are considerably less vigorous in summer than in winter. In summer, gentle to fresh breezes prevail, and winds come from a variety of directions with almost equal frequency. But in winter they are like their counterparts in the Southern Hemisphere—strong, boisterous, and westerly more often than anything else.

The poleward margins of the westerlies near the subpolar troughs of low pressure are particularly subject to great equatorward surges of polar air in the colder months. The sinuous line of discontinuity known as the polar front, which separates polar from tropical air, is the zone of origin for a great many middle-latitude cyclones and anticyclones. It follows that the poleward margins of the westerlies are much more subject to stormy, variable weather than the subtropical margins. Since this polar front and the accompanying belt of storms migrate with the sun, retreating poleward in summer and advancing equatorward in winter, it is logical that stormy weather in the middle latitudes is much more pronounced during the winter season.

Convergence and divergence in surface winds

Earlier sections have discussed the climatic significance of horizontal convergence and divergence in airflow and the associated vertical movements. A more integrated representation of horizontal convergence and divergence in the earth's mean surface winds is provided by Figs. 3.36 and 3.37. From them it can be seen that the strongest and longitudinally most continuous convergent zone is the one between the trades—the well-known ITC. As would be expected, Figs. 3.36 and 3.37 also show that divergence is particularly well developed in the latitudes of the subtropical anticyclones. These features have been emphasized throughout the chapter.

But an almost equally striking fact is that the distribution of surface-wind convergence and divergence shown in these maps is not nearly as simple, and certainly not as zonal, as it may appear to be in an idealized atmospheric circulation which accentuates zonal trades, westerlies, horse latitudes, and doldrums. The many nonzonal elements, or

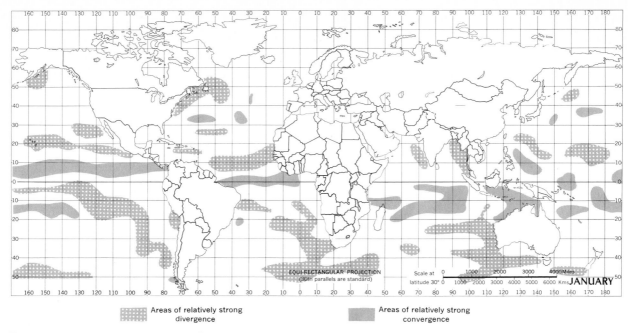

Figure 3.36 Distribution of areas of strong divergence and convergence of the mean surface wind in January over the oceans. The equatorial latitudes stand out as the main extended zone of convergence. Subtropical latitudes show more consistent divergence, characteristically in centers rather than in belts, and with the centers concentrated on the eastern side of an ocean. (*After Mintz and Dean.*)

Figure 3.37 Distribution of areas of strong divergence and convergence of the mean surface wind in July. Equatorial convergence and subtropical divergence are apparent, but centers are more conspicuous than belts. Note particularly the strong center of divergence in the easternmost North Pacific, off the west coast of North America. (*After Mintz and Dean.*)

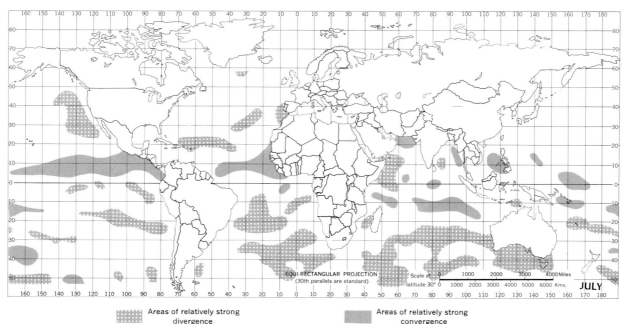

local areas of convergence and divergence, shown in Figs. 3.36 and 3.37 are just as conspicuous and greatly complicate the general world pattern of climates. This book will not catalogue in detail the numerous nonzonal elements in the distribution patterns of convergences and divergences. The important point to remember is that they are a major factor in world climates.

TERRESTRIAL MODIFICATIONS OF SURFACE WINDS

Among the terrestrial modifications of surface winds are several which were mentioned earlier, such as seasonal reversal of wind direction (monsoon systems), and the latitudinal shifting of wind belts with the seasons. Some terrestrial modifications are planetary in magnitude, others are of subcontinental size, and still others are essentially local. They result from (1) the inclination and parallelism of the earth's axis, which causes a seasonal north-south shifting of the belts of solar energy, (2) the earth's nonhomogeneous surface, on which land and water areas have contrasting temperature, pressure, and wind characteristics, and (3) the variable configurations and altitudes of land areas.

Latitudinal shifting of wind belts

It was explained in Chapter 1 that during half the yearly period of the earth's revolution about the sun, the sun's vertical noon rays shift from $23\frac{1}{2}°$ N (summer solstice) to $23\frac{1}{2}°$ S (winter solstice)—a total of 47°. The belt of maximum solar energy actually shifts as much as 70°, and temperature belts, which are largely sun-controlled, follow the latitudinal migration of solar energy. Pressure and wind belts, which are in part thermally controlled, also shift north and south with the sun (Figs. 3.8 and 3.29).

This north-south shifting of the wind belts is not a simple zonal phenomenon, however; it varies in the amount and rapidity of the shift from one part of the earth to another. In general it lags 1 month, or possibly 2, behind the sun. Over the oceans and along coasts, where the migration is more readily observable, it is fairly small—usually not much over 10 to 15°. But over continents, where seasonal temperature changes are greater, the total latitudinal shift is also greater. In addition, the time lag is considerably less on land than over oceans. Surface-wind systems are more confused over land masses, however, owing to the effects of terrain irregularities and strong seasonal and diurnal temperature contrasts. Consequently an orderly migration of wind belts is less evident over continents.

Latitudes affected by more than one wind belt.

Latitudinal shifting of the wind belts is climatically most important in regions occupying an intermediate position between two unlike wind systems. Such a region is encroached upon at the opposite seasons of the year by contrasting air masses and their strikingly different weather conditions. Theoretically, three such transition zones should be present in each hemisphere; and there is evidence that these actually do exist, although in imperfect form and certainly not as continuous zonal belts (Fig. 4.29).

Latitudes about 5 to 15° lie between the wet equatorial convergence zone (the ITC and the equatorial westerlies) on the one hand, and the dry subsidence-divergence zone of the trades and the subtropical anticyclones on the other. With the north-south seasonal shifts of pressure and wind belts, these latitudes mainly get the effects of the ITC and its rainbringing disturbances at the time of high sun (summer), and the effects of subsidence in the subtropical anticyclones and the dry trades at the season of low sun (winter). One wet and one dry season are the result. This seasonal pattern usually is muted along the eastern side of a land mass, where the subtropical anticyclone is ordinarily weak (Fig. 3.35). See also Fig. 4.29.

Latitudes about 30 to 40° fall between the fair-weather subtropical anticyclones and the middle-latitude stormy westerlies. Drought associated with subsidence and divergence should therefore be characteristic of summer, while the winter season should have adequate precipitation from traveling cyclonic storms and fronts in the westerlies. Actually, however, these dry summers and wet winters are found only in certain restricted longitudes, mainly on the eastern sides of oceans and the adjacent western margins of the continents. Such a location is dominated by the stable eastern flanks of an oceanic subtropical anticyclone in summer, so that subsidence is strong and the temperature inversion low. But the anticyclone is weaker in the western parts of subtropical oceans and the adjacent eastern sides of continents, as well as in their interiors. During summer in these regions atmospheric disturbances are numerous and any inversions are high or weak, so that it may be actually the wettest season.

Latitudes 60 to 70°, which are the approximate positions of the subpolar lows, are located between the stormy westerlies and winds of polar origin, so that latitudinal shifting of winds should allow this region to experience both during the course of a year. However, the numerous cyclones and anticyclones which infest these latitudes tend to complicate and obscure any simple migration of wind belts. This is much more a region of alternating thrusts of cold polar and warm tropical air. Nevertheless, there is a greater prevalence of cold polar air in winter and of warmer southwesterly currents in summer, a fact which suggests a semblance of wind-belt and storm-belt migration (Fig. 4.29).

Monsoons

Concepts of a monsoon wind system.   Unfortunately, "monsoon" is a term which has come to have a variety of meanings among geographers and meteorologists. It seems to have been derived either from the Arabic "mausin" or from the Malayan "monsin." As first used it was applied to southern Asia and the adjacent waters, and it referred to the seasonal surface airstreams there which reverse their directions between winter and summer. In the original meaning of monsoon, cause was not implied. With such a simple descriptive definition involving only a seasonal wind reversal approaching 180° in magnitude, quite a few regions of the earth actually qualify as having monsoon wind systems. As described in the previous section, seasonal wind reversal is to be expected in the transition zones between unlike wind belts.

But in 1686 Halley of England burdened the original descriptive concept of monsoon with a theory concerning the origin of the Asiatic monsoon, a theory which subsequently was applied to other continental areas as well. He proposed that monsoons are more than just seasonal winds—that they are also directly thermally induced, and so are gigantic convectional systems stemming from differential seasonal heating of continental and oceanic areas. The chain of events is from temperature through pressure and winds to rainfall. In summer, according to Halley, the land surface becomes warmer than that of the surrounding seas. This thermal contrast generates surface pressure differences, resulting in a low center over the warm land and higher pressure over adjacent seas. As a consequence the summer monsoon is a sea-to-land wind, with the tropical maritime air normally bringing to the land an abundance of moisture and warmth.

Conversely, the winter monsoon is a wind of land origin. Because the land is colder than the sea then, a shallow, thermally induced high-pressure cell develops over the land, with lower pressure over the adjacent ocean. The result is a land-to-sea pressure gradient, which causes cold, dry continental air masses to flow seaward.

It is debatable whether or not Halley's definition of monsoons, which involved origin and explanation as well as description, was an improvement. However this may be, throughout geographic and meteorological literature the most common definition of a monsoon involves not only the fact of a seasonal wind reversal, but also the idea of thermal origin arising from differential heating of extensive land and water surfaces. According to this more restricted point of view, monsoons develop because the earth's surface is not homogeneous, for they could not arise if the earth were composed of all land or all water.

Still, authorities use this somewhat standard definition in different ways. To some authors, monsoon is always singular, and the term applies only to the sea-to-land surface circulation in summer. Also, some insist that a genuine monsoon must have a complete 180° reversal of wind direction between winter and summer, while others are satisfied with a smaller directional variation. A number of meteorologists apply the term monsoon to seasonal variations in winds aloft as well as at the surface. As noted in the previous section, it can be that monsoonal wind reversals, both at the surface and aloft, will originate from the normal latitudinal migration of the planetary wind zones. Some authors, therefore, have applied the term monsoon to what appear to be trade winds which are deflected as they cross the equator into the summer hemisphere so that they become equatorial westerlies.

Modifications of the classical concept. There is some doubt whether monsoons as purely thermal systems of winds actually are common and widespread. What seems probable is that the recognized monsoon winds are far more complex than was previously supposed. For example, monsoon lows over the summer continents are not always persistent features on the daily weather charts, even though they are conspicuous on mean summer charts of surface pressure. On daily weather maps, the monsoon lows show great variations in position and intensity from day to day, variations which may be attributed to different cyclonic systems moving through the area. Many of these moving cyclonic storms seem to come in from the sea fully formed, so that they cannot be described as heat lows. Such local disturbances are characteristic of the summer season of monsoon areas like India, China, and northern Australia.

Certainly also, the winter anticyclones over cold continents, which are conspicuous on average sea-level charts, are only partly a consequence of radiation cooling from the cold land surface. Even the thermal winter anticyclone over Siberia, for example, does not have great depth except on certain occasions. Significantly also, the seaward movement of cold continental air is not at all continuous. It occurs in intermittent surges of considerable vertical depth, which must originate in part from great outpourings of Arctic air drawn toward lower latitudes on the rear of eastward-moving upper troughs. These invasions also feed the surface anticyclone.

As it happens, most of the land areas that show a seasonal reversal of wind direction are not in latitudes with sufficiently severe winters to develop thermal anticyclones on the continents. Therefore the offshore land winds of the low-sun season in tropical regions must result directly from nonthermal causes. The fact is that the monsoons of the constantly warm low latitudes, such as those of southern Asia, West Africa, and Northern Australia, are not a consequence of seasonal temperature extremes. Instead, they come from a normal latitudinal migration of wind belts following the course of the sun. As described in the preceding section, tropical easterlies, or trades, prevail at the time of low sun, and equatorial westerlies at the period of high sun.

To a greater extent than was formerly thought, monsoons appear to be modifications of the general planetary wind system. Any thermally induced monsoon circulation is always superimposed upon the original planetary winds, which usually form the dominant framework. If the monsoon were a convectional system of gigantic proportions, the direction of airflow aloft above the surface current should be opposite to that at low levels, and in many instances this is not the case.

Active discussion of monsoons has developed in recent years. Not only have many of the widely held opinions regarding their origins and wind systems been questioned, but some authors have also disputed the efficacy of the summer monsoon itself as an originator of precipitation. Undoubtedly the summer monsoon does transport great quantities of evaporated moisture from sea to land, and hence creates an atmospheric environment favoring the rainmaking processes. However, this is not sufficient to guarantee precipitation. The weather in the summer monsoon is actually much more variable than most textbook accounts would lead one to assume. It does not rain continuously throughout the season of general onshore winds; there are frequent spells of fair weather, some of them lasting several days, during which the skies may be relatively clear and no rain falls. Except in highlands, most spells of cloud and showers during the summer monsoon appear to come with atmospheric disturbances in the form of waves, depressions, or cyclones.

Climatic implications of monsoons. Since a seasonal wind reversal usually involves a change in air masses, it naturally has important effects on precipitation and temperature. Most monsoon regions, tropical as well as middle-latitude, have more rainfall in summer than in winter; in fact, winter is often quite dry. Summer wetness reflects the abundance of moisture in the tropical-equatorial air masses which invade the land at that season. Winter's lesser precipitation, in some parts even drought, is associated in middle latitudes with the dominance of dry, cold, stable air of continental origin. In the tropics the winter circulation obviously cannot be composed of cold air, but it is derived from a subtropical anticyclone, usually of land origin, so that

it is stable and dry. However, if the winter monsoon air crosses a large expanse of ocean before coming onshore, as it does in western Japan and the eastern Philippines, winter precipitation may also be plentiful.

Important temperature effects of monsoons are mainly confined to the middle latitudes, where genuine cold sources exist over the large continents in winter. There the monsoon circulation advects tropical heat in summer and subarctic continental cold in winter. The consequence is abnormally cold winters, sultry heat in summer, and large annual ranges of temperature. No such cold sources exist in the tropics to produce cold winters and large ranges.

Asiatic monsoons. Eastern and southern Asia together have the earth's largest and most perfectly developed monsoon circulation (Figs. 3.30, 3.31). In most parts of this region, the general direction of the seasonal airflow is from the continent in winter and from the sea in summer (Fig. 3.38). The magnitude and full development of the monsoon are largely a consequence of Asia's size and arrangement of mountains, two factors which combine to

Figure 3.38   Principal elements of the low-level circulation patterns over eastern and southern Asia in the cold and the warm seasons. (*After Thompson, Watts, Flohn, and others. From Trewartha, The Earth's Problem Climates, University of Wisconsin Press.*)

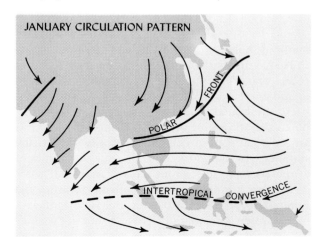
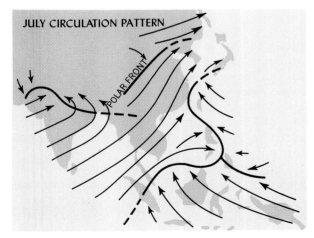

accentuate the winter and summer temperature extremes. These in turn magnify the latitudinal migration of winds as well as their seasonal reversal.

The huge land mass which is the Asian continent naturally intensifies all weather effects associated with continentality. And because of Asia's high east-west cordillera, the grain of the continent is hostile to a meridional circulation of air. This makes the seasonal contrasts in temperature and pressure even more dramatic. Consequently, the winter anticyclone is unusually strong and is positioned well to the east of the continent's center. In summer, the intertropical convergence reaches its maximum northern migration in the hot lowlands of northern India-Pakistan where, because of the barrier of high mountains, air from the north cannot penetrate. By comparison North America, with its smaller size and triangular shape, as well as its mountain system extending north-south, has no such seasonal contrasts in pressure centers, for a meridional exchange of air masses is relatively easy.

East Asia monsoon. This circulation mainly affects China, Taiwan, Japan, and Korea. Since it is located in middle latitudes, seasonal temperature contrasts between continent and ocean play a significant role in producing the wind reversal between winter and summer. In both seasons, local cyclones and anticyclones produce numerous interruptions in the prevailing wind direction, though it is somewhat more constant in winter than in summer. The winter monsoon of East Asia is also much stronger than that of summer, the velocities of the mean resultant surface wind along the coasts being several times greater in January than in July.

Percentage of wind frequency in North China

	N	NE	E	SE	S	SW	W	NW
Winter	17	8	5	6	6	8	18	*32*
Summer	10	9	12	*26*	16	10	7	10

During the *warm season,* a weak continental low, at least partly of thermal origin, is positioned over eastern Asia. Three basic air currents dominate the region (Fig. 3.38, July).

1. North of about 40° are the deep zonal westerlies, with a widely fluctuating jet stream aloft whose average position is about 40 or 45°N. Locally over northern Japan this circulation may be from the north.

2. South and west of the zonal westerlies and dominating most of central and southern China and Korea is a deep and humid southwesterly current, which is a poleward extension of the same equatorial westerlies (called the southwest monsoon) that prevail over tropical South Asia in summer. It is this current which supplies much of the moisture for the abundant summer rainfall of East Asia, since during the warm season the mean flux of water vapor over most of that region appears to be from west to east. Separating the zonal westerlies to the north from the equatorial southwesterlies farther south is the fluctuating polar front convergence. Its disturbances generate much of the copious summer rainfall of North China, including the Northeast (Manchuria).

3. The third basic current is the tropical easterlies, whose circulation around the western end of the North Pacific High causes them to arrive in East Asia as a southeasterly, southerly, or even southwesterly current. Its moisture is mainly of Pacific origin. Being somewhat more stable and less deeply humidified than the equatorial southwesterly current, this air from the Pacific does not provide as favorable an environment for the rainmaking processes. Moreover, since the Pacific air normally does not enter deeply into the continent, it is not such an important moisture source in summer, except in eastern subtropical China and in Japan. Significantly, rainfall declines in these regions in mid- and late-summer, at the time when Pacific air reaches its greatest dominance and the polar front is positioned farthest north.

The summer rainfall of eastern Asia does not appear to be directly a result of the warm continent being flooded by humid tropical air, in which surface heating produces numerous random convective showers and thunderstorms. Actually, eastern Asia has many fewer thunderstorms than the southeastern United States in similar latitudes, even though the much rougher East Asian terrain should act to trigger convective systems. Most of the lowland summer rainfall of eastern Asia appears to be associated with extensive atmospheric disturbances, some of which develop along fronts. There are other nonfrontal disturbances which are in the nature of pressure waves, speed convergences, and typhoons.

A Chinese climatologist has pointed out that the stronger and steadier the summer monsoon, in South China at least, the less the precipitation is in that region. This again suggests that the rainfall is not directly monsoonal, but rather is caused by disturbances many of which are embedded in the southerly flow. In this respect, it is highly significant that the probability of rainfall in subtropical China is least with southerly winds and highest with northerly winds—a fact which tends to link rainfall with a disturbed monsoon flow. As noted previously, the migration northward of the polar front and its disturbances to a mean position in North China–Manchuria in midsummer brings those regions a brief but heavy rainfall. Since North China–Manchuria is about the northern limit of penetration of the summer monsoon, and since its northern frontier is a pulsating, migrating border, it is logical that these northern parts of East Asia should have a precarious and variable rainfall.

The surface *winter monsoon* of East Asia is a wind which appears to originate in a cold anticyclone centered in the Baikal area of Siberia. The seaward flow is not a steady one, but occurs in the form of great nonperiodic surges, a fact which suggests that the anticyclone is fed at higher levels by huge expulsions of Arctic air which move southward on the rear of long troughs in the upper atmosphere. At its source in Siberia the northwest winter monsoon is very cold, dry, and stable, with a well-developed inversion, so that it provides a poor environment for precipitation. Over eastern Asia the winter monsoon's temperature varies considerably, depending on whether it has had a land or a sea trajectory, and also on the nature of the terrain over which

it has traveled. Terrain-induced turbulence tends to disrupt the surface inversion and bring down warmer air from above.

Normally the winter monsoon prevails over all eastern Asia, including China, Korea, and Japan, and extends at times into the northern part of the Indochina Peninsula. A widely fluctuating polar front separating cold monsoon air from that of the maritime trades is positioned along the eastern margins of Asia and the adjacent seas (Fig. 3.38, January). Along this front pass many weather disturbances. Zonal westerlies prevail above the not-so-deep monsoon current. The westerlies have two branches, the main one to the north of the central Asian highlands and the other to the south. Each branch is overlain by a jet stream, and these tend to steer cyclonic storms along their routes.

As a result of its well-developed winter monsoon, eastern Asia has the lowest sea-level winter temperatures for its latitude anywhere on the earth. The dry, stable air is not conducive to precipitation. Where the winter monsoon has had a land trajectory and is especially cold and dry, weather disturbances are weak and winter precipitation is meager, as it is in North China–Manchuria. But where it has been humidified by traveling some distance over water, as is the case in Japan and parts of southern China, winter precipitation may be considerable. Over the North Atlantic, for example, studies have shown that polar continental air becomes relatively mild and wet through a thickness of about 4 km (2.5 miles) after a sea trajectory of 600 km (375 miles). Except where highlands cause forced ascent of the air, almost all winter precipitation in the winter monsoon is generated by cyclonic storms, many of them frontal in origin.

Monsoons of tropical southern Asia. The tropical monsoon circulation of southern Asia, including India-Pakistan and Southeast Asia, differs significantly from that of East Asia. The Indian monsoon is effectively separated from that of China by the highlands of the Himalaya-Tibet system (Fig. 3.38). In Southeast Asia, however, this separation of tropical and middle-latitude monsoons is less complete. Among the elements in which the monsoons of

India and China differ is that of the strength and constancy of the seasonal winds. While the winter monsoon in eastern Asia is stronger and more constant in direction than the summer monsoon, the opposite is true in South Asia. There the winter monsoon is frequently only a gentle drift of air (2 or 3 miles an hour). But the average velocity of the summer monsoon at Bombay is 14 miles an hour, and it is even stronger over the sea.

A further difference between the two Asiatic monsoons is in the locations of their centers of action. In *summer* a deep and widespread surface pressure trough, to some degree thermal in origin, extends across northern India–Pakistan into Southeast Asia. This is part of the planetary intertropical convergence zone, which here reaches its maximum poleward displacement on the earth. To the south of the trough is a deep, humid, unstable current of equatorial maritime air, called the southwest monsoon (equatorial westerlies). This unstable southwesterly current crosses India, continues eastward over the Indochina Peninsula, and then moves northward over much of eastern Asia (Fig. 3.38, July). It is the great humidity source for most of southern Asia.

In *winter* much of southern Asia is dominated by weak subtropical anticyclones, so that the northerly winds of that season are essentially variable subtropical winds and trades (Fig. 3.38, January). In the far northwest of the Indian subcontinent, some modified polar air enters from the northwest, while in northeastern Indochina there are also modest invasions of greatly modified polar air from China. But except at these two extremities, the high mountains effectively separate the winter monsoon of South Asia from that farther north. A southern branch of the zonal westerlies, together with a subtropical jet stream, is positioned just south of the Himalayas in winter, and at times this westerly flow may reach down to the surface in the northern lowlands of the subcontinent. In summary, the seasonal reversal of winds in South Asia represents only a latitudinal migration of planetary winds, with northeast trades prevailing in winter and equatorial westerlies in summer.

The summer period of southwest monsoon (equatorial westerlies) dominance is, of course, the rainy season for the region as a whole. Where these warm, humid, unstable air masses of equatorial origin are forced upward over highlands, copious rains result. Over lowlands, most of the showery summer rainfall accompanies the passage of extensive atmospheric disturbances, a great majority of which have weakly developed wind systems. A steady and uninterrupted southwesterly flow is usually associated with fair weather over lowlands. The typical weather sequence during the summer months is one of spells of disturbed weather with organized cloud and rain areas, alternating with intervals of relatively fair weather. In the Indian subcontinent, the advance northward of the ITC and southwest monsoon in late May and June is accompanied by particularly turbulent weather. This is known as "the burst of the monsoon."

During the period of the winter monsoon, the weather element is relatively feeble because of the prevailing anticyclones, and the air is stable, nonbuoyant, and disinclined to rise. Precipitation is meager except along elevated windward coasts and in highlands, and clear skies and fair weather are the rule. The very modest winter rainfall in northern India–Pakistan is associated with weak cyclonic disturbances, which take an easterly course approximating that of a southerly jet stream across the northern part of the subcontinent, across Southeast Asia, and even into southern China.

Other monsoon-like areas. Nowhere else in the world is such an extensive land area affected by seasonal wind reversal as the eastern and southern rimlands of Asia. Yet as the previous section showed, even Asia's monsoons depart from the textbook model in a number of respects. And there are other regions of the earth whose wind systems and consequent weather bear a degree of resemblance to those associated with monsoons.

One such area is tropical West Africa, which fronts on the Gulf of Guinea and the Atlantic Ocean from Nigeria to Sierra Leone. Here during the Northern Hemisphere summer a southwesterly flow of equatorial maritime air prevails, while during the season of low sun it is replaced by northeast trades (Figs. 3.30 and 3.31). This resembles the situation in South Asia, except that along a narrow coastal belt in West Africa the southwesterlies persist in greatly weakened form even in winter, with northeasterly winds on the average not quite reaching to the littoral. The West African wind reversal, like that of South Asia, functions indirectly to produce wet summers and dry winters.

Northern Australia also has an onshore northwesterly flow of humid equatorial maritime air in summer and offshore dry southeast trades in winter, so that the wind reversal is nearly 180° (Figs. 3.30 and 3.31). In most ways it resembles the monsoons of South Asia and West Africa. A wet summer and

Percentage of wind frequency in North Australia

	N	NE	E	SE	S	SW	W	NW
Winter	4	4	14	62	11	2	1	2
Summer	17	9	10	9	3	5	8	39

dry winter result. A similar wind reversal in equatorial East Africa is likewise referred to as monsoonal by the local meteorologists. Keep in mind, however, that seasonal wind reversals in the tropics are rarely induced by the differential heating of land and water, for the winter circulation, the trade wind, is not derived from a cold anticyclone but from a warm subtropical anticyclone.

In middle latitudes, where a convective type of seasonal wind reversal is more likely, North America is the only continent other than Asia whose size might permit the development of a thermally induced monsoon circulation along its eastern and southern margins. South America is too narrow in its middle-latitude parts to produce such a continental system of winds. And even North America, because of its smaller size and meridional mountain systems, cannot compare with Asia in the creation of a thermally induced wind system. Still there are semblances of a monsoon pattern, for the central and eastern United States show an average

Percentage of wind frequency in the eastern United States

	N	NE	E	SE	S	SW	W	NW
Winter	11	15	6	6	7	18	14	23
Summer	8	12	6	11	13	28	9	13

winter circulation of cold northerly surface winds which is fairly anticyclonic. Along the Atlantic and Gulf Coasts, prevailing winter winds are definitely offshore. In summer, by contrast, this same region is fed by a southerly circulation of warm tropical-subtropical air (Figs. 3.30, 3.31). This southerly flow of tropical Gulf air is probably not drawn to the interior by a thermal low, however; rather, it mainly represents the normal circulation around the western end of a subtropical anticyclone. An average onshore-wind component is not characteristic of the Atlantic Coast in summer. Nor does the seasonal march of rainfall everywhere in central and eastern North America have a strong monsoonal character. A wet summer and drier winter are mainly features of stations well inland, especially west of the Mississippi. Much of New England and Maritime Canada have no important seasonal accent in precipitation; parts even show a slight winter maximum, and an extensive region in the subtropical southeastern United States receives more precipitation in the winter half-year than in the summer half.

Minor terrestrial winds

Land and sea breezes.   Just as there are seasonal wind reversals (monsoons) caused at least in part by seasonal temperature contrasts between land and water, so there are diurnal, or daily, monsoons resulting from similar temperature contrasts within the 24-hr period. These are called land and sea breezes. Along coasts, there is often a drift of cool air from land to water at night (corresponding to winter) and a wind from sea to land during the heat of the day (corresponding to summer). These diurnal monsoons represent a convectional circulation (Fig. 3.39).

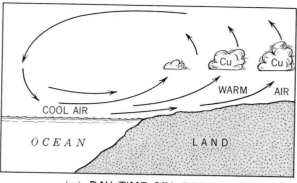

(a) DAY TIME SEA BREEZE

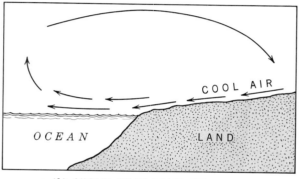

(b) NIGHT TIME LAND BREEZE

Figure 3.39 One type of diurnal wind change.

Usually the sea breeze begins between 10 and 11 A.M., reaches its maximum speed between 1 and 3, and gradually subsides between 3 and 8, after which it is replaced by the weaker land breeze. The height to which the sea breeze extends varies with different climates and with local conditions. Sea breezes off large lakes may reach a depth of 200 to 500 m. Along coasts in the subtropics and tropics this may increase to 1,000 to 2,000 m. The land breeze is much shallower.

The distance on land to which the sea breeze penetrates also varies greatly. Distances of 15 to 50 km are typical of the middle latitudes; in the tropics 50 to 65 km are more characteristic, and maximum distances of over 100 km have been recorded. The land breeze has a much smaller seaward range.

Velocities of the sea breeze also show great variations. In the middle latitudes, normal speeds do not exceed 8 to 12 miles per hour, and maximum values go up to 15 to 25 miles per hour. But along some tropical coasts, the sea breeze may reach storm intensity. It is particularly strong along dry tropical coasts which are paralleled by cool ocean currents, for there the afternoon temperature contrasts between land and water are unusually strong.

The maximum development of strong and regular sea breezes occurs on tropical coasts; the climatic importance of these winds normally decreases in middle latitudes. It is particularly in tropical areas where skies are clear that the sea breeze is a full-grown phenomenon, for such conditions favor large daily temperature contrasts between land and sea. The generally weaker prevailing winds in the low latitudes likewise interfere less with the development of the sea breeze than the winds in stormier regions farther poleward. In the middle and higher latitudes, the sea breeze is more or less confined to the warmer seasons. In the Baltic Sea region, land and sea breezes are characteristic of only about 20 percent of the days even in summer. In Jakarta, Java, on the other hand, they are a year-round phenomenon and occur on 70 to 80 percent of the days. Along a dry coast at Karachi, Pakistan, they blow on 100 percent of the days from May through September.

The sea breeze creates a more livable and healthy climate along tropical littorals than they otherwise would have. When it begins to blow, it may cause a drop in temperature 15 to 20° within ¼ to ½ hr. At a coastal station in Senegal, the temperature at 12:30 P.M. on one April 14 was 100°, with a land wind from the northeast and a relative humidity of 3 percent. But at 12:45 the wind direction was northwest from the sea, temperature had dropped to 82°, and the relative humidity had risen to 45 percent (Hann). Frequently the most uncomfortable part of the day on a tropical coast is in the forenoon before the arrival of the sea breeze. Coasts with well-developed sea breezes are inclined to have modified marine climates, with the daily temperature maxima much reduced. In cities like Milwaukee and Chicago, the residents recognize two distinct belts of summer climate: a cooler marine type lying within a mile or so of Lake Michigan, and a markedly warmer one farther in the interior.

Rainfall effects of the sea breeze are less than its thermal effects and do not seem t everywhere. But along certain coasts in the and subtropics, the sea breeze appears to p speed or directional wind convergence which lifting. Such lifting results in increased s activity, particularly in the warm seasons and warmer hours of the day, when the sea bre at its maximum. The strong summer maximu precipitation in peninsular Florida and along Gulf Coast in the United States is believed t partly due to sea breeze convergence. This sum maximum is especially prominent in penins Florida, where there is a convergence of two breezes—an easterly one from the Atlantic and westerly one from the Gulf.

Mountain and valley winds.[7]   Like land and sea breezes, mountain and valley winds have a distinct diurnal periodicity. Local thermal differences in mountain areas tend to create a convectional circulation which in the daytime consists of a lower current of air directed toward and up the mountain, and an upper current in the opposite direction. This circulation pattern reverses at night (Fig. 3.40).

The daytime surface current directed toward the mountains and upslope is composed of two elements: the thermal upslope wind and the valley wind. The *thermal upslope wind* is the result of temperature differences between the air over the intensely heated exposed slopes and that over the center of the valley at the same altitude. The result is a strong rising of air along the mountain slopes. The upslope

[7]The foehn or chinook wind is discussed in Chapter 11 on highland climates.

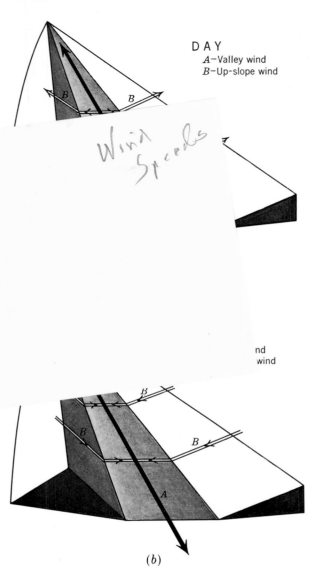

DAY
A—Valley wind
B—Up-slope wind

(b)

Figure 3.40   Another type of diurnal wind change.

winds start shortly after sunrise, reach their maximum at midday, and reverse their direction shortly after sunset. As a result of the stronger solar radiation, they are particularly well developed on southern slopes (in the Northern Hemisphere) and are weak or absent on northern slopes. A consequence of upslope winds is the development of daytime cumulus clouds over peaks and summits. These are the "visible tops of invisible ascending air currents." Daily afternoon cloudiness and showers are therefore common in mountains, and visibility is restricted during the warm hours of the day because of the cloud masses and rain.

Closely associated with slope winds are the *valley winds,* which blow along the axes of the larger valleys. They are especially well known in the wide and deep valleys of the Alps, where they have been studied in detail. They are best developed and most frequent during clear anticyclonic weather in summer. During the day, an up-valley or valley wind blows from about 9 or 10 A.M. until sunset, after which an opposite down-valley or mountain wind blows (Fig. 3.40).Within large mountain valleys, the diurnal temperature variations are more than twice as large as those in a similar air layer over the adjacent plains. This results in a pressure gradient from plains to mountain valleys during the day, and from valleys to plains at night. Weather and climate in mountainous regions are of course greatly influenced by these local winds.

Diurnal variation in wind velocity.   This is another common terrestrial modification of surface winds. It often happens that calm nights and early mornings in the warmer months are followed by windy middays, the maximum surface wind velocity corresponding with the time of greatest heat (Fig. 3.41). By sunset the atmosphere usually becomes calm again. The boisterous midday winds are associated with convectional overturning, which effects an interchange of air between upper and lower strata, at the time of greatest surface heating. Under those conditions the surface air, which becomes linked with the faster-moving air aloft as a result of the ascending and descending currents, is dragged along at a rapid rate. Cumulus clouds are often numerous during the windiest hours when convection is at a maximum, a coincidence that has gained them the name of "wind clouds." During the night, when the

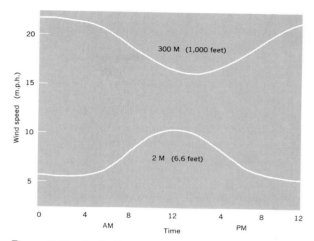

Figure 3.41 Typical variation in wind speed near the ground and at about 1,000 ft (300 m). (*After Riehl.*)

lower air is colder and heavier, there is no tendency for it to rise. The interchange between the upper and lower air is then at a minimum, so that the surface air is relatively undisturbed by the fast-moving currents aloft.

Gravity winds.   Along the coasts of Greenland and Antarctica, which are backed by ice plateaus, unusually strong winds are caused by the descent of the cold, dense air from the seaward-sloping icecap plateau to the shore below. On the edge of the Antarctic Continent, Mawson reported an average wind velocity of 50 miles per hour, while velocities of over 100 miles an hour were frequent. Similar gravity winds on a smaller scale are characteristic of all regions where cold air develops on uplands and along slopes at higher elevations and drains downslope to lower elevations. Such a wind is the *bora* type, characteristic of the Yugoslav coast. Here a narrow coastal plain is backed by a plateau where, under anticyclonic conditions in winter, the air becomes very cold. Because of its density, this plateau air slips through the passes and down the slopes of the plateau as a cold, gusty wind whose temperatures, even after warming by descent, are markedly colder than those along the coast below.

OCEANIC CIRCULATION AND ITS CLIMATIC EFFECTS

In general it can be said that the atmosphere directly and indirectly controls the oceanic circulation, and that the movements of ocean waters in turn have a significant influence upon world climates. The interaction between atmospheric and oceanic circulation is so complicated, however, that it is often impossible to separate cause and effect.

Drifts and currents

Although the oceanic circulation is three-dimensional, it is mainly the surface parts of the system, or those in contact with the atmosphere, that will be considered here. Much the larger part of the surface movement of ocean waters is a slow, relatively inconspicuous transfer (at an average rate of $2\frac{1}{4}$ miles per hour) that affects only shallow depths. This movement is more correctly spoken of as a drift, in contrast to the deeper and more rapidly flowing currents that attain velocities two to three times that of drifts. Currents are much less common than drifts and are usually confined to localities where discharge takes place through restricted channels—as, for instance, the Florida Strait, through which a portion of the Gulf Stream, or Florida Current, emerges from the Gulf of Mexico.

Origin of ocean-water movements. The atmosphere and the oceans are unlike in the nature of their energies. Since the atmosphere is heated chiefly from below, a great amount of kinetic energy (energy of motion) is generated, as is shown by winds and by the general atmospheric circulation. But an ocean is heated from above, so that only about 2 percent as much kinetic energy is imparted to it as the atmosphere possesses. Hence the ocean circulation depends for its driving energy upon the atmosphere.

There are two primary causes for the general circulation of ocean waters. (1) Owing to the frictional effect of winds on the ocean surface, a relatively thin layer of top water about 100 m deep is driven slowly in the general direction of the air movement. If the winds prevailingly are from one direction, a steady drift of surface ocean water, moving in the same general direction, is usually the result. This fairly close correspondence between the great surface drifts of ocean water and the prevailing winds immediately suggests a causal connection and has led some authors to conclude that winds are almost the sole cause of sea drift. (2) The second basic reason for ocean-water circulation lies in contrasts of density within the water itself—contrasts which are due to differences both in temperature and in salinity. Since atmospheric circulation is associated with differences in density, it would seem logical to believe that oceanic circulation is also.

In all probability, the system of water movement as we know it is largely the result of the combined forces of wind friction and contrasting densities. Density differences are responsible chiefly for the great primary circulations, both vertical and horizontal, that take place throughout the entire hydrosphere. Winds largely determine the more readily observable features of surface currents and drifts. But although there is a general agreement between atmospheric and oceanic circulations, the flow patterns of the two fluids are not identical. Modifying agencies affecting the general water circulation are depth, degree of enclosure, shapes of coastlines, and the deflective force of earth rotation.

Mechanics of ocean-water circulation. Observational data on oceanic circulation, especially on the differences of temperature and salinity at various depths in all the seas, has not yet been accumulated in enough detail for a complete understanding of this subject. In general it appears that temperature has more to do with the circulation than salinity.

Temperature. The ocean waters in high latitudes are colder and therefore denser than those in the tropics, which makes for a constant exchange of water between polar and equatorial regions. This exchange is one part of the mechanism, involving

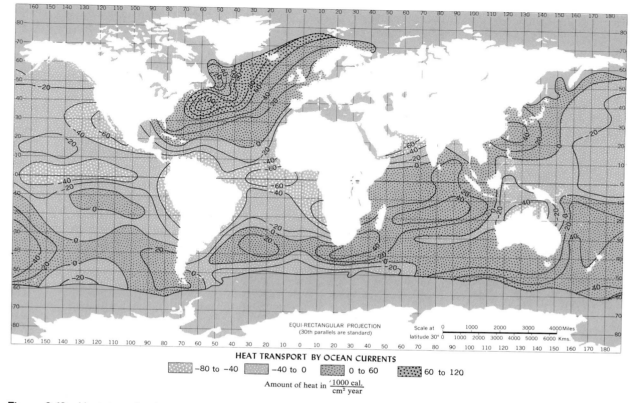

HEAT TRANSPORT BY OCEAN CURRENTS

-80 to -40   -40 to 0   0 to 60   60 to 120

Amount of heat in $\frac{1000\ cal.}{cm^2\ year}$

Figure 3.42 Heat transfer by ocean currents. Shadings indicate the annual amount of heat gained or lost by the atmosphere in contact with the ocean surface. Compare this with Fig. 3.44 showing the general scheme of ocean currents. Note the remarkable heat gain in the northern North Atlantic from the powerful Florida–North Atlantic Current. (*After map by Rudolph Geiger, Justes Perthes, Publishers.*)

circulations of both atmosphere and oceans, for maintaining the earth's latitudinal heat balance (Fig. 3.42). The Antarctic Ocean, having much wider connections with warmer seas than the more enclosed Arctic, no doubt provides more cold water for the general circulation. The principal southern connection between the Arctic and warmer seas to the south is the route through the North Atlantic between Greenland and Eurasia. Here the warm surface waters flow poleward on the eastern side of the basin, and cold return currents move equatorward in the western North Atlantic, passing both to the east and to the west of Greenland (Fig. 3.44). In the North Pacific there is no such free connection

with the Arctic Ocean. As a result the circulation in the North Pacific is less vigorous, and both the warm and cool currents are much smaller than they are in the North Atlantic. In the Atlantic, the easternmost projection of Brazil so divides the equatorial current that much the larger part of its warm water is shunted northward to form the extraordinarily powerful Florida Current (Gulf Stream) and the North Atlantic Drift (Figs. 3.42, 3.44).

Salinity. Differences in salinity tend to complicate the thermally induced circulation further. In the latitudes of the trades and the great permanent subtropical high-pressure centers, where rainfall is meager and evaporation great, salinity of the surface water is relatively high, and accordingly its density is greater. A constant sinking of the denser, saltier surface water of these latitudes is the result. On the other hand, both to the south and to the north of these dry belts with high evaporation are regions of relatively abundant rainfall and decreased evaporation—the wet tropics and the belts of westerlies. In these latitudes the surface waters are constantly being freshened, by direct precipitation as well as by runoff from the lands. The less salty and less dense surface waters have little tendency to sink.

General pattern of surface drifts and currents

Except for the polar seas, all the great oceans have broadly similar general circulations of surface drifts and currents (Fig. 3.43). This follows naturally from the general similarity of atmospheric circulation patterns. The scheme of ocean currents as developed in the North Atlantic ocean will be taken here as an example (Fig. 3.44).

Figure 3.43 Surface currents of the oceans in the Northern Hemisphere winter. (*After Schott and Dietrich.*)

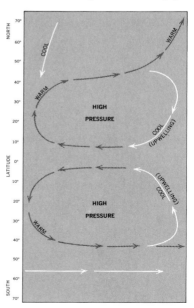

The most conspicuous element of the North Atlantic circulation is the great closed elliptical whirl about the subtropical Azores High. The trade winds on the equatorward sides of the subtropical anticyclones in both hemispheres tend to drift the surface waters before them across the ocean. The deflective force of earth rotation (right in the Northern and left in the Southern Hemisphere) acts to make this a westward-flowing current, moving somewhat at an angle to the direction of the trades. This is the Atlantic Equatorial Current. (In fact, there are really two equatorial currents, separated in the eastern part of the ocean by a minor countercurrent setting toward the east.)

Checked in its westward progress by the eastward thrust of the South American continent, the Equatorial Current is divided so that the larger part of it flows northwestward and the smaller part southwestward. Partly because of the deflective force of earth rotation and the trend of the coastline, and partly because of the wind direction around the western end of the subtropical Azores High (Fig. 3.43), the warm northward-moving current gradually is bent more and more to the east. A part of it enters the Caribbean Sea and passes through the Straits of Yucatán into the Gulf of Mexico. This water returns to the Atlantic through the Florida Strait, where it joins the major part of the warmwater drift, which has kept eastward of the West Indies. Together they form the Gulf Stream (more appropriately called the Florida or Caribbean Current), which parallels the American Atlantic seaboard.

At about latitude 40°N, westerly winds and deflection cause the warm surface waters of this current to turn slowly eastward across the ocean in the form of a west-wind drift. In the eastern Atlantic the drift divides. A part of it is carried by the subtropical anticyclone's northerly and northwesterly winds southward along the coast of southwestern Europe and northwestern Africa, until it again joins the Equatorial Current and thus completes the low-latitude circuit. This is the relatively cool Canaries Current (Fig. 3.43).

A large portion of the west-wind drift, however,

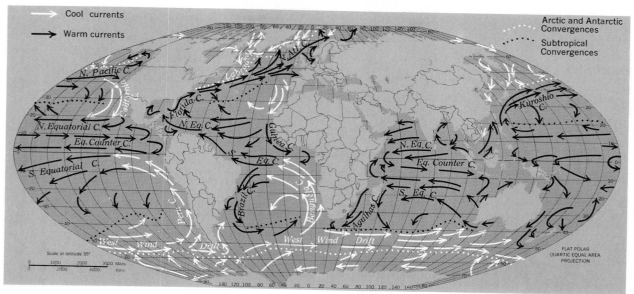

Figure 3.44 Generalized scheme of ocean currents.

is carried northeastward by the stormy southwesterlies toward Europe to form the mighty North Atlantic Drift, which eventually enters the Arctic Ocean. Nowhere else in the Northern Hemisphere is there a comparably large heat transport by an ocean current. The Arctic, compensating for this receipt of warm water from the eastern Atlantic, produces an outward flow of cold water in the western Atlantic on either side of Greenland. The western branch, called the Labrador Current, may reach as far south as New England.

The circulation pattern sketched above is a greatly simplified one, of course. Superimposed upon this generalized average flow are numerous eddies and surges, together with changes in direction and strength of currents following the seasonal shifting and reversals of winds.

Convergences, divergences, and vertical movements. For both hemispheres oceanographers recognize three general contrasting water masses separated by relatively distinct lines of discontinuity. In the high latitudes are the polar waters, which are characterized by low temperatures, low salt content,

a greenish color, and high plankton content. In the low latitudes are the contrasting tropical or subtropical waters, high in temperature and salt, low in organic forms, and blue in color. Between the two extremes are the "mixed waters" of the middle latitudes. Therefore both hemispheres must have two principal lines of discontinuity where waters differing in temperature, life forms, saltiness, and color are in contact. Those two lines, which separate the westward-drifting tropical waters from the eastward-moving drifts of the middle latitudes, are called the subtropical convergences, sometimes spoken of as the "polar fronts" of the oceans.

The simple schematic diagram of ocean currents and drifts (Fig. 3.44) discloses a number of locations where there is typically a divergence or convergence of flow. For example, convergence occurs in the eastern equatorial oceans and in the western parts of middle-latitude oceans, for in both locations poleward-moving warm currents from the tropics meet equatorward-moving cold currents from the polar seas. Slow convergence also occurs near the centers of the subtropical anticyclones. Divergent flow characterizes the western parts of tropical oceans, where the bifurcated equatorial current spreads to the north and south. Divergence is also typical of the eastern parts of middle-latitude oceans, where the west-wind drift divides the waters.

Where surface drifts converge, surface water must sink. If the converging waters are significantly unlike in temperature and density, the colder, denser water sinks beneath the less dense in a manner similar to that which occurs between unlike air masses along an atmospheric front. If density differences are slight, the sinking is slow and diffuse. Surface divergence is accompanied by a slow rising of waters from below. The climatic effects of surface convergence and divergence are discussed in a later section.

Warm and cool currents. In general poleward-drifting surface waters, since they come from lower latitudes, are inclined to be relatively warm, while those from higher latitudes are likely to be cooler than the surrounding waters. In the lower latitudes (equatorward from about 40°) warm ocean currents typically parallel the eastern sides of continents, while cool ocean currents parallel their western sides (Figs. 3.42, 3.43, and 3.44). For example, along the west coast of the United States at about 40°N, the average surface-water temperature near the coast is on occasions as low as 50°, while in the same latitude off the east coast of Japan the sea temperature is 70° or higher. At 2°S along the Peruvian coast the water temperature is as low as 62°, whereas in a corresponding latitude on the coast of New Guinea the surface temperature is about 83°. In the middle and higher latitudes this pattern is usually reversed; warm ocean currents parallel the western sides of land masses, and cool ones the eastern sides. Along east coasts (western sides of oceans), therefore, there is likely to be a convergence of contrasting currents, while along west coasts such currents tend to diverge.

Upwelling. It should be added that a part of the cool water along west coasts in lower latitudes (Peru and northern Chile, northwest and southwest Africa, southern California and adjacent northwestern

Mexico, among others) is the result of coastal divergence and a consequent upwelling from somewhat greater depths along the coast. Such regions of upwelling occupy positions along the eastern margins of oceanic subtropical anticyclones. There, paralleling equatorward-moving winds of the subtropical whirls drive the surface waters toward lower latitudes. Owing to the deflective force of earth rotation, the ocean currents along these cool-water coasts have a component of movement away from the land. Colder water from below, therefore, rises to replace the diverging surface water. Usually this overturn takes place within the upper 500 to 600 ft, so that upwelled water does not represent Arctic or Antarctic deep water.

Climatic influences of ocean drifts and currents

It is difficult to formulate simple principles concerning the effects of ocean drifts and currents upon the climate of adjacent land areas. Obviously the temperature effects of a cool current will differ from those of a warm. But whether a current actually has much of an influence in chilling or warming a land also depends on the comparative temperatures of the land and water surfaces. Clearly the same current may affect land temperatures quite differently in winter, when the land is cold, than in summer. The warm North Atlantic Drift makes a much greater contribution to tempering western Europe's winter cold than it does to moderating its summer heat. By contrast, the cold California Current is much more effective in mitigating subtropical coastal California's summer heat than in tempering its mild winter chilliness.

The influence of the water offshore also depends on how consistently the temperatures it generates are carried landward by winds. On windward coasts the temperature influences are far stronger and are carried inland a greater distance than on leeward coasts. Climatic effects become more complicated when, as a result of seasonal wind reversal, a coast that is windward in one season becomes leeward at the opposite time.

When cool currents border tropical coasts, they

greatly moderate the tropical heat. The coast of Peru, which is paralleled by the cold Peru Current, is 10° or more cooler than the coast of Brazil in similar latitudes, where a warm current prevails. In subtropical and middle latitudes, the effects of cool currents are chiefly restricted to the warmer months. Santa Monica on the California coast, where cold water lies offshore, has a cool July temperature of only 66°. This is 10° or more lower than east-coast stations in the subtropical United States, where warm water prevails. Kushiro in eastern Hokkaido (Japan), which is paralleled by the cool Okhotsk Current, has a chilly August temperature of only 63.7°. But the same month in Sapporo on the west side of the island averages about 70°.

Warm ocean currents are particularly effective in moderating the winter temperatures of windward littorals in middle and high latitudes. A most remarkable example is the case of western Europe, parts of which are 30 to 40° warmer in January than the world average for the latitude. This represents the combined effects of the unusually large and warm North Atlantic Drift, and an atmospheric circulation whose westerly winds persistently advect this warmth to the continent (Fig. 3.44). Warm waters also wash the shores of much of the eastern United States and Asia, but they are much less important in tempering the winter cold, for the prevailing winter winds are offshore.

As the previous section said, there is a tendency for cool and warm ocean currents to converge along the western sides of middle-latitude oceans and to diverge along the eastern sides. Where contrasting currents converge, the general effect is to squeeze the isotherms of both sea and air closer together. This makes for marked latitudinal contrasts in ocean temperatures, steep thermal atmospheric gradients, and advection fogs—all of which are found, for instance, along parts of the east coasts of Asia and North America. Where contrasting currents diverge, they tend to spread the isotherms, creating milder temperature gradients in both water and air, as are found in the eastern Atlantic and western Europe. This does not mean, however, that ocean currents are the principal cause of these temperature-gradi-

ent phenomena on the opposite sides of oceans; at best they are only auxiliary to the more powerful overall oceanic and continental influences.

Cool-water coasts in low latitudes are often characterized by both fog and aridity—on first thought an apparently contradictory combination. The surface fog is the result of warm air from over the open ocean being chilled by blowing over the cool current lying alongshore and mixing with the cool air above it. Fog may then be drifted in over the land, though usually it is confined to a narrow coastal fringe. During daytime hours the sea fog over land may be prevented by turbulence from reaching down to the surface. Frequent high fog or stratus cloud is also a consequence of subsidence and a low-level inversion in the strong anticyclonic circulation that prevails. The intense aridity of cool-water tropical coasts with their stable air masses is well known. To what extent the stability of the air is the result of the cool water is controversial. At the present time the stabilizing influence of the water seems to be viewed as an auxiliary process rather than as the primary one. The more effective of the two aridifying agents is probably the strong above-surface subsidence in the subtropical anticyclonic cell, whose eastern margins are located over these tropical and subtropical west coasts.

In summary, the following generalizations may be made concerning the climatic characteristics of coasts paralleled by warm and cold currents (Fig. 3.45).

1. West coasts in subtropical and tropical latitudes which are bordered by cool waters have relatively low average temperatures and small annual and diurnal ranges. They are foggy even though arid.

2. West coasts in the middle and higher latitudes which are bordered by warm waters have strikingly marine climates. Unusually mild winters and small annual ranges of temperature are the rule.

3. East (and therefore leeward) coasts in the lower middle latitudes, even though bordered by warm waters, have modified continental climates with relatively cold winters and warm-to-hot summers.

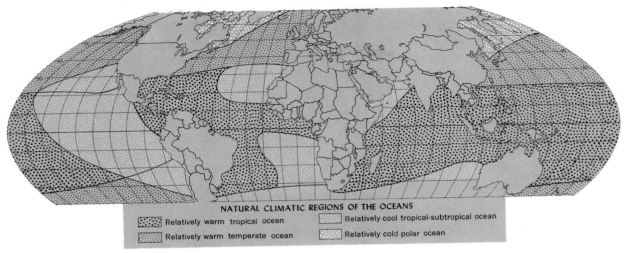

NATURAL CLIMATIC REGIONS OF THE OCEANS

- Relatively warm tropical ocean
- Relatively warm temperate ocean
- Relatively cool tropical-subtropical ocean
- Relatively cold polar ocean

Figure 3.45   Classification of oceanic regions relative to their surface temperature. (*After Markus.*)

4. East coasts in the higher middle latitudes which are paralleled by cool ocean currents are noteworthy for their cool summers.

Indirect climatic effects of ocean currents.   Indirectly, ocean currents may affect the general climatic and weather conditions of a region by their influence upon the locations of the great frontal zones, and consequently upon the tracks of cyclonic storms. Two of the main frontal zones and associated regions of cyclone formation in winter are situated off the eastern coasts of North America and Asia, roughly between latitudes 25 and 50°. In both places, an abundance of tropical water is being transported into the middle latitudes, and as a result energy is being supplied to the atmosphere in great amounts. It is in these same locations that temperature gradients are unusually steep, for the cold eastern sides of the great continents and the warm waters offshore create marked thermal contrasts. The fact that frontal zones appear to occur where unusual amounts of energy are being supplied to the atmosphere suggests that the cyclones which develop along these fronts are at least partly of thermodynamic origin. It has been pointed out also that the two principal hurricane paths in the Caribbean region coincide with warm waters. One follows the route into the Gulf of Mexico, and the other the route of the warm waters (Caribbean or Florida Current) off the northern and eastern coasts of the Greater Antilles and Florida (Fig. 3.42).

SELECTED REFERENCES FOR FURTHER STUDY OF TOPICS IN CHAPTER THREE

1. **Atmospheric pressure: relation of winds to pressure.**   Blair and Fite, pp. 21–36, 122–128; Petterssen, pp. 145–163, 174–181; Hare, pp. 44–56; Blüthgen, pp. 213–225; Riehl, *Tropical Meteorology;* Riehl, *Introduction to the Atmosphere*, pp. 18–20, 117–136; *Aviation Weather*, pp. 19–25; Critchfield, pp. 70–88.

2. **General circulation of the atmosphere.**   Blair and Fite, pp. 135–150; Blüthgen, pp. 338–378; Byers, pp. 260–286; Critchfield, pp. 88–92; Gentilli, pp. 37–65; Hare, pp. 91–103; Jerome Namias, "The Jet Stream," *Scientific American*, Vol. 187, pp. 27–31, October, 1952; Willett and Saunders, pp. 118–151;

Compendium of Meteorology, pp. 541–567; H. Flohn, "Studien zur allgemeine Zirkulation der Atmosphäre," *Ber. Deut. Wetterdienstes in der U.S. Zone,* no. 18, 1950; H. Flohn, "Neue Anschauungen über allgemeine Zirkulation der Atmosphäre und ihre klimatische Bedeutung," *Erdkunde,* Vol. 4, pp. 141–162, 1950; Hermann Flohn, "Die Revision der Lehre von Passatkreislauf," *Meteorologische Rundschau,* Vol. 1, pp. 1–60, 1953; Riehl, *Tropical Meteorology,* pp. 358–382; H. Riehl, M. A. Aeaka, C. L. Jordan, and R. J. Renard, "The Jet Stream," *Meteorological Monographs,* Vol. 2, No. 7, 1954; Reiter, *Jet-Stream Meteorology,* pp. 324–409; Reiter, *Jet Streams,* 55–178; Riehl, *Introduction to the Atmosphere,* pp. 203–214; Willett and Saunders, pp. 152–195.

3. **Surface winds and their characteristics.** Conrad, pp. 252–277; Haurwitz and Austin, pp. 44–63; Yale Mintz and Gordon Dean, "The Observed Mean Field of Motion of the Atmosphere," *Geophysical Research Papers,* No. 17, Air Force Cambridge Research Center, Cambridge, Mass., 1952; *Atlas of Climatic Charts of the Oceans,* U.S. Weather Bureau, 1938; J. S. Sawyer, "Memorandum on the Intertropical Front," *Meteorol. Repts.,* No. 10, British Meteorological Office, 1952; *Compendium of Meteorology,* pp. 859–887; Hare, 104–115; Petterssen, 169–189; Blüthgen, 227–232; Riehl, *Introduction to the Atmosphere,* pp. 229–232.

4. **Monsoons.** Blüthgen, pp. 354–365; Hare, pp. 139–154; "Monsoons of the World," *Symposium on Monsoons of the World,* New Delhi, 1960; P. Pédelaborde, *The Monsoon* (translated by M. J. Clegg), Methuen & Co., Ltd., London, 1963.

5. **Local winds.** *Compendium of Meteorology,* pp. 655–672; Blüthgen, pp. 232–248; Petterssen, pp. 164–168.

6. **Ocean drifts and currents.** William S. von Arx, *An Introduction to Physical Oceanography,* Addison-Wesley Pub. Co., Reading, Mass., 1962; Cuchlaine A. King, *Oceanography for Geographers,* E. Arnold, London, 1962.

7. **General references on wind and pressure systems.** Jen-Hu Chang, "The Indian Summer Monsoon," *Geog. Rev.,* Vol. 57, no. 3, pp. 372–396, July, 1967; F. K. Hare, "The Westerlies," *Geog. Rev.,* Vol. 50, pp. 345–367, 1960; "Energy Exchanges and the General Circulation," *Geography,* Vol. 50, pp. 229–241, 1965; H. H. Lamb, "Representation of the General Atmospheric Circulation," *Met. Mag.,* Vol. 89, pp. 319–330, 1960; P. Koteswaram and N. S. Bhaskara Rao, "The Structure of the Asian Summer Monsoon," *Australian Met. Mag.,* Vol. 42, pp. 35–56, 1963; J. E. McDonald, "The Coriolis Effect," *Scientific American,* Vol. 186, pp. 72–78, 1952; Jerome, Namias, "Interactions of Circulation and Weather between Hemispheres," *Monthly Weather Rev.,* Vol. 91, No. 10–12, pp. 482–486, 1963; E. Palmén, "The Role of the Atmospheric Disturbance in the General Circulation," *Quar. Jour. Roy. Met. Soc.,* Vol. 77, pp. 337–354, 1951; Herbert Riehl, "General Atmospheric Circulation of the Tropics," *Science,* Vol. 135, pp. 13–22, 1962; "Jet Streams of the Atmosphere," *Tech. Paper No. 32,* Colorado State University, 117 pp., 1962; V. P. Starr, "The General Circulation of the Atmosphere," *Scientific American,* Vol. 195, pp. 40–45, 1956; G. B. Tucher, "The General Circulation of the Atmosphere," *Weather,* Vol. 17, pp. 320–340, 1962; E. W. Wahl, "Singularities and the General Circulation," *J. Meteorology,* Vol. 10, pp. 42–45, 1953.

CHAPTER FOUR

ATMOSPHERIC MOISTURE AND PRECIPITATION

HUMIDITY

Humidity refers to the water in gas form, or water vapor, in the atmosphere—but not to liquid droplets of cloud, fog, or rain. Water vapor has nothing to do with wetness, for it is dry like any other gas. Also like the other atmospheric gases, it exerts a pressure, which is known as the *vapor pressure* of the air. Air is saturated when its vapor pressure equals that over a smooth water surface.

Although water vapor comprises less than 2 percent of the atmosphere's total volume, it is the single most important component of the air from the standpoint of weather and climate. As noted in the Introduction, the proportions of most of the atmosphere's gases are relatively constant from place to place. It is chiefly water vapor that is highly inconstant, varying from nearly zero up to a maximum of about 3 percent in the middle latitudes and 4 percent in the humid tropics. This variability, in both place and time, is of outstanding importance for several reasons:

1. Water vapor is the source of all forms of condensation and precipitation. The amount of water vapor in a given mass of air is one indicator

of the atmosphere's potential for precipitation.

2. As the air's principal absorber of solar energy, and of radiant earth energy as well, water vapor operates as a heat regulator and so has a great effect on air temperature.

3. The latent heat contained in water vapor is an important energy source both for the circulation of the atmosphere and for the development of many atmospheric disturbances, especially those in tropical latitudes.

4. Also because of its latent heat, the amount and vertical distribution of water vapor in the atmosphere indirectly affects the buoyancy of air and hence its tendency to ascend. This in turn is closely related to the formation of cloud and precipitation.

5. Humidity is an important factor in the rate of evaporation, a process that greatly influences plant and animal life. By affecting the human body's rate of cooling, evaporation is one determinant of its sensible temperature, or feeling of heat and cold.

6. Unlike other gases comprising the air ocean, water vapor can be changed to liquid and solid forms well within the range of atmospheric temperatures.

Evaporation

The process by which water in its liquid or solid form is converted into water vapor, a gas, is called evaporation.[1] Water vapor enters the atmosphere by evaporation, chiefly from the ocean surfaces, but also from smaller bodies of water, from the moist ground, from the transpiring leaves of plants, and from raindrops falling through the air.

Evaporation takes place when the vapor pressure at the earth's surface exceeds that in the air above. At an open water surface, molecules escape from the liquid water into the air, while at the same time

---

[1]As used here, evaporation includes transpiration from plants, sometimes called evapotranspiration. Water loss from plant surfaces (mainly the leaves) by transpiration is a complex process. It occurs when the vapor pressure in the leaf cells is greater than the atmospheric vapor pressure. Evapotranspiration losses from natural surfaces cannot be measured directly. Only indirect methods and theoretical formulas are available (see Chapter 12, Dry Climates).

some gaseous molecules return to the liquid. If the temperature of the liquid is higher than that of the air, more molecules escape from the water than return to it, and evaporation occurs. But as the number of water vapor molecules in the air increases, so also does the partial vapor pressure, and consequently a larger number of molecules returns to the liquid. Eventually equilibrium is reached, with the number of escaping molecules equaling those returning. Evaporation then ceases; the air is saturated. Under existing temperature conditions it can hold no more water vapor. But saturation pressure changes rapidly with temperature. Over the earth's range of temperature, it varies from under $\frac{1}{10}$ mb to over 100 mb.

There must be an expenditure of energy to change liquid water into the vapor state. This energy is usually supplied from the evaporating surface, which as a consequence is cooled. Thus one experiences the chilling effect of evaporation upon stepping out of a bath. Evaporation from falling raindrops chills the air, since the air provides most of the energy for evaporating the water.

The cooling effect of evaporation is involved in the functioning of a wet-bulb thermometer. As Chapter 2 described, the instrument consists of an ordinary thermometer whose bulb is covered by gauze saturated with water. Evaporation from the wet gauze causes the wet-bulb thermometer to record a lower temperature than the ordinary dry-bulb thermometer does. The amount of cooling is directly proportional to the dryness of the air. From the differential temperature readings of the wet- and dry-bulb thermometer, one can determine from prepared tables the dew point, vapor pressure, and relative humidity of the air.

The rate of evaporation depends mainly upon the temperature of the evaporating surface compared with that of the air, but also upon the relative humidity of the air or its degree of saturation, and the speed of the wind. Evaporation is highest when the evaporating surface is warmest, when the air is relatively dry, and when the winds are strongest. Consequently evaporation is greater in summer than in winter, and at midday than at night. When frigid air with low humidity is advected over a relatively warm water surface, evaporation is rapid, since the vapor pressure of the cold atmosphere is always less than that over the warmer water. Moreover, under these circumstances the air is heated from below by contact with the warm water so that it becomes unstable. This increases turbulence and therefore evaporation.

Moisture entering the air as water vapor is widely distributed by winds and vertical air currents. Often it is carried great distances before being eventually removed as liquid or solid condensation and precipitation.

Distribution of evaporation. Since the rate of evaporation depends mainly on temperature, obviously evaporating power, or *potential evaporation*, will decline from tropical to polar latitudes, given the presence of unlimited amounts of water (see the following table and Fig. 4.1).

Variation of average annual *potential* evaporation by latitude zones*

Latitude	North		South	
	cm	in.	cm	in.
70–60°	18	7	?	?
60–50°	40	16	30	12
50–40°	70	27	60	24
40–30°	110	43	100	40
30–20°	130	51	130	51
20–10°	140	55	130	51
10–0°	120	47	120	47

* Evaporating power given in height of evaporated water column.

Source: After Landsberg.

Figure 4.1   World distribution of average annual effective evapotranspiration. (*After Rudolph Geiger, Die Atmosphäre der Erde, Justus Perthes, Darmstadt, Germany.*)

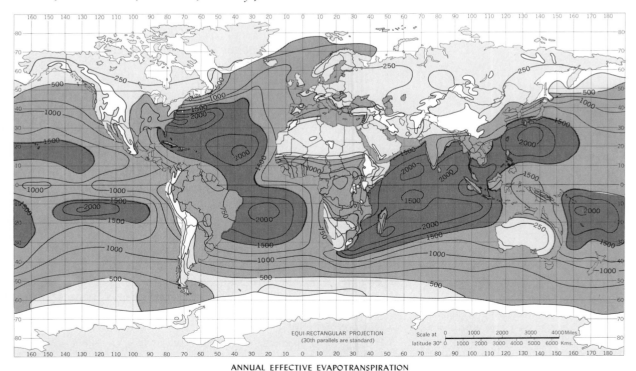

ANNUAL EFFECTIVE EVAPOTRANSPIRATION

0 to 500   500 to 1500   1500 to 2500 mm

The *actual* evaporation (as distinguished from *potential*), however, depends not only upon air and surface temperatures, air motion, and degree of air saturation, but also on the available supply of moisture. On nonliquid surfaces, the evaporating power of the air may be several times greater than the amount of water available.

Available data on measured evaporation are scanty and often unreliable. They are usually supplemented by estimates for actual evaporation from various natural surfaces, compared with amounts measured from evaporating pans, or calculated by the use of various formulas. Evaporation from large water surfaces is estimated to be about 70 to 75 percent of that from a pan exposed to the same temperature and wind conditions. Evaporation from a wet bare soil may be about 90 percent of that from an open water surface.

The following table illustrates two main points concerning the distribution of actual evaporation.

1. In most latitudes, actual evaporation is greater over oceans than over continents. This reflects the facts that the supply of water is unlimited at the ocean surface, while over many land areas water is scarce. But in the very low latitudes, between about 10° N and S, there is more evaporation over land than over water, because of the abundant supply of water available in these rainy regions and the unusually great transpiration from the dense equatorial forests.

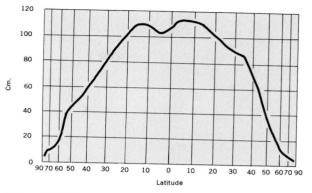

Figure 4.2 Distribution of average annual effective evaporation by latitude zones. Since evaporation is mainly a function of temperature, highest values are in the low latitudes, and there is a decline toward either pole. (*Data from G. Wüst.*)

2. The mean zonal or meridional pattern of actual-evaporation distribution is chiefly a function of air temperature, since there is a maximum in low latitudes and a steady decline toward either pole (Fig. 4.2). Perhaps as much as 60 percent of the earth's evaporation is furnished by the belt 25°S to 25°N, and 80 percent occurs in the zone bounded by the 35° parallels.

Latent energy in water vapor

It is common knowledge that energy is required in the form of heat to melt the solid, ice, into the liquid,

Zonal distribution of actual mean annual evaporation (Inches)

	Latitude					
	60–50°	50–40°	40–30°	30–20°	20–10°	10–0°
Northern Hemisphere						
Continents	14.2	13.0	15.0	19.7	31.1	45.3
Oceans	15.7	27.6	37.8	45.3	47.2	39.4
Mean	15.0	20.1	28.0	35.8	42.9	40.6
Southern Hemisphere						
Continents	(7.9)	(19.7)	20.1	16.1	35.4	48.0
Oceans	9.1	22.8	35.0	44.1	47.2	44.9
Mean	8.8	22.8	35.5	39.0	44.5	45.7

Source: After Wüst.

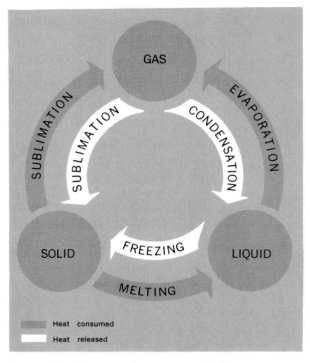

Figure 4.3 Changes of state as applied to water.

it is used only in producing a change in state from liquid to gas. One reason why the sea surface or a vegetation cover never becomes really hot on a sunny day—as a dry, sandy beach does, for example—is that so much energy is consumed in evaporation at their surfaces.

If energy is consumed in the process of evaporation, then conversely, energy will be released during condensation, when water vapor returns to a solid or liquid state. This released heat, known as the latent heat of condensation, is an important source of atmospheric energy. On a night when large-scale cloud condensation and precipitation take place, cooling is retarded by the liberation of so much latent heat. Condensation, then, is the mechanism for releasing latent heat gained during evaporation. Through cloud formation, heat of condensation is mixed through the troposphere—in the tropics, up to a height of 50,000 ft. Evaporation-condensation probably account for about two-thirds of the heat exchange between the land-water surface and the atmosphere. To maintain the heat balance, it is estimated that on the average, 30 to 40 in. of water a year must be evaporated at the land-water surface into the atmosphere and subsequently precipitated.

The hydrologic cycle

A never-ending round of evaporation from sea and land creates the air's reservoir of water vapor, which then condenses to form clouds and subsequently returns to land and sea as precipitation. This hydrologic cycle involving evaporation → condensation → precipitation has two main subdivisions: a meridional cycle which involves a transfer of moisture between different latitude belts, and another which involves an exchange between land and sea. From the previous section on latent heat, it must be clear that any exchange of water vapor must also involve a transfer of energy (Fig. 4.4).

Meridional moisture exchange. As the following table indicates, some latitude zones have a surplus of precipitation over evaporation, while in others the reverse is true. Such an arrangement requires

water, and to evaporate water into vapor, a gas (Fig. 4.3). The unit of heat energy, the calorie, is the amount of heat required to raise the temperature of a gram of water 1° centigrade. But it takes 79 calories to convert a gram of ice into a gram of water at freezing temperature, and 607 calories to evaporate the gram of water at freezing temperature and convert it into water vapor at the same temperature.

Since energy is required to change the solid into a liquid, and likewise the liquid into a gas, it follows that water vapor contains more potential energy than liquid water, and water in turn more than ice. This stored-up energy in water vapor is known as latent heat, or latent energy. It is transformed sun energy, which has been employed in evaporating water or ice and converting them into water vapor. The energy which the evaporating surface loses during evaporation does not warm the water vapor gas;

Precipitation excess and deficiency by zones (In 1,000 cu km)

North latitude		Precipitation—evaporation	South latitude		Precipitation—evaporation
90–80°		+ 0.4	90–80°		+ 1.0
80–70°	Precipitation excess	+ 2.3	80–70°	Precipitation excess	+ 2.5
70–60°		+ 5.0	70–60°		+ 3.6
60–50°		+ 8.1	60–50°		+12.0
50–40°		+10.1	50–40°		+10.7
40–30°		− 7.1	40–30°		+ 0.1
30–20°	Precipitation deficiency	−19.0	30–20°	Precipitation deficiency	−16.5
20–10°		−16.2	20–10°		−16.0
10–0°	Precipitation excess	+19.3	10–0°		− 0.2

Source: After Wüst, Gentilli, and others.

a meridional exchange of moisture, and hence energy, between zones of surplus and deficiency.

The table shows that there are three main belts of precipitation excess, and hence of net moisture and energy gain: the two middle-latitude zones (40° to 70° in each hemisphere), and the equatorial zone, especially 0° to 10°N. Winds advect vast amounts of moisture, and consequently of latent heat of condensation, to these regions. In contrast, there are two extensive belts of precipitation deficiency, located at about 10° to 40° in each hemisphere where evaporation exceeds rainfall. These are the zones of the subtropical anticyclones and the trade winds. In them there is an enormous energy loss by evaporation. This energy, in the form of latent heat in water vapor, is advected by winds to the equatorial and middle latitudes, where precipitation exceeds evaporation.

Within the high-evaporation trade-wind belt, the moist convective layer gradually deepens in the direction of the airflow (equatorward and to the west). In the trades, most of the moisture is retained in vapor form rather than rained back to the ocean. The trades' accumulated water vapor is transported toward the equatorial trough, where much of it is condensed, releasing latent heat that is carried to great heights in cumulonimbus clouds. Such large clouds are concentrated in atmospheric disturbances, which are relatively numerous within the equatorial trough. Thus the upward transfer of energy, both latent and sensible heat, from the land-sea surface to the atmosphere is significantly increased by synoptic disturbances. Such traveling disturbances form a vital link in the moisture transfer between the tropical land-sea surface and the atmosphere.

Land-sea moisture exchange. A moisture exchange between land and sea is required because continents have more precipitation than evaporation, and oceans more evaporation than precipitation.

	Cu km per year
Precipitation on continents	122,000
Evaporation from continents	97,000
Surplus on continents	25,000
Precipitation on oceans	259,000
Evaporation from oceans	384,000
Deficiency on oceans	25,000
Transferred from continents to oceans	25,000

Some of the water vapor added to the atmosphere by evaporation over the sea is transported by winds to the continents where, together with land-evaporated moisture, it condenses and falls as precipitation. A part of this land precipitation enters the ground to form an important groundwater resource. A part is returned to the sea by runoff

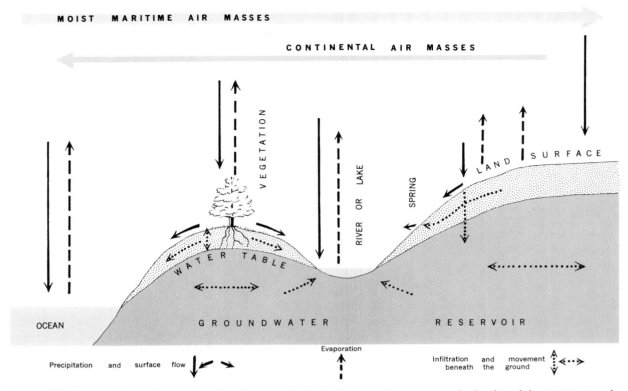

Figure 4.4 Some of the major relationships between the hydrologic cycle, waters on the land, and the air-mass cycle.

in the form of streams and glaciers. And still another part is transported back to the oceans by land winds containing land-evaporated moisture (Fig. 4.4). Water evaporated in one locality is usually precipitated hundreds or even thousands of miles distant; the hydrologic cycle is rarely completed locally. Of course, much of the sea-evaporated moisture is returned directly to the ocean by rains over the sea.

Through careful observation it has been discovered that the amount of water precipitated upon the continents is well in excess of the runoff from those same areas in the form of rivers and glaciers. Since only roughly 20 percent of the land precipitation is removed by runoff, it follows that close to 80 percent must be returned to the atmosphere by evaporation from the ground, inland bodies of water, and the vegetation cover. From these data

it might be concluded that the oceans contribute to continental precipitation only that proportion included in the runoff, while local continental evaporation must provide the other 80 percent. But this reasoning neglects a fundamental consideration—that if sea winds can transport water vapor from the oceans to the land, then land winds can also transport it from land to the oceans (Fig. 4.4).

In the general circulation of the atmosphere in the middle latitudes, and in the cycle of atmospheric exchange between land and sea, humid tropical maritime air masses traveling poleward become cooled, precipitate much of their moisture, and are ultimately converted into polar continental and polar maritime air masses. Conversely, polar continental air masses in moving equatorward over land areas become warmed, absorb much land-evaporated moisture, and eventually are transformed into warm, moist, tropical maritime air masses over the oceans. Thus it is not only rivers and glaciers that carry land-precipitated moisture back to the oceans, but also these dry polar continental air masses, which may pass entirely across the continent from north to south without any re-precipitation.

The amount of evaporation by North American polar continental air masses has been calculated by measuring their mositure content when they cross the Canadian border and when they arrive at the Gulf of Mexico. It was found that the water content of southward-moving polar continental air at Ellendale, North Dakota, was equivalent to 0.44 in. of rainfall, and at Pensacola, Florida, to 0.97 in.—a gain of 0.53 in. If these figures are representative, the water removed by polar continental evaporation from the Mississippi drainage basin reaches the astounding figure of 5,909,000 cu ft per sec, which is nine times the average discharge of the Mississippi River. Air masses and their various characteristics will be treated in more detail in Chapter 5. The important point here is that polar continental air masses moving southward across a continent resemble great invisible rivers transporting vast amounts of water vapor from the land to the ocean. There it is precipitated, later to be evaporated from the sea and returned to the continents by tropical maritime air masses.

Water balance of individual continents. From the accompanying table, it can be seen that the greatest amounts of precipitation, evaporation, and runoff are in South America, which contains large regions

with tropical humid climates and few areas with dry climates. Runoff is least in Australia, which is predominantly a continent of dry climates.

Distribution of humidity

The actual moisture content of the atmosphere in any locality may be expressed in various ways. As noted earlier, *vapor pressure* is that part of the whole atmospheric pressure which is due to the water vapor. It is expressed in the same units as total air pressure—i.e., either in millibars, or in inches or millimeters of mercury. *Specific humidity* is defined as the weight of water vapor per unit weight of air. It is usually expressed as the number of grams of water vapor contained in 1 kg of air. Specific humidity is nearly proportional to the water vapor pressure, so that the two terms are used interchangeably in the following discussion on distribution of humidity. *Absolute humidity* refers to the weight of water vapor per unit volume of air—as, for example, grains per cubic foot or grams per cubic meter. This concept is less frequently used by meteorologists because, since volume changes as an air mass is elevated or subsides, absolute humidity varies with contraction or expansion of the air.

The capacity of the air for water vapor, expressed in any of the three ways mentioned above, depends upon its temperature. Not only does the capacity increase at higher temperatures, but it increases at an increasing rate (see the following table and Fig. 4.5). Thus a 10°C increase from 0° to 10° advances the capacity only 4.6 g, but an equivalent increase of

Water balance of the continents (In mm per year)

Continent	Precipitation	Evaporation	Runoff
Europe	600	360	240
Asia	610	390	220
North America	670	400	270
South America	1,350	860	490
Africa	670	510	160
Australia	470	410	60

Source: After Budyko.

Maximum water vapor capacity of a cubic meter of air at different temperatures

Temperature			Temperature		
°C	°F	Grams	°C	°F	Grams
−5	23	3.261	20	68	17.300
0	32	4.847	25	77	23.049
5	41	6.797	30	86	30.371
10	50	9.401	35	95	39.599
15	59	12.832	40	104	51.117

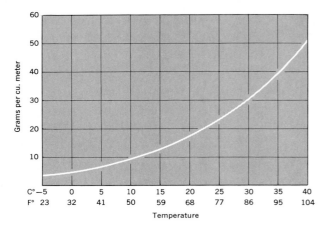

Figure 4.5 Variations in the capacity of air at different temperatures for water vapor. Not only does the capacity increase as the air temperature rises, but equally important, it increases at an accelerating rate.

10°C from 25° to 35° results in an increased capacity of 16.6 g, or three to four times as much. Obviously warm summer air is able to contain much more moisture than cold winter air, and it therefore has greater potentialities for precipitation. When air contains all the water vapor it can hold at a given temperature and pressure, it is said to be *saturated*. It has reached its *dew-point* temperature and also its *saturation vapor pressure*.

Average vertical distribution of water vapor in middle latitudes

Height (km)	Water vapor (vol. in %)
0.0	1.3
0.5	1.16
1.0	1.01
1.5	0.81
2.0	0.69
2.5	0.61
3.0	0.49
3.5	0.41
4.0	0.37
5.0	0.27
6.0	0.15
7.0	0.09
8.0	0.05

Source: From Landsberg.

Because the main source of atmospheric humidity is the earth's land-sea surface, much of the water vapor in the air is concentrated in the lower troposphere, and there is a rapid decrease with height. Half the total water vapor is below 6,500 ft. Above about 6 miles (10 km), the atmosphere contains almost none. (See preceding table.)

The zonal distribution of water vapor in the air is shown in Fig. 4.6, where the scale has been adjusted to indicate the relative size of the latitude belts. The atmosphere's water vapor content, expressed as specific humidity or vapor pressure, is highest near the equator and lowest in high latitudes. This reflects the distribution of air temperature, since temperature largely determines the air's capacity for water vapor. It is temperature also which causes the specific humidity of any latitude to be higher in summer than in winter, so that the values shown in Fig. 4.6 are displaced northward in July and southward in January. But distribution of water vapor is not simply zonal in character. Along the same parallel, a dry continental area is likely to have a lower specific humidity than the oceans. It is these contrasts which cause the minor irregularities in the relatively symmetrical zonal curve of Fig. 4.6. However, the water vapor content

Figure 4.6 Distribution by latitude zones of the water vapor content of the air. Specific humidity is highest in equatorial latitudes and decreases toward the poles. (*After Haurwitz and Austin.*)

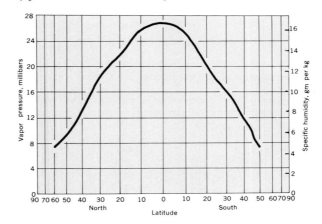

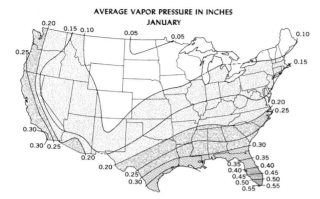

Figure 4.7

of desert air is not necessarily low in an absolute sense; the superheated air of the Sahara in summer may contain more moisture than cold maritime air farther poleward.

For most regions of the earth, the *annual* variation of specific humidity closely parallels that of temperature, being highest in summer and lowest in winter. Over the north central United States, July air contains three to six times as much water vapor as January air (Figs. 4.7 and 4.8). The *diurnal* variation of specific humidity is more complicated, for while over the oceans it follows the temperature march, over land there are two maxima and two minima within the 24-hr period. This is because a secondary minimum occurs at midday. It is the

result of convective turbulence, which transports enough water vapor upward away from the earth's surface to offset the effect of the high afternoon temperatures.

Relative humidity differs from the parameters of humidity described earlier, for it represents the amount of water vapor actually present in the air compared with the maximum that could be contained under conditions of saturation at a given

Figure 4.9 Each cube represents 1 cu ft of air. The top row shows the changing capacity for water vapor of a saturated cubic foot of air under three different temperature conditions. The cubic-foot samples in the bottom row have the same temperatures as those in the top, but they are unsaturated—they do not contain all the water vapor possible at those temperatures. The different amounts of water vapor which they hold at the different temperatures cause them to have different relative humidities, as represented by the variations in shading.

Figure 4.8

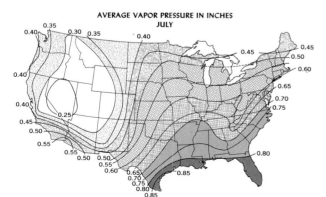

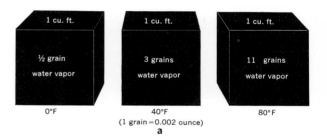

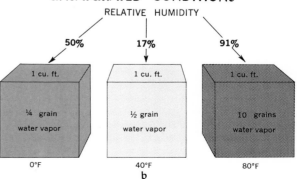

temperature and pressure. It is always expressed in the form of a ratio, fraction, or percentage. When relative humidity reaches 100 percent, the air is, of course, saturated. For example, if at constant pressure and a given temperature, 1 kg of air can hold 12 g of water vapor, but actually contains only 6 g, it has a relative humidity of 50 percent. Relative humidity can be altered either by changing the amount of water vapor or by varying the capacity of the air, i.e., changing its temperature.

Relative humidity is an important determinant of the amount and rate of evaporation. Consequently it is a critical climatic factor in the rate of moisture and temperature loss by plants and animals, including human beings. Various humidity relationships are illustrated by Fig. 4.9.

Comparison of Figs. 4.10 and 4.6 indicates that the zonal distribution of relative humidity is quite different from that of specific humidity. Relative humidity has its highest maximum at the equator

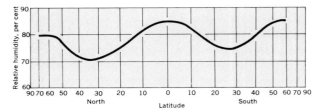

Figure 4.10 Distribution of relative humidity by latitude zones. Note that zonal distribution of relative humidity is quite different from distribution of specific humidity (see Fig. 4.6).

and decreases poleward toward the latitudes of the subtropical anticyclones. It is in these subsiding and diverging air masses that the lowest values occur. From about 30° poleward, relative humidity increases as a result of the decreasing temperature. The belts of highest and lowest relative humidity shift northward in July and southward in January, following the course of the sun (also see Fig. 4.11).

Figure 4.11 World distribution of average relative humidity in July. Two zones of low values approximately coincide with the latitudes of the subtropical highs. Higher values are characteristic of the oceanic parts of the middle-latitude westerlies, the equatorial latitudes, and the monsoon coasts of southern and eastern Asia. (*After J. Száva-Kováts.*)

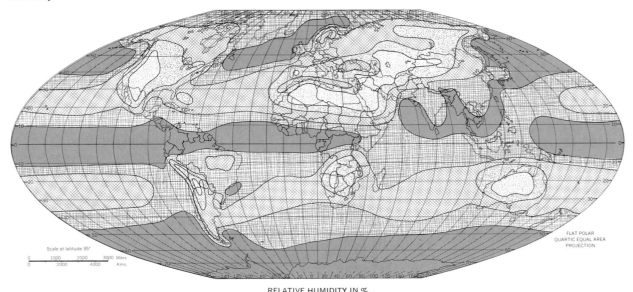

RELATIVE HUMIDITY IN %

<50   50–70   70–80   80–85   >85

The seasonal distribution of relative humidity differs with latitude. In the low latitudes from about 30°N to 30°S, where seasonal temperature variations are small, average relative humidity is higher in the wet summer than in the drier winter. But in higher latitudes the highest humidities coincide with the cold winters, especially over the frigid continents. The diurnal variation in relative humidity is generally the reverse of the march of temperature; relative humidity declines from a maximum in the early morning to a minimum around mid-afternoon.

CONDENSATION

If air that is not saturated is sufficiently cooled, so that its capacity for moisture is reduced, it eventually reaches a temperature at which its relative humidity is 100 percent, even though the amount of water vapor has not been altered. As noted earlier, this critical temperature at which saturation is reached is called the dew point. If air is cooled below the dew point, then the excess of water vapor over what the air can contain at that temperature is given off in the form of minute droplets of water or particles of ice. Thus condensation has taken place.

The only method known whereby significant amounts of water vapor in the atmosphere can be converted into the liquid (condensation) or solid (sublimation) state is to reduce the temperature of the air close to or below the dew point. When the air is cooled, its capacity for water vapor is lowered, and if this capacity is sufficiently reduced, condensation must result. The dew point of any mass of air is closely related to its relative humidity. When the relative humidity is high, the air is close to the saturation point. Only a slight amount of cooling will be required before the dew point is reached and condensation begins. On the other hand, when relative humidity is low, as it usually is over the hot deserts, an astonishingly large amount of cooling is required before the dew point is reached. Condensation therefore depends upon two variables: the amount of cooling, and the relative humidity of the air. If the dew point is not reached until the temperature falls below 32°, some of the condensed water vapor may be in the form of tiny ice crystals (white frost, snow, and some clouds). If condensation occurs above the freezing point, it will be in the liquid state (dew, fog, and cloud). Observations seem to indicate that while solid condensation may occur at any temperature below 32°, actually much cloud condensation down to about 5°F (−15°C) is still in the form of supercooled liquid droplets.

Enough cooling of the atmosphere to cause condensation may occur in several ways. Such processes as radiation and conduction from the overlying air to the cold earth, and the mixing of two unlike air masses of contrasting temperatures and humidities, can lower the temperature of relatively shallow layers of air below condensation level. But the condensation forms (dew, white frost, and fog) resulting from these methods of cooling usually represent small-scale condensation which is confined to air layers close to the earth's surface. Appreciable rainfall probably never develops from such cooling. The one process that can reduce the temperature of deep and extensive masses of air enough to bring about cloud condensation which results in abundant precipitation is the expansion associated with rising air currents—adiabatic cooling.

Condensation apparently is a continuous process, for strongly hygroscopic particles begin to attract water around them even at humidities as low as 75 percent. Haze is the first evidence of such sub-saturation condensation. Visibility then decreases, and the landscape takes on a soft blue-gray appearance. As the relative humidity approaches 100 percent, the condensation droplets rapidly grow larger, so that the haze thickens and gradually develops into fog. Thus haze (with the exception of the ordinary dust haze of dry weather) is often the forerunner of fog. Each cloud and fog droplet forms around a condensation nucleus consisting of a water-attracting substance such as common salt, derived from evaporated sea spray, or sulphuric acid, a product of city smoke and motorcar exhausts.

Minor condensation: dew, white frost, rime, fog

Dew, white frost, and rime are relatively unimportant forms of condensation which develop at or near the ground. Dew and white frost result from radiational cooling at night, usually with clear skies and little or no wind. Rime consists of ice crystals formed when droplets of supercooled fog or cloud touch cold objects. Dew is unimportant in humid climates, but in dry and subhumid climates frequent night dews may significantly contribute to the meager water supply for plant life. In subhumid Palestine, the maximum annual dew water amounts to as much as 55 mm (2.17 in.), and most of it occurs in the dry summer. Since plants are able to absorb dew water on their leaves, this moisture source has considerable agricultural usefulness.

Fog, although far less important than cloud and precipitation, is more prominent in human affairs than the other forms of surface condensation. For the most part its consequences are negative. For example, fog at sea makes navigation difficult or hazardous; fogged-in airports are temporarily put out of use; and fog slows highway motor traffic and can cause serious accidents. On the other hand, through its retarding effects on radiational cooling, a night fog may save a region from killing frost in fall or spring.

Unlike thick, rain-producing clouds, which are characteristically formed by the expansional cooling of rising air masses, fog is usually caused by the cooling of humid surface air below the dew point. This cooling can result from radiational processes or from the mixing of warm and cool air masses.

Radiation fogs. A very common type of land fog, known as *radiation* or *ground-inversion* fog, is produced by radiation cooling of relatively shallow layers of quiet, humid air overlying a chilled land surface. Nighttime conditions which favor radiation fogs are a surface inversion of temperature, a cloudless sky, and slight air movement (but not an absolute calm). The inversion is necessary to prevent the fog from rising. Cloudless skies promote fog because a low cloud cover absorbs a large part of the earth radiation and reradiates it back to the earth's surface, where it is absorbed and consequently acts to reduce the net loss of heat. Slight air movement is a requirement because fogs of the radiation type cannot form when the wind velocity is greater than 6 to 10 miles an hour; at higher speeds, the turbulence associated with the wind carries heat downward and thus prevents formation of an inversion. Some slight turbulence and mixing appear to be necessary, however, in order to develop a fog of moderate density.

Radiation fogs build up by gravitation. For this reason they are most prevalent in valleys and depressions, where air drainage causes the colder, heavier air to collect. Often the slopes and uplands of a region are clear while the adjacent lowlands are fogged in, so that from an elevation one sees many "lakes" of fog occupying the surrounding depressions. On a convex land surface, air drainage reduces the likelihood of temperature inversion and radiation-fog formation.

Radiation ground fog has a distinct diurnal periodicity and is usually short-lived. It grows and deepens from the ground upward as the night progresses, only to burn off and disappear as the temperature inversion is dissipated and the fog droplets are evaporated by the warming air during the following morning. Radiation fog is at its worst in the vicinity of large cities, where the air is rich in hygroscopic particles derived from coal and oil smoke and the exhaust fumes from cars. These abundant condensation nuclei result in tinier fog particles, which make for greater density of the fog, or smog.

Another variety of radiation fog is known as high-inversion fog. In this type not only is the inversion layer much deeper than in the radiation ground fog, but in addition the inversion commonly is located 400 to 2,000 ft above the earth's surface. With warmer, drier air overlying cooler air at the surface, upward movement of air is suppressed. Continued cooling of the surface air for a succession of nights results in a fogging of the surface strata. Turbulence and convective overturning by day usually lift the fog layer above the surface, so that it appears as a deck of low stratus at the inversion base.

Ideal conditions for high-inversion fogs are similar to those for a ground fog except that they are caused by cooling which results not from one night's radiation, but from a cumulative net loss of heat by radiation. Polar maritime air that has become stagnant over a continent provides ideal temperature and moisture conditions for such fogs. Thus western Europe, which is open to direct invasion of polar maritime air, has many high-inversion fogs in winter. The stratus deck formed at the inversion level is an excellent reflector of solar radiation by day and an equally good radiator of earth radiation at night. This tends to intensify the inversion, so that the high fog, which descends to ground level at night, may persist for days or even weeks. Under such conditions airports may be closed for days at a time.

Advection-radiation fogs. Formation of an advection fog always involves the movement of air, and usually warm moist air, over a cold surface, with a consequent loss of heat by radiation. Here the emphasis is upon *moving* air, rather than quiet air. Favored locations for such fogs are along seacoasts and the shores of large inland bodies of water such as the Great Lakes, where temperature contrasts are great. Advection fogs are most common over oceans and large inland bodies of water in summer, and over lands in winter. This is because the horizontal contrasts in temperature are more marked over continents in winter. Over oceans, on the other hand, temperature contrasts are greater in summer, since the northern waters are kept cool by melting ice. The diurnal periodicity so conspicuous in most types of radiation fogs is not characteristic of the advection fog as a class.

A common variety of advection fog is the sea fog, which forms when warm, moist air moves over colder waters. It is especially prevalent in the vicinity of cool ocean currents. Another form of advection-radiation fog results when a moist tropical air mass is transported poleward over a progressively cooler surface. This type is most frequent in the middle latitudes, where latitudinal temperature contrasts are most marked. Such fogs are common over both sea and land, and since they are related to an extensive air mass, they usually cover large areas. Mist or even drizzle often accompanies them. They occur frequently in the north central United States in winter when a cyclone traveling on a northern track draws tropical air northward over a snow-covered surface.

Frontal fogs. These fogs are associated with the passage of the boundaries separating unlike air masses. Such boundaries are known as fronts. Frontal fog is formed chiefly through saturation of the cold surface layer of air by rainfall from the overrunning warm air aloft.

Distribution of fog. For the earth as a whole, fogs are much more common over the ocean than over land. A study of fog distribution over oceans obviously also involves a study of ocean temperatures and the system of ocean currents. In the low latitudes, including the subtropics, cool ocean currents are concentrated along the eastern sides of oceans (west sides of continents), making these regions some of the foggiest of the earth. In the middle latitudes, much ocean fog develops as a result of warm tropical air moving poleward in the warm sectors of cyclonic storms. Such fogs are common in the North Atlantic, especially its eastern parts, and in the North Pacific near the Aleutian Islands and the Gulf of Alaska. In the Newfoundland area of the North Atlantic and in the region northeast of Japan in the North Pacific, summer fogs are prevalent because of the convergence of warm and cool ocean currents (Fig. 4.12).

For the continents, as might be expected, fog is strongly concentrated on the coastal margins. In the United States there is a definite regional pattern of

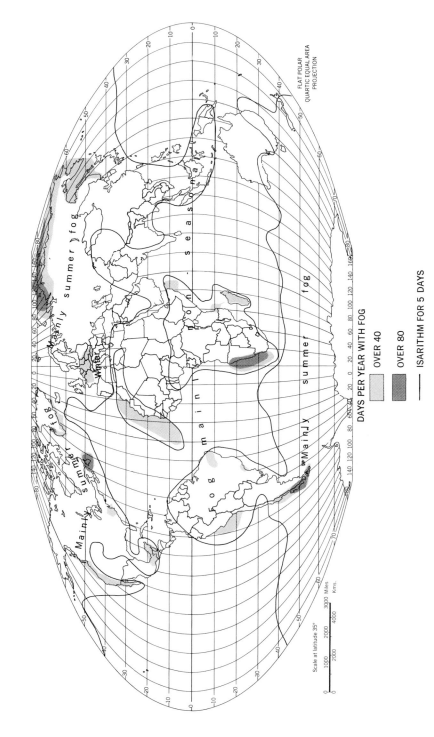

Figure 4.12 World distribution of days with fog. In the high latitudes fog is most prevalent in summer, when it is caused by advected warmer air being cooled over ice surfaces and cold seas. In tropical latitudes fog is less seasonal. It is most frequent in the vicinity of cool currents and their upwelled waters. (*After J. Blütgen in Allgemeine Klimageographie.*)

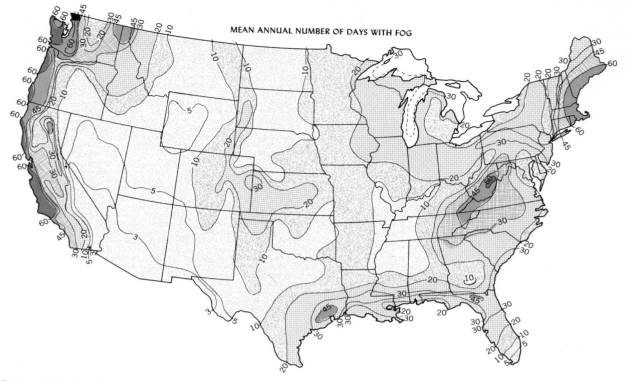

Figure 4.13 Fog frequency in the United States. (*After Arnold Court and Richard D. Gerston in Geog. Rev., vol. 56, October, 1966.*)

fog frequency. Three regions have the most days with heavy fog: the coast and west-facing slopes of the Pacific states; eastern New England; and sections of the Appalachian Highlands (Fig. 4.13). Secondary concentrations are found in parts of the Rocky Mountains, the Gulf Coastal Plain, and restricted coastal sections of the upper Great Lakes. The dry West and southernmost Florida are areas of minimum fog.[2]

Seasonal variations in fog frequency differ for continental and marine locations. Continental stations usually have a maximum of fog in winter and a minimum in summer, while marine stations have the opposite. Radiation ground fogs tend to develop

over land areas in winter because of the greater frequency of large anticyclones then. The clear, calm air of the anticyclone and the long winter nights provide ideal conditions for strong radiational cooling and hence for fog formation. Advection fogs also are more frequent over land in winter because moist maritime air masses entering a continent are chilled at the base from moving over the relatively cold land surface. Over oceans, radiation fog is infrequent, and advection fog is at a maximum in summer since the water surface is cooler than the land at that season.

In both continental and marine locations, the low latitudes have less fog than the middle and high latitudes. Moreover, fog in the tropics is not seasonally concentrated, while in the middle and high latitudes it is most prevalent in summer.

[2]Arnold Court and Richard Gerston. "Fog Frequency in the United States." *Geog. Rev.*, Vol. 56, pp. 543–550, 1966.

Formation of clouds in ascending air currents

Adiabatic temperature changes.   As air is moved upward away from the land-water surface or downward toward it, very important changes occur in the air temperature. Air moving upward away from the surface comes under lower pressures because there is less weight of atmosphere upon it, so it stretches or expands. Air moving downward toward the surface from higher elevations encounters higher pressures and shrinks in volume. Even when there is no addition or withdrawal of heat from surrounding sources, the temperature of the upward- or downward-moving air changes because of its expansion or contraction. This type of temperature change which results from internal processes alone is called adiabatic change.

According to kinetic theory, a gas such as air is composed of molecules which are in a state of constant motion. Individual molecules are continually colliding, and the impact of these collisions produces the pressure of the gas. The pressure therefore depends upon the number and mass of the molecules and the speed at which they are moving, and the speed in turn is determined by temperature. If the temperature is high, the molecular velocity is greater and the number of collisions more frequent. Consequently the gas pressure is greater at high temperatures than at low temperatures, when the volume remains constant.

Some principles indicating the relationship between pressure and volume (hence also density) in gases may be expressed in the form of physical laws as follows:

1. In a gas kept at constant temperature, the volume is inversely proportional to the pressure. If pressure on the gas is doubled, its volume is reduced by one-half.

2. In a gas kept at constant pressure, the volume varies directly with the absolute temperature. Therefore, as the temperature of a gas increases, its volume also increases, and as its temperature is lowered, its volume is decreased.

When a mass of air is lifted, it comes under decreased pressure, so that its volume is increased.

The rising and expanding air does work in making room for its increased volume. This requires an expenditure of energy by the expanding air, with a consequent lowering of its temperature. Stated in a different way, the increased volume of rising air results in fewer molecular collisions. Conversely, when air descends, its volume is decreased by the greater air pressure, molecular collisions increase, and its temperature and density are increased.

The rate of adiabatic heating and cooling of dry or nonsaturated air masses through vertical movement is constant, no matter what the temperature of the air (Fig. 4.14). It is approximately 5.5°F per 1,000-ft change in altitude, or 10°C per 1,000 m. This is known as the dry adiabatic rate, for it applies only to air which is not saturated. The rate of cooling of ascending air, therefore, is considerably more rapid than the normal vertical decrease in temperature (lapse rate), which averages about 3.5°F per 1,000 ft, or 6.5°C per 1,000 m.

These two rates should be clearly differentiated. The lapse rate, which is variable as to time and place, is simply a record of vertical temperature distribution in the atmospheric environment and has nothing to do with rising or subsiding air and associated temperature changes. It can be measured by thermometer readings at fixed levels in the atmosphere. By contrast, the adiabatic rate in nonsaturated air is a fixed rate, and it expresses the temperature changes occurring within unsaturated air which is itself moving upward or downward.

The moist or retarded adiabatic rate.   Since the adiabatic rate of cooling for unsaturated air is the same no matter what the temperature, the three dry adiabats in Fig. 4.14 are all parallel. But as the unsaturated air continues to rise and cool, it is likely to reach a temperature at which condensation begins. This is called the condensation level. In Fig. 4.14 the condensation levels represented for the three rising air masses, each of which has a different temperature, are not the same, because their relative humidities probably are not the same. Thus the cooler air C, which normally would have the highest relative humidity, is shown reaching condensation level first, or at the lowest elevation. The

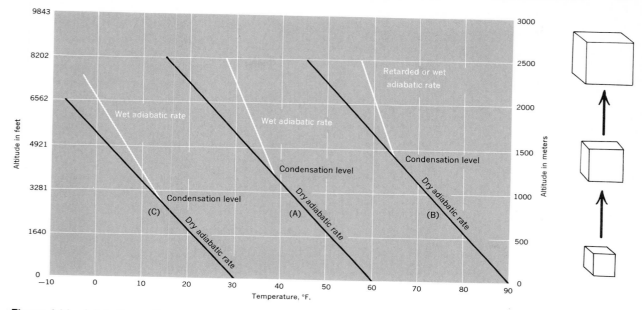

Figure 4.14 Adiabatic cooling in rising air of different temperatures.

warmest air *B*, which normally would be the lowest in relative humidity, is represented as having to rise higher and undergo a greater drop in temperature before the saturation point is reached and condensation begins.

After the rising air reaches the condensation level, where clouds begin to form, the now-saturated air as it continues to rise no longer cools at the previous dry adiabatic rate, but at one which is somewhat slower. This is called the retarded, or moist adiabatic, rate. Above the condensation level, water vapor is being converted into clouds composed of liquid and ice particles, and heat of condensation is released into the atmosphere. It is this added source of energy which slows up the rate of cooling in the rising air. Thus two counteracting processes, one resulting in cooling and the other in heating, are proceeding simultaneously; for while the saturated air continues to cool by expansion, the heat liberated as a result of condensation acts as a brake on the cooling action. Of the two processes, the cooling due to stretching is the primary one, so that the rising saturated air continues to cool but at a slower rate.

It may be observed from Fig. 4.14 that the retarded, or moist adiabatic, rates are not identical for the three air masses, but vary with the temperature of the air. The same amount of cooling will lead to the condensation of a greater amount of moisture when air temperature is high, because the specific humidity of warm air is higher than that of cool air. As a result, the amount of liberated heat of condensation usually is much greater at high temperatures, so that the moist adiabatic rate of cooling is slower in 90° air than in 30° air. In very cold air, the moisture content is so low that the moist adiabatic rate is similar to the dry adiabatic rate. With increasing altitude there is less and less moisture to condense, and the rate of cooling in saturated air approaches the dry adiabatic. In the wet tropics, the rising warm humid air cools more slowly than that farther poleward and hence ascends to greater heights. At high temperatures the moist (or wet) adiabatic rate may be about half the dry adiabatic rate, or about 3° per 1,000 ft (0.5°C per 100 m).

Adiabatic cooling and precipitation. Air which rises voluntarily because of its instability and buoyancy will continue to rise until it reaches an atmospheric environment of its own temperature or density. At this point buoyancy vanishes, and upward movement decelerates. To repeat a statement made earlier, this process of cooling by expansion in rising air is the only one capable of reducing the temperature of extensive and thick masses of air below the dew point. Therefore it is the only one capable of producing condensation on such a large scale that abundant precipitation results. Nearly all the earth's precipitation is the consequence of this expansion and cooling in rising air. The direct outcome of cooling due to ascent is clouds, a form of condensation characteristic of air at altitudes usually well above the earth's surface—just as dew, white frost, rime, and fog are forms characteristic of the surface air. Though all precipitation has its origin in clouds, not all clouds generate precipitation, which is the result of additional processes beyond those causing condensation.

Stable and unstable atmospheric conditions

Since practically all precipitation is caused by the upward movement of air, the conditions that tend to promote or hinder such movement are extremely important. Air is said to be stable, and consequently antagonistic to precipitation, if it is nonbuoyant and resists vertical displacement. Voluntary vertical motions are largely absent in stable air. On the other hand, if displacement results in buoyancy and a tendency for further movement away from the original position, the air is unstable. Under such conditions vertical movement is prevalent. Whether an air mass is stable or unstable can be determined by comparing its lapse rate, a variable, with the dry adiabatic rate, which is fixed. When the dry adiabatic rate (about 5.5°F per 1,000 ft) is exceeded by the lapse rate—as may possibly happen on a hot, sunny afternoon, for example—the surface air is buoyant, unstable, and inclined to rise. But when the lapse rate is lower than the adiabatic rate—as it may be, for example, over a cold middle-latitude continent in winter—the air is nonbuoyant and stable, and it resists vertical displacement. Remember that the lapse rate is the variable one; the dry adiabatic rate is the same at all times and under all conditions.

In Fig. 4.15, the black lines *A* and *X* represent conditions of stability and instability, respectively.

Figure 4.15 Atmospheric stability and instability: When the lapse rate exceeds the adiabatic rate, instability prevails. When the reverse is true, the air is stable.

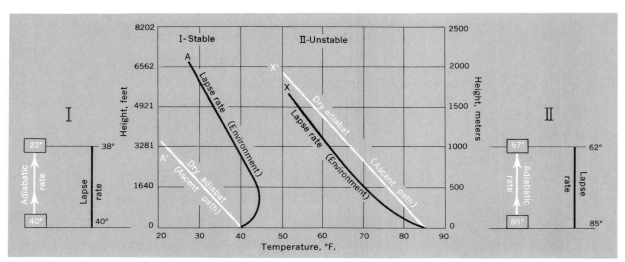

The white lines *A'* and *X'* indicate the dry adiabatic rate—which, of course, is the same for unsaturated air, no matter what the temperature. The white lines are therefore parallel.

Instability.   If in Fig. 4.15 a parcel of unsaturated air with a surface temperature of 85° is caused to rise, its rate of cooling will be represented by the white dry adiabat line *X'*. At any point along this line, the parcel of air will be warmer, and therefore less dense, than the surrounding atmospheric environment, represented by the black-line lapse rate *X*. The parcel will therefore continue to rise freely because of its buoyancy after an initial upward impetus has been given it.

This condition, where the lapse rate throughout is greater than the dry adiabatic rate of 5.5°F per 1,000 ft, illustrates what is known as absolute instability. Moderately unstable air will not begin to rise voluntarily, but if upward movement is once started, or "triggered," either by convergence or by a highland barrier, the ascent will continue because of the buoyancy forces involved (Fig. 4.16).

Stability.   When an air layer has a lapse rate that is less than the dry adiabatic rate of 5.5° per 1,000 ft, the air is stable and opposed to vertical movement. If in Fig. 4.15 an air parcel with a surface temperature of 40° is forced to ascend, then at any point along dry adiabat *A'* the parcel will be cooler and hence denser than the surrounding atmospheric environment, whose vertical temperature structure is represented by black line *A*. Consequently the parcel of air will tend to sink back to its former position. It would never have risen in the first place except under compulsion (Fig. 4.16).

If the lapse rate is less than the *moist* adiabatic rate, which is a variable but on the average is about 2.5°F. per 1,000 ft, the air is said to be absolutely stable. This air resists vertical displacement even when condensation takes place. A common instance of absolute stability is the temperature inversion. An inversion is so stable that it acts as an aerial "lid" to halt ordinary rising currents. Rising columns of smoke or growing clouds are forced to spread

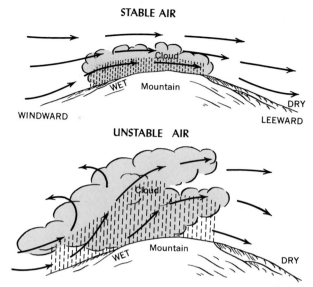

Figure 4.16   When stable or nonbuoyant air is forced upward, as over a terrain obstacle, the resulting clouds are more likely to be of the stratus variety with limited vertical thickness, so that the resulting precipitation may be fairly light. But in unstable or buoyant air, cloud forms are likely to be of the cumulus variety with great vertical thickness, and the resulting showery rainfall will ordinarily be heavier.

out horizontally beneath its base. When nonbuoyant, stable air is forced to ascend—as, for example, over a highland barrier or along a convergence zone or a front—a dense cloud cover without great vertical thickness results. If any rain falls from this cloud cover, it is likely to be light and fairly steady.

Conditional instability.   The information on stability and instability discussed in the previous sections leads to the conclusion that a lapse rate greater than 5.5°F per 1,000 ft represents instability, while a lapse rate of less than about 2.5°F indicates absolute stability. If an air mass has a lapse rate between the dry adiabatic (5.5° per 1,000 ft) and the moist adiabatic (over 2.5° per 1,000 ft) rates of cooling, its instability is conditional upon the saturation of the air at all levels. The lapse rate shown in Fig. 4.17 represents a state of conditional instability.

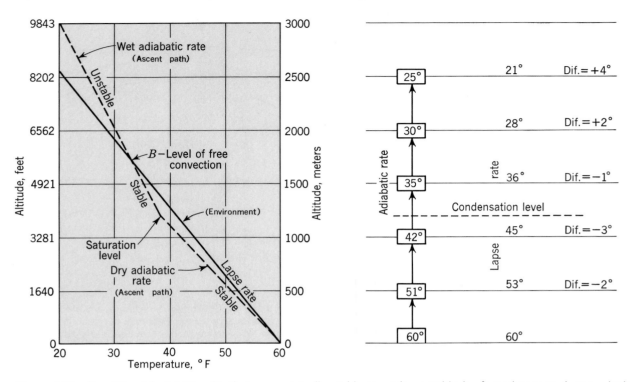

Figure 4.17 Conditional instability: Air that was originally stable is made unstable by forced ascent, during which heat of condensation is added.

While still unsaturated, the rising air parcel follows the dry adiabatic rate of cooling. The lapse rate is less than the dry adiabatic rate, so that during this early phase the rising air is cooler than its environment. It is therefore stable and resists lifting, and any rise is the result of forced ascent. But when the rising air reaches condensation level and latent heat is added, any further cooling takes place at the retarded, or moist adiabatic, rate, which is less than the lapse rate. As a result of the lapse rate being greater than the wet adiabatic rate, from point B upward the rising air is warmer than its atmospheric environment. Consequently it is buoyant and unstable, so that it rises freely of its own accord. In other words, conditionally unstable air performs like stable air until it rises somewhat above condensation level, whereupon it behaves like unstable air.

Summary. The concepts of atmospheric stability, instability, and conditional instability are fundamental to an understanding of the origin of precipitation and the relationship between air masses and precipitation. The following points summarize the basic temperature data connected with these concepts (Fig. 4.18).

1. Absolute stability prevails when the temperature lapse rate is less than about 2.5°F per 1,000 ft.

2. Stability prevails in a relatively dry air mass when its lapse rate is less than the dry adiabatic rate (lapse rate between about 2.5° and 5.5° per 1,000 ft).

3. Conditional instability prevails in a relatively moist air mass when its lapse rate lies between the dry and the wet adiabatic rates (lapse rate between about 2.5° and 5.5° per 1,000 ft).

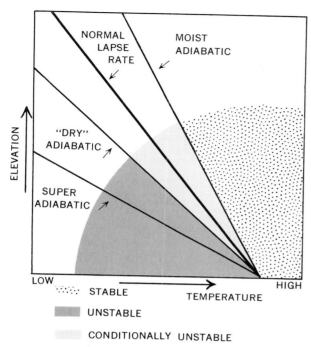

Figure 4.18 Degree of air stability in relation to the rate at which temperature changes with height.

4. Absolute instability prevails in an air mass when its lapse rate is greater than the dry adiabatic rate (lapse rate greater than 5.5° per 1,000 ft).

Vertical displacements of deep and extensive air masses. For the sake of simplicity, the previous discussion dealt with the rising or subsidence of isolated parcels or currents of air. But the lifting or sinking of an entire air mass several thousand feet thick, and covering hundreds of thousands of square miles, is a different story. It involves vertical stretching or contraction of the air mass as a whole, and there are consequent changes in the vertical thickness of the mass which greatly affect condensation processes. In addition to the general cooling of the air mass with ascent and heating with descent, there are differential rates of cooling and heating at different levels within the thick layer.

In the lifting process, accompanied by vertical

stretching, the upper layers of a thick air mass expand and stretch more than the lower ones and as a result are cooled more. This differential cooling tends to steepen the lapse rate within the air mass. Conversely, during subsidence the lower strata are less involved in the vertical shrinking than those aloft, so that adiabatic heating is not so marked at the lower levels. The result of this differential heating at the different levels is to decrease the lapse rate of an air mass and make it more stable. These effects upon the lapse rate of lifting and subsidence are illustrated by Fig. 4.19. When the layer of air lying between the 1,000- and 900-mb surfaces is lifted sufficiently so that pressure decreases by 300 mb and its bounding isobaric surfaces are 700 and 600 mb, the vertical thickness of the air mass has increased and stretching has taken place. This is because the density of air is less at higher altitudes, and consequently the isobaric surfaces are farther apart. In the lifting process graphed in Fig. 4.19, the lower part of the layer at A will cool to A'; the upper part with the temperature B will cool to B'. The result is that the lapse rate has been steepened

Figure 4.19 Effects upon lapse rates of the lifting and subsidence of a thick air mass.

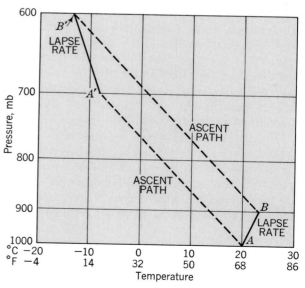

from *AB* to *A'B'*, and the air mass is less stable. If the layer shown at *A'B'* is caused to sink, the reverse process occurs, and the lapse rate is made more stable (*AB*). Actually, the lapse rate *AB* indicates an inversion condition, so that absolute stability prevails. As a general rule, then, lifting and stretching of a thick air mass steepens its lapse rate and therefore increases instability. Conversely, subsidence or sinking, accompained by horizontal spreading, decreases the lapse rate and so increases stability.

Convective instability.   It frequently happens that the lower layers of an air mass have a much higher relative humidity than those aloft. When lifting takes place in an air mass in which relative humidity decreases rapidly with elevation, the lower strata reach condensation level first. Added heat of condensation then causes them to cool more slowly than the drier upper layers, which continue to follow the dry rather than the wet adiabatic rate of cooling. A greatly steepened lapse rate thus develops within the air mass, and it becomes unstable. Such an air mass is said to be convectively unstable.

In Fig. 4.20, the air mass lying between the 1,000- and 900-mb levels has a lapse rate indicated by line *AB*. Because the air at *A* has a higher relative humidity than that at *B*, after ascent begins it reaches condensation level sooner than *B* and therefore cools at a retarded rate. If the air mass with lapse rate *AB* is lifted so that its pressure is reduced 300 mb, *A* will cool to *A'* and *B* to *B'*. The new lapse rate of the ascending air layer, *A'B'*, is considerably steeper than the old lapse rate *AB* and represents a condition of greater instability.

The importance of convective instability lies in the fact that even air that was originally stable may be converted into a buoyant, unstable air mass by a modest forced ascent—provided the surface layers are humid and the water vapor content decreases markedly with elevation. Such forced vertical dis-

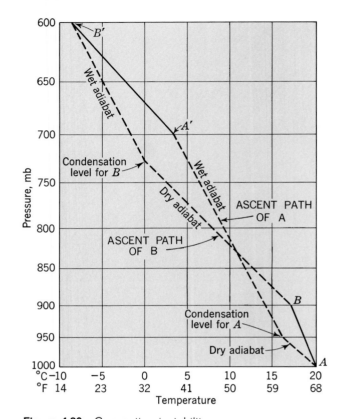

Figure 4.20 Convective instability.

placement of extensive air masses takes place along the windward sides of mountain ranges, and also along frontal surfaces in cyclonic storms where cold and stable air masses act as the obstacle to produce forced ascent. When potentially unstable (convective or conditional) air is lifted, the resulting clouds are usually a mixture of the layer clouds and convective cumulonimbus forms. The cumulonimbus types are associated with the buoyant unstable later stages of the ascent. Rainfall is likely to be heavier than is the case with stable air, as well as more showery and localized.

CLOUD TYPES

Classification of clouds. Since clouds clearly reflect the physical processes taking place in the atmosphere, they are good indicators of weather conditions and so are studied intensively by weather analysts. They are also very important to geographers for a number of reasons. As described in the previous sections, they are the originators of the earth's precipitation. Chapters 1 and 2 discussed them in connection with the earth's heat budget, since clouds greatly influence air temperature by their reflection and scattering of solar radiation and their absorption of earth radiation. In addition, clouds account for a large part of the upward heat transport at heights above the shallow turbulent layer. And of course, cloud forms and illumination are important features of the natural landscape.

The detailed study of cloud types which is useful to the weather forecaster lies outside the province of this book. Below are brief descriptions of the 10 principal types recognized in the international classification of clouds that has been adopted by most countries, including the United States (Fig. 4.21).

Family A. High Clouds. (Mean lower level, 20,000 ft)

1. Cirrus (Ci)—thin featherlike clouds with a fibrous structure and a delicate, silky appearance. When detached and arranged irregularly in the sky, they are harbingers of fair weather. On the other hand, when they are systematically arranged, as in bands, or connected with cirrostratus or altostratus, they usually foretell bad weather. They are always composed of ice crystals.
2. Cirrostratus (Cs)—a thin whitish sheet of cloud covering the whole sky and giving it a milky appearance. These clouds commonly produce a halo around the sun and moon and usually are the sign of approaching storm.
3. Cirrocumulus (Cc)—small white flakes or small globular masses, usually without shadows. They are usually arranged in groups, lines, or ripples resulting from undulation of the cloud sheet. Called a mackerel sky.

Family B. Middle Clouds. (Mean upper level, 20,000 ft; mean lower level, 6,500 ft)

4. Altostratus (As)—a uniform sheet cloud of gray or bluish color, frequently showing a fibrous structure. It is like thick cirrostratus and often merges gradually with it. Through it the sun and moon shine wanly, with a faint gleam. Altostratus commonly is followed by widespread and relatively continuous precipitation.
5. Altocumulus (Ac)—flattened globular masses of cloud, arranged in lines or waves. Differs from cirrocumulus in that it has larger globules, often with shadows.

Family C. Low Clouds. Such clouds commonly cover the whole sky, causing a gray overcast. (Mean upper level, 6,500 ft; mean lower level close to earth's surface)

6. Stratocumulus (Sc)—large globular masses or rolls of soft gray clouds with brighter interstices. The masses are commonly arranged in a regular pattern.
7. Stratus (St)—a low uniform layer of cloud resembling fog, but not resting on the ground.
8. Nimbostratus (Ns)—a dense, shapeless, and often ragged layer of low clouds from which continuous precipitation commonly falls.

Family D. Clouds with Vertical Development. Such clouds usually create a broken sky with clear areas between individual clouds. (Mean upper level, that of cirrus; mean lower level, 1,600 ft)

9. Cumulus (Cu)—a thick, dense cloud with vertical development. The upper surface is dome-shaped with a cauliflower structure, while the base is nearly horizontal. Much cumulus cloud is of the fair-weather type, although towering cumulus may develop into cumulonimbus or thunderheads.
10. Cumulonimbus (Cb)—heavy masses of cloud with great vertical development whose summits rise like mountains, towers, or anvils. They are accompanied by sharp showers, squalls, thunderstorms, and sometimes hail (Fig. 4.22).

Distribution of cloudiness. Cloudiness is expressed in terms of the total area of sky covered by clouds and is given in either tenths or percent. A cloudiness of 0 indicates a cloudless sky; a cloudiness of 10 represents a sky completely overcast, and so does a cloudiness of 100 percent.

The zonal distribution of cloudiness parallels that of rainfall fairly closely. Thus in equatorial latitudes, where horizontal convergence is strong, there is a modest secondary maximum of cloudiness (see the table on page 146 and Fig. 4.23). It is not as strong as the equatorial primary maximum in rainfall,

Figure 4.21 A very generalized vertical arrangement of cloud types. (*From Atmosphere and Weather Charts, published by A. J. Nystrom and Company.*)

Figure 4.22   A cumulonimbus cloud, a type in which convective showers originate. Photographed from an airplane. (*Courtesy of T. Fujita.*)

however, because the tropics have a predominance of convective clouds of the cumulus type. While vertically deep, these are localized clouds which usually cover only a part of the sky, even though

Figure 4.23   Distribution by latitude zones of the annual means of cloudiness, in percent. Note that the greatest cloudiness is not in equatorial latitudes, where rainfall is heaviest, but in the middle and higher latitudes, where stratus clouds associated with cyclonic storms and their fronts reach a maximum development. Cloudiness is least in the latitudes of the subtropical anticyclones and the trade winds. Compare with Fig. 4.27. (*Data from C. E. P. Brooks.*)

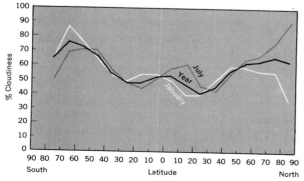

they yield heavy showery precipitation. In the latitudes of the subtropical anticyclones and dry trades (20°–30° N and S), there are minima of cloudiness, as there are of rainfall. The subtropical minima are much more striking on the continents, with their extensive low-latitude deserts, than on the oceans. Poleward from the subtropics, both cloudiness and rainfall increase. But while rainfall shows only secondary maxima in the middle latitudes of both hemispheres (Fig. 4.22), cloudiness reaches its primary maxima there. This is because much of the middle-latitude cloudiness and rainfall is associated with cyclones and fronts. In these weather disturbances, clouds tend to be of the stratus type which covers the whole sky, but a good part of the precipitation that falls is only light or moderate.

The *annual variation* in cloudiness is likely to roughly parallel that of rainfall. Close to the equator, the difference in cloudiness between individual months may be small. In the tropical belts located between 10° and 20°, the cloud maximum coincides with the summer rainfall maximum (see table on page 146). But toward the western sides of continents in the subtropical latitudes (between 30° and 40°),

Average monthly cloudiness (Percent)

Latitude	J	F	M	A	M	J	J	A	S	O	N	D	Yr	Continents	Oceans
90–80°N	36	47	56	46	76	87	90	85	84	64	45	41	63		63
80–70°N	56	56	55	63	70	74	75	76	78	75	63	50	66	63	70
70–60°N	57	56	54	59	65	66	66	68	71	72	67	60	63	62	72
60–50°N	59	57	57	59	64	63	63	62	62	67	67	64	62	60	67
50–40°N	59	57	57	57	56	56	54	49	49	54	58	61	56	50	66
40–30°N	50	49	49	48	48	43	42	39	39	43	45	48	45	40	52
30–20°N	41	41	41	39	41	43	45	44	40	39	38	40	41	34	49
20–10°N	40	39	39	40	47	53	59	58	54	46	44	44	47	40	53
10– 0°N	50	48	49	53	54	56	57	55	53	53	53	53	53	52	53
0–10°S	54	53	53	52	50	50	50	52	53	53	53	55	52	56	50
10–20°S	54	52	52	49	46	45	43	44	43	47	49	54	48	46	49
20–30°S	49	50	50	47	48	48	47	45	48	47	49	50	48	38	53
30–40°S	53	52	54	53	55	56	56	54	55	56	55	52	54	48	57
40–50°S	64	65	63	64	64	64	69	64	66	67	67	66	66	58	67
50–60°S	76	69	71	74	83	82	70	69	68	71	71	75	72	70	72
60–70°S	86	80	80	72	72	66	68	74	75	77	83	80	76		76
70–80°S	64	80	69	69	64	47	49	59	65	74	62	63	64		64
Continents	47	47	47	48	49	50	49	48	48	49	49	50	49		
Oceans	59	58	58	57	58	58	59	58	58	59	58	59	58		

Source: After C. E. P. Brooks.

it is the cool season that has the most cloud and also the most rainfall. Poleward from about 50°, summer is cloudier than winter, especially in the Northern Hemisphere. This contrast between summer and winter strengthens in the higher latitudes as the winter continental anticyclones become increasingly dominant. The interiors of large middle-latitude continents not uncommonly show annual variations of cloud and precipitation that are out of phase with each other—a summer minimum of cloudiness, but a maximum of rainfall. This situation reflects the greater amounts of local convective cloud in the warm season and of stratiform frontal cloud in winter.

The *daily variation* in cloudiness is complicated because of the contrasts in the origin of cumulus and stratiform cloud types. Since surface heating tends to produce cumulus clouds, they often reach a maximum in the early and middle afternoon. Stratiform clouds, on the other hand, are usually created by stable atmospheric conditions, so that their maximum occurs in the early morning hours and their minimum in midafternoon.

PRECIPITATION

Precipitation processes.   It is possible for ascending air which reaches and passes beyond the condensation level to form cloud but still yield no precipitation. Cloudy condensation probably does not take place without at least a slight degree of supersaturation. As relative humidity exceeds 100 percent, condensation first occurs on the largest hygroscopic nuclei. But if it continues, condensation collects on almost innumerable smaller nuclei as well, with the result that the individual cloud droplets are too small to fall to earth as rain. Even a very slight ascent of air is sufficient to keep such minute droplets in suspension. And even if they did succeed in falling below condensation level, probably they would be evaporated before reaching the earth. It is understandable, then, why there can be so many days with cloud on which no precipitation occurs.

Thus rainfall is not the result of a simple continuation of the condensation processes which produce clouds. The essential feature of the precipitation process is the combining of myriads of small cloud droplets into fewer larger drops capable of falling to earth. Two precipitation mechanisms are believed to cause most of this combining that forms raindrops.

The first mechanism is the result of cloud instability brought about by the coexistence of water droplets and ice particles in a cloud at temperatures below 32°F. For an appreciable amount of precipitation to occur in this instance, the ascending air must rise above the freezing level, where some of the liquid droplets will be changed into ice, and where direct sublimation (water vapor to ice) may also take place. The difference in vapor pressure around the liquid droplets and around the ice particles leads to evaporation of the liquid droplets and then to condensation around the ice particles. As water continues to accumulate around an ice nucleus, the drop eventually reaches such a size that it can no longer be held up by the ascending currents. It begins to fall, first within the cloud, and then through the lower air to the land-water surface. Unless it is broken up by the speed of its descent, the drop continues to grow by accretion until it leaves the cloud. The best evidence that this mechanism operates as described here is the sudden release of precipitation when a cumulus cloud grows into a cumulonimbus after it reaches the glaciation, or ice-nuclei, level. Glaciation of the upper part of the cloud changes its form from the typical boiling, cauliflower top to one that is anvil-shaped.

The second mechanism believed to cause precipitation occurs without the presence of ice particles. It involves simply the coalescence of droplets of various sizes as they collide because they are falling at different rates within the cloud. Thus the rate of growth by this process depends upon the size, size distribution, and concentration of the drops.

A good-sized raindrop contains as much water as 5 to 10 million cloud particles and falls 200 times as fast. The maximum size to which a raindrop can grow is about $\frac{1}{5}$ in. (5 mm) in diameter, for at that size its rate of fall is about 18 miles per hr. Above that size and speed, drops begin to break up. In gently ascending air currents condensation is relatively slow. Even small drops can fall to earth through the slowly rising air, so that light rain or drizzle results. On the other hand, the strong upward surges in a violent thunderstorm produce vigorous condensation and are able to support drops of great size. Consequently the first few isolated drops that fall preceding a general downpour make sizable splashes. Raindrops are usually larger than $\frac{1}{50}$ in. in diameter, and in still air they fall faster than 10 ft per sec. Drizzle droplets are less than $\frac{1}{50}$ in. in diameter.

Solid forms of condensation and precipitation. In rising air, solid condensation forms may appear in the atmosphere as a result of either the freezing of liquid condensation, or the direct condensation from the vapor to the solid state. Liquid condensation is possible at temperatures well below freezing. Down to 15°F, clouds composed of supercooled water droplets are much more common than ice clouds, and they have been observed at far lower temperatures.

Snow condensation is in the form of intricately branched, flat hexagonal crystals in an almost infinite variety of patterns. A snowflake is simply an agglomeration of snow crystals matted together because of a film of water on the individual crystals. Since very cold air contains little moisture, heavy snowfalls are usually associated with surface temperatures not much below freezing. Large wet flakes fall when temperatures are comparatively high, while fine hard snow is characteristic of very cold regions and periods. As a rough approximation, it requires about 12 in. of snow to equal 1 in. of rain, although the ratio may vary from 5 to 1 to 50 to

1, depending on the density of the snow. If snow forms at fairly high levels in the atmosphere while surface temperatures are well above freezing, it will melt before reaching the ground and therefore will arrive at the surface as rain. A considerable proportion of the rainfall reaching the ground in middle latitudes originated as snow.

Data on the amount of snowfall and the duration of the snow cover are scanty and fragmentary for much of the earth, so that a satisfactory description of the distribution of snow is difficult to provide. Snow falls occasionally near sea level even in subtropical latitudes, but it does not remain on the ground long. At low elevations a durable snow cover in winter lasting for a month or more is characteristic only of the interior and eastern parts of Eurasia and North America poleward of about 40°. Köppen indicates that a snow cover of appreciable duration is typical of regions where the temperature of the cold month is 27°F (−2.8°C) or below. A permanent snow cover exists at very high altitudes even in the tropics. The height of the snow line declines poleward. Thus at 68°N in Norway, permanent snow is found at an elevation of about 3,500 ft, while on Mount Kilimanjaro in equatorial East Africa at 3°S, the snow line is at about 18,400 ft. Obviously the height of the snow line depends not only on temperature, but also on the amount of snowfall, for if the snowfall is sufficiently heavy a permanent snow cover may exist even where the average summer temperatures are somewhat above freezing (Fig. 4.24).

Sleet is frozen, or partly frozen, rain and appears as particles of clear ice. Glazed frost is actually not a form of precipitation but is the accumulation of a coating of ice on surface objects. Fortunately it is not of common occurrence, for the so-called ice storm that produces glazed frost is one of the most destructive of the cool-season weather types. It occurs when supercooled rain or drizzle strikes surface objects and is immediately converted into ice. The weight of the ice accumulation may become so great that trees are often damaged; telephone, telegraph, and electric wires are broken and their poles snapped off. Rime is composed of white layers of ice crystals deposited on windward edges or points of objects, generally in supercooled fog or mist. Hail, although the heaviest and largest unit form of solid precipitation, is exclusively the product of vigorous convection such as characterizes thunderstorms. These in turn are mainly features of the warm regions and warm seasons.

Precipitation types

Since most precipitation results from expansion and cooling in ascending air, it is important to analyze the conditions under which large masses of air may be caused to rise. Three common types of ascent and the precipitation characteristics associated with each are described below. The three are not mutually exclusive, however, and any particular rainfall is not necessarily the result of only one. In reality, most of the earth's precipitation is caused by the joint action of several types of atmospheric lifting.

Convectional precipitation. This type of precipitation originates with the adiabatic cooling of buoyant air currents whose ascent is truly vertical with respect to the land-water surface. It is not an oblique upgliding of air such as occurs along a hill slope or an atmospheric front. Convection currents, usually of limited diameter, may attain considerable speed. The whole convective system consists of numerous cells of local up currents and down currents, with the rapid vertical chimneys of updraft separated by broader, slower, down currents. When the chimneys of buoyant rising air reach condensation level, they are capped by the dramatic cauliflower-topped cumulus cloud. Such a cloud is the visible top of an invisible ascending convectional current. If the air is deeply humidified and unstable, the released heat of condensation may be so great as to cause some of the cumuli to burgeon into vertically deep cumulonimbus clouds. These are capable of producing heavy showery precipitation, lightning and thunder, and perhaps even hail.

But even in conditionally and convectively unstable air, some initial upward displacement of the surface air is usually required in order to trigger

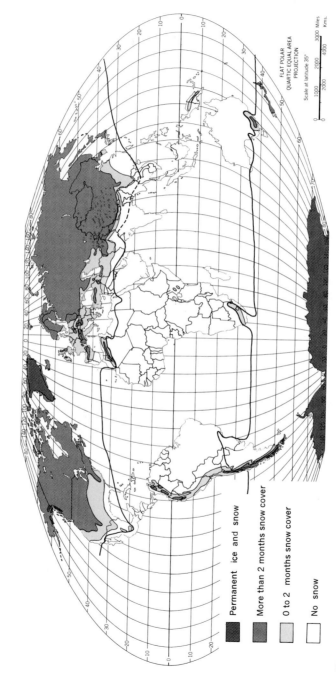

Permanent ice and snow

More than 2 months snow cover

0 to 2 months snow cover

No snow

Figure 4.24 Duration of the snow cover and equatorial limits of snowfall. (*After Sekiguchi and Blütgen.*)

FLAT POLAR
QUARTIC EQUAL AREA
PROJECTION

Scale at latitude 35°

1000 2000 3000 Miles
2000 4000 Kms.

the buoyant convectional currents. This trigger action may be supplied in several ways—by uplift over a terrain obstacle such as a mass of hills; by mechanical turbulence resulting from ground friction or from internal eddies, which are always present in moving air; by lifting along fronts where less dense air glides up over colder denser air; and by directional convergence in the airflow. Some authorities would add that trigger action can be supplied by a warming effect from below, as when land surfaces heat up in the daytime, or when cold air is advected over warm water. This is controversial, however. There can be no doubt that heating from below makes an air mass more buoyant and unstable. But it is uncertain whether surface heating in itself is sufficient to trigger ascent.

In fact, weather specialists appear to be increasingly skeptical of the notion that widespread thermal convection is random in occurrence and stems from surface heating of air. If heating from below were a main trigger for convectional activity, then convective showers should develop every afternoon over land surfaces in the tropics, which they do not. Convection and its associated shower activity appear to occur in organized (not random) regional patterns, with the shower regions related to one or more of the types of trigger action mentioned above. And since many of the shower regions change their shapes and locations from day to day, the evidence suggests that intensified convection is often connected with extensive moving atmospheric disturbances in which convergence in airflow is present. Thus it is clear that much convectional precipitation is complex in origin.

Rainfall associated with convective overturning has certain distinctive characteristics. Since a single convective unit is composed of several columns or cells of rapidly ascending air, each often capped by a cumulus or towering cumulonimbus cloud, cooling is rapid, and the resulting rain commonly falls as a pelting downpour. But since the ascending chimneys of air and their cumulonimbus clouds are not horizonally extensive, the rain is often spotty, local, and brief, for the raincloud quickly drifts by.

We usually speak of convective *showers* (or thundershowers) rather than of convective *rains*.

Much convective shower activity has a seasonal and diurnal periodicity reflecting the influence of solar heating. It tends to follow the sun and occurs at the time of greatest heat and greatest atmospheric instability. In the middle latitudes this means the warmer months, when the daily forecasts refer much more frequently to showers than to general rains. In the low latitudes, also, convective shower activity follows the sun and coincides with the meridional shifting of wind belts. Over the 24-hr period, shower activity generally reaches a maximum in afternoon and early evening, when the surface air is warmest and most unstable.

It has been discovered recently that within the tropics (and probably middle latitudes as well), half the total rainfall falls within 10 percent of the time intervals, where the latter range from 24 hr to $\frac{1}{10}$ hr.[3] Obviously this demonstrates the importance of convective-scale motions in the origin of tropical rainfall. Three scales of circulation or disturbance are recognized: (1) the synoptic or cyclonic scale (time scale, 24 hr), (2) the convective scale (time scale, hours), and (3) individual convective cells (time scale, tenths of hours). The essential point is that synoptic-scale disturbances provide a favorable environment for the development and organization of convective-scale motions, within which the smaller individual convective cells originate; and it is the latter that are responsible for producing most of the precipitation. Thus synoptic-scale disturbances are just as essential for rainfall in the tropics as in the middle latitudes, but precipitation occurs in synoptic systems as a consequence of enhanced convection, characteristically in organized patterns. Some authorities are of the opinion that all tropical rainfall has its immediate origin in convective cells. But towering cumuli require convergence in the

[3]N. E. La Seur. *Proc. 1964 Army Conference on Tropical Meteorology.* Fort Monmouth, N.J. May 14–15, 1964. Pp. 20–25. Michael Garstang. "Atmospheric Scales of Motion and Rainfall Distribution." *Proc. 1966 Army Conference on Tropical Meteorology.* Miami Beach, Fla., May 26–27, 1966. Pp. 24–35.

large-scale flow such as is provided by macro-scale disturbances.

Because it comes frequently in the form of heavy showers, convectional precipitation is less effective for crop growth than steady rain, since much of it runs off in the form of surface drainage instead of entering the soil. This is a menace to plowed fields, where soil removal through slope wash and gullying is likely to be serious. On the other hand, for the middle and higher latitudes convectional precipitation comes at the time when it is most effective: in the warm season of the year when vegetation is active and crops are growing. Moreover, it provides the maximum rainfall with the minimum duration of cloudiness.

Orographic precipitation. This may be defined as precipitation which is caused or intensified by the upthrust effects on air of highlands. Total annual precipitation in highlands is strikingly larger than that normal to the surrounding lowlands, so that highlands stand out on a rainfall map as centers of precipitation. One of the commonest effects of a mountain chain is the upslope motion given to broad and deep currents of air which lie athwart its course. This motion is called forced ascent, because an airstream can only reluctantly be made to surmount a highland barrier.

Since water vapor is largely confined to the lower layers of atmosphere and rapidly decreases upward, heavy orographic rainfall is often the result of such forced ascent of air associated with the blocking effect of terrain obstacles. Examples include the very abundant precipitation along the western, or windward, flanks of the Cascade Mountains in Washington and Oregon; along parts of the mountainous east side of Madagascar, which lies in the trades; and bordering the abrupt west coast of India, where the westerly currents of the summer monsoon meet the escarpment of the Western Ghats practically at right angles. The leeward sides of such mountain barriers, where the air is descending, warming, and becoming more stable, are characteristically drier (Fig. 4.25). This is called the rain

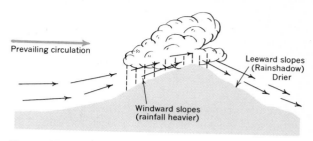

Figure 4.25 Rainfall contrasts on windward and leeward slopes.

shadow. The blocking effect of a mountain upon an airstream is normally felt at some distance out in front of the abrupt change in slope. Heaviest orographic rainfall along a mountain front usually is not far above the point where precipitation begins, although its elevation varies with the season, exposure, and latitude.

It seems likely, though, that a considerable part of the increased precipitation associated with highlands is not solely the result of the forced upslope ascent of prevailing winds, but is complex in origin. Other indirect effects of highlands are also important: (1) their production of strong turbulence of both a mechanical and a convective nature; (2) their obstructing and slowing effect upon the progress of cyclonic storms; (3) orographically conditioned convergence in horizontal currents; and (4) the trigger effect of highlands that gives the initial upthrust to conditionally or convectively unstable air masses. Sometimes only a slight amount of lifting is necessary to bring unstable air masses to the condensation level. After that, added heat of condensation makes them unstable and buoyant and causes them to continue rising, yielding abundant showery convective rainfall. Thus highlands less than 3,000 ft in elevation, although they probably induce no great amount of rain through simple blocking action, may by these indirect means become distinctly wetter than adjacent lowlands.

As might be expected, orographic rain has less seasonal and daily periodicity than that of convectional origin, and it varies from region to region.

Such features as the strength of the winds, the angle at which they meet the mountain barrier, and the degree of contrast between land and water temperatures may determine the season of maximum orographic rainfall. The effects of highlands upon precipitation cannot be judged exclusively by their height, for factors such as temperature and humidity of the upslope air, as well as speed and direction of the wind, are also involved. The cloud types and the rainfall intensity and extensiveness associated with orographic effects are highly variable in character. Upslope movement of stable air favors the development of sheet clouds with accompanying light and long-continued precipitation. But in potentially convective air, cumulus cloud forms and showery rainfall will also be present.

Disturbance precipitation.   Whenever surface airstream convergence occurs, lifting of the opposing air masses results, and atmospheric instability increases. Extensive, mobile atmospheric disturbances of many types are one of the most common situations where widespread convergence and upward movement of air masses occur. Lifting of the air in such a convergent system is often rather general and slow, so that widespread instability is attained gradually and pervasively. In warm latitudes and warm seasons, areas of organized convection and associated shower activity commonly develop within the convergent air of these extensive atmospheric disturbances. Obviously such rainfall is a combination perturbation-convection type.

In contrast with the tropics, middle-latitude air masses differ in temperature and density, so that much lifting of air within extensive atmospheric disturbances results from a slow and gradual upglide of warm and less dense air over colder air (Fig. 4.26). Such lines or zones of convergence between contrasting air masses are known as fronts. Frontal precipitation is very common in the middle latitudes, and since fronts are a characteristic feature of cyclonic vortex disturbances, most frontal rainfall is also cyclonic in origin.

Frontal and cyclonic clouds and precipitation

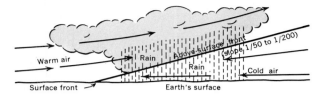

Figure 4.26   The origin of precipitation along an atmospheric front. Here the warmer and less dense air is forced upward over a mildly inclined wedge of cooler and denser air, with the consequence that the rising air is cooled by expansion.

have no fixed character. Because a considerable amount of the upward movement of air is a slow upglide, cooling may be slow and the resulting clouds may be the stratus type providing a gray overcast and steady long-continued precipitation. But in conditionally and convectively unstable air, a slight amount of uplift along a front may provide the trigger effect necessary to start vigorous convective overturning, especially in the warmer months. Cumulus clouds and showery precipitation result.

Distribution of precipitation

Salient features.   Among the data concerning precipitation which are of special importance in describing regional climates are the average annual amount or depth, the average number of rain-days, the seasonal distribution or annual march of precipitation, its dependability or reliability, and its probability.

It is estimated that if the *total average precipitation for the year* were spread evenly over the earth's surface, it would form a layer about 39 in. deep. In reality, the annual precipitation is spread very unevenly, for there are great desert areas that receive less than 5 in., while a few spots receive over 400 in. (33 ft). Since continental air is lower in humidity than oceanic air, the average yearly precipitation over land is estimated to be only about 60 percent of that over the oceans.

The number of rain-days, or days on which 0.01 in. or more of precipitation falls, can be compared with the total annual or monthly amount to give some indication of the rate of fall or *intensity* of rain in a region. This measure also provides a general impression of the dampness or dryness of a climate (see the following table). For example, Marquette, Michigan, receives about 32.5 in. annually which, since there are 165 rain-days, is 0.20 in. per rain-day. By contrast, Pensacola, Florida, has nearly 58 in. of precipitation, which falls in 114 rain-days, making 0.51 in. per rain-day. In the United States the number of rain-days is especially large along the North Pacific Coast and in the Great Lakes–New England region. It is particularly small in the dry Southwest. Figures on the *probability* of rain-days for an area are valuable for such people as farmers and resort owners. Probability may be computed by dividing the number of rainy days in a month or year by the total number of days.

The *seasonal distribution* of precipitation is as important as the total amount. From a geographer's standpoint, the fact that Los Angeles, California, annually receives about 15 in. of precipitation must be evaluated in the light of the fact that 60 percent of this falls in the three winter months and almost none in the three summer months. Seasonal distribution of precipitation becomes especially significant in latitudes which have a dormant season for plant growth imposed by low temperatures, i.e., the winter season. In the tropics, where frost is practically unknown except at higher elevations, rainfall is effective for plant growth no matter at what time of year it falls. In the middle latitudes, however, only the part of the annual precipitation which falls during the freeze-free season may be effective. In the more severe climates, a strong concentration of rainfall in the warmer months when plants can use it is desirable.

The *dependability*, or *reliability*, of the annual or seasonal precipitation is an expression of its variability (Fig. 4.35). Variability may be defined as the deviation from the mean computed from 35 years or more of observations. In humid climates the annual variability is usually not greater than 50 percent on either side of the mean; i.e., the driest year may have about 50 percent of the normal value, while the wettest year may have 150 percent. In dry climates these values vary between about 30 and 250 percent. Thus it is a general rule that variability increases as the annual amount of rainfall decreases: an inverse ratio prevails.

Precipitation variability must be taken into consideration by the farmer, for he must expect that there will be years when the precipitation is less than average. In semiarid and subhumid climates, where crop raising ordinarily depends on a small margin of safety in rainfall, variability is crucial. Moreover, the agriculturist in such regions must bear in mind that negative deviations from the mean are more frequent than positive ones. This means that a greater number of dry years are com-

Rain-day information for Philadelphia, Pennsylvania, and Cherrapunji, India

	J	F	M	A	M	J	J	A	S	O	N	D	Yr
Precipitation, mm													
Philadelphia	86	88	88	85	85	88	112	118	87	75	77	84	1,078
Cherrapunji	17	59	268	809	1,495	2,632	2,729	2,069	1,255	426	58	7	11,824
Rain-days													
Philadelphia	12	11	12	11	11	10	11	11	8	8	9	10	124
Cherrapunji	1.4	3.3	8.4	17.3	21.0	24.8	27.2	26.0	18.7	8.4	1.4	0.6	158.5
Mm per rain-day													
Philadelphia	7.1	8.0	7.4	7.7	7.7	8.8	10.2	10.8	10.9	9.5	8.6	8.4	8.7
Cherrapunji	12.1	17.9	31.9	46.7	70.7	100.0	101.0	79.5	67.0	50.6	41.4	11.6	69.5

Source: After Landsberg.

pensated for by a few excessively wet ones. Variability of seasonal and monthly rainfall amounts is characteristically greater than that of annual values.

Average annual amounts. Even a glance at the rear endpaper shows that precipitation is distributed very unevenly over the earth's surface and that the patterns of distribution are complex. No single explanation is widely applicable. Basically, the amount of precipitation which falls in a region depends on two groups of factors—those which influence the vertical motions of the atmosphere, and those relating to the nature of the air itself, particularly whether it is dry or moist, warm or cold, stable or unstable. The first of these, which emphasizes lifting and subsidence, is closely related to the distribution of (a) the principal zones of horizontal convergence and divergence, (b) atmospheric disturbances, and (c) highland barriers. The second group of factors, the nature of the air, chiefly depends on its place of origin and its subsequent trajectory, both of which determine whether it is maritime or continental, tropical or high-latitude, in its characteristics.

Some of these controls of precipitation—for example, certain belts of horizontal convergence and divergence—are fairly zonal in arrangement. Others—such as the distribution of land and water, and the arrangement of highlands and lowlands—are nonzonal and hence operate to produce a modification of the zonal patterns. Certainly the complicated character of annual rainfall distribution on the individual continents (see rear endpaper) makes generalizations difficult to perceive. In order to simplify the situation, some of the more significant facts of world distribution are portrayed by a graph of a meridional profile of average annual precipitation in Fig. 4.27, and by a diagram of the main rainfall type-regions on a hypothetical continent in Fig. 4.28. Other features of distribution unique to individual continents will be discussed in Part Two, which deals with climatic types and regions.

Figure 4.27 Distribution of average annual precipitation by latitude zones. (*Data from Meinardus, Brooks, and Hunt.*)

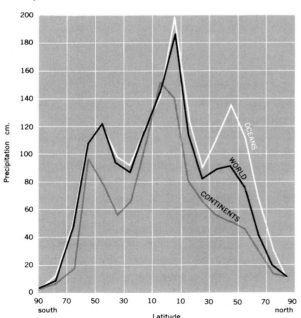

Zonal features of annual averages. A meridional profile of average annual rainfall amounts, showing the means for different parallels around the entire earth, suggests some of the most fundamental facts of rainfall distribution (Fig. 4.27). Of course, much is omitted in such a profile, for it indicates none of the variations in amount of rainfall in an east-west direction, i.e., along a parallel.

As seen in Fig. 4.27, there is a strong primary maximum of precipitation (±1,600 mm) in a belt about 10 to 20° wide in the vicinity of the equator. In this belt, directional convergence (ITC) of surface winds results in large-scale lifting of warm, humid, unstable air. Numerous weak, rain-generating disturbances develop along the ITC. (The main exception within the equatorial belt of heavy rainfall is the narrow dry zone along the equator in the eastern central Pacific Ocean, as will be discussed in Chapter 8. Recent satellite photographs confirm this dramatic climatic anomaly.) From the equator, annual amounts of precipitation decline poleward in each hemisphere. Belts of lower rainfall, providing secondary minima, exist at latitudes about 20 to 35° N and S. Here the average rainfall is only 800 to 900 mm. These are the latitudes of the great subtropical anticyclones, with their diverging wind systems and associated vertical subsidence. Since some of the earth's great deserts are located here, it may seem odd that the average zonal rainfall is not lower than Fig. 4.27 indicates. It must be kept in mind, however, that the western sides of the subtropical highs are rainier, and these heavier rainfalls tend to raise the average for the subtropical latitudes as a whole.

Poleward from the subtropics, rainfall again increases. Secondary maxima with an average annual precipitation of 1,000 to 1,200 mm are indicated for latitudes about 40 to 55° N and S. These are the zones of the middle-latitude convergences (polar front), with which the belts of maximum cyclonic activity are associated. The average zonal rainfall of these belts is lower in the Northern Hemisphere than in the Southern, because of the extensive interior middle-latitude deserts in Eurasia and North America. (If rainfall over land alone is considered, there is no middle-latitude maximum in the Northern Hemisphere.) Poleward from about latitude 50 to 55° in both hemispheres precipitation declines sharply, so that primary zonal minima of less than 150 mm are reached in the cold regions of the high latitudes beyond about the 75° parallel.

Although the total annual precipitation is about the same in the Northern and Southern Hemispheres, some contrasts in zonal distribution may be observed. The most conspicuous difference is that latitudes 0 to 10° N have more precipitation than latitudes 0 to 10° S. This reflects the fact that the intertropical convergence is positioned north of the equator over more extensive longitudes and for a greater part of the year than it is to the south. Another difference is that the secondary maximum of precipitation at about latitudes 40 to 60° is considerably higher in the Southern Hemisphere than in the Northern, because of the larger proportion of ocean in these latitudes in the Southern Hemisphere.

Another principal factor affecting zonal rainfall amounts is the distribution of oceans and continents. For the earth as a whole, the average amount of precipitation is much higher over the oceans than over the continents: about 44 in. for the oceans, compared with 26 in. for the land. Taking into consideration the fact that 71 percent of the earth's surface is water and only 29 percent land, it has been computed that 19 percent of the earth's total annual precipitation by weight falls on land surfaces and 81 percent on water. It is in the Northern Hemisphere, where there is a more equitable distribution of land and water areas, that the zonal patterns of rainfall over oceans and continents are most in contrast. Continental rainfall, for example, shows no secondary middle-latitude maximum north of the equator, while this zonal maximum is very definite over the oceans (Fig. 4.27).

Precipitation regions on a hypothetical continent. Figure 4.28 is a highly generalized representation of rainfall distribution, in both its zonal and its nonzonal aspects, on a hypothetical continent which is broad in the Northern Hemisphere and tapers southward. In equatorial latitudes in the vicinity of the intertropical convergence zone, wet climates with abundant rainfall extend across the entire breadth of the continent. Explanations for this equatorial rain belt have been provided in the preceding section on zonal distribution of rainfall. The wet belt is broader along the eastern side of the continent, for this is the windward side, which is paralleled by warm ocean currents and lies adjacent to the more unstable western margins of the subtropical anticyclones. Here inversions are high or absent. The wet belt is narrower along the west side, which is dominated by the stable eastern sides of subtropical anticyclones and by cold currents with upwelling. Wet regions are also found along the elevated windward western margins in middle latitudes, both north and south of the equator. These wet areas are far smaller than the equatorial rain belt.

Figure 4.28 shows that areas characterized by below-average precipitation, including both dry and subhumid types, are to be found in continuous

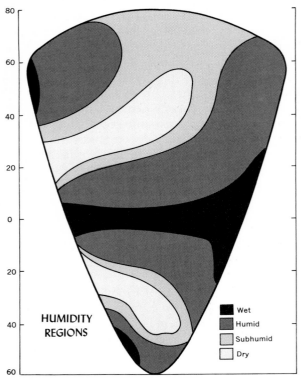

Figure 4.28 Very generalized patterns of distribution for four levels of average annual rainfall, as they might appear on a hypothetical continent.

HUMIDITY
REGIONS

Wet
Humid
Subhumid
Dry

As diagrammed in Fig. 4.28, humid conditions, reflecting moderate rainfall, prevail in two locations in each hemisphere. Both are situated between the wet and the dry-subhumid types. One occupies a zonal belt lying just poleward of the equatorial wet region and then continues into the middle latitudes along the eastern side of the continent, so that the whole has roughly the form of a boomerang. The low-latitude part of this boomerang-shaped area lies in the region of alternating ITC and trades. The part which continues into the middle latitudes is affected both by frequent cyclonic storms and by tendencies toward monsoon systems. The other location of humid conditions, or moderate rainfall, is in middle latitudes along the western side of the continent, inland from the wetter oceanic margins. Here cyclonic activity accounts for much of the precipitation.

These patterns of rainfall distribution represented on the hypothetical continent in Fig. 4.28 should be compared with rainfall patterns of actual continents on the rear endpaper and with the index-of-humidity map, Fig. 4.29.

Seasonal rainfall variations. One factor which greatly influences the seasonal march of precipitation, especially in the tropics and subtropics, is the north-south migration of the wind belts. As explained in Chapter 3, these, together with their associated zones of convergence and divergence, follow the course of the sun. In the low latitudes, evidences of a zonal pattern of seasonal rainfall distribution are moderately conspicuous. By contrast, the middle latitudes show more evidence of a nonzonal pattern, imposed in considerable degree by the effects of continents and oceans. Oceanic areas not only have a greater total depth of annual precipitation than the lands do, but their precipitation is also less seasonal in its fall. Large continents in middle latitudes are likely to have more of their annual precipitation concentrated in the warmer months, both because of their cold winters and warm summers, which greatly influence the stability and moisture content of seasonal air masses, and because of their tendency to develop a monsoonal wind system.

bands in both hemispheres, reaching from low into middle latitudes. In the low latitudes the dry regions are asymmetrically developed, for they are concentrated in the western and central parts of a large land mass. Such locations are under the influence of the stable eastern ends of subtropical anticyclones, where subsidence and horizontal divergence prevail. Cool ocean currents and upwelling along the subtropical and tropical west coasts may intensify the aridity. In the middle latitudes the dry-subhumid areas are located toward the center of the continent, since this region is farthest removed from the oceanic sources of moisture. Among the actual continents in the Southern Hemisphere (see rear endpaper), only South America extends far into the middle latitudes, and there the land mass is so narrow that dry climates reach to the east coast.

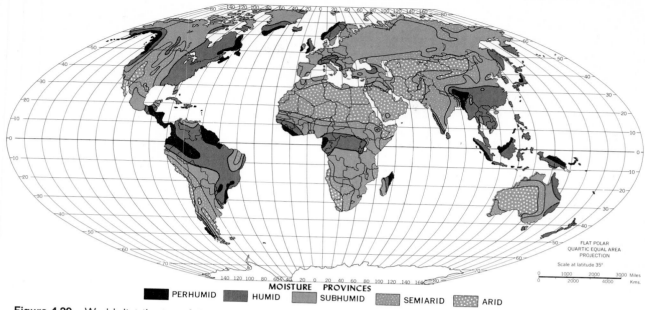

Figure 4.29 World distribution of the index of humidity. The five humidity-index subdivisions are defined as follows:

A	Perhumid $\geqq$	16.2
B	Humid	8.7–16.1
C	Subhumid	4.7–8.6
D	Semiarid	2.5–4.6
E	Arid $<$	2.5

The humidity index *EP* (effective precipitation) is calculated after the following formula: $EP = P/1.025\,T + X$, where *P* is annual rainfall in inches, *T* is the annual temperature in °F, and *X* is the correction supplied by a nomogram. (*After Harry P. Bailey, Geog. Ann., Vol. 40, pp. 196–215, 1958. For the formula, see his Fig. 2, p. 200.*)

Zonal variations in seasonal rainfall. Figure 4.30 is a diagram of the principal zones of convergence and divergence associated with the entire atmospheric circulation. Parts *A* and *B* indicate the maximum latitudinal displacement in the extreme seasons. Part *C* provides a general description of the zonal belts of seasonal rainfall that would result from a regular north-south migration of the zones of convergence and divergence. This diagram does not show the important seasonal variations in precipitation which are the result of nonzonal factors: longitudinal differences in the atmospheric circulation pattern, and terrestrial influences such as land-water distribution and the arrangement of highlands. Consequently Fig. 4.30 should be studied together with Fig. 4.31, which illustrates the general arrange-ment of the types of seasonal rainfall concentration that might appear on a hypothetical composite continent. Here zonal and nonzonal features are combined (see also Fig. 4.32).

Tropics. In the very low latitudes near the equator, where the trades converge and weak disturbances are numerous, rainfall is abundant and falls throughout most of the year, so that there is either no dry season or only a very short one (Fig. 4.30, zone 1). Farther away from the equator, from about 5 or 10° out to 15 or 20°, rainfall becomes more seasonal as it decreases in amount. There is a marked dry season in low sun, or winter, while summer, or the high-sun period, is wet (zone 2). This feature of high-sun rainfall and low-sun drought is related to the north-south shifting of the

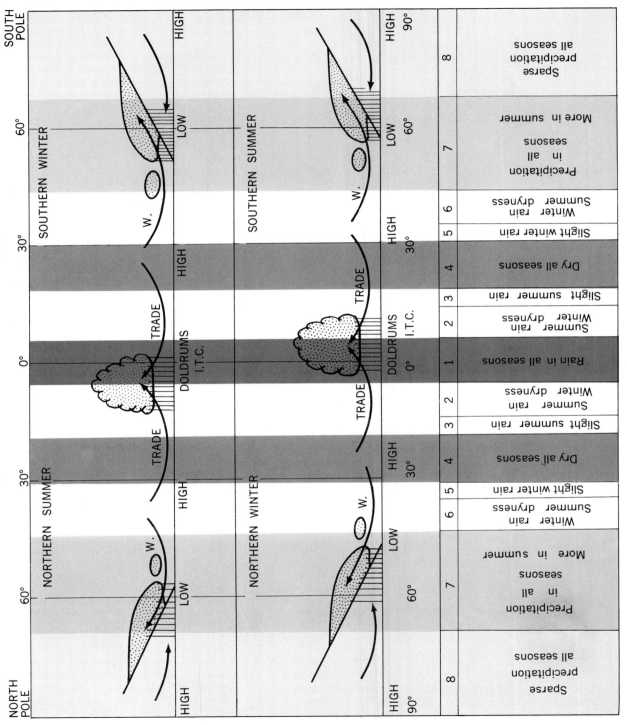

Figure 4.30 Main zones of atmospheric horizontal convergence and ascent, and of divergence and subsidence, together with associated seasonal characteristics of precipitation; *A*, during the Northern Hemisphere summer; *B*, during the Northern Hemisphere winter; *C*, zones of seasonal precipitation. But remember that many nonzonal features of precipitation distribution cannot be adequately represented on this type of schematic latitudinal cross section. (*From Petterssen, Introduction to Meteorology, McGraw-Hill Book Company, New York.*)

158

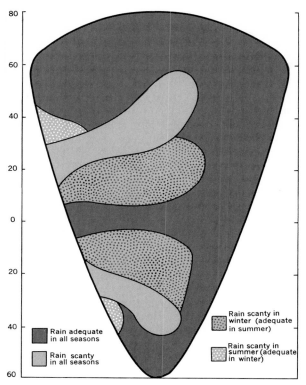

Figure 4.31 Very generalized patterns of seasonal rainfall concentration as they might appear on a hypothetical continent.

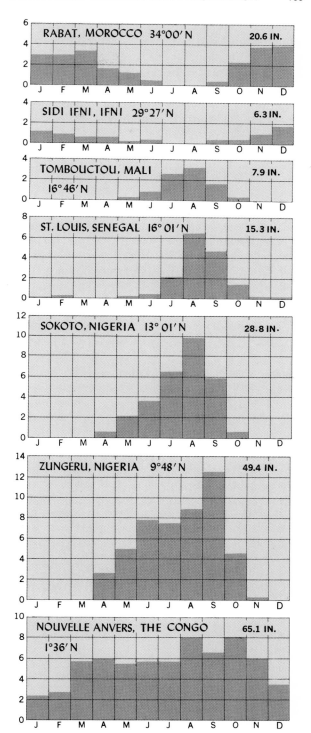

wind belts and zones of convergence following the course of the sun. Such latitudes are influenced by the ITC and its disturbances at the time of high sun, and by the subtropical anticyclone with its subsidence and divergence in the low-sun period. And as Fig. 4.31 shows, the tropical winter-drought regime does not ordinarily extend to the east side of the continent, where the subtropical anticyclone is weaker and prevailing winds are onshore.

Within the area (shown in Fig. 4.30) where rainfall is both meager in annual total and scanty in all seasons, (zone 4) seasonal variation is not of

Figure 4.32 Illustrating seasonal rainfall regimes in the low latitudes by means of a series of stations in Africa north of the equator. The stations are arranged according to latitude, with Nouvelle Anvers closest to the equator and Rabat farthest away from it.

great consequence. In tropical dry climates, the equatorward margins have least rain in winter (zone 3). On the poleward western margins, summer is the driest season (zone 5). Where the dry type extends well poleward into the middle-latitude heart of the hypothetical continent, winter is usually the season of scantiest rainfall (Fig. 4.31).

Subtropics (about 30 to 40° N and S). Located along the subtropical *western* side of the hypothetical continent, in an intermediate position between the middle-latitude convergence zone on its poleward side and the subsident western end of a subtropical anticyclone on its equatorial side, is a restricted area where rain is scanty in summer and moderately abundant in winter (Fig. 4.31). The wet winters coincide with the equatorward migration of middle-latitude westerlies and their cyclonic storms following the sun (Fig. 4.33a). The aridity of summer is associated with the poleward retreat of the storm belt and the reestablishment of anticy-

clonic control with its strong upper-air subsidence (Fig. 4.30, zone 6).

The eastern sides of the continents in these same subtropical latitudes (about 30 to 40°), however, are very different in their precipitation characteristics (Fig. 4.33b). Normally, they have more rainfall than the western side, and in addition the winter maximum and summer drought have entirely disappeared. Here it is the western margin of a subtropical anticyclone with its unstable air masses which is in control, so that rain falls throughout the year, and summers, with their tendency to onshore monsoon winds, are likely to have more precipitation than winters.

Latitudes poleward of about 40°. On the hypothetical continent, most of the middle and higher latitudes are represented as having adequate rain in all seasons (Fig. 4.31). This is the latitude of the cyclonic westerlies, whose disturbances bring precipitation both in winter and in summer. But this

Figure 4.33 Subtropical rainfall regimes: *left,* west side of continent; *right,* east side of continent.

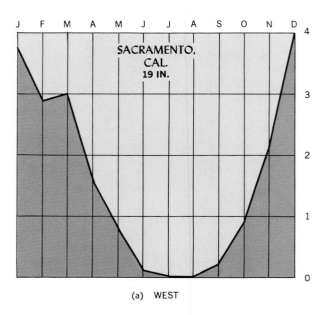

(a) WEST

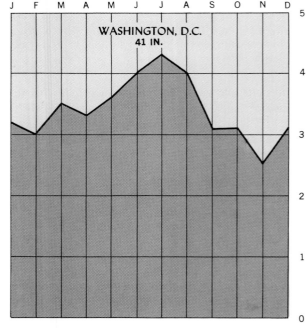

(b) EAST

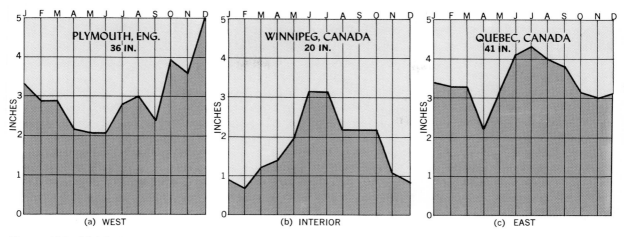

Figure 4.34 Middle-latitude rainfall regimes.

does not mean that seasonal contrasts are entirely absent.

On the *western* or windward side of the continent, it is not uncommon for the most marine locations to show a slight winter maximum (Fig. 4.34*a*). This is probably a widespread feature of oceanic climates in middle latitudes. It is associated with the intensified cyclonic activity in the cooler seasons and the greater temperature contrasts between sea and land at those times.

The *interior* of a large continent in middle latitudes, where total annual rainfall is on the scanty side, is likely to show varying degrees of summer maximum, even though winter is by no means dry (Fig. 4.34*b*). The summer maximum is a distinctive feature of land-controlled climates. This fact is associated with the low winter air temperatures which reduce the possible water vapor capacity, the greater prevalence of anticyclones in winter, and the tendency toward monsoonal outflows of continental air in the winter season. In summer, on the other hand, surface heating causes greater convective activity, the warm air has a higher moisture capacity, strong anticyclones are less pronounced, and there is a tendency toward a monsoonal indraft of warm, humid air from the oceans.

The humid *eastern* side of a large continent in middle latitudes usually shows varying degrees of summer concentration of precipitation. In eastern Asia, where the monsoon circulation is particularly strong, the summer maximum is more pronounced than in eastern Anglo-America, where winter cyclones are more active (Fig. 4.34*c*). Dry Patagonia, in the rain shadow of the Andes, is the exception to humid eastern sides of middle-latitude continents.

In the very high latitudes, the maximum precipitation probably occurs during the season of least cold, since at that time there is more moisture in the air, and cyclonic influence can penetrate deeper (Fig. 4.30, zone 8).

Annual precipitation variability. Annual precipitation is most variable in dry and subhumid regions. As described earlier, there is usually an inverse ratio between rainfall amount and rainfall variability. In Fig. 4.35 two types of regions stand out: the deserts and steppes of the tropics and middle latitudes, and the cold regions of high latitudes. The highest variability is in the deserts. But the economic consequences of high variability in the earth's sparsely populated dry regions and its cold regions may not be nearly so serious as a smaller variability in more populous humid areas, where agriculture is a major industry.

Small rainfall variability is characteristic of two types of humid regions—those of the wet tropics,

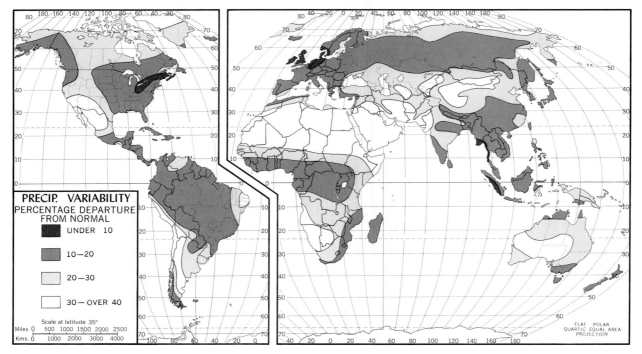

Figure 4.35   Variability or undependability of annual precipitation is usually at a maximum in dry and subhumid climates. (*After Biel, Van Royen, and others.*)

and those of the cyclonic middle latitudes. The least variability is found in maritime western Europe and an east-west belt along the United States–Canadian border in eastern Anglo-America.

Diurnal variation in precipitation.   This is a relatively complicated phenomenon, although two gen-

eral types may be recognized; a continental type and a marine one. Over large land masses, more of the rainfall is likely to occur in the warmer hours of the day, when solar heating of the land surface is at a maximum and a steepened lapse rate prevails in the lower air, causing it to be more buoyant. Oceans tend to have a maximum of precipitation

Diurnal variation in precipitation (In thousandths of total amount)

Place	Hours					
	12 P.M.–4 A.M.	4 A.M.–8 A.M.	8 A.M.–12 M.	12 M.–4 P.M.	4 P.M.–8 P.M.	8 P.M.–12 P.M.
Valentia, Ireland, 52°N	181	183	160	149	162	163
Pavlovsk, U.S.S.R., 60°N	148	156	144	201	185	157
San José, Costa Rica, 10°N	13	6	44	342	485	84

Source: After Köppen.

during the night or early morning, when the maritime air is most unstable (Fig. 2.18). The largest midday maximum over land in middle latitudes occurs during the summer season. The surface temperature of oceans, unlike those of continents in summer, changes very little from day to night, and as a result the surface air varies little in temperature. At some distance above the ocean surface, where the air temperature is largely controlled by radiation, the atmosphere is warmer by day and cooler by night. Consequently the lapse rate is steepest at night and convective activity is greater. These generalizations are illustrated in the table on page 162, which shows the diurnal variation in the amount of precipitation for Valentia, Ireland, a marine station in middle latitudes; Pavlovsk, U.S.S.R., a continental station in middle latitudes; and San José Costa Rica, located in the tropics.

SELECTED REFERENCES FOR FURTHER STUDY OF TOPICS IN CHAPTER FOUR

1. **Atmospheric humidity and condensation.** Riehl, *Tropical Meteorology*, pp. 144–155, 156–192; Riehl, *Introduction to the Atmosphere*, pp. 67–89; Blair and Fite, pp. 36–59, 100–110; Blüthgen, pp. 108–157; Critchfield, pp. 37–46; Hare, pp. 17–25; Gentilli, pp. 80–86, 112–114; Petterssen, pp. 61–69; Haurwitz and Austin, pp. 83–96; *Compendium of Meteorology*, pp. 165–175, 192–197, 199–205, 1179–1189; Willett and Sanders, pp. 82–101.

2. **Precipitation: origin and kinds.** *Compendium of Meteorology*, pp. 175–179; Blair and Fite, pp. 110–120; Blüthgen, pp. 158–186; Riehl, *Tropical Meteorology*, pp. 105–124; Riehl, *Introduction to the Atmosphere*, pp. 91–101; Critchfield, pp. 46–69; Hare, pp. 25–37; Petterssen, pp. 40–60; Willett and Sanders, pp. 101–117.

3. **Precipitation distribution.** Haurwitz and Austin, pp. 64–83; Blüthgen, pp. 186–210; Riehl, *Introduction to the Atmosphere*, pp. 218–244; H. Flohn and J. Huttary, "Zur Kenntnis der Struktur der Niederschlags Verteilung," *Zeitschrift für Meteorol.*, Vol. 6, pp. 304–309, 1952; Gentilli, pp. 87–108; Fritz Möller, "Vierteljahrskarten des Niederschlags für die ganze Erde," *Petermanns Geograph. Mitt.*, Vol. 95, pp. 1–7, 1951.

4. **General references on atmospheric moisture and precipitation.** Harry P. Bailey, "A Simple Moisture Index Based upon a Primary Law of Evaporation," *Geog. Ann.*, Vol. 40, pp. 196–215, 1958; C. E. P. Brooks and T. M. Hunt, "The Zonal Distribution of Rainfall over the Earth," *Mem. Roy. Meteorol. Soc. (Great Britain)*, no. 3, pp. 139–158, 1930; M. I. Budyko, *Evaporation Under Natural Conditions* (translated from Russian), Office of Technical Services, Washington, D.C., 1963; J. J. George, "Fog," In *Compendium of Meteorology*, pp. 1179–1189, 1951; International Cloud Atlas (2 vols.), World Meteorological Organization, Geneva, 1956; B. J. Mason, *Clouds, Rain and Rainmaking.* Cambridge University Press, London, 1962; J. E. McDonald, "The Evaporation-Precipitation Fallacy," *Weather,* Vol. 17, pp. 168–177, 216, 1962; W. Meinardus, "Niederschlagsverteilung auf der Erde," *Meteorol. Zeitschrift*, Vol. 51, pp. 345–350, 1934; H. L. Penman, "Weather, Plant, and Soil Factors in Hydrology," *Weather,* Vol. 16, pp. 207–219, 1961; H. L. Penman, "Estimating Evaporation," *Trans. Amer. Geophysical Union*, Vol. 37, pp. 43–50, 1956; R.C. Sutcliffe, "Water Balance and the General Circulation of the Atmosphere," *Quart. Jour. Roy. Meteorol. Soc.*, Vol. 82, no. 354, pp. 385–395, 1956; *Symposium on Monsoons of the World*, Indian Meteorological Department, New Delhi, pp. 177–234, 1960; C. W. Thornthwaite and J. R. Mather, "The Water Balance," *Publications in Climatology*, Vol. 8, no. 1, Drexel Institute of Technology, Centerton, N. J., 1955; R. C. Ward, "Measuring Potential Evapotranspiration," *Geography*, Vol. 48, pp. 49–55, 1963.

CHAPTER FIVE

AIR MASSES AND FRONTS

An air mass is an immense body of air that moves over the earth's land-sea surface as a recognizable entity, with temperature and humidity characteristics which are relatively uniform in a horizontal direction at different levels. Its horizontal homogeneity is not complete, of course, for since an air mass may extend through 20 to 30° of latitude and cover up to several million square miles, it is bound to have some internal differences. But these are always small compared with the much more rapid rates of temperature change across the air-mass boundaries. Air masses show up on weather maps as broad-scale atmospheric currents of polar or tropical origin. As earlier chapters have said, these currents are the vast units of exchange by which the general circulation effects a transfer of energy between tropical and polar latitudes. Weather is likely to be relatively uniform throughout an area covered by the same air mass. It changes rapidly along the air-mass margins.

The air-mass concept is mainly a refinement and amplification of the general circulation and planetary winds described in Chapter 3, for both trades and westerlies have unlike properties at different places and times. The concept of air masses is also related to the discussion of atmospheric disturbances in Chapter 6, since a majority of disturbances in middle latitudes appear to originate along the boundary zones separating unlike air masses. The prevailing weather is determined to a large degree by the nature of the air masses moving over a region. This is especially true in the zones of the disturbed westerlies, where the succession of weather events is intimately related to the passage of cyclones and anticyclones and the sequence of air masses accompanying them. Although the air-mass concept may not provide a very sophisticated tool for the climatologist, it continues to be useful and revealing. It has even served as the basis for a number of climatic classifications.

Weather characteristics of an air mass depend on two basic properties: vertical temperature distribution, and moisture content and its vertical distribution. The first property indicates not only the warmness or coldness of an air mass, as reflected in its surface temperatures, but also its stability, which in turn influences the extent of vertical movement. The moisture content partly determines the presence or absence of condensation forms, while the vertical stratification of moisture also influences vertical movement.

Source regions.   Air masses originate under conditions which promote the development of vast bodies of horizontally uniform air. An ideal air-mass source region has two essential characteristics: an extensive physically homogeneous surface, and a sufficient stagnation of atmospheric circulation so that the air has time to acquire the temperature and moisture properties of the underlying surface. A region with highly uneven terrain, or one composed of both land and water surfaces, is not satisfactory. Nor is one in which there is strong convergent airflow, for in most parts of the earth outside the inner tropics, converging surface winds usually advect unlike temperatures toward the zone of convergence, resulting in steep temperature gradients. It is anticyclonic circulations, with their light winds and calms, and their prevailing subsidence and horizontal divergence, that favor development of the horizontal temperature uniformity required in an air

mass. Examples of good source regions are the snow-covered arctic plains of North America and Eurasia in winter, the extensive subtropical and tropical oceans, and the superheated Sahara in summer.

The depth to which an air mass is affected by its source region depends upon both the length of time it remains in that region, and the kind and degree of temperature differential between the air and the underlying surface. When the air is initially colder than the surface of the source region, it is heated from below. This produces convective ascending currents which carry heat and moisture aloft, so that the air mass is modified up to considerable heights. But when the air is initially warmer than the underlying surface, it is cooled from below and made nonbuoyant. In such stable air convective currents do not develop, and the modification of the air is much slower and shallower. Since the properties of source regions vary from season to season, so also do the temperature and moisture properties of air masses originating in them.

Once having formed over its source region, an air mass does not long remain stationary. The general circulation of the atmosphere, which acts to compensate for the energy imbalance between low and high latitudes, ensures that an air mass will soon move outward from its source region. When this happens, the air itself is slowly modified by its new environment, while it in turn modifies the weather of the region it invades. An extensive and horizontally homogeneous body of air is conservative in character: it can travel for long distances over the earth's surface and still retain many of the properties acquired at its source. Because of its great size and the slowness with which it is modified, the movement of an air mass can be traced from day to day, and any changes which are being induced by its new environment can be measured. Usually such changes are small and gradual compared with those which occur along the boundary zones that separate unlike air masses.

In addition to being influenced by the surfaces over which it moves, and the different amounts of radiation and moisture which it receives, an air mass is also greatly modified by the lifting and subsidence of thick air layers within it.

Surfaces of discontinuity and fronts. When air masses with different temperature and moisture properties, and therefore with density contrasts, are brought together, they do not mix freely with each other. They tend to remain separate, with more or less distinct sloping boundary surfaces, called surfaces of discontinuity or fronts, between them (Fig. 5.1). Where such sloping surfaces of discontinuity in the free atmosphere intersect the earth's land-sea surface, surface fronts are formed. It is these intersections which are charted on a surface weather map. Since there is an upward movement of air along fronts, many of them are accompanied by an organized system of clouds and precipitation extending over immense horizontal distances. Fronts are active weather breeders, and their location in the atmosphere and the nature of the contrasting air masses on either side of them are of great significance in weather analysis.

Two conditions are essential for the formation of fronts, and unless both are operating simultaneously, frontogenesis cannot occur. (1) The two adjacent air masses must have contrasting temperatures, so that one is colder and denser than the other. (2) The atmospheric circulation must have a convergent flow strong enough to transport the air masses toward each other at the front line. But even when these two conditions are present, fronts still would not develop in the absence of Coriolis force. If the earth suddenly stopped rotating, the cold denser air would settle quietly under the warm air, like water under oil in a tank. Then the surface of discontinuity separating the two would be horizontal, not sloping, and there would be no surface front.

The decay and disappearance of fronts, called frontolysis, is caused whenever (1) the temperature contrast between the air masses weakens and eventually disappears or (2) the wind system no longer brings about convergence of the air masses.

Surfaces of discontinuity, or fronts, are not mathematically abrupt features, but rather transition zones of appreciable width, where changes in the weather elements are much more rapid than within the air masses themselves. The width of the frontal transition zone of pronounced temperature gradient separating cold polar from warm tropical air is most commonly between 50 and 150 miles. Marked shifts in temperature, and also in humidity and wind direction, usually can be observed upon crossing an air-mass boundary zone. The front represents a temperature discontinuity most of all, for it is chiefly through temperature contrasts that differences in air-mass densities are maintained.

The discontinuity surface separating unlike air masses rarely remains very long in a stationary position. Usually one air mass begins to advance into the domain of the other, so that the front itself begins to move. Two principal types of fronts are recognized: (1) a zone where there is an active upglide of lighter warm air over cold dense air; and (2) a zone where there is a forced ascent of warm air over cold because the cold air is actively underrunning and lifting the warm. These are the well-known warm front and cold front (Fig. 5.2). Chapter 6 on atmospheric disturbances will discuss their characteristics and the types of weather each brings.

It is the middle latitudes, lying as they do between cold polar and warm tropical source regions, that have the greatest air-mass contrasts. Consequently these latitudes also have the best-developed fronts and the most pronounced weather changes associated with them. In fact, the concepts of air masses and

Figure 5.1 Three-dimensional representation of an atmospheric front.

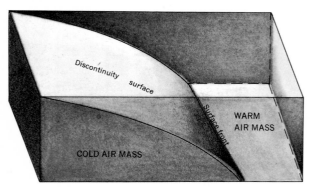

Discontinuity surface

Surface front

WARM AIR MASS

COLD AIR MASS

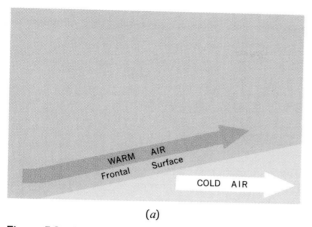 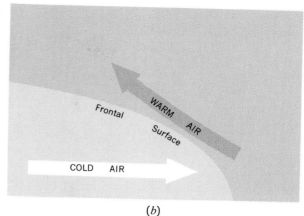

Figure 5.2 Arrangement of air masses along a warm front (*a*) and a cold front (*b*).

fronts developed out of weather analysis in the middle latitudes and are not very useful when applied to the tropics. Even though a zone of strong wind convergence in the tropics may be accompanied by a wind shift and an organized system of clouds and precipitation, usually the converging air masses do not have significant temperature and density contrasts. Thus a front does not exist. As a result, there is no upgliding of the less dense air along an airmass surface of discontinuity, but instead sporadic convectional ascent of a vertical nature.

CLASSIFICATION OF AIR MASSES

The attributes of an air mass which determine its characteristic weather features are chiefly acquired from its source region. To a lesser degree, an air mass is also affected by other influences which continue to modify it as it moves away from its source region. Any useful classification of air masses must express their weather characteristics and consequently must be based upon the nature of their source regions and the subsequent modifications which they undergo.

Polar and tropical. The primary air-mass source regions are polar (*P*) and tropical (*T*). It is mainly in high latitudes and low latitudes that extensive homogeneous surfaces characterized by light air movement can be found. Since the middle latitudes are the scene of intense interaction between air masses arriving from polar and tropical latitudes, they lack the uniformity of conditions essential to a source region. Nearly all air masses in middle latitudes can be traced back to polar or tropical source regions.

Some authorities differentiate cold air from high latitudes into arctic (*A*) and polar (*P*) subgroups, the former originating in the polar areas and the latter over the cold continents of Eurasia and North America. Similarly, warm air from low latitudes is sometimes separated into equatorial (*E*) and tropical (*T*), the equatorial air, which originates in the inner tropics, being more unstable and more deeply humidified. Although these refinements have some usefulness, they are not included in the simplified scheme of air-mass classification presented here.

Maritime and continental. Polar and tropical air masses may each be subdivided into *maritime (m)* and *continental (c)* subgroups, depending upon whether they originated over oceans or over continents. The *m* or *c* symbol indicates whether the air mass left the source region with moist maritime characteristics, so that condensation is likely, or whether it originated as a dry continental air mass in which condensation will be meager. When a dry continental air mass moves out over an ocean surface, moisture evaporates into the dry air and changes it relatively rapidly into a humid maritime air mass. The transformation from maritime to continental air is ordinarily much slower, since there is not necessarily a prompt and large-scale removal of moisture by precipitation from a maritime air mass that moves inland over a continent. This classification, then, includes four principal types of source regions and four corresponding principal air masses—*cP, mP, cT,* and *mT.*

Modifications of polar and tropical air masses

Modifications of an air mass after it leaves its source region may partly determine the nature of the weather which occurs within the air mass. Two principal types of air-mass modification are recognized: those which are thermodynamic in origin and those which arise from mechanical causes. They may occur separately or in combination.

Thermodynamic modifications. These result from the transfer of heat between the bottom of an air mass and the surface over which it moves. The degree of modification depends upon the original character of the air mass, the nature of the underlying surface, the trajectory or path of the air mass as it leaves its source, and the number of days it has traveled in arriving at the observation point.

If the air mass moves over a surface that is warmer than its own ground temperature, the con-

Basic classification of air masses

Major group	Subgroup	Source region	Properties at source
Polar (including arctic)	Polar continental (*cP*)	Arctic Basin; northern Eurasia and northern North America; Antarctica	Cold, dry, very stable
	Polar maritime (*mP*)	Oceans poleward of 40 or 50°	Cool moist, unstable
Tropical (including equatorial)	Tropical continental (*cT*)	Low-latitude deserts, especially Sahara and Australian Deserts	Hot, very dry, stable
	Tropical maritime (*mT*)	Oceans of tropics and subtropics	Warm, moist; greater instability toward west side of ocean

sequent warming of its basal layers will result in an increased lapse rate and added buoyancy and instability. This condition favors ascent of the heated lower air and thus creates the possibility of condensation and precipitation. The maximum vertical change in temperature occurs in the lowest layers. Conversely, when an air mass moves over a surface that is colder than its own ground temperature, the basal air is chilled, and a surface inversion develops which increases the stability of the mass. Such a condition is opposed to the ascent of air and consequently also to the formation of clouds and precipitation. Naturally, polar air masses will most frequently undergo the first type of modification, tropical air masses the second.

These modifications of the source properties of an air mass are represented by the letters *W* (warm) or *K* (*kalt,* or cold). *W* indicates that the air mass is warmer than the underlying surface and so is in the process of being chilled at the base and made stable or nonbuoyant. *K* means that the air is colder than the underlying surface and is therefore being warmed at the base and made unstable and buoyant. Note that these letters do not specify whether the air mass itself is either hot or cold; they only indicate its *relative* temperature with respect to the surface beneath.

A *K* air mass typically has these traits: turbulence up to a height well above the friction layer; an unstable lapse rate; cumuliform clouds; generally good visibility except in rain; showers, thunderstorms, and possibly snow flurries. By contrast, a *W* air mass is more often characterized by these features: little turbulence above the friction layer; a stable lapse rate; poor visibility at lower levels, where dust and smoke are concentrated; stratiform clouds; fog and precipitation, if there is any, in the form of drizzle.

The height to which the surface thermal influence extends up into the air mass depends on whether the air has a *W* or *K* character. Since surface cooling tends to produce stability, vertical turbulence and convective transport are restricted. Thus the *W* character, or chilling of the lower air, is confined

to a fairly shallow layer—usually the first few thousand feet. But the *K* characteristic, signifying an air mass warmed at the surface, is usually carried to a much greater height by convective turbulence. In an air mass which is moderately unstable aloft, there is almost no limit to which the surface convection and its attendant transport of water vapor can extend into the troposphere.

Further thermodynamic modification of an air mass results from the addition of moisture to it by evaporation—either from a moist underlying surface (particularly a water surface), or from raindrops which fall through the air mass out of an overrunning air current. Turbulence and vertical convection act to distribute the moisture through a moderately thick layer of the atmosphere. Thus in addition to a direct increment of sensible heat, evaporation is another way of adding energy to the lower atmosphere and increasing its instability, for the evaporated moisture represents actual or potential energy in the form of latent heat.

Mechanical modifications.   Besides thermal modifications of air masses, others of a mechanical nature can affect stability. One of these is turbulence resulting from the frictional effects of the earth's surface. The effect of turbulence is to mix the atmosphere vertically, so that heat and moisture are carried upward from the surface and a layer of atmosphere several thousand feet deep may be modified.

Much more important modifications of an air mass result from large-scale horizontal convergences and divergences that may occur near, as well as above, the earth's land-sea surface. Such circulations, which produce slow upward and downward movements of thick and extensive masses of atmosphere, thereby affect air-mass stratification and possibly cloud and precipitation. An air mass that is part of an anticyclonic circulation undergoes above-surface subsidence and so becomes more stable. Likewise, air which descends on the leeward side of a mountain barrier is made more stable by the subsidence. On the other hand, an air mass which is involved in a cyclonic circulation, where convergence, lifting, and stretching are dominant, is made more unstable. Lifting over highlands has a similar effect (Fig. 4.19).

Also, an air mass moving toward higher latitudes, where the earth's circumference shrinks, tends to undergo horizontal convergence, while increasing divergence characterizes equatorward-moving air. In addition, other things being equal, the relative warmth of a tropical current moving poleward promotes upward movement, while the relative coldness of a polar current moving equatorward favors subsidence. As a consequence the development of stable characteristics is favored in polar air masses reaching low latitudes, and unstable characteristics tend to appear in tropical air attaining higher latitudes.

The mechanical changes described here mainly affect those parts of an air mass above the surface, or friction, layer. They are likely to be independent of the other modifications resulting from surface heating or cooling which were treated in the previous section. In order to represent air-mass qualities associated with stability and instability conditions aloft, a fourth pair of letters may be added to the air-mass classification system: s = stable air aloft; u = unstable air aloft. Under most conditions the s symbol is associated with anticyclonic circulation and the u symbol with cyclonic circulation. Note that the s or u characteristic of an air mass is a feature unrelated to the surface qualities of its source region, or in many instances, to its subsequent trajectory. Instead, this feature is associated with the local or broad-scale circulation pattern in which the air mass originates or subsequently migrates through.

Figure 5.3 Air masses and fronts in January. (*After Willett.*)

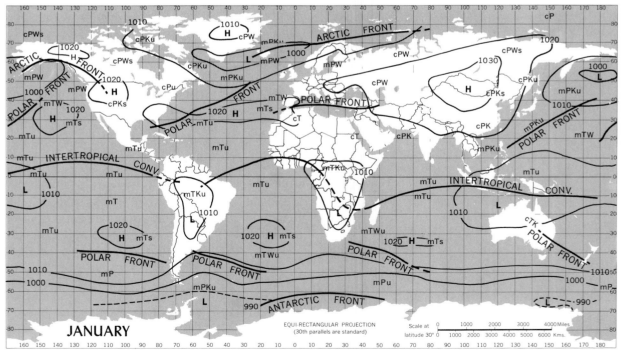

JANUARY

EQUI-RECTANGULAR PROJECTION
(30th parallels are standard)

Scale at latitude 30°

- - - - - Front
— mb — Isobar

AIR MASS SYMBOLS: T = Tropical P = Polar m = Maritime c = Continental
W = Warmer than, K = Colder than, underlying surface. s = Stable aloft u = Unstable aloft

WORLD PATTERN OF FRONTS AND AIR MASSES

Figures 5.3 and 5.4 offer a generalized picture of the distribution of the principal air masses and the average locations of the major fronts and zones of convergence at the two extreme seasons. (Compare them with Figs. 3.30 and 3.31 showing resultant surface winds, and review the discussion of lines of convergence under the general topic of surface winds in Chapter 3.) At best, of course, these seasonal charts of air masses are a bird's-eye view; only the most conspicuous elements are represented. In addition, they can show only the mean positions of frontal zones, which in reality are constantly shifting. At any particular time, both frontal zones and air masses may occupy positions well removed from those given on the charts. The air-mass symbols and their positioning on the maps represent mean conditions only.

Generalized mean frontal zones

Chapter 3's discussion of the planetary winds said that because of the nature of the general circulation there are bound to be certain extended zones of convergence in wind direction (Fig. 5.5; also Fig. 3.25). And when the converging circulations have distinctly unlike temperatures, and hence densities, a more or less permanent, though fluctuating, frontal zone must develop. For the world's weather, the

Figure 5.4 Air masses and fronts in July. (*After Willett.*)

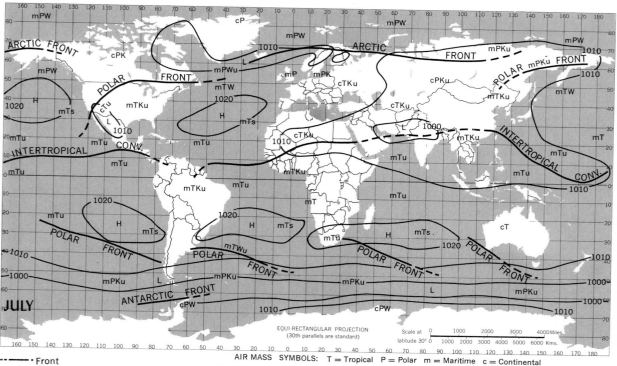

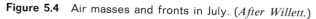

------- Front
—mb— Isobar

AIR MASS SYMBOLS: T = Tropical   P = Polar   m = Maritime   c = Continental
W = Warmer than,   K = Colder than,   underlying surface.   s = Stable aloft   u = Unstable aloft

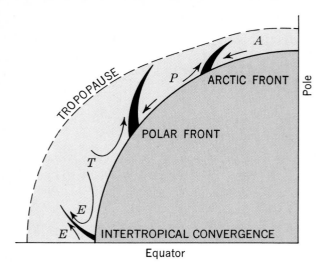

Figure 5.5 Diagrammatic cross section of the atmosphere showing principal frontal zones and air masses in the Northern Hemisphere. (*From Sverre Petterssen, Introduction to Meteorology, McGraw-Hill Book Company, New York.*)

most important of these extended frontal zones of wind and air-mass convergence is the polar front of middle latitudes, which marks the zone of conflict between polar and tropical air masses, either continental or maritime (Fig. 5.6). As would be expected, temperature gradients are steep across the polar front, and it is along certain parts of this convergence zone that numerous cyclonic storms develop. The mean latitudinal position of the polar front shifts periodically following the seasonal course of the sun—poleward in summer and equatorward in winter. But since it also has wide short-time fluctuations of a nonperiodic character, any mean position on the maps is only an average of a feature that is extremely fickle in location. Along certain parts of the polar front, cold air surges equatorward; in other parts, warm air pushes poleward. The total effect is one of alternating tongues of cold and warm air, giving the front a sinuous or wavelike shape. Usually it is not continuous around the earth. At times the polar front and its polar air deeply invade the tropics.

Along the high-latitude Arctic and Antarctic fronts, the converging circulations are both of polar (or arctic) origin. In some sectors polar continental and polar maritime air masses are in conflict; in others it is two continental or two maritime air masses. Normally the temperature contrasts are weaker than along the polar front, and cyclogenesis is also.

As mentioned earlier, zones of wind convergence within the tropics, like the ITC, are rarely density fronts with upgliding air, since the opposing air masses are too similar in temperature. In certain localities the polar front may occasionally break through into the poleward margins of the tropics—as it does, for example, in Caribbean America, northeastern Southeast Asia, and perhaps tropical West Africa—but only rarely do these invasions reach equatorial latitudes.

General properties of the main air masses

cP **air.** Polar continental air originates over the higher latitudes of Eurasia and North America,

Figure 5.6 Hemispheric view of the polar front.

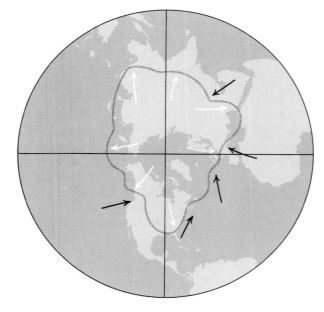

Greenland, and the Arctic Basin in the Northern Hemisphere, and over Antarctica in the Southern Hemisphere.

In *winter* the snow-covered source regions grow bitterly cold because of prolonged earth radiation and because they are remote from warm oceans. The circulation is anticyclonic and divergent, with light air movement. Hence winter *cP* air, a consequence of both intense surface cooling and outpourings of polar air aloft, consists of a layer of extremely cold, dry air reaching at times up to 3,000 m (10,000 ft). There is a persistent strong temperature inversion, with the lowest temperatures at or near the surface. Weather is predominantly clear and frigid; clouds and precipitation are meager. The air-mass symbol at the heart of the source region is *cPWs*, indicating that the air is warmer (*W*) than the surface and therefore is being chilled by it, while anticyclonic subsidence produces above-surface stability (*s*).

As winter *cP* surges outward from its source region, it carries its cold and dryness into the areas which it invades. Simultaneously the *cP* is itself modified—a process that is likely to be slow as long as the air's route is over a cold snow surface. But as soon as the *cP* moves out over extensive warmer surfaces of open water, it is rapidly heated, humidified, and made more unstable. Then it is on its way to being converted into *mP* air.

If the outward-moving *cP* encounters highlands, the resulting turbulence tends to destroy the surface inversion and so causes the low-level temperatures to be less cold. Along the seaward eastern margins of Asia and North America, the modified polar air may warrant the air-mass symbol *cPK*—or even *cPKu*, if the circulation has become cyclonic (Fig. 5.3). Conversion from *cP* to *mP* goes on with greatest speed over the western parts of middle-latitude oceans lying leeward of the great Northern Hemisphere continents, where both heat and moisture are rapidly acquired.

In *summer* the principal northern source regions of *cP* are free of snow, the land surfaces are much warmer, and anticyclonic circulation is less prevalent. Consequently *cP* in July at its continental sources has moderate surface temperatures and so is less stable than in winter. Bright, pleasant weather is characteristic. *cPK*, and even *cPKu*, are the typical air-mass symbols in the vicinity of the source regions. But air from other sources frequently invades the summer *cP* source regions, so that fronts develop, with associated cloud and precipitation. Obviously *cP* source regions are not as ideal for air-mass development in summer as in winter. Thus summer *cP* is much less homogeneous and clearly identifiable than its winter counterpart. Nor does it invade regions surrounding the source region to the same degree.

mP **air.**   Polar maritime air acquires its distinguishing properties over the great oceans of the higher middle latitudes. These are not regions where air can lie stagnant for long, so that *winter mP*, in the Northern Hemisphere at least, is characteristically polar-arctic air which has had a sea route of variable length. During its fairly prolonged period of contact with a *relatively* warm sea surface, the air has been heated and humidified in its lower layers and converted into a milder, relatively moist and, many times, unstable air mass. On the January map (Fig. 5.3) its usual symbol is *mPKu*, the *K* indicating that it often is cold *cP* air in the process of being modified by additions of heat and humidity from below. Its *u* character is acquired because it becomes involved in transient or semipermanent cyclonic circulations over the northern Pacific and Atlantic Oceans. These circulations produce lifting and consequently instability. Toward the eastern side of the oceans, the *K* may be replaced by *W* as a result of the air's long southerly sea trajectory, while the *u* disappears because the air becomes more neutral as regards stability aloft. Weather accompanying *mP* air is variable. Fresh *mPKu* air on the rear of a cyclonic storm is unstable both at the surface and aloft. It is full of cumulus clouds, which often bring rain showers over the sea and over onshore lands as well. But warmer *mPW* air arriving from the southwest may be warmer than the sea or the land. In this case it is stable at low levels, so that in winter it provides only stratus cloud and drizzle.

Summer mP is apt to be more stable at low levels than the same air in winter. If it was originally warm land air, subsequently chilled over a cooler sea, its air-mass symbol is likely to be *mPW*. And since the circulation over the ocean is only weakly cyclonic, or even anticyclonic, the air aloft is usually neutral or actually stable. Such air often provides relatively clear weather (see Fig. 5.4).

cT **air.** Tropical continental air, formed over the tropical and subtropical deserts, does not spread widely beyond its source regions and hence is fairly unimportant except in those locations. In both winter and summer the air is very dry, and in summer it is exceedingly hot. Subsidence and stability aloft are characteristic of a majority of the source regions.

mT **air.** Tropical maritime air, which includes equatorial air, is the world's most extensively developed air mass. It is also the tropical air mass which most frequently and deeply invades the middle latitudes. It has source regions both in the great tropical oceans and in the extensive parts of tropical continents where lush forests yield a bountiful supply of moisture for the atmosphere. Typically, *mT* surface air is humid and fairly hot (though not as hot as summer *cT*), and in the absence of wind it produces a sultry and oppressive environment.

But while *mT* air is characterized by uniformly high temperatures and humidity (so much so that convergence rarely results in density fronts within the low latitudes), the vertical structure often varies so widely that *mT* air masses may show great contrasts in weather and in amounts of precipitation. Two principal contrasting types are (1) tropical maritime in which subsidence and divergence prevail and hence produce stability aloft (*mTs*); (2) tropical maritime in which convergence and upward movement prevail, with resulting instability aloft (*mTu*). In addition there is tropical maritime air which is neither markedly stable (*s*) nor unstable (*u*) aloft, but intermediate or neutral in stability. This neutral condition (*mT*) is very common.

Source regions of stable *mTs* air are found most often in the eastern and central parts of the sub-

tropical anticyclones over oceans, and in the poleward and eastern parts of their tropical circulation (trade winds). Here, above-surface subsidence is marked and the inversion level low, so that the air is very stable and rainfall is meager. Such *mTs* air is dry and warm aloft.

Source regions of convergent unstable tropical maritime air (*mTu*) are characteristic of equatorial latitudes over both oceans and continents (Figs. 5.3 and 5.4). In some classification systems this air is designated as equatorial (*E*). Here in the ITC between the trades, the prevailing upward movement of air has two consequences: (1) the moisture evaporated at the surface is carried to great heights, so that a deep layer is charged with humidity, and (2) the inversion is dissipated, and general instability exists aloft. Such air has a high potential for producing abundant cumulonimbus clouds and heavy showery rainfall. In addition to the equatorial latitudes, *mTu* air may develop at the western extremities of the oceanic anticyclones and their trades.

Neutral *mT* air, as its name suggests, forms in regions where neither subsidence nor convergence is marked. It is common in parts of the trades where *mTs* air has had a long journey over warm waters, traveling equatorward and westward in the trade-wind circulation. Hence *mT* air occurs most often along the equatorward and westward flanks of the oceanic subtropical anticyclones. As a consequence of its long trajectory over a warm ocean, the lower air is thoroughly warmed and moistened, and turbulence and convection have carried the heat and humidity upward to considerable heights. In addition, as the air has progressed farther from the poleward and eastern parts of the oceanic anticyclone, subsidence has abated, and the temperature inversion has weakened and lifted. Thus clouds extend up to greater heights, and resulting rainfall is more abundant.

From this description of stable, unstable, and neutral *mT* air, it follows that *mT* air reaching western Europe (east side of ocean) is likely to be more stable than that entering eastern North America and eastern Asia (west side of ocean). *mT* air in Atlantic Europe occurs either in the warm sectors of passing cyclonic storms or as a broad flow on the north and west side of a fairly stationary anticyclone. It is not only fairly stable aloft, but is undergoing low-level stabilization through cooling at the surface. As a result its symbol may well be *mTs*. Stratus clouds and sea fog are common in such air.

The *mT* air affecting the West Coast of the United States is quite similar to that in Atlantic Europe. But the *mT* air which enters the eastern United States and eastern Asia in summer is warmer, moister, and less stable (i.e., either unstable or neutral), for its course has been over warmer waters. The warm land surface then further increases surface instability (*mTK*). In winter, however, the colder land causes the air to take on *mTW* characteristics, so that it becomes stabilized in its lower layers and is not inclined to ascend convectively. But if it is forced to rise along fronts or over highlands, widespread stratus clouds may form and sometimes may yield considerable rain or snow.

NORTH AMERICAN AIR MASSES

Winter

cP **air masses.** North American *cPWs* air in winter originates over the snow and ice-covered areas of interior Canada north of about latitude 50 or 55°, as well as over Alaska and the frozen Arctic seas (Figs. 5.7 and 5.8). It enters the United States from Canada, usually between the Rocky Moun-

Figure 5.7 North American air masses and their source regions.

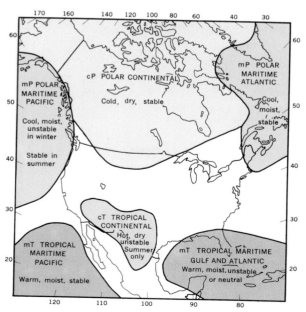

tains and the Great Lakes. On the weather map it commonly takes the form of a rapidly moving polar-outbreak anticyclone. Because the continent has no major highlands between the arctic plains and the Gulf of Mexico, it is relatively easy for these outbreaks of polar air to move rapidly southward, carrying abnormally cold weather to the subtropical latitudes of the east central United States (Fig. 5.9). The first freeze in fall, the last in spring, and the cold waves over most of the country all are associated with *cP* air. The openness of the continent also makes central and eastern North America an ideal location for the clash of polar and tropical air masses. In winter the middle latitudes of North America are the world's most active continental region of cyclogenesis, an area whose winter storminess is proverbial. (See also Fig. 6.15.)

Modified cP air in the southern and eastern United States. As winter *cP* moves southward from its source, it is modified by the addition of heat and moisture at the surface and by mechanical turbulence. Surface heating becomes more marked after the polar air leaves the snow surface, whose southern boundary, on the average, is about at the latitude of central Illinois. Continuing subsidence aloft serves to maintain the upper-level stability, so that *s* persists in the air-mass symbol. Turbulence acts to weaken the surface inversion and raise the surface temperature. As the wind dies down, however, a nighttime surface inversion is reestablished, and ex-

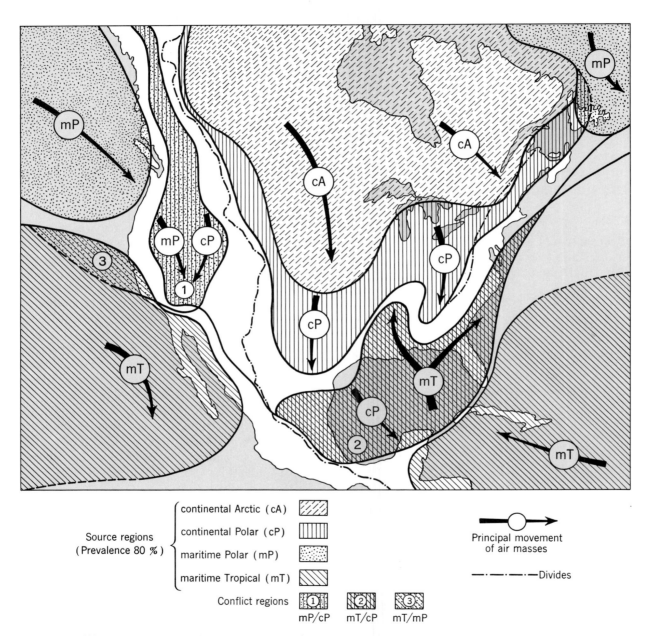

Source regions
(Prevalence 80 %)
{
continental Arctic (cA)
continental Polar (cP)
maritime Polar (mP)
maritime Tropical (mT)
}

Conflict regions

① mP/cP   ② mT/cP   ③ mT/mP

→○→ Principal movement
of air masses

—·—·— Divides

Figure 5.8 Winter air masses in North America. (*After Brunnschweiler.*)

Classification of North American air masses

Classification by local source regions			General classification after Bergeron (international)	Local air-mass names
Source by		Local source regions		
Latitude	Nature	Local source regions	Season of frequent occurrence		
Polar	Continental	Alaska, Canada, and the Arctic	Entire year	cP or cPW, winter cPK, summer	Pc, polar continental
		Modified in southern and central United States	Entire year	cPk	
	Maritime	North Pacific Ocean	Entire year	mPK, winter mP or mPK, summer	Pp, polar Pacific
		Modified in western and central United States	Entire year	cPW, winter cPK, summer	
		Colder portions of the North Atlantic Ocean	Entire year	mPK, winter mPW, spring and summer	Pa, Polar Atlantic
		Modified over warmer portions of the North Atlantic Ocean	Spring and summer	mPK	
Tropical	Continental	Southwestern United States and northern Mexico	Warmer half of year	cTK	Tc, tropical continental
	Maritime	Gulf of Mexico and Caribbean Sea	Entire year	mTW, winter mTW or mTK, summer	Tg, tropical Gulf
		Modified in the United States or over the North Atlantic Ocean	Entire year	mTW	
		Sargasso Sea (Middle Atlantic)	Entire year	mTW, winter mTW or mTK, summer	Ta, tropical Atlantic
		Modified in the United States or over the North Atlantic Ocean	Entire year	mTW	
		Middle North Pacific Ocean	Entire year	mTW, winter mTW or mTK, summer	Tp, tropical Pacific
		Modified in the United States or over North Pacific Ocean	Entire year	mTW	

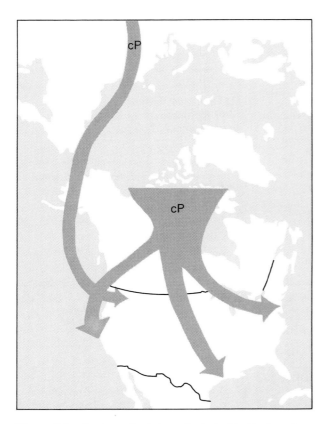

Figure 5.9 Routes of winter *cP* air in North America.

ceptionally low minimum temperatures, associated with clear skies, are the rule. Surface heating and humidification are by no means sufficient to overcome the stability (*s*) aloft, so that the air mass remains stable and largely cloudless (*cPKs*) even over the southern United States.

After crossing the Gulf of Mexico, a polar outbreak may arrive in Central America and Mexico in the form of a strong "norther." By then it has become either cool *mTKu* air accompanied by heavy convective precipitation, or anticyclonic *mTKs* air with drizzle or no rain at all.

If the *cP* air takes a more eastward course, somewhat different modifications occur (Fig. 5.10). One is that the stability aloft weakens as a result of a waning anticyclonic circulation within the polar cur-

rent eastward from the Great Lakes. Consequently upper-level subsidence is diminished, and the *s* element gradually fades from the air-mass symbol. Another modification takes place over the open water of the Great Lakes in winter, where the lower air is humidified and warmed, producing some surface instability. Heavy snow showers are characteristic of this *cPK* air on the lee side of any of the Great Lakes. Still another modification is created by the rough terrain of the Appalachian country, which causes a thorough mixing of the lower few thousand feet of air and so eliminates the surface temperature inversion and raises surface temperatures. On the western slopes of the eastern highlands, the forced ascent of modified *cP* air from over the Great Lakes produces general cloudiness and heavy local snowfall as long as there is a strong flow of cold air. Heat of condensation then acts to warm the air mass, and as it descends the eastern slopes of the highlands, it is further warmed as a result of subsidence (see the data for Boston in the table on page 179). Thus the cloud cover disappears and the snowfall ceases.

Modified cP air west of the Rockies. Because of the Rocky Mountain barrier and the general west-to-east atmospheric circulation in middle latitudes, *cP* air less frequently invades the intermontane region lying between the Rockies and the Pacific Coast mountains. And it is only occasionally that these air masses reach the Pacific Coast lowlands. When *cP* air does arrive at the intermontane plateaus and basins, it has been considerably warmed from turbulent mixing in the lower layers, which has brought

Figure 5.10 Modifications induced in North American winter *cP* air by the Great Lakes and the Appalachian Highlands.

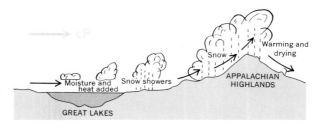

Winter characteristics of some American air masses

Air mass	Station	Weather element	Elevation above sea level				
			Surface	1 km	2 km	3 km	4 km
cP	Ellendale, N.D.	Temp., °F Sp. humid., g/kg Rel. humid., %	−15 0.32 82	−13 0.35 80	−4 0.60 75	−7 0.50 63	−13 0.45 71
	Boston	Temp., °F Sp. humid., g/kg Rel. humid., %	21 0.9 43	6 0.6 50	0 0.5 50	−9 0.3 44	−20 0.2 48
mP	Seattle	Temp., °F Sp. humid., g/kg Rel. humid., %	46 4.4 66	32 2.7 64	18 1.5 64	7 0.8 52	−2 0.4 35
	Ellendale, N.D.	Temp., °F Sp. humid., g/kg Rel. humid., %	30 3.0 83	45 3.0 43	34 2.2 44	19 1.5 48	7 1.1 60
mT	Miami	Temp., °F Sp. humid., g/kg Rel. humid., %	77 16.3 82	68 13.3 82	55 9.8 83	46 6.2 66	37 5.2 67
	Boston	Temp., °F Sp. humid., g/kg Rel. humid., %	57 8.8 88	57 6.5 59	48 6.2 70	36 4.6 75	25 2.9 65
mT	San Diego	Temp., °F Sp. humid., g/kg Rel. humid., %	68 11.9 86	59 9.8 81	50 6.8 70	43 4.0 51	34 2.1 33

Source: After Willett.

down warmer air from aloft (Fig. 5.11). If cP air continues on to the Pacific Coast in Washington and Oregon, occasionally it produces subfreezing temperatures and snow flurries. In rare cases this weather type may extend southward to coastal California. In years when cP air is more prevalent in the region west of the Rockies, the Pacific Coast may have relatively severe winters as a result of transmontane spillovers of the air mass. Air from cold northern Asia which has crossed the ice-covered Arctic Ocean, Alaska, and the Gulf of Alaska also occasionally brings very cold weather to the Pacific Northwest (Fig. 5.9). Asian air is sufficiently warmed and humidified over the Gulf of Alaska to yield

Figure 5.11 In North America the western mountains tend to obstruct and modify winter cP air masses moving westward toward the Pacific Coast. (*After Phil Church in Weatherwise, April, 1962.*)

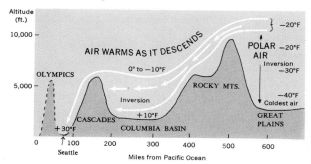

showers and squalls when it is forced to ascend the coastal highlands.

mP-**Pacific air masses.**   It is *mP*-Pacific air which largely controls the Pacific Coast weather of North America in winter (Fig. 5.8). The source region of these air masses is the northern part of the North Pacific Ocean, an area which in winter is dominated by the cyclonic circulations of the Aleutian Low and its traveling disturbances. The eastern part of the Pacific in these latitudes is a region of relatively warm surface waters. Most of the air that enters the *mP*-Pacific source region originates in cold continental areas to the north and west and is drawn into the source region by the strong northwesterly flow converging on the Aleutian cyclone. Warmth and humidity from the ocean surface are added to the stable *cPWs* air, and it becomes involved in the prevailing cyclonic circulation, which acts to lift the air mass. Thus it is gradually transformed into unstable *mPKu*.

As it arrives on the Pacific Coast of North America, *mPKu* air is relatively mild, with temperatures well above the freezing point. It has a steep lapse rate and is accompanied by low clouds with frequent showers. Convective instability extends up to about 10,000 ft. As this humid, unstable air advances inland, its shower activity is greatly increased when it is forced to ascend over the cooler land air along the coast and also over the coastal ranges. Rain in the lowlands and heavy snowfall in the mountains are characteristic of the winter season.

Modified mP-Pacific air east of the Pacific Ranges.   As the maritime air continues inland over successive mountain ranges and plateaus of the intermontane region, the cold land lowers the air's surface temperature, while condensation on the highlands causes the low levels of the air mass to become drier. This *mP* air becomes involved in an anticyclonic circulation over the intermontane plateau region and also to the east of the Rocky Mountains, so that its stability at upper levels increases and the *u* gradually changes to *s*. Thus the mild, moist, unstable air (*mPKu*) of the West Coast is slowly transformed into a cold, dry, stable continental air mass

(*cPWs*) east of the Rockies in the central United States. But is not nearly so cold or so dry as fresh *cP* air from arctic Canada.

The weather associated with this greatly modified *mP* air over the central part of the country is some of the finest that winter has to offer—clear skies, light winds, and moderate temperatures. But along frontal zones, where it makes contact with *mT* air from the Gulf, or with fresh *cP* air from Canada, widespread bad weather conditions may result.

mP-**Atlantic air masses.**   This polar maritime air originates over the abnormally cold North Atlantic waters (35 to 40° in winter, and 50 to 60° in summer) off Newfoundland, Labrador, and Greenland (Fig. 5.7). It was originally *cP* air that moved eastward off the continent and subsequently was modified over the cold waters, which acted as a secondary source region. Owing to the prevailing west-to-east air movement in middle latitudes, the North Atlantic is not an important source region for air masses affecting North America. This is especially true in winter. Normally, *mP*-Atlantic air moves eastward across the ocean toward Europe. Under certain synoptic weather conditions the westerly flow is temporarily reversed, and Atlantic air may enter the continent shallowly. Normally in such cases it keeps east of the Appalachians and north of Cape Hatteras.

In winter, *mP*-Atlantic air (*mPKs*) is dry and stable aloft as a result of subsidence, while the surface layer is moderately unstable, moist, and chilly. Temperatures are raw rather than bitterly cold. Compared with *mP*-Pacific air at Seattle, the low-level *mP*-Atlantic air is colder, drier, and more stable, because of both the colder waters in the North Atlantic and the shorter trajectory of *mP*-Atlantic air over the water (Fig. 5.8).

The weather associated with an invasion of *mP*-Atlantic air is distinctly unpleasant. Surface temperatures near freezing or somewhat below, high relative humidity, strong northeast winds, low, thick clouds, and variable amounts of precipitation, depending on the synoptic pattern—such is a typical "northeaster" of the New England coast.

mT-Gulf-and-Atlantic air masses. These air masses greatly affect the weather and climate of much of Anglo-America east of the Rockies. Their source regions are the exceptionally warm ocean surfaces of the Gulf of Mexico, the Caribbean Sea, and the subtropical western Atlantic (Fig. 5.7). Consequently mT-Gulf air is characteristically warm and humid, with specific humidities and temperatures higher than those of any other American air mass in winter. Its properties aloft—absence of pronounced subsidence, and a neutral or unstable vertical structure (mTK[u])—reflect the source region's dominance by the weak western limb of the North Atlantic subtropical anticyclone (Fig. 5.8).

During the colder months, polar air prevails in the central eastern United States well to the south. As a result the polar front separating mT-Gulf and cP air lies close to the northern and western portions of the mT-Gulf source region. Such a condition does not favor easy entrance of mT air into eastern and central North America in winter.

In the American South, mT air masses are associated with mild, humid winter days that have a good deal of cloudiness of the stratus type, especially at night and in early morning. As mT air moves northward, the ground strata are cooled by contact with a progressively colder land, and the surface stability of the air mass is increased (mTW). Convective showers consequently are uncommon in winter. But if the northward-flowing mT air is drawn into a cyclonic circulation or encounters a relief barrier, its forced ascent may cause heavy and widespread precipitation. Much of the winter rain and snow over central and eastern North America are the result of mT-Gulf air being lifted along fronts in cyclonic circulations. If nothing forces it to rise, however, northward-moving mT air frequently develops a dense advection fog because of cooling at the surface.

mT-Pacific air masses. Chiefly the extreme southwestern United States and northwestern Mexico are affected by these air masses. They originate in the subtropical eastern Pacific Ocean, in almost identical latitudes to those in which mT-Gulf air develops (Fig. 5.7). But although the latitudes of their source regions are similar, their other characteristics are not. The eastern subtropical Pacific westward from Mexico and California is dominated by the circulation around the *eastern* end of an oceanic subtropical high, where strong upper-level subsidence produces marked stability (mTs). Therefore mTs air which arrives along the North American Pacific Coast is considerably cooler, drier, and more stable than the mT-Gulf air which affects the southeastern United States.

As mT-Pacific air moves northward along the coast of California, occasionally reaching the latitudes of Oregon and Washington, it is cooled at the surface by the colder water and is further stabilized, acquiring W characteristics. Fog is often the result. When, as frequently happens, this mT air is drawn into a frontal or cyclonic convergence, its s property changes to u, so that moderate precipitation may be caused by lifting along fronts or topographic obstacles. Ordinarily mT-Pacific is not as good a rain-bringer as mT-Gulf air, which is potentially more unstable. As an mT-Pacific air mass moves inland aloft over the western states, it continues to yield moderate precipitation as it ascends over colder air or over mountains. East of the Rockies, it can rarely be identified.

Summer

cP air masses. In summer this polar continental air mass has much the same geographical source region as in winter, but its properties are different (Fig. 5.12). In the absence of snow, the land surface is moderately warmed by solar energy during the long summer days. Air entering the continent from cooler seas is warmed at the base and given a K character. Evaporation from numerous lakes, as well as from vegetation and moist soil, provides a fair amount of moisture. A weak anticyclonic circulation causes summer cP aloft to be mildly stable (s), but less so than in winter. Surface temperatures are moderate. Strong solar heating by day and rapid radiational cooling at night create large diurnal temperature variations. At its source cP air is typically cloudless,

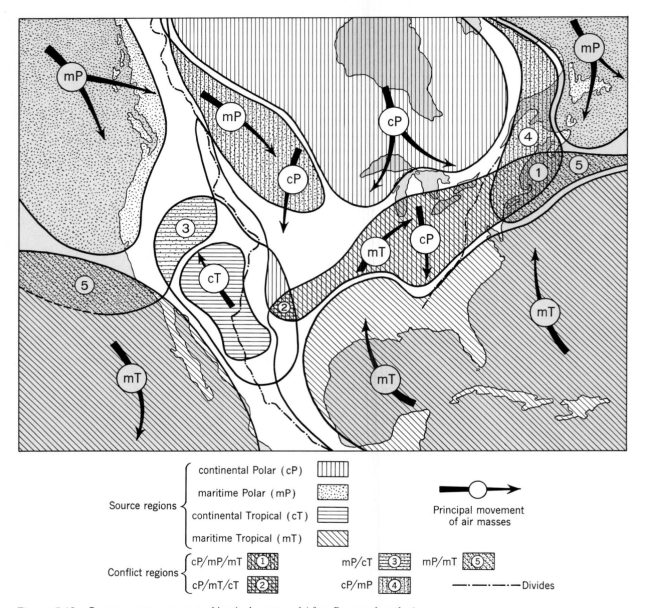

Figure 5.12   Summer air masses in North America. (*After Brunnschweiler.*)

for in spite of its daytime surface instability and the resulting convective turbulence, the condensation level is so high that a few scattered high cumulus clouds are usually the most that can develop.

In the northern half of the central and eastern United States, summer heat waves are normally broken by a southward advance of fresh *cPKs* air which, for a day or two, substitutes cool, clear weather for the sultry or desiccating heat. Southward-moving outbreaks of summer cP air in the United States are much less frequent than in winter, and they are also weaker and slower. As a rule they move eastward more than southward, so that the cool air rarely reaches the southern states. As cP air advances eastward and southward from the Canadian border, it is slowly warmed at all levels, and some moisture is added at the base. Its general upper-level stability persists, however, except when it enters a cyclonic circulation, which causes instability to develop aloft and may lead to scattered light thundershowers in the afternoon (see Figs. 3.13, 3.14, and 3.15).

mP-Pacific air masses. In summer this polar maritime air has the same geographical source region as in winter, but the properties of the summer and winter mP-Pacific air masses are somewhat different (Fig. 5.12). During the warm season, the waters of the North Pacific are relatively cooler than the sur-

Summer characteristics of some North American air masses

Air mass	Station	Weather element	Elevation above sea level				
			Surface	1 km	2 km	3 km	4 km
cP	Ellendale, N.D.	Temp., °F	66	61	50	39	27
		Sp. humid., g/kg	6.3	5.6	3.9	3.1	2.9
		Rel. humid., %	42	45	43	44	57
	Royal Center, Ind.	Temp., °F	63	55	43	36	28
		Sp. hum., g/kg	8.3	5.8	4.5	2.6	1.4
		Rel. humid., %	68	57	64	43	27
	Pensacola, Fla.	Temp., °F	73	68	54	45	
		Sp. humid., g/kg	13.4	9.8	7.2	5.0	
		Rel. humid., %	79	65	67	57	
mP	Seattle	Temp., °F	63	48	41	34	28
		Sp. humid., g/kg	7.1	6.3	3.9	2.3	1.7
		Rel. humid., %	62	91	60	42	33
	San Diego	Temp., °F	64	72	70	57	43
		Sp. humid., g/kg	10.3	7.7	3.5	3.6	3.4
		Rel. humid., %	79	40	20	27	36
mT	Miami	Temp., °F	75	68	59	48	41
		Sp. humid., g/kg	17.3	14.9	9.3	6.3	4.3
		Rel. humid., %	93	88	74	58	48
	Royal Center, Ind.	Temp., °F	84	77	65	51	
		Sp. humid., g/kg	15.9	13.9	11.5	8.6	
		Rel. humid., %	61				
	Ellendale, N.D.	Temp., °F	84	81	72	55	
		Sp. humid., g/kg	16.5	13.3	8.7	5.7	
		Rel. humid., %	66	54	42	43	
cT	El Paso, Tex.	Temp., °F	75	81	75	64	
		Sp. humid., g/kg	11.0	9.7	9.9	7.6	
		Rel. humid., %	52	37	43	43	

Source: After Willett.

rounding lands, so that air entering the source region may be chilled at the base and stabilized (*W*). And since the summer circulation over the eastern North Pacific is less cyclonic than in winter, the air aloft is also less unstable.

During summer the strengthened Pacific anticyclone shifts poleward, so that the northeastern axis of the high along the Pacific Coast is positioned as far north as 45 to 50°N. A southward flow of cool, stable air along almost the entire United States Pacific Coast is the result. Originally, a part of this circulation is *mP*, but gradually cool, stable *mT* comes to dominate the northerly flow around the eastern end of the subtropical high.

After *mP* air has reached the interior of the continent, it cannot be distinguished from *cP*. Therefore the designation *cP* is satisfactory for both *mP* and *cP* air in summer in the North American interior.

mP-Atlantic air masses. These polar maritime air masses occur most frequently during late spring and early summer. At that time the chilly ocean surface from Cape Cod to Newfoundland becomes a genuine source region for cool air (Fig. 5.12). As *cP* air moves eastward from the continent over the cold waters, it is chilled and stabilized (*W*) at the base, while the predominantly anticyclonic circulation produces upper-level stability (*s*). The circulation on the southerly side of this high carries chilly *mP*-Atlantic air to New England, and also occasionally to the coastal areas east of the Appalachians as far south as Cape Hatteras.

Along the North Atlantic Coast in summer, *mP*-Atlantic air brings weather with low temperatures, clear skies, and good visibility. Because of its dryness, surface fog is uncommon even over coastal waters. Thin and broken clouds are relatively frequent at the top of the turbulence layer, but precipitation never comes from these clouds in summer.

mT-Gulf-and-Atlantic air masses. During summer these air masses largely control the weather of much of the United States east of the Rockies. Because of the prevailing sea-to-land pressure gradients over the eastern part of the country in summer, *mT* air is able to enter much deeper into the continent than in winter (Fig. 5.12). A strengthening of the Bermuda High over the western Atlantic Ocean, together with the tendency toward a thermally maintained low over the warm continent, produces a fairly persistent indraft that flows northward even into Canada. Not only do the *mT*-Gulf air masses cover much wider areas in summer than in winter, but they are present a much greater share of the time. They are responsible for the oppressive, humid summer heat that characterizes so much of the central and eastern United States. Outbreaks of polar air are comparatively weak and infrequent at this season, so that the belt of maximum frontal activity, which in winter normally lies somewhere over the Gulf States, is shifted northward to the Great Lakes or beyond.

As it leaves its source region, summer *mT*-Gulf air is relatively similar to winter *mT*, though it is somewhat warmer and more humid. Although the general circulation in which the *mT*-Gulf air is involved is anticyclonic, this is the western end of the cell, where subsidence is much less marked. Therefore upper-level instability (*mTu*) is not uncommon. And as summer *mT* air moves inland over the warm continent, its surface temperature rises (*mTKu*), with a consequent increase in surface instability. (During winter, in contrast, a cold continent has the reverse effect.) Because of the high relative humidity of this air, only a small amount of vertical lifting is required in order to start active convection and shower activity. The marked potential instability, plus the effect of daytime heating of the surface air, favors the development of cumulus clouds which produce showers and thunderstorms—features characteristic of summer *mT* air in the Gulf states. Much the greater share of this precipitation appears to be of an organized variety associated with passing synoptic disturbances.

Even in the arid southwestern United States, warm-season thunderstorms appear to be related to westward thrusts of *mT*-Gulf air aloft. These summer showers are frequently so intense that the name "cloudburst" has been applied to them. They are usually localized with respect to mountain masses. On rare occasions *mT*-Gulf air even reaches southern

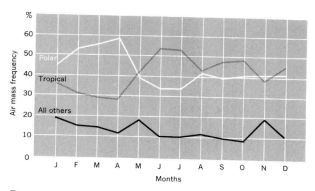

Figure 5.13 Relative frequencies of polar, tropical, and other air masses in central Pennsylvania. (*From H. Landsberg in Physical Climatology.*)

California, bringing that region its rare summer showers. Severe local thunderstorms are likewise characteristic of *mT* air as it rises over the Appalachian Mountains. Such storms may continue for 100 miles or more eastward over the coastal plain (see Figs. 5.13, 5.14, and 5.15).

mT-**Pacific air masses.** It is not easy to distinguish *mT* and *mP* along the Pacific Coast of the United States in summer. The circulation around the eastern end of the poleward-displaced anticyclone makes for a steady transport of cool, stable air southward along the entire American coast to southernmost California and beyond (Fig. 5.12). The result is unusually low summer temperatures along the whole Pacific littoral. Abnormally cold upwelled water along the coast of California tends to cool the northerly airflow still

further. As a consequence the maritime air at San Diego has surface temperatures similar to those at Seattle. In the north, a good part of this cool airflow probably should be classed as *mP*, but southward it gradually changes to cool *mT*. The *mT* air is stable both at the base because of the cool ocean surface, and aloft (with low inversions prevalent) because of anticyclonic subsidence. Thus lowlands throughout the entire Pacific Coast from 45 or 50°N to 20 or 25°N have a dry summer. Coastal fog is widespread. In both temperature and stability, summer *mT*-Pacific is very different from *mT*-Gulf.

cT **air masses.** Owing to the marked narrowing of the continent to the south and the prevalence of highlands there, North America has no extensive source region for tropical continental air. Only in northern Mexico and adjacent parts of the United States is there even a small source region (Fig. 5.12). For the most part *cT* air probably consists of *mT*-Pacific air that has stagnated over this secondary source region for several days, becoming intensely heated and desiccated in its lower strata. *cT* air is a feature of the warm season; it is absent in winter, for the source region has no distinctive characteristics then. Very high daytime temperatures, extremely low humidities, and an almost complete lack of precipitation characterize *cT* air in summer. Outside its source region *cT* air is found chiefly in the southern Great Plains, where it is associated with severe drought. It rarely moves east of the Mississippi River or north of Nebraska.

Figure 5.14 Relative frequencies of maritime and continental air masses in central Pennsylvania. (*From H. Landsberg in Physical Climatology.*)

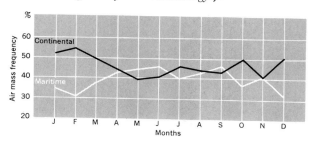

Figure 5.15 Percentage of days annually with air-mass changes in central Pennsylvania. (*From H. Landsberg in Physical Climatology.*)

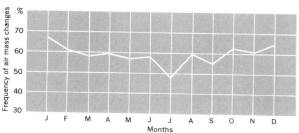

AIR MASSES OF EUROPE

In accordance with the general circulation patterns Europe, which forms the western part of the great Eurasian continent, has air masses that often resemble those of western North America in similar latitudes (Figs. 5.8 and 5.12). In many ways *mP*-Atlantic air in western Europe resembles *mP*-Pacific air along the Pacific margins of Anglo-America, while *mT*-Atlantic air in Portugal-Spain resembles *mT*-Pacific air in California. There are several significant differences, however. Mainly, the trend of highlands in Europe is approximately east-west, or parallel with the westerly circulation, while those of North America lie athwart the westerlies. In Europe, maritime air masses from the west therefore enter the continent more readily and more deeply, and retain their maritime properties much longer, than their counterparts in western Anglo-America. Because of the Pacific Coast highlands in North America, moderate *mP*- and *mT*-Pacific air is rapidly transformed into more severe and dry continental air.

It would seem as if these differences in terrain arrangement between the two continents would make it easier for *cP* air to reach the Baltic countries and Germany than to invade the Pacific Coast of Anglo-America. However, Europe thrusts far westward in its lower latitudes and withdraws eastward in its higher latitudes, a configuration which is opposite to Anglo-America's. This means that most polar-air invasions of western Europe are necessarily of maritime origin, a feature much less true of western North America east of the Pacific highlands.

It might be expected also that Europe's air masses would be strongly influenced by the large subtropical bodies of warm water along the continent's southern borders. But the high east-west mountain chains of Europe tend to confine the effects of these water bodies chiefly to the great southern peninsulas. The Gulf of Mexico, in contrast, has a marked influence on North American weather. Still, the Mediterranean Sea does play a great part in the generation and routes of cyclonic storms within its basin and hence in its weather as well.

cP **air masses.** Polar continental air masses in *winter* chiefly affect the weather of Russia and central Europe. It is only occasionally that they reach western Europe, for except under certain synoptic conditions, the west-to-east atmospheric circulation hinders a reverse movement of air from the source regions farther east. Moreover, since Europe narrows in a poleward direction, the source region for *cP* air is limited in extent. Most *cP* air affecting central and western Europe originates in Fenno-Scandia or western Russia. It has temperatures less severe than air from Siberia or subarctic Canada, so that it shows more resemblance to the modified *cP* air found in the central United States. A European invasion of Russian *cP* air usually develops in connection with a synoptic condition involving an anticyclone over northern Russia and a deep low over Scandinavia. Cold air then enters central-western Europe as a thrust of anticyclonic air from the east and north. Clear skies lead to relatively low night temperatures, although the minima decrease rapidly in intensity toward the coast as the *cP* air is swiftly modified.

cP air in *summer,* though it is quite different from that of winter, still has its source in northern Russia. Summer *cP* is fairly warm and humid at the surface and cool aloft, so that it is much less stable than winter *cP*. Since it has probably developed from *mP* air, its moisture content is higher than that of its counterpart in the United States. With daytime heating cumulus clouds and thundershowers may develop, especially if the circulation is cyclonic. European *cP* in summer is only slightly cooler than *cT* air in western Europe, but it is somewhat warmer than *mP*. Summer air masses of central and western Europe lack important temperature contrasts. Consequently summer weather in central and western

Europe is not characterized by the amount of temperature variability that it has in central and eastern North America, where air masses show greater contrasts. In Europe the diurnal range is large, however, with cool nights and warm days. Summer *cP* is uncommon in westernmost Europe.

mP **air masses.**   Polar maritime air, because of the westerly circulation, tends to dominate the weather of western and central Europe in all seasons. Since in *winter mP* is sea-modified *cP*, its lower layers have been warmed, humidified and made unstable, while instability aloft may have been induced through circulation in the North Atlantic Low. Instability showers are therefore common over the ocean and along the coasts of Europe. As *mP* progresses eastward, surface instability decreases owing to loss of heat to the underlying cold land surface.

Depending on the ocean route traveled by the air mass, *mP* in western Europe brings a variety of weather. If it comes from a northerly quarter with a relatively short trajectory over open water, it arrives in Europe as some of the earth's coldest maritime air, with a surface temperature around freezing. Winds are strong and gusty, the lapse rate steep, and humidity in the lower layers high, so that brief, heavy snow showers are common. If the *mP* air has had a longer trajectory, surface temperatures are likely to be 10 to 15° warmer, and the humidification reaches up to higher levels. In such air the lapse rate is not quite so steep, but rain and snow showers are still common. On occasion *mP* air may enter Europe from a southwesterly direction, having had a long trajectory into southerly latitudes. Its surface temperature may then be warmer than that of the winter ocean (*mPW*), so that it is stable near the surface. Stratus cloud and drizzle are usual in such air. Fronts between different *mP* air masses are common, and many cyclonic disturbances that affect western Europe have warm sectors of warm *mPW* air rather than of *mT*.

mP air in Europe in *summer* shows considerable variation in character, depending on where it originates within its extensive source region and how long it has traveled over the ocean. If it forms in the southern part of the source region and reaches Europe after a long trajectory over the water, it is generally stable. This air mass gives rise to cool fair weather and is a rain producer only when it undergoes a marked lift along fronts in well-developed cyclones. On the other hand, air that reaches Europe more directly from the northern part of the source region is mildly unstable and may show some convective activity. Heavier showers result when this air is lifted along frontal surfaces. Summer *mP* air is distinctly cooler than the land it invades, so that its general effect is to give western Europe a cool, marine summer climate.

cT **air masses.**   Tropical continental air in *winter* is not of great importance in Europe, at least not at the surface. Mild and dry in character, it originates over the Sahara in North Africa and over dry southwestern Asia. Only occasionally does it invade southern and central Europe at the surface. But frequently, after being shallowly humidified over the Mediterranean Sea, it is drawn into atmospheric disturbances where it is caused to rise over frontal surfaces, with resulting moderate precipitation. No doubt a part of the cool-season cyclonic precipitation in southern and central Europe is from *cT* air modified over the Mediterranean.

cT air masses are found over Europe mainly in *summer*. Since they originate in the Sahara, southern Russia, and Asia Minor, they are extremely dry and hot. Although lapse rates are steep, the unmodified air is too arid to produce much rainfall. Under the influence of low-pressure centers poleward of 40 or 45°N, *cT* air occasionally flows northward across the Mediterranean Sea in the form of a hot sirocco wind. These winds are very desiccating in areas of southern Europe where the southerly airstream has crossed only a narrow expanse of water, but they are more humid and sultry where the trajectory has been long. Such modified *cT* air masses provide a fair share of the moisture for the modest summer rainfall of parts of southern and central Europe.

Winter air masses in Europe

	Elevation, m	Temperature, °F	Specific humidity, g/kg
		1. cP air	
	——	1	1.0
	800	7	1.0
Moscow	1,650	3	1.0
	2,600	− 9	0.56
	3,700	−26	0.39
	4,950	−42	0.20
	800	12	1.4
	1,700	1	1.0
Munich	2,650	− 6	0.60
	3,750	−15	0.38
	5,050	−33	0.16
		2. cP (arctic) air, modified over the Norwegian Sea	
	——	36	3.7
	800	28	2.9
Berlin	1,700	16	2.0
	2,700	5	1.1
	3,850	− 9	0.56
	5,150	−26	2.26
		4. mT-Atlantic air	
	——	39	4.4
	850	34	3.5
Hamburg	1,750	23	2.2
	2,750	14	1.3
	3,950	0	0.8
	5,250	−17	0.5
		3. mP-Atlantic air	
	——	48	6.2
	850	41	5.4
Hamburg	1,800	36	3.7
	2,850	25	2.2
	4,000	10	1.4
	5,400	− 2	0.8

Source: After Petterssen.

mT air masses. Tropical maritime air masses in Europe strongly resemble mT-Pacific air along the West Coast of the United States. They are cooler, drier, and more stable than mT-Gulf air in the southeastern United States. mT air that arrives in Europe comes from the northern and eastern flanks of the Atlantic subtropical high-pressure cell, where subsidence is strong. Although relatively warm and moist compared with polar air, it is markedly stable aloft. A pronounced inversion is present at relatively low levels. After leaving its source in the subtropical anticyclone, this air moves over a colder surface as it travels progressively northward, so that surface stability likewise evolves (mTWs). Widespread sheets of stratus cloud develop underneath the inversion and penetrate well inland. Some drizzle may fall from such clouds, but owing to the pronounced stability aloft, rainfall from this air is modest even along fronts. mT air reaches western Europe chiefly in the form of warm sectors of cyclonic storms. As an element of the weather of that region, it is more conspicuous in the cooler seasons and relatively unimportant in summer.

In summary, European *winter* air masses cover a much smaller range of temperature than those of North America, a feature which reflects the greater marine influence in Europe and greater continentality in America. As a consequence fronts are less sharp and less active in Europe, and weather less changeable and extreme. Also, *mT* winter air plays a much less significant role in European weather. The *mT* that reaches Europe, being from the eastern side of an oceanic subtropical anticyclone, is definitely more stable than winter *mT*-Gulf air in North America.

In *summer,* it is Europe's cooler air masses that are most unstable and bring showery convective precipitation, while warmer *mT* is stable. In North America the opposite is true, for *mT*-Gulf air is rich in heat energy and moisture and is relatively unstable. Europe has no tropical air-mass conterpart to *mT*-Gulf. In general, Europe's air masses are characterized by greater uniformity in lapse rates, the main contrasts being in temperatures. As might be expected, the North American air masses show greater differences both in vertical structure and in temperatures.

Summer air masses in Europe

	Elevation, m	Temperature, °F	Specific humidity, g/kg
		1. cP air	
		63	9.6
	850	57	7.5
	1,850	46	5.7
Berlin	2,900	36	3.9
	4,050	21	2.3
	5,450	7	1.3
		2. mP air	
		59	8.5
	850	50	6.3
	1,800	39	4.7
Hamburg	2,850	28	3.1
	4,050	16	1.8
	5,400	1	1.0
		3. mT air	
		68	8.9
	900	57	7.5
	1,850	48	6.2
Hamburg	2,900	37	4.6
	4,100	25	3.3
	5,500	12	2.1
		4. cT air (source region in Europe)	
		68	9.5
	900	64	8.5
	1,850	52	6.5
Berlin	2,925	39	4.6
	4,150	27	2.7
	5,500	7	1.6

Source: After Petterssen.

AIR MASSES OF EAST AND SOUTH ASIA

Because of the direction of the general circulation, eastern Asia's air masses are more like those of eastern North America than those of western Europe and western North America (Figs. 5.8 and 5.12). The larger continent of Eurasia tends to create greater air-mass temperature extremes than exist in North America. But because of the east-west mountain chains of Asia, polar and tropical air masses do not come into conflict there as readily as in Anglo-America, which has no zonal terrain barriers in its central and eastern parts.

cP **air masses.**   Polar continental air in *winter* originates in the cold anticyclone centered over central eastern Siberia. At its source, it is bitterly cold and very dry, even more so than its North American counterpart. As *cP* surges seaward in the form of great thrusts of cold, dry air, spaced an average of about a week apart, at least two important types of modification occur. First, the air is forced to travel over extensive highlands before descending to the plains of China. Mechanical turbulence caused by the highlands dissipates the strong surface inversion and increases the temperature of the lower air, as well as raising its humidity somewhat. Second, before reaching Japan this continental air is forced to traverse several hundred miles of open sea. Here such large amounts of heat and moisture are added at the surface that instability results. Over the sea and along the windward western side of Japan, this modified *cP* air gives rise to heavy snow showers.

cP air reaching western Korea also has been modified in different degrees by longer or shorter sea journeys. The same is true for subtropical China south of the Yangtze. In this case the flow of air is from a northeasterly direction, for it originated in more northerly latitudes on the back side of a cold anticyclone centered east of the mainland and moving eastward. The arid winters of northern China, including Manchuria, are largely caused by the dominance of dry, cold land *cP* in these parts. Likewise sea-modified *cP* is at least partly responsible for the wetter winters of Japan, Korea, and subtropical southern China.

Along the sea margins of Asia, each fresh surge of *cP* air comes into conflict either with air of genuinely maritime origin or with sea-modified *cP*, giving rise to atmospheric disturbances and fronts. It is in such disturbances that most of the winter precipitation of Japan, Korea, and southern China is generated. If the outbreak of polar air is unusually strong, it may carry with it a large amount of fine dust picked up from the interior dry lands. This is the source of the extensive loess deposits of North China. *cP* air usually does not invade southern Asia deeply, for the high mountains and plateaus tend to isolate the south from the north. Only at the eastern and western extremities of the highlands—in North Vietnam in Southeast Asia and in northwest India-Pakistan—are there significant intrusions of modified *cP*. In India-Pakistan these intrusions aid in the generation of weak winter cyclones which bring modest amounts of cool-season precipitation, chiefly to the northern and western parts of the Indian subcontinent.

cP air masses in *summer* are relatively unimportant in eastern and southern Asia, where *mT*, the summer monsoon, dominates the weather. Usually the summer *cP* is felt only in northernmost China and to some degree in northern Japan. In both regions it makes fronts with the tropical air, giving rise to precipitation. Some of the *cP* that enters northern China, and certainly the *cP* in Japan, has been more or less modified over cool waters and hence is more humid in its lower layers than most summer *cP* in North America.

mP **air masses.**   Polar maritime air masses which have had a long trajectory over oceans play a minor role in the weather of eastern Asia. In *winter,* the strong westerly circulation in eastern Asia almost precludes the possibility of genuine *mT* reaching the coast, although as mentioned earlier, sea-modified *cP* does have some influence. In *summer, mP* air from the Sea of Okhotsk takes on modest importance, for the summer monsoon north of about 40° probably is chiefly *mP* air. In Japan modified *mP*-Okhotsk air moves southward in early summer, forming a front with *mT* air. The disturbances along this zone of conflict produce overcast, rainy weather and so create the first of the two summer maxima of summer precipitation.

mT **and** *cT* **air masses.** Tropical maritime air masses in *winter* are largely excluded from East Asia by the Siberian anticyclone and its land-to-sea monsoon winds, so that they play only a modest role in the cool-season weather there. Surface winter *mT* is rarely found as far north as the Yangtze Valley; farther south it occurs somewhat more frequently at the ground. Characteristically it is to be observed aloft in cyclonic storms, where it overruns *cP* air and causes precipitation. Since *mT* air is warmer and moister than other winter air masses in East Asia, fog, low stratus cloud, and misting rain may occur when it moves northward over a cooler land surface. South Asia in *winter* is chiefly dominated by the northeast trades, which have *mT* characteristics in the insular and coastal parts but take on *cT* attributes over the broader parts of the two large peninsulas of Southeast Asia and India-Pakistan. Rainfall is slight except along the windward eastern sides of these land masses.

mT (tropical maritime, including equatorial) air masses in *summer* largely dominate the weather of eastern and southern Asia, for they represent the monsoon circulation of that season. Except in extreme northwest India-Pakistan, where the main air masses are *cT*, South Asia is under the influence of a deep southwesterly circulation of equatorial origin that is warm, humid, and convectively unstable to a remarkable degree. Naturally it has a high potential for precipitation when it is lifted over highlands or in convergent circulations of atmospheric disturbances. A lift of only a few hundred meters is sufficient to saturate the air and bring about heavy showery rainfall. As *mT* air moves inland over the heated land surface, instability is further increased.

The equatorial current of southwesterly air continues northward over humid eastern China clear into North China and Manchuria, where it meets *cP* air along the polar front. Heavy summer precipitation is the result. This same current carries its heat and moisture to Korea and Japan. Still another branch of the summer monsoon in East Asia is derived from the western limb of the North Pacific subtropical anticyclone. It may arrive in Japan and eastern China as a southeasterly, southerly, or even southwesterly current. Because of its anticyclonic origin, even though it comes from the anticyclone's moister western flanks, this Pacific air may not be quite as humid and unstable as the equatorial southwesterlies from the Indian Ocean.

SELECTED REFERENCES FOR FURTHER STUDY OF TOPICS IN CHAPTER FIVE

1. **World.** H. C. Willet, *Descriptive Meteorology*, 1944 ed., pp. 181–196; Petterssen, *Weather Analysis and Forecasting*, 2d ed., McGraw-Hill Book Company, New York, 1956, pp. 138–204; Petterssen, *Introduction to Meteorology*, pp. 189–203; John R. Borchert, "Regional Differences in the World Atmospheric Circulation," *Ann. Assoc. Amer. Geographers*, Vol. 4, pp. 14–26 1953; Dieter H. Brunnschweiler, "Die Luftmassen der Nordhemisphäre," *Geographica Helvetica*, Vol. 12, pp. 164–195, 1957; *Aviation Weather*, pp. 62–72; Byers, *General Meteorology*, 3d ed., pp. 325–342, 1959; Hare, pp. 57–69; Blair and Fite, pp. 169–177, Haurwitz and Austin, pp. 101–105.

2. **North America.** H. C. Willet, *Descriptive Meteorology*, 1944 ed., pp. 196–224; Willet, "Characteristic Properties of North American Air Masses," in *Air Mass and Isentropic Analysis*, Milton, Mass., 1940, pp. 73–108; Byers, *Synoptical and Aeronautical Meteorology*, pp. 121–136, Byers, *General Meteorology*, 1944 ed., pp. 255–277; Dieter H. Brunnschweiler, "The Geographic Distribution of Air Masses in North

America," *Vierteljahresschr. naturforsch. Ges. Zürich,* Vol. 97, pp. 42–49, 1952.

3. **Europe.** Hare, pp. 157–161; Petterssen, *Weather Analysis and Forecasting,* numerous scattered references to European air masses on pp. 138–204; Byers, *General Meteorology,* 1944 ed., pp. 278–287; J. E. Belasco, "Characteristics of Air Masses in the British Isles," *Geophys. Mem., no.* 87, Great Britain Meteorological Office, 1952. Rudolph Treidl, *European Air Masses,* Canada, Meteorological Division, circular 2298.

4. **Eastern and southern Asia.** H. Arakawa, "The Air Masses of Japan," *Bull. Amer. Meteorol. Soc.,* Vol. 18, pp. 407–410, 1937; Chang-Wang Tu, "Chinese Air Mass Properties," *Quart. J. Roy. Meteorol. Soc.,* Vol. 65, pp. 33–51, 1939; Byers, *General Meteorology,* 1944 ed., pp. 287–291. A. Liu, "The Winter Frontology of China," *Bull. Amer. Meteorol. Soc.,* Vol. 26, pp. 309–314, 1945.

5. **General references on air masses and fronts.** J. Gentilli, "Air Masses of the Southern Hemisphere," *Weather,* Vol. 4, pp. 258–261, 292–297, 1949; H. H. Lamb, "Essay on Frontogenesis and Frontolysis," *Meteorol. Magazine,* Vol. 80, pp. 35–36, 65–71, 97–106, 1951; A. Miller, "Air Mass Climatology," *Geography,* Vol. 38, pp. 55–67, 1953; J. Namias, "An Introduction to the Study of Air Mass Analysis and Isentropic Analysis," *Amer. Meteorol. Soc.,* Milton, Mass., 1940; I. J. W. Pothecary, "Recent Research on Fronts," *Weather,* Vol. 12, pp. 147–150, 1956.

CHAPTER SIX

ATMOSPHERIC DISTURBANCES AND THEIR ASSOCIATED WEATHER

If solar control of the earth's weather were complete, climate would exhibit perfect diurnal and annual rhythms. As the previous chapters have shown, however, the rhythms are there but are far from perfect, for the nonperiodic effects of another control—that of atmospheric disturbances—are superposed on the regular daily and yearly march of solar energy. Atmospheric disturbances are extensive waves, eddies, or whirls of air embedded in the earth's major wind systems. It is they that create the great diversity of weather from day to day and week to week, for they are irregular in their occurrence and have time scales of variable magnitude.

The geographer's interest in atmospheric disturbances stems from the powerful influence they have on weather, and therefore on the distribution of the main climatic elements—solar energy, cloud-precipitation, and temperature. Except where highlands cause an ascent of air, traveling disturbances, or perturbations, furnish the chief mechanisms for elevating deep and extensive masses of the earth's atmosphere. Since these mechanisms in turn generate most of the earth's precipitation, the present chapter concentrates especially on types of disturbances that influence rainfall, either positively or negatively.

Other climatic effects of atmospheric disturbances are also important, however. Through their cloud systems, these storms substantially affect the distribution of solar energy and so are among the determinants of air temperature. And their wind systems carry out the advection which leads to major latitudinal, as well as land-sea, exchanges of heat and moisture.

The atmosphere's traveling disturbances cannot be observed on average monthly, seasonal, and annual charts of pressure and winds; they are entities of shorter, although variable, duration. On the other hand, the seasonal isobaric and wind-flow patterns described in Chapter 3, and shown in Figs. 3.30 and 3.31, are not so easy to see on any particular day's weather map, such as that of the United States or that of the Northern Hemisphere. This is because of the great variety of transient local wave and vortex disturbances which are superimposed on the broader hemispheric patterns of the general circulation (Fig. 6.1). These disturbances, sometimes called the *secondary* circulation, are so numerous that they tend almost to mask the general circulation, just as

Figure 6.1   A series of three vortex disturbances and their associated fronts over the North Pacific Ocean and North America, with TIROS cloud pictures, superimposed on a conventional surface weather chart, May 20, 1960. Such traveling disturbances are especially characteristic of the middle-latitude westerly wind belts and are responsible for much of the weather of those latitudes. (*Prepared by V. J. Oliver, U.S. Weather Bureau.*)

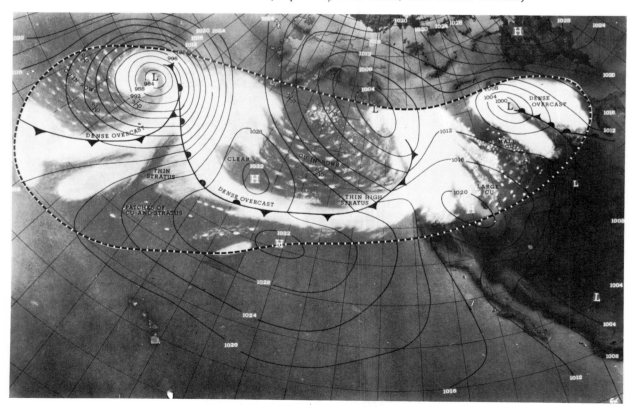

eddies, whirls, and crosscurrents of a river can nearly obscure the general downstream flow of the main current. The disturbance provides the dynamic element in weather and cannot be divorced from climate.

MIDDLE-LATITUDE TRAVELING CYCLONES AND ANTICYCLONES

Although the many disturbed pressure and wind fields shown on the daily surface weather charts have various shapes and patterns, the two most common, and most important in weather and climate, are the cyclone and the anticyclone. In the middle latitudes a succession of alternating low- and high-pressure systems moves from west to east in the stream of the prevailing westerlies (Fig. 6.1). These disturbances, with their accompanying wind, cloud, rainfall, and temperature phenomena, infest the belts of westerly winds and cause the variability and fickleness of weather characteristic of the middle latitudes. Here synoptic- or disturbance-scale systems dominate the distribution of pressure, temperature, and wind at any time; larger- or smaller-scale systems are less important. It is in the middle latitudes that weather forecasting services are most necessary and best developed.

General features.   No two cyclones or anticyclones exactly resemble each other. They vary in size, shape, and intensity, as well as in their cloud and rainfall. So the following generalizations must not be expected to apply in all respects to any individual disturbance. Atmospheric disturbances also differ to some extent from one part of the world to another. Even within an area the size of the continental United States, they look and act somewhat differently from region to region.

As can be seen on a surface weather map, cyclones have centers of relatively low pressure surrounded by concentric closed isobars, while anticyclones have cores of relatively high pressure surrounded by closed isobars (Figs. 6.2 and 6.3). A middle-latitude cyclone, therefore, frequently goes by the name of low, depression, or trough if it is elongated. An anticyclone is called a high or, if elongated, a ridge. Since its surface pressure is low, the cyclone must be a column of air with less mass than an anticyclone. While the isobars of lows and highs are roughly concentric, the shapes of individual storms vary greatly.

In a cyclone the pressure is lowest at the center and increases toward the margins of the storm. In an anticyclone the highest pressure is at the center. However, the terms low and high as applied to pressure in these storms are not precise but relative. No specific amount of pressure separates lows from highs; it is always a comparative matter. Pressure at the center of a weak or moderate low is about 1,000 mb or a little more. It may be as low as 940 to 980 mb in a strong cyclone. Normally the pressure difference between the center and margins of a low is 10 to 20 mb, or several tenths of an inch. In an unusually large and deep winter low, the pressure difference may be as much as 35 mb, or 1 in. An intense high may possibly show an equivalent pressure difference between center and circumference, although normally the difference will be half that amount. There are many more large and vigorous

Figure 6.2 Model cyclone (*A*) and anticyclone (*B*), Northern Hemisphere, showing arrangement of isobars, winds, air masses, and surface fronts.

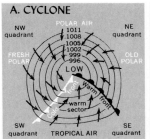
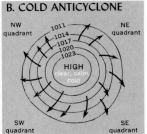

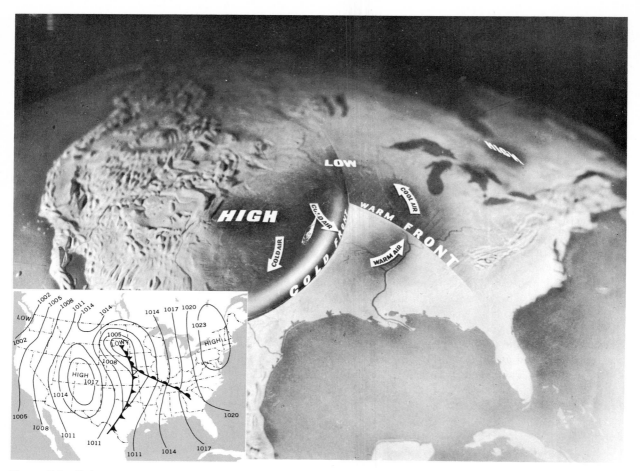

Figure 6.3 If the air masses and fronts were actually visible, the cyclonic disturbance shown on the small inset weather map might appear as represented in the three-dimensional illustration. (*U.S. Weather Bureau.*)

cyclones and anticyclones in winter than summer, so that the cooler seasons have more variable weather.

Because the cyclone is a low-pressure storm, it has a converging surface-wind system of moderate velocity. Rarely are its winds violent or destructive, however. Earth rotation causes the converging winds to approach the low center obliquely, so that the surface-wind arrows make an angle of 20 to 40° with the isobars. Because surface airflow in a low is converging, there must be upward movement and a compensating outflow aloft, commonly with cloud

and precipitation. In the Northern Hemisphere such a converging system of surface winds usually results in a warm sector to the south and east of the center, where the winds are from a southerly direction, and a more extensive cool or cold section to the north and west, where the winds are northerly. The surface winds in a high move obliquely outward from the center toward the circumference; in other words, an anticyclone is a diverging system, with inflow aloft and vertical settling or subsidence feeding the surface currents.

Note, however, that a cyclone or an anticyclone

is not composed of the same air at different times as it shifts its position. Such disturbances are simply pressure systems through which the air moves.

There are great variations in the size of these storms, but typically they spread over extensive areas. Sometimes they cover as much as one-third of the conterminous United States, or 1,000,000 square miles, although most of them are smaller. They are extensive rather than intensive, so that a normal cyclone with a vertical thickness of 6 to 7 miles probably will have a diameter one hundred times as great. Cyclones tend to be elliptical or egg-shaped, with the narrow end toward the equator. The long axis, extending in a northeast-southwest direction, is commonly nearly twice the length of the shorter one. Thus a typical well-developed winter cyclone might have long and short diameters of 1,200 and 650 miles, respectively. Anticyclones are apt to be somewhat larger, with diameters roughly 25 percent longer than those of lows.

Direction and rate of movement. As mentioned earlier, middle-latitude cyclones and anticyclones are carried along by the deep westerly circulation, so that they travel in a general west-to-east direction. This is not to say that such storms always move due east, for they follow different routes, concentrating in some regions and spreading out in others. But in spite of their vagaries in direction, and also in rate of movement, their general progress is eastward.

The direction and rate of movement of cyclones are approximately those of the winds in the free atmosphere. This is why a weather forecaster in the middle latitudes bases his prediction upon weather conditions to the west, rather than to the east, of his station. Disturbances to the east already have gone by; those to the west are approaching. Storms have a tendency to follow certain very general tracks, whose worldwide patterns will be described later in the chapter. Keep in mind that the movement of an atmospheric disturbance involves a pressure system, not a particular body of air. In this respect atmospheric disturbances resemble ocean waves.

Rates of storm movement are variable both as to season and as to individual disturbances. In the United States, cyclones move eastward across the country at velocities averaging 20 miles an hour in summer and 30 miles an hour in winter. Highs are somewhat slower than lows. Speeds are reduced in summer because the whole atmospheric circulation is slowed down, and as a consequence warm-season weather is less changeable. In the winter a well-developed low characteristically requires 3 to 6 days to cross the United States.

Just as temperature, pressure, and wind belts follow the movements of the sun's rays, shifting poleward in summer and equatorward in winter, so also do the storm tracks. This fact helps to explain why storms are fewer and weaker over the lower middle latitudes in summer than in winter.

Origin and structure of middle-latitude cyclones

Middle-latitude depressions appear to consist of a variety of disturbances, not all of which have the same origin. A few seem to develop as eddies on the lee side of highlands—for example, in the eastern foothill region of the North American Rockies. Occasionally small, short-lived depressions form within homogeneous, unstable polar maritime air. Such depressions are fairly common in the British Isles. Other weak disturbances appear to develop as a consequence of intense and prolonged solar heating. But the great majority of surface cyclonic storms of middle latitudes which appear on the daily weather map seem to begin as waves along surface fronts separating cold and warm air masses.

Doubtless the genesis of many surface cyclones and anticyclones is closely linked to conditions in the upper-troposphere waves and jet streams. On occasions the flow patterns aloft favor cyclogenesis at low levels; at other times they are neutral or oppose it. However, the same patterns do not produce the same results in all locations. Figures 6.4 and 6.5 represent a fairly common situation in the case of a developing cyclone. Since surface barometric pressure varies only modestly, mass convergence at the surface must be nearly compensated for by divergence aloft. Conversely, divergence at

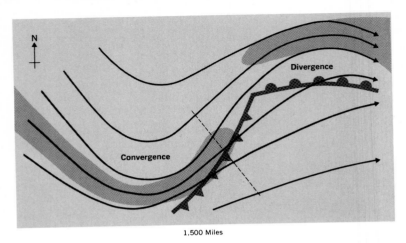

1,500 Miles

Figure 6.4 Model of flow pattern in troposphere above a developing wave cyclone. Cold and warm surface fronts are shown. Shaded bands indicate location of jet streams. (*From Herbert Riehl, Introduction to the Atmosphere.*)

the surface must be replaced by convergence aloft.

Although many times the upper flow provides the key for a cyclone's beginning, yet there is also clear evidence of instances where surface disturbances build upward from the earth's surface to create the wave aloft. Since historically it was the surface disturbances that were studied first, and since it is these which are most easily observed and about which most is known, cyclones will be analyzed here in relation to fronts at the lower levels.

Preceding cyclogenesis, there is a quasi-stationary front separating a colder and a warmer air mass. The winds in each are parallel to the surface front, and they may blow from opposite directions (Fig. 6.6A), or even from the same direction provided they have different speeds. As noted in Chapter 5, because of Coriolis force the colder air underlies the warmer in the form of a gently sloping wedge, whose inclination may be on the order of 1 to 100. At this stage neither air mass has a significant sustained component of motion toward the front; any upglide of the warm air over the cold wedge is minor. That there is some, however, is indicated by a fair amount of cloud—usually stratus—within the warm air just above the frontal surface. Since well-developed fronts rarely remain stationary or undisturbed for long, the existence of such a front portends imminent wave and cyclone development.

Stages in the life history of a cyclone. Often the most easily detected first evidence of wave development along a front is an expansion and intensification of the cloud and precipitation areas, together with a general fall of pressure, around the nascent wave. When this occurs, closer inspection usually reveals that a shallow indentation is forming along the front (Fig. 6.6B). Such an incipient wave is observable both on the above-surface discontinuity and along the surface front.

Exactly how a frontal wave gets its start is not always clear. Some may result just from the temperature contrast or the opposing air motion along a front. Others appear to develop in what may be called favorable type locations—for example, on the lee side of highlands, or in areas where steep temperature gradients exist between continents and oceans. Still others may be induced by a neighboring disturbance, or by the approach of an upper-level trough. In its initial stage, where the warm air thrusts only a shallow salient into the colder air mass, there is no great difference between the warm and cold fronts.

The incipient wave moves rapidly eastward along the front. If the length of the wave is less than about 300 miles, it is said to be stable, for it does not evolve into a full scale cyclone, although it usually intensifies the bad weather. Larger waves, on the other hand, ordinarily grow rapidly as the wedge of warm air becomes increasingly more deeply embedded in the mass of cold air. Pressure falls, weather worsens, and a young cyclone with closed isobars forms at the apex of the wave.

Anywhere from 30 to 60 hours after the first appearance of the wave, the cyclone reaches the stage of early maturity (Fig. 6.6C). The depression is roughly circular, the warm sector is broad and deep, clearly defined warm and cold fronts have evolved, and cloud and precipitation are widespread.

From this stage on through full maturity, decline, and finally old age, the more rapidly advancing cold front begins to overtake the slower-moving warm front. Commencing at the warm-sector apex, the warm air is gradually pinched out at the surface, and *occlusion* is said to occur (Fig. 6.6E). With oc-

Figure 6.5 Coupling between waves in the upper westerlies and high- and low-pressure systems near surface. *Top:* Horizontal view of upper jet-stream core in relation to surface systems. *Bottom:* Vertical cross section showing the convergent and divergent parts of the atmospheric motion in the horizontal plane, and vertical motion (white arrows). (*From Herbert Riehl, Introduction to the Atmosphere.*)

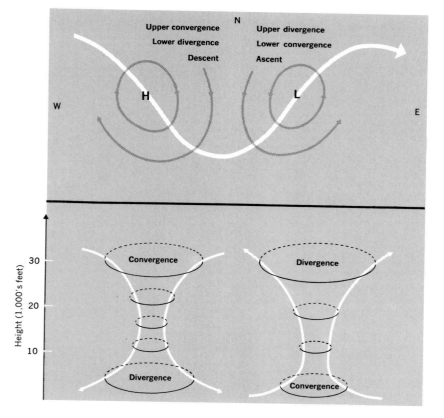

During the life cycle of a middle-latitude cyclone, the warm air which initially existed side by side with the cold air is lifted to higher and higher levels, and the cold and heavier air comes to occupy the lower levels. By this process the center of gravity of the whole cyclone system is lowered and potential energy is released. Such vertical movement provides the main source of energy for cyclone growth as represented in the kinetic energy of the cyclone's wind system. A subordinate source of energy is latent heat of condensation.

The cyclone model. A well-developed cyclone in a stage of early maturity is portrayed in Fig. 6.7. The middle drawing, (b), represents the cyclone as it appears on the surface weather map. The upper part of the figure (a) is a vertical cross section through

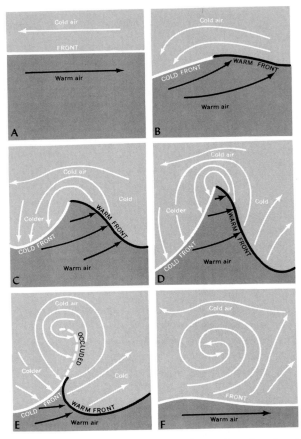

Figure 6.6 Six stages in the life cycle of a frontal cyclone. Drawing B shows the beginning of a small horizontal wave along the front. In drawing C the wave development has progressed to the point where there is a definite cyclonic circulation with well-developed warm and cold fronts. Because of the more rapid movement of the cold front, drawing D shows a narrowed warm sector as the cold front approaches the retreating warm front. In drawing E the occlusion process is occurring, the cyclone has reached its maximum development, and the warm sector is being rapidly pinched off. In drawing F the warm sector has been eliminated; the cyclone is in its dying stages and is represented only by a whirl of cold air.

clusion the total area of the warm sector shrinks. Eventually as the warm air disappears at the surface, the cyclone at lower levels becomes an immense vortex of relatively homogeneous air (Fig. 6.6 F).

Figure 6.7 Cyclone model: Ground plan (b) and vertical sections (a) and (c) of a Northern Hemisphere cyclone in a stage of early maturity, before occlusion has commenced.

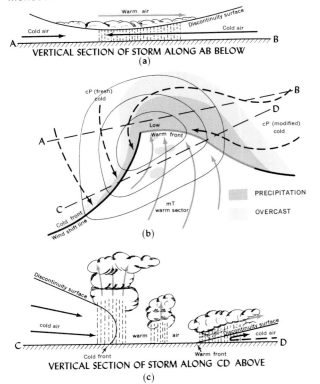

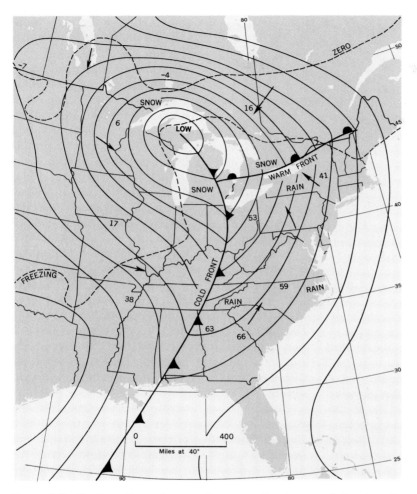

Figure 6.8 Section of a surface weather map showing a well-developed wave cyclone over central and eastern North America. This representation of a particular cyclone should be compared with the ground plan of the model cyclone in Fig. 6.7*b*.

the cyclone north of its center. The lower part (*c*) is a similar vertical cross section south of its center. In the cross section *AB* to the north of the storm center, the warm air is present aloft but does not reach the earth's surface. It is the continued lifting of the warm air aloft that produces the precipitation shown. South of the center, warm air does reach down to the earth's surface. As the vertical cross section along line *CD* shows, this warm sector is bounded by two surface fronts.

Thus a cyclone consists of two essentially different air masses separated from one another by fronts (Figs. 6.7 and 6.8). To the south and east is the warm sector—a poleward projection of warmer air, many times of tropical origin, which is enveloped on its western, northern, and northeastern sides by colder air, usually of polar origin. The nature of this polar air varies with different locations and source regions. In North America east of the Rocky Mountains, the fresh polar air to the west and

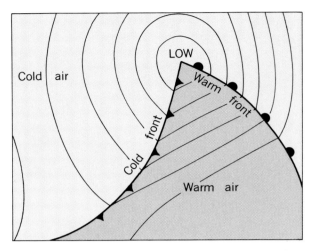

Figure 6.9 A cold front and a warm front on a surface weather map.

northwest is usually colder than the old polar air to the east and northeast, which has been considerably modified by several days' travel over the middle latitudes.

The discontinuity surface separating the polar and tropical air masses may be classified as a warm or a cold front, depending on whether the warm or the cold air is the aggressor (Fig. 6.9). Where warm air leaves the earth's surface and actively moves upward over a retreating wedge of cold air, as it does ahead of the eastward-shifting tongue of tropical air, a warm front develops. On the other hand, a cold front forms when advancing cold air aggressively undercuts warm air and brings about its forced ascent, which happens to the rear, or west, of the warm sector. Along a warm front, warm air replaces cold air at the ground; along a cold front, cold air replaces warm. Warm fronts in the central and eastern United States most commonly develop as a consequence of *mT*-Gulf air overrunning modified polar air with a southerly component. The latter is returning poleward on the rear of a retreating anticyclone (Fig. 6.6*D*). Note that there is an upward movement of air in at least two separate areas: (1) along the warm front, where the faster-moving warm air upglides over the eastward-retreating

wedge of cold air, and (2) along the cold front, where the more rapidly advancing cold air thrusts under the warm air ahead of it.

Warm fronts. The inclination of the surface of discontinuity in a warm front is relatively gentle, usually on the order of 1:100 to 1:200. In other words, if an aircraft were flying 5,000 ft above the ground and at right angles to the front, it would cross the boundary separating the two air masses 100 to 200 miles ahead (east) of the surface front (Figs. 6.9 and 6.10). As the warm air slowly glides up the gently inclined surface of the cold-air wedge, adiabatic cooling results and may lead to cloud condensation and precipitation. Any vertical component of the warm air along the warm-front surface of discontinuity is likely to be small, except where convection develops. Frictional drag tends to thin out and flatten the wedge of retreating cold air near the surface front. This, together with the mixing of air that occurs along the front, tends to blend the two air masses within the contact zone so that it may be difficult to locate the warm front at the surface with any precision. The change in temperature and wind direction is transitional rather than abrupt.

Because of the west-to-east movement of the cyclone and the arrangement of fronts within it, warm-front weather phenomena are the first indications of an approaching cyclonic storm. Since the discontinuity surface has such a mild inclination, signs of it in the form of certain high-level cloud types may be seen as far as 500 to 1,000 miles ahead of the surface front. The first clouds heralding the storm are the high cirrus and cirrostratus, the latter creating the familiar halos around sun and moon. If the overrunning warm air is unstable, cirrocumulus will appear, producing the well-known "mackerel sky" recognized by sailors as a harbinger of stormy weather. As the front approaches and the wedge of cold air thins, the cloud deck becomes lower and thicker. Altostratus and altocumulus deepen into stratocumulus and finally nimbus. Seen from below, the warm-front cloud deck is a formless gray overcast, known as nimbostratus. Precipitation usually begins with the altostratus and continues until the surface front has passed. Soon after the rain starts, clouds and fog commonly form within the cold air mass below the discontinuity surface. These are largely the result of evaporation from raindrops and

Figure 6.10 Warm fronts in vertical cross section: (a) A warm front with stable warm air. (b) A warm front with unstable warm air. (c) A wavy warm-front surface.

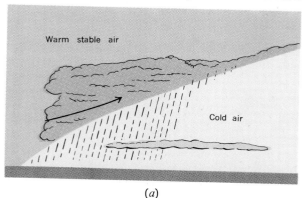

Warm stable air

Cold air

(a)

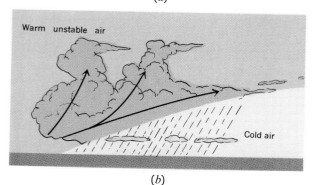

Warm unstable air

Cold air

(b)

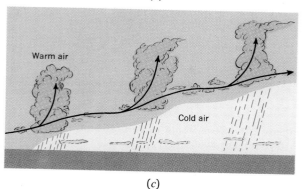

Warm air

Cold air

(c)

surface water. The suddenness with which these lower clouds and fog appear makes them a hazard to air navigation.

The character of the weather at the warm front depends mainly upon the nature of the overrunning warm air. If it is dry and stable, considerable ascent will be necessary to produce any condensation forms, and precipitation will be light if it occurs at all. On the other hand, if the warm air is conditionally or convectively unstable, ascent over the cold wedge may be vigorous enough to trigger showers, and even thunderstorms in the warmer months (Fig. 6.10). To the observer at the ground, the rain appears to be coming from the cold air. Most of it, however, has its origin in the overrunning warm air aloft. Because of the cold air at the surface, the weather is likely to be chilly, damp and disagreeable, with low visibility. It was such weather that a New England poet had in mind when he wrote

The day is cold and dark and dreary;
    It rains and the wind is never weary.
The vine still clings to the moldering wall,
    And with every gust the dead leaves fall,
And the day is dark and dreary.

Because of the slight inclination but great horizontal extent of the surface of discontinuity, warm-front precipitation is inclined to be fairly steady, long-continued, and widespread. It is not unusually heavy, however; unless the ascending air is so unstable that convective showers result from the lifting along the front.

With the passage of a warm front at the surface, the following weather changes are likely: (1) marked clearing, (2) a distinct temperature rise, (3) a rapid increase in specific humidity, (4) a slight barometric trough, (5) a small shift in wind direction—usually about 45°.

Warm sector. Within the warm sector of a cyclone, which is situated south of the center and between the warm and cold fronts, the weather depends on the nature of the air mass and the season (Fig. 6.7). Here there is no large-scale upglide of air along fronts. Still, the wedge of warm air, gradually shrinking in area because of the scissors action of the two bounding fronts, is a region of convergence which has some upward air movement and cloud. In summer over the central and eastern United States, the unstable *mT* air in the warm sector may give rise to a broken cloud deck, with cumulus clouds and occasional convective showers and thunderstorms. In western Europe, warm-sector air is more stable in summer, and a stratus overcast with drizzle is characteristic. Warm-sector winter weather in the United States, especially where the surface has a snow cover, brings more stratus cloud, drizzle, and advection fog.

Cold fronts. In this type of discontinuity the warm air is forced to rise by the aggressive action of the eastward-moving wedge of cold air (Fig. 6.11). Along such a front, cold air replaces warmer air. When cold fronts are well developed, they are easily recognized on the weather map by their marked wind discontinuities, which are known as wind-shift lines.

Under normal conditions the slope of a cold-front surface of discontinuity is considerably steeper than that of a warm front, being usually on the order of 1:50 to 1:100. This difference between cold and warm fronts is largely due to the contrasting effects of frictional forces. Along the warm front, a drag effect upon the retreating wedge of cold air tends to thin it out and reduce its slope. Along the cold front, the advancing cold air at the ground is held back by friction, which gives it a tendency to push forward aloft, thereby increasing the discontinuity slope. Characteristically the forward edge of a rapidly moving cold front is steepened and blunted, as shown in Fig. 6.7c. Here the cold air aloft may even push out ahead of the front at the ground, obviously creating a highly unstable condition and tumultuous weather. Where the front moves more slowly, the air aloft has less tendency toward overrunning, and the weather is less turbulent.

Since the surface of discontinuity slopes upward away from the direction of storm movement, the approach of the cold front is not heralded far in

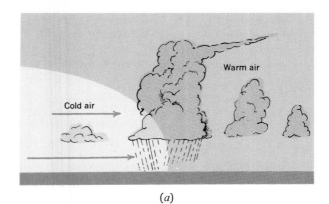

(a)

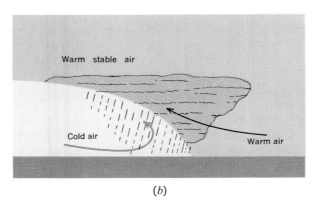

(b)

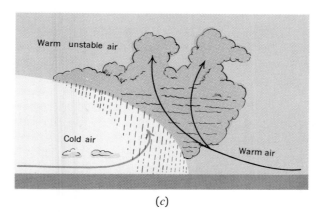

(c)

Figure 6.11 Cold fronts in vertical cross section: (*a*) A fast-moving cold front, with unstable warm air. (*b*) A slow-moving cold front, with stable warm air. (*c*) A slow-moving cold front, with unstable warm air.

advance by cloud forms or other weather phenomena. Moreover, the zone of bad weather associated with rising air is much narrower at the cold than at the warm front and so passes more rapidly. To some extent the cold front's comparative narrowness is due to its greater speed and steeper slope. Mainly, however, this difference exists between the two fronts because warm air moves in opposite directions with respect to them. If the cold front moves rapidly and steadily, clearing takes place quickly after it has passed. If it is retarded for any reason, the sky may remain cloudy for some time.

The nature of the weather along the cold front depends in a large degree upon the structure of the warm air that is being displaced. If it is dry and stable, the front may be accompanied only by broken clouds, with no precipitation (Fig. 6.11). On the other hand, if it is warm, humid mT-Gulf air, cumulonimbus clouds and frontal thunderstorms are likely. But on the whole, owing to the steeper slope of the cold-front surface of discontinuity, and to the blunter forward nose of the rapidly advancing wedge of cold air, the upthrust of warm air along the cold front is more vigorous than along the warm front. This results in more tumultuous weather and in pelting rains, although their duration is shorter than warm-front rains. Cloud forms along a cold front tend to be of the cumulus and cumulonimbus types, but stratus types may also appear. In some cases most of the precipitation falls in advance of the front; in others it occurs in the cold air behind the front.

If an unusually cold and vigorous air mass back of a cold front passes over a warmer water surface, the addition of heat and moisture in the lower layers often leads to development of heavy rain or snow showers, with the precipitation originating in the cold air. This phenomenon is most common in late fall and early winter, when temperature contrasts between air and water are most marked. Similar snow flurries may occur over land areas where the land is distinctly warmer than the southward-advancing cold air mass.

With the passage of a cold front, the following weather changes are characteristic: (1) a marked clearing and improvement in the weather, usually fairly rapid, (2) an abrupt fall in temperature, (3) a well-defined barometric trough, (4) a pronounced decrease in specific and relative humidity, and (5) a pronounced wind shift of 45 to 180°.

Occluded fronts. Since the cold front of a cyclone normally travels at about twice the speed of the warm front, it gradually approaches and eventually overtakes the warm front. Occlusion is a normal development in the life history of middle-latitude cyclones (Fig. 6.6). It begins at the apex of the warm sector near the center of the storm and gradually progresses equatorward until the warm sector has been completely obliterated at the surface. Thus the cyclone is partly occluded during the major part of its life history. Actually it usually reaches its maximum intensity 12 to 24 hr after the beginning of the occlusion process. Although the two polar air masses involved in the occlusion may have had a similar origin, their routes since leaving the source regions have usually been different enough that their temperatures, humidities, and densities are no longer the same. Consequently as the cold front overtakes the warm front, a new discontinuity—this one between the two polar air masses—is created. When the air behind the cold front is colder than that ahead of the warm front (which is usually the case in the central and eastern United States) the resulting discontinuity will resemble a new cold front to some extent, although one without sharp temperature contrasts (Fig. 6.12). But when the air ahead of the warm front is colder than that behind the cold front, the occlusion, and the weather, at the surface will be more of the warm-front type (Fig. 6.12*b*).

Precipitation along occluded fronts results from both the continued lifting of the warm air aloft and the vertical displacement of the less dense of the cold air masses. As occlusion continues, the depth of cold air along the occluded front increases, and the warm air is crowded further upward until it is high enough to spread out laterally over the cold air masses enclosing it. When it does spread lat-

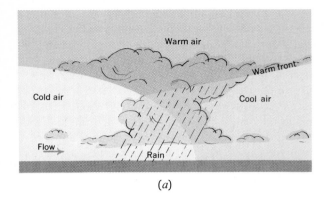

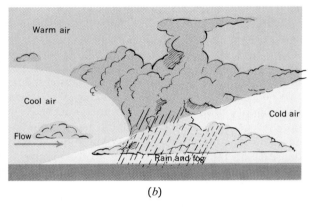

Figure 6.12 An occluded cyclone in vertical cross section: (*a*) Cold-front occlusion. (*b*) Warm-front occlusion.

erally, further lifting of the warm air ceases, and that particular air mass plays no further part in cloud and rain production. Subsequently the weather conditions are governed by the contrasts in the characteristics of the two cold air masses, one on either side of the occluded front.

In its later stages of occlusion a cyclone weakens, and precipitation is less abundant than earlier in the storm's life cycle. The temperature and density contrasts between the cold air masses comprising the cyclone are not marked, and the heat of condensation derived from lifting the cool air is not great, so that the storm's energy sources are reduced. There may be an extensive cloud cover, but usually the clouds are not thick and the precipitation is

light. Lifting of the cool air mass does not ordinarily lead to heavy rains, for the air's moisture content is relatively low and the air mass is likely to be stable. Regions with a predominance of occluded cyclones, such as western Europe, have a great deal of low cloud, usually only modest amounts of precipitation, and small daily as well as nonperiodic temperature changes.

Stationary fronts. A front between two opposing air masses which shows little or no movement, with neither air mass replacing the other, is a stationary front. The weather along such a front resembles that accompanying a warm front, but is usually less intense. A feature of stationary fronts is that their bad weather conditions may persist over a region for a number of days.

Families of cyclones. Toward the terminal part of a cyclone's occluded stage, and along its southwesterly-trailing cold front, a new wave often develops. And as this second wave matures, still a third and a fourth may develop behind it. Thus there comes to be a whole family of four or five cyclones strung out along a single frontal system, each progressively smaller than the preceding one, and each developing farther equatorward. In such a family of cyclones the easternmost unit is likely to be fully occluded, while that farthest to the southwest is still in an incipient stage. The last wave in such a series is commonly followed by a great outbreak of fresh polar air, which may surge far southward, even into the tropics.

Origins and structure of middle-latitude anticyclones

Chapter 3 on pressure and wind systems described the great seasonally permanent anticyclonic systems: both the cold anticyclones of high latitudes and the warm subtropical anticyclones of lower latitudes. This chapter will deal with moving anticyclones of the middle latitudes. These are by no means, however, completely divorced from the larger more permanent systems of Chapter 3.

Anticyclonic disturbances have received less attention from students of the atmosphere than cyclones, so that less is known concerning their origin and structure. Probably there are various types of moving anticyclones, but the usual classification recognizes two main groups: the cold and the warm. This obviously follows the division made for Chapter 3's more durable and stationary high-pressure systems. Some authorities would also specify a third type which includes the sluggish systems which seem to fill the spaces between the more vigorous individual cyclones. These somewhat inert wedges of high pressure bring clearing weather after the cyclone's cloud and rain. Their clear skies in turn promote large diurnal ranges in temperature, with marked night cooling. Although they probably have various origins, superficially they come to resemble a mild form of cold anticyclone over middle-latitude continents.

Cold anticyclones, which have abnormally cold air in the lower troposphere, are sometimes called "polar-outbreak highs." They are rapidly moving masses of cold air that emerge from the subpolar regions and are drawn toward lower latitudes on the rear of a well-developed depression, often the last in a cyclone family. Such cold polar-outbreak highs slow down as they move equatorward and eastward across the middle latitudes. In the subtropics they are gradually transformed into warm anticyclones. Occasionally, however, they break through into the inner tropics. In the Northern Hemisphere there are two general major routes, one from northern Canada southward and southeastward across the central eastern United States, and the other from intensely cold eastern Siberia across northern China and Japan. Such cold anticyclones are dome-shaped pools of cold air whose high pressure is due to the weight of this dense air.

The exact ways by which a cold anticyclone is formed are still somewhat obscure. One element seems to be the strong radiational cooling of the lower air as it rests on a snow-covered polar or subpolar surface. A strong southward surge of cold air in the surface anticyclone, however, is usually initiated by an expulsion of cold Arctic Basin air aloft, acting in conjunction with an upper-troposphere long wave. Cold anticyclones are commonly con-

fined to the lower 5,000 to 10,000 ft of atmosphere. There is usually marked subsidence within them, so that a pronounced inversion is present and the air is very stable. When they have little wind, a surface inversion likewise develops. Such anticyclones typically bring clear, dry, cold weather.

Warm anticyclones, while certainly arising from dynamic (not thermal) effects, remain something of a mystery as far as origin is concerned. They are characterized by slow rates of movement, great vertical depth, and unusual warmth throughout the troposphere (except in the lowest layers, where there may be some radiational cooling). Subsidence is marked throughout, inversions often develop, and clear, dry weather is the rule. Precipitation is uncommon, although along certain subtropical west coasts paralleled by cool ocean currents, a deck of stratus cloud associated with a strong low-level inversion may produce some drizzle.

Warm anticyclones are most common in the subtropical latitudes, especially over the oceans and adjacent parts of continents. This is the realm of the great semipermanent highs. From these extensive subtropical anticyclones, local buds or cells in the form of slow-moving warm highs frequently emerge and travel eastward and poleward into the middle latitudes. There, if their trajectory is over continents in winter, they are gradually transformed into cold anticyclones. In reality, some of the so-called permanent subtropical high-pressure cells seem to be just monthly and seasonal averages of traveling warm anticyclones, especially in the Southern Hemisphere.

Compared with cyclonic weather, weather associated with an anticyclone is much more autonomous or independent of influences outside its own system. The clear skies, weak and divergent air movement, and general subsidence of anticyclones make for weather that is strongly regional and diurnal in character. It is little affected by advection from outside sources. By contrast, cyclonic weather, in which the emphasis is on advective processes, is powerfully influenced by outside conditions, that are brought by a converging circulation into the cyclone's system. This advection, together with the cloud cover characteristic of cyclones, results in weather that is more nonperiodic than diurnal.

Wind systems

Cyclones.   Paramount features of cyclones are the surface convergence and ascent of air within them (Figs. 6.2 and 6.13). Various modifications of a simple converging spiraling system of winds result from such effects as terrain irregularities, as well as from the fact that the depression is not a stationary but a moving system traveling at speeds of 500 to 1,000 miles a day. Also, in any converging system of winds in middle latitudes, air masses of contrasting temperatures and densities are almost bound to be drawn into the vortex system on its northern and southern sectors in a manner which departs from a symmetrical in-spiral. In Fig. 6.7b, for example, although a convergent airflow is still conspicuous, there are two contrasting air masses: a warmer one to the south and a colder one enveloping it on the north, west, and northeast.

As a result of the converging circulation about a cyclonic center, surface winds on the front of the center (east, or in the direction the storm is traveling) are from a general easterly and southerly direction—southeast, east, and northeast in the northeastern quadrant, and south and southeast in

Figure 6.13   Circulations in a cyclone and anticyclone in the Southern Hemisphere. Note that the cyclonic circulation here is clockwise and that of the anticyclone is anticlockwise—just opposite to those in the Northern Hemisphere.

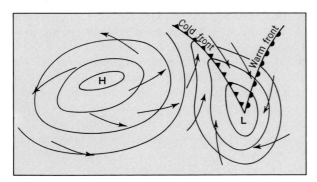

the southeastern quadrant. These easterly and southerly winds on the front of the cyclone can be either polar or tropical air, depending on the directional trend of the warm front. One cannot be certain of the primary origin of the air merely by observing wind direction, for the modified polar air on the front of a cyclone is frequently a southeasterly current. This is polar air that was brought southward in a preceding anticyclone and has had a relatively long trajectory over the middle latitudes.

Obviously all these easterly winds on the front of a cyclone are blowing opposite to the deep planetary westerlies in which they exist, and also opposite to the direction of storm movement. Thus while the storm travels from west to east, the winds on its front typically have an easterly component. This characteristic of a cyclone, which as a whole is an eddy or vortex in the westerly circulation, has a counterpart in the whirlpool in a river. The whirlpool of water is carried downstream by the major current just as the cyclonic vortex is carried eastward by the westerlies. On the whirlpool's downstream side, the water flows into the whirl in a direction opposite to that of the general current, and also opposite to the downstream movement of the whirlpool itself.

Wind shift with the passing of a cyclone.   When a cyclonic center approaches and passes by an observer, he will have general easterly winds as long as the low center is to the west of him and he is on the front of the vortex. As the center passes by, leaving him in the western, or rear, part of the low, the winds shift to the west. Easterly winds, therefore, frequently indicate the approach of a cyclone with its accompanying rain and cloud, while westerly winds more often foretell the retreat eastward of the storm center and the approach of clearing weather accompanying the following anticyclone.

In many cyclones this shift from easterly to westerly winds is gradual. But in others—especially in storms that have a marked equatorward elongation, so that the isobars are roughly in the form of the letter V—the wind shift is likely to be abrupt (Fig. 6.8). Along the wind-shift line, which is approximately a line joining the apexes of the V-shaped isobars south of the center, winds of contrasting temperature, humidity, and density meet at a sharp angle, and in the warmer months vigorous thunderstorms and turbulent weather conditions often result. This wind-shift line is the cold front described earlier.

Veering and backing winds. If the center of a cyclone passes to the north of the observer, so that he is in the southern quadrants of the storm, the succession of winds will be from southeast to west and northwest by way of south (Fig. 6.14). This is called a veering wind shift. If the storm center passes south of the observer, so that he is on the north side of the cyclone, his winds will shift from northeast to northwest by way of north. This is known as a backing wind shift.

Backing and veering winds have important climatic significance. A veering wind shift is likely to involve a succession of modified polar air, tropical air, and fresh polar air, as well as one or more fronts. With a backing wind shift, the observer is continuously in air of polar origin, and surface fronts are less likely. Therefore regions which commonly lie on the equatorward sides of passing cyclones and experience veering winds advecting some warm air are likely to have relatively higher average temper-

Figure 6.14 Illustrating veering and backing wind shifts with the passing of a cyclonic disturbance to the north or to the south of a station. Numerals 1, 2, 3 represent a time sequence.

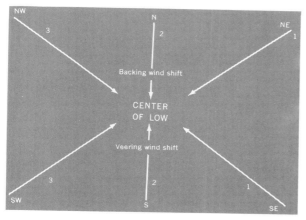

atures, greater variability of temperature, stronger convection, and more cumulus clouds than they would if they were located poleward of the cyclonic centers. Backing winds with their cold-air advection are associated with generally lower temperatures, less variability in temperature, a more continuous cloud cover, and greater likelihood of snow in winter.

Anticyclones. Key elements of anticyclonic circulation are subsidence and surface divergence. The wind systems of highs and lows, therefore, are opposite—in direction of gradient-induced flow, in direction of spiral rotation, and in the nature of the vertical movement.

Anticyclonic wind systems are usually less well developed than those of cyclones, so that no characteristic wind shift is forecast as they pass by. In general, however, winds on the front (east) of an eastward-advancing high are northwesterly, while those on the rear are southeasterly (Figs. 6.2B and 6.15). Since lows and highs often alternate with one another as they move across the country, it is evident that the westerly winds on the front of a high and the rear of a low have a similar origin and are much alike in character.

Pressure gradients are usually less steep, and wind velocities lower, in anticyclones than in cyclones. Weak pressure gradients are particularly conspicuous toward the centers of slow-moving or stagnant highs, where there is much light wind and calm. The strongest winds are likely to be found on the front margin of a rapidly advancing cold high, for there the high is merging with the preceding low. In this location between a cyclonic and an anticyclonic system, the isobars, which are closely spaced, tend to become nearly parallel straight lines trending in a general north-south direction. Winds here are vigorous northwest ones, a feature stemming from a steep surface pressure gradient and a strong westerly circulation aloft. Cold waves and blizzards over central and eastern North America are associated with these strong outpourings of cold air in the transition areas between the rears of lows and the fronts of cold highs. (See Fig. 6.15.)

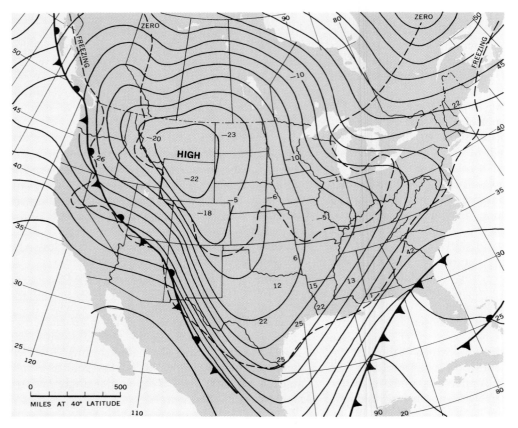

Figure 6.15 Polar-outbreak cold anticyclone advancing rapidly southward as a mass of unusually cold and stable *cP* air. On this occasion Minot, North Dakota, had a minimum temperature of 23° below zero, Denver −18°, and Jackson, Mississippi, 15°. Below-freezing night temperatures were common along the margins of the Gulf of Mexico. (*From Weatherwise, June, 1962.*)

Warm highs, which are subtropical in origin, normally have weaker pressure gradients and less well developed wind systems than cold highs, and their movement is likely to be much slower (Fig. 6.16). They are apt to remain stagnant over an area for days at a time.

Precipitation

Since the convergent circulation and upward air movement of cyclones are conducive to condensation, they are commonly accompanied by large areas of cloud, and often by precipitation as well. To be sure, not every cyclone brings rain, but many of them do. Anticyclones, in contrast, are likely to be fair-weather areas with fewer clouds and meager rainfall. This is because the subsidence and lateral spreading of the air in an anticyclone operate to create stability and to prevent a conflict of air masses and the formation of fronts.

Even though anticyclones are in general fair-weather systems, it is not correct to assume that rainfall is never associated with them. If the easterly and southerly circulation on the rear of a warm anti-

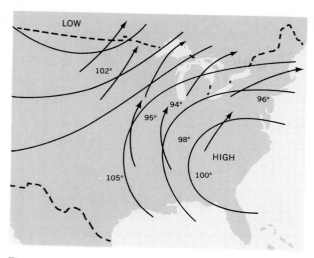

Figure 6.16 A warm anticyclone with clear skies, relatively stagnant over the southeastern United States. Such a synoptic feature produces unseasonably hot weather in the central and eastern parts of the country.

cyclone comes from a warm water surface, it may provide an environment modestly conducive to some shower activity. Similarly, the cold northwest winds on the front of a polar-outbreak anticyclone in winter may generate snow showers on the windward shores of a large lake. And occasionally the circulation aloft may be opposite to that at the surface—for example, a surface anticyclone may be capped by a cyclonic circulation at higher levels—creating circumstances which produce uncertain weather conditions.

Precipitation areas within a cyclone. In both summer and winter, cyclones are the great generators of precipitation within the middle latitudes. Even most of the convective shower activity and thunderstorms of the warmer months appear to be of an organized type, patterned in distribution and located within the generally convergent wind systems of passing extensive depressions. It should be stressed that precipitation in any disturbance would be minor and brief if new supplies of water vapor were not constantly being imported by winds to any area where rain is falling.

Neither the probability of precipitation nor its nature and origin are the same in all parts of a cyclone. In general, the front or eastern half has more cloud and rain than the rear or western half, although the rear is not completely rainless. Clouds usually extend much farther ahead of the center than to the rear, partly because of the location and arrangement of fronts within the cyclone and partly because the front or eastern half of a low is an area of more pronounced and extensive horizontal convergence than the rear. In well-developed winter lows, snow is more common in the cooler northern and northeastern parts, while rain occurs more frequently in the warmer southerly sections. Heavy snows over the central and eastern United States usually arrive when storms travel the more southerly routes, so that the central and northern states are on the poleward sides of the cyclones.

As discussed in an earlier section, two of the main areas of precipitation within a cyclone develop in conjunction with the broad-scale lifting of air that occurs along both the warm and the cold fronts (Fig. 6.7). As occlusion proceeds within the maturing or aging depression, the occluded front as well may tend to localize precipitation. In addition to frontal shower activity, non-frontal rain resulting from general convergence occurs in other parts of a cyclone. In Anglo-America and eastern Asia, showery rains unrelated to fronts are most common within the warm sector of lows. A feature of the warm sector of certain cyclones with especially well-developed cold fronts is the prefrontal squall line, which is accompanied by strong winds, heavy showers, and thunderstorms (Fig. 6.29). Such prefrontal squall lines may precede the actual cold front by 100 to 200 miles. Their origin is not well understood.

Surface temperatures

In themselves cyclones and anticyclones are neither always warm nor always cold. Their surface temperatures depend upon (1) the source and nature of the air masses involved, (2) the season of the year with their varying lengths of day and changing angle of sun's rays, and (3) the degree of humidity and cloudiness. To a large degree an analysis of surface-air temperatures in cyclones becomes a study of the air masses that comprise individual disturbances, for both advection and the heat exchange between atmosphere and ground are major determinants of temperature. In addition, temperature is almost as much a function of the cloud deck (or lack of it) as of the air mass. Winter night minima between nearby stations, one cloudy and the other clear, may differ by as much as 20 to 30°.

Anticyclones. In the winter season a vigorous, well-developed high, advancing rapidly toward the central and eastern United States from northern Canada, moves as a mass of cold, dry polar continental air with clear skies (Fig. 6.15). Such a cold anticyclone brings the cold waves and bitterest winter weather. This type of high is cold for two reasons: (1) it travels rapidly southward from subarctic regions as a mass of cold polar continental air, and (2) its dry, clear air fosters rapid and prolonged losses of the earth's heat during long winter nights. An anticyclone composed of modified polar maritime air brings much less severe winter weather to the interior of the country, although its night temperatures also are likely to be below normal. Even in summer, a well-developed polar continental high advects unseasonably low temperatures, providing several days of clear, cool delightful weather. Pleasantly warm days are followed by cool nights, creating large diurnal ranges. Thus in the middle latitudes many people associate highs with low temperatures for the season.

In summer, however, when a large, relatively stagnant warm high composed of air of subtropical or tropical origin spreads over the south central part of the country, it is likely to produce excessively high temperatures, called hot waves, over the central and eastern United States (Fig. 6.16). The same clear skies and dry air that make for rapid terrestrial radiation during the long winter nights are conducive to maximum receipts of strong solar radiation during long summer days. Moreover, as the tropical air from this anticyclone moves northward over the country, the south winds carry heat absorbed in the lower latitudes. These same stagnant warm highs over the south central part of the country can also bring clear, mild days in the cooler seasons, such as October and November's much-cherished Indian Summer weather.

In summary, anticyclones may be composed of either polar or tropical air masses and so may be responsible for either cold or warm weather. Vigorous moving cold highs arriving from higher latitudes are likely to bring lower-than-normal temperatures, especially in winter. Weak, stagnant warm highs produce abnormally high temperatures, especially in summer. In both winter and summer the clear skies associated with anticyclones are conducive to large diurnal swings in the temperature curve.

Cyclones. A well-developed cyclone, accompanied by an extensive cover of low clouds and perhaps precipitation, is likely to bring temperatures that are higher than average in winter and somewhat lower than average in summer—just the opposite of those induced by an anticyclone. During the long winter nights, the cloud cover and humid air of a cyclone tend to prevent rapid loss of earth heat. The same conditions in summer, when days are long and the sun stronger, tend to reduce incoming solar radiation.

Temperature contrasts within different parts of a cyclone. These general rules concerning lows and seasonal temperatures cannot be accepted too literally, however. Within its several parts or quadrants a cyclone usually has marked temperature contrasts, depending upon the air masses that are represented. Thus in central and eastern North America the south and southeast part of a low, which is the warm sector of tropical air, is considerably warmer than the north and west portion, where the air movement is from cooler higher latitudes. These temperature importations cause the isotherms in cyclones to be skewed in a counterclockwise fashion, so that they assume a north-northeast by south-southwest alignment instead of the normal east-west direction. Warm-air advection by southerly

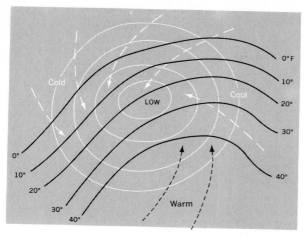

Figure 6.17 Characteristic arrangement of isotherms in a cool-season cyclone over the central and eastern United States.

winds on the front of the storm pushes the isotherms poleward, and cold-air advection by northwest winds on the rear bends them equatorward (Fig. 6.17). To the east and north of the storm center, where the air mass is of a modified polar character—and therefore colder than the tropical air but less severe than the fresh *cP* air to the rear—the isotherms follow the parallels more closely.

In general, the average rise of temperature above seasonal normal in front of a winter storm in the central eastern United States is not far from 10°, although it may be as high as 20 or 30°. Between the front and rear of a well-developed winter cyclone in the same part of the country, temperatures may differ by as much as 30 to 40° or even more. If the temperature of the center of a low is taken as a standard, the average departures of the four quadrants of well-developed winter cyclones in the eastern United States are as follows: northwest, −8.7°; northeast, −5.6°; southeast, +6.3°; southwest, +2.6°.[1] Stations on the southern side of a passing low therefore experience a greater change in temperature than those to the north of the center,

[1]Robert De C. Ward, *Climates of the United States,* Ginn and Company, Boston, 1925, p. 56.

even though the average temperatures are not so low to the south.

Cyclonic-anticyclonic control

The essence of the cyclonic-anticyclonic control of weather is its irregularity or nonperiodic character. Marked nondiurnal weather changes are characteristic of regions and seasons where cyclones and anticyclones are numerous and well developed—as they are, for instance, over central eastern North America. Throughout the middle latitudes temperature changes with passing storms are especially pronounced in winter, when latitudinal temperature gradients are steepest and cold and warm advection by winds is consequently most striking. In summer, with the poleward migration of the frontal zones, cyclones are fewer and weaker, temperature gradients are milder, and sun control, with its daily periodicity, is more influential.

Cyclonic control is not identical, however, in different parts of the earth. Although they all conform to the same general laws, storms differ in character with regions and seasons. Even within the United States, for example, storms act differently over the Pacific Coast or the western plateaus from the way they behave east of the Mississippi. If northwest winds on the rear of a cyclone come from the open ocean, as they usually do in northwestern Europe and on the Pacific Coast of the United States, they obviously cannot advect severe low temperatures as they do in the central and eastern United States.

Typical cyclone and anticyclone: eastern United States. This section summarizes weather changes that might be expected during the passage of a well-developed cyclone and a following anticyclone over the eastern United States (Figs. 6.18 and 6.19). As the cyclone approaches from the west and the barometer falls, skies gradually cloud up. The first evidence of ascent along the warm-front surface of discontinuity is fine veils or films of cirrus and cirrostratus, which produce circles around the moon or sun. These distant heralds of the storm appear as

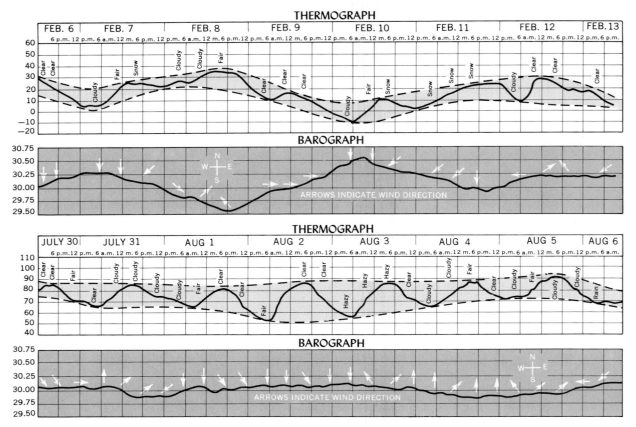

Figure 6.18 Barograph and thermograph traces for a week of winter (*top*) and summer (*bottom*) weather. Note that the barograph trace for winter shows much wider fluctuations than that of summer, indicating much stronger cyclonic-anticyclonic control in the cooler months. The two temperature traces demonstrate that diurnal sun control is stronger in the warm season, when the rise and fall in temperature is diurnal and regular. If a temperature "belt" (represented by shading) is created for each of the two temperature traces, by connecting the diurnal crests and troughs, it will be noted that the temperature belt of winter shows wide nonperiodic oscillations, while that of summer is relatively flat. The wide winter oscillations of the belt are induced by the air masses of passing cyclones and anticyclones. The flat belt of summer indicates weaker air mass and cyclonic-anticyclonic control in that season.

much as 500 to 1,000 miles ahead of the surface warm front, where the inclined wedge of cold air may be several miles thick. Together with the arrival of the cirrus and cirrostratus clouds, the wind sets in from an easterly direction, and it will continue easterly until the surface warm front passes.

As the surface warm front moves closer and the wedge of cold air becomes thinner, the clouds gradually thicken, darken, and lower. Precipitation usually begins several hundred miles ahead (east) of the surface warm front and continues until the front has passed. Temperature increases somewhat as the surface warm front draws nearer, but since the air on the front of the storm may be modified polar in character, there is no abrupt temperature rise until the warm front at the surface has gone by. The passage of the warm front at the surface, with an associated shift from polar to tropical air,

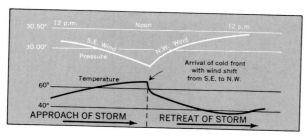

Figure 6.19 Barograph and thermograph traces registered during the approach and retreat of a middle-latitude cyclone in the central and eastern United States.

is marked by a number of weather changes (described on page 203).

Weather conditions within the warm air mass may vary considerably depending on the nature of the air and the season of the year. In summer, air from the Gulf of Mexico creates hot, sultry weather with showers and thunderstorms. Tropical air from a drier source brings heat that is less oppressive but more desiccating. In winter, a tropical air mass produces mild weather and rapid thaws, often associated with fog.

The arrival of the cold front at the surface, with an accompanying shift from the warm to the cold air mass, frequently results in strong turbulence, particularly if the cold front is well developed. In summer, severe thunderstorms with strong squall winds are common. Other weather changes associated with cold-front passage are noted on page 205. The anticyclone which follows the cyclone moves southeastward as a mass of cold polar continental air with northwest winds. It continues to reduce the temperature until its center has passed. Toward this center winds are light and calms are prevalent, and in winter surface temperatures may be extremely low. As the anticyclone retreats, the cycle is complete: winds again become easterly, and another approaching cyclone begins a new sequence.

Obviously storms can vary considerably from this generalized description. The weather phenomena noted in any individual storm depend on the nature of its air masses and on its route with respect to the observer.

Distribution of middle-latitude disturbances

All parts of the middle latitudes, including the adjacent margins of the low and high latitudes, are affected by the traveling cyclones and anticyclones of the westerlies. Not all parts are affected to the same degree, however. While there is no rigid system of clearly defined *storm tracks,* there are certain broad belts over which storm centers travel more frequently than elsewhere. There are also regions where cyclones originate more frequently and are more numerous than elsewhere. Of course, the effects of a disturbance are felt far beyond the path of its center. In a general way, the path of a disturbance follows the upper-flow pattern. In all parts of the world cyclones move chiefly toward higher latitudes, while anticyclones move toward lower latitudes.

Cyclones. In the following discussion Figs. 6.20 to 6.22 showing world distribution of cyclones should be compared with Figs. 5.3 and 5.4, on which the main frontal zones are charted. Although locations of the frontal zones are indicated by single lines, these show only mean positions of convergences that oscillate within wide limits, depending on the season and on actual air-mass distribution.

In the Southern Hemisphere, cyclonic storms are vigorous in all seasons as a result of a very stable and intense cold source over the Antarctic Continent throughout the year. These storms appear to concentrate along the subpolar low-pressure trough, and their belt of maximum frequency is continuous around the earth. The northerly margins of the cyclones, however, affect the poleward parts of all the Southern Hemisphere continents, especially in the low-sun season. The Cape Horn region of South America, extending as it does nearly to latitude 55°S, is stormy the year round. Winter cyclones are also relatively numerous over the Argentine Pampa.

In the Northern Hemisphere, the arctic and subarctic cold sources which provide the southward thrusts of cold air essential to the mechanism of cyclone formation are strong in winter but relatively weak in summer. Accordingly, while winter cyclonic

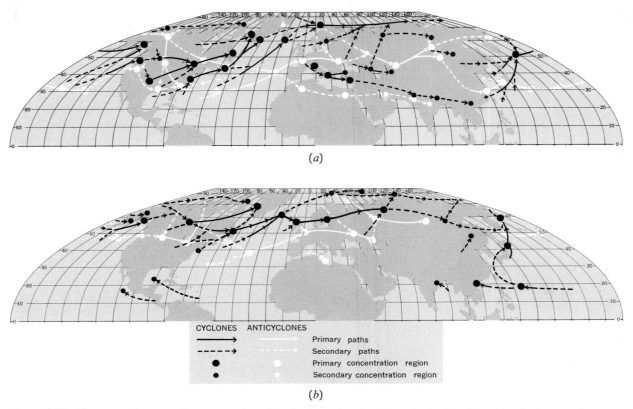

(a)

(b)

CYCLONES   ANTICYCLONES
⟶   ⟶   Primary paths
- - - ➤   - - - ➤   Secondary paths
●   ○   Primary concentration region
●   ○   Secondary concentration region

Figure 6.20   Routes and areas of concentration of cyclones and anticyclones in the Northern Hemisphere: *a*, January; *b*, July. (*After W. Klein.*)

activity is vigorous, that of summer is comparatively mild. In addition, there is a general poleward shift of the cyclone belt in the warmer months.

Although *winter* cyclones originate in almost any and all parts of the middle latitudes, two of the most important cyclogenetic regions are the frontal zones of steep temperature gradients positioned off the east coasts of Asia and North America. Other cyclones form inland over China and Siberia, but most western Pacific storms appear to develop along the polar front over the relatively warm waters around Japan. These Pacific depressions move northeastward toward the Aleutian Low, which is a center of maximum cyclone frequency, and then along a route south of the Aleutian Islands into the Gulf of Alaska. Because of the concentration of storm

tracks, the Gulf of Alaska has one of the largest number of lows on earth. A second center of cyclone frequency is located near Vancouver Island, which receives some depressions from the Gulf of Alaska besides being the terminus of many storms that develop in the eastern Pacific (Figs. 6.20*a*, 6.21).

Unlike interior Asia, interior North America is a region both of major cyclogenesis and of frequent vigorous cyclonic storms. Most of the continent's storms either originate or redevelop in two centers just east of the Rockies—one in the Alberta region and the other in Colorado–western Nebraska. The primary Alberta storm track, reaching from southern Alberta to a center of maximum frequency in the vicinity of the Great Lakes, is probably the best defined in all North America (Fig. 2.20*a*).

In the Great Lakes region the Alberta track is joined by cyclonic tracks from other areas of the United States. The most important of these is the route followed by the Colorado lows, which move northeastward across the central part of the country. Others from Texas and the Ohio Valley also converge on the Great Lakes center, from which they move eastward through the St. Lawrence Valley to a center of maximum frequency in the vicinity of Newfoundland. This center also receives many lows from the region of cyclogenesis lying close off the Atlantic seaboard from Cape Hatteras northward, where the thermal contrasts between cold land and warm water favor cyclone development (Fig. 6.23).

Interior and eastern North America is the world's most cyclonic continental area. This unique distinction derives in part from the numerous disturbances which originate on the lee side of the Rocky Mountain barrier and whose development is subsequently promoted by the free clash of contrasting air masses over the great central lowlands which reach from the Arctic to tropical latitudes.

From the western Atlantic, Newfoundland cyclones and those from south and east of that island move northeast toward the Iceland Low, some being shunted into Davis Strait by the Greenland ice plateau. Cyclones emerging from the eastern end of the Iceland Low travel northeastward to a strong center of maximum cyclone frequency in the vicinity of the Barents Sea north of Scandinavia (Fig. 6.20a).

Figure 6.21 Percentage frequency of cyclone centers in winter. (*From Sverre Petterssen, Introduction to Meteorology.*)

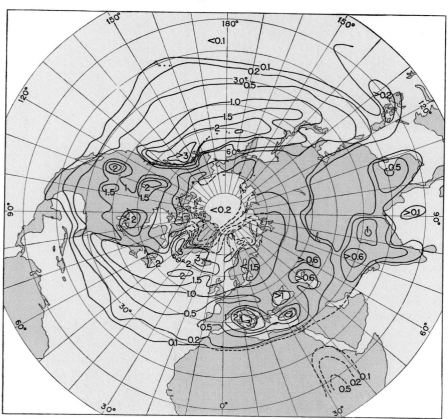

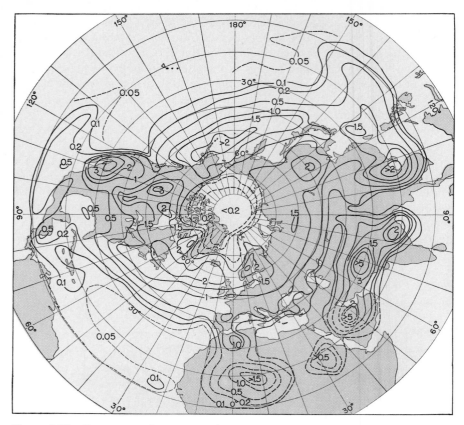

Figure 6.22 Percentage frequency of cyclone centers in summer. (*From Sverre Petterssen, Introduction to Meteorology.*)

Within Europe, cyclones move in a generally eastward direction along two primary routes located along the northern and southern margins of the continent, and a lesser intermediate track across the North and Baltic Seas. The Mediterranean Basin is one of the earth's great regions of winter cyclogenesis and cyclone frequency. From the eastern Mediterranean, secondary tracks extend eastward across Iraq and Iran into northern India-Pakistan, which also acts as something of a center of cyclogenesis. A few of these storms may even reach China. Other Mediterranean cyclones travel northeastward across the Black and Caspian Seas, and a few move north into Russia. Siberia, because of its prevailing cold anticyclone, is markedly free of deep winter cyclones—a situation which contrasts with that of interior North America.

By *midsummer* in the Northern Hemisphere there has been a considerable northward shift of most features of the general circulation, and with it a corresponding poleward displacement of the principal storm tracks and centers of cyclone concentration. As a consequence, the important storm tracks in the subtropics and lower middle latitudes have generally disappeared. Now they concentrate in the high latitudes of the Bering Strait and in the American and Russian subarctic and Arctic. The western and southeastern parts of the United States are free of major storm tracks, and the same is true of China and the Mediterranean Basin (Fig. 6.20*b*).

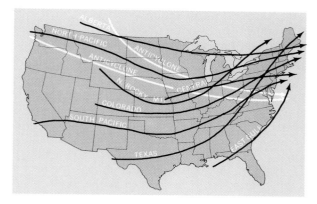

Figure 6.23 Main tracks of cyclones and anticyclones in the United States. Note how the cyclone tracks converge toward the northeastern part of the country. Cold anticyclones have a dual origin. Those which move south from northern Canada (*cP* air) bring severe cold in winter; those from the Pacific Northwest (modified *mP* air) bring only moderate cold. Many of the latter are in the nature of weak highs separating traveling cyclones. Warm anticyclones are not shown.

Anticyclones.   Cold and warm anticyclones have quite different regions of origin and routes of progression. Unlike extratropical cyclones, which have a northward component of movement, cold anticyclones, since they originate in high latitudes, usually move southward. But the warm anticyclones of middle latitudes have more of a west-to-east progression—as many cold anticyclones do also, after they have reached about latitude 40°. Like cyclones, the prevailing tracks of highs and their areas of maximum genesis and occurrence are generally farthest south in late winter and farthest north in late summer.

The highest frequency of individual cold anticyclones in *January* is found in two major centers: northwestern Canada and east central Siberia (Fig. 6.20). From the North American center the cold highs push rapidly southward along a primary track into a center of maximum in the Dakotas (Figs. 6.20 and 6.23). In about the latitude of the Ohio Valley the main track gradually bends eastward, so that its highs contribute to the center of maximum frequency in the Middle Atlantic states. It is these

Canadian anticyclones that bring cold waves, blizzard conditions, and minimum temperatures to the Mississippi Valley. A few of the Canadian anticyclones, instead of taking an eastward course, push farther south into the Gulf states, where their advance is heralded by the notorious "norther." In the western Gulf states, the track of these southward-displaced highs merges with that of highs arriving from the Great Basin. Strong westerlies appear to shunt some anticyclones onto an eastward course, which carries them along a route north of the Great Lakes.

From the Siberian center, cold highs move south and east over northern China, usually crossing the seaboard between Shantung and the mouth of the Yangtze River. There a strong maximum of anticyclone occurrence is located. These highs continue eastward over Japan.

A less severe variety of cold anticyclone originates in the Great Basin area of the intermontane western United States. In this region the frequency of both anticyclones and anticyclogenesis is higher than anywhere else in the Northern Hemisphere. Most of the air in such highs is of Pacific origin, although there has been some addition of colder air from Canada west of the Rockies. Stagnation of air masses within the Great Basin gradually gives them a degree of continentality, but they never approach the severity of the Arctic anticyclones composed of fresh *cP* air. After crossing the Rockies, they redevelop farther east over the southern Plains states.

As for the European continent, it has some winter anticyclones which move southward from the Scandinavian centers. Most European highs, however, are in the form of warm cells emerging from the subtropical Azores High which cross the continent on a main track located at about 50°N.

In *July* the main anticyclonic tracks and centers of maximum occurrence are displaced northward. One major zonal track across the northern United States is followed by cells derived from both the Pacific and northernmost Canadian sources. This track has pronounced centers of maximum anticyclone frequency positioned over the Great Lakes and northern Great Plains (Fig. 6.20*b*). In eastern Asia, anticyclone frequency is greatly reduced in the summer monsoon, so that there are only secondary tracks across China. In Europe, one primary track located at about 50°N is fed principally by cells from the subtropical high and to a lesser degree by anticyclones from high latitudes. A second route of warm highs is through the Mediterranean Basin.

Large-scale weather situations, singularities, and the weather calendar

In addition to the local circulation patterns of migratory and transient cyclones and anticyclones, there are others of a broad-scale, quasi-stationary type which account for the longer weather episodes of a region. In response to a particular and persistent type of extensive circulation pattern, the weather of a region may remain fairly consistent for as much as several weeks, although minor changes will occur. According to some atmospheric scientists, such spells of fairly homogeneous weather tend to recur annually at about the same times in the calendar year. In the United States, examples of such large-scale weather episodes, sometimes called natural seasons, are the Indian Summer of autumn and the January Thaw of midwinter. Each of these is controlled by a particular synoptic pattern. In the case of Indian Summer, a large, stagnant warm anticyclone is positioned over the southern and southeastern part of the United States, with the circulation around its western end advecting warm air from the southwestern dry areas of the country.

If the transition from one large weather episode to another occurs within a relatively short time range, and if it is associated with recognizable broad-scale synoptic changes, it is spoken of as a *singularity* (described in Chapter 2). A chronological table of the year's natural seasons, or longer weather episodes, is designated as the *weather calendar*. Unfortunately, no one has yet developed a widely accepted classification of large-scale weather situations, singularities, and weather calendars that can be applied to the whole earth, or even to the Northern Hemisphere. By far the greatest progress in this direction has been made in the application of these concepts to European weather.

If large-scale weather patterns are genuine entities, physically conditioned, that tend to recur at particular times of the year, it might be expected that their weather calendar could be useful in long-range forecasting. One group of weather scientists does hold this point of view. Others are inclined to look upon the larger weather episodes as somewhat fortuitous circumstances, the result of chance fluctuations caused by the coincidence of many minute elements not readily observable.

TROPICAL DISTURBANCES

As discussed in Chapter 3, the increase in information about tropical weather disturbances since World War II has tended to undermine earlier ideas about the monotonous, uneventful, and generally fair weather of the low latitudes. Today there is a general awareness that a considerable variety of weather disturbances exist within the tropics. At the same time, the literature on low-latitude weather is confusing when it attempts to distinguish and classify the various types of disturbances, describe their distinctive weather features, and trace their geographic distributions.[2]

Most types of extensive tropical disturbances appear to be relatively mild phenomena as far as pressure gradients and winds are concerned. Their primary climatic significance lies in the fact that

[2] See N. E. LeSeur, "Synoptic Models in the Tropics," *Symposium on Tropical Meteorology, Rotorua, New Zealand, 1963,* New Zealand Meteorological Service, Wellington, 1964.

they generate clouds and precipitation. In the low latitudes, temperature changes and variations associated with the advective effects of disturbances are negligible. Their cloud decks, on the other hand, do perceptibly lower the average temperature as well as shrink the diurnal range.

Unfortunately for the climatologist, it is the violent and spectacular hurricane type of tropical storm which has been studied most intensively. Yet there are relatively few of these storms compared with the milder and more beneficent types, and they occur in much more restricted areas within the tropics. Genuine hurricanes cannot develop over extensive land masses or even invade them for any distance, so that in the low latitudes it is exclusively the weaker disturbances which affect the weather of any land but islands and the margins of continents.

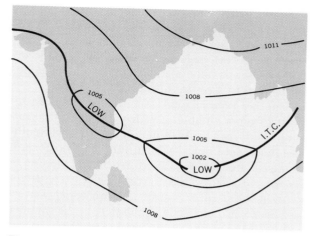

Figure 6.24 Weak tropical disturbances of the summer monsoon period over the Bay of Bengal and India. Some such monsoon lows appear to be associated with a fluctuating ITC.

Weak tropical disturbances

It is not uncommon for a spell of rainy, cloudy weather to approach and pass by a weather station in the tropics without the instruments registering any significant changes in wind and pressure that might indicate the nature and origin of the disturbed weather. On occasions it is suspected this is because the disturbance is too weak to register preceptible changes on the instruments. Moreover, many weak tropical disturbances do not exist in the lower atmosphere, but are phenomena of the mid- and upper troposphere and hence are much more difficult to detect. Then too, weather stations throughout much of the tropics are very widely spaced, and many are not well equipped with modern instruments, so that cooperative efforts in the detection and plotting of weak disturbances are rendered ineffective. No adequate classification of types of weak tropical disturbances now exists; there are probably many more varieties than this discussion indicates. As for their origins, these are uncertain also.

One type of extensive tropical disturbance which appears to be widespread throughout the humid tropics (and in summer, within parts of the humid subtropics as well) is the weak, shallow low-pressure

system. It develops chiefly within the humid, unstable equatorial air that follows behind the ITC when this zone is displaced well to the north or south of the equator within the summer hemisphere. On the surface weather map such a system may be detected by its one or two widely spaced closed isobars (Fig. 6.24). Pressure gradients are weak, its convergent wind system is poorly developed, and its rate and direction of movement are sluggish and erratic. But in spite of its being weak, such a disturbance is able to induce a great deal of showery convective precipitation. When one remains fairly stationary over an area for several days, rainfall may be excessive.

While these weak tropical disturbances appear to be widespread, naturally they are best known in regions with well-developed weather services, such as India and Australia. In India, where they go by the name of "monsoon depressions," they are responsible for a large part of the summer rainfall. Originating over the Bay of Bengal, perhaps as waves along the diffuse ITC, they enter India from the east with an average frequency of about one every 6 days. Their course is northwestward, which tends to concentrate their weather effects over the northern, and especially the northeastern, parts of the subcontinent. While monsoon lows are extensive—usually several hundred miles in diameter—they are not as large as most middle-latitude cyclones. Occasionally one of these weak depressions develops into a severe hurricane or typhoon.

Another type of weak disturbance has been reported from a number of widely dispersed tropical-subtropical regions (including India, parts of Southeast Asia, subtropical China, interior Brazil south of the equator, Tropical West Africa, and the Congo Basin) and is suspected of existing elsewhere. It appears to be a surge within the equatorial westerlies, or tropical monsoon circulation, of the summer hemisphere. In India, for example, the summer southwest monsoon is not a steady current. Instead it is frequently interrupted by pulsations which move downstream with the main current, traveling in a general easterly direction. During these surges, which are only apparent aloft, wind velocities may increase to twice their normal rates. Such wind surges result in speed convergences, with the accelerated air acting to force the air ahead of it upward. The result is a spell of bad weather with cloud and rain.

A somewhat better-known weak disturbance is called an *easterly wave*. It is a feeble trough of low pressure extending in a general north-south direction, and hence lying athwart the trade-wind circulation. It moves slowly from east to west in the trade winds, mainly those of the summer hemisphere. Such weak transverse waves in the trades are easier to detect through their deformation of the wind field than of the pressure field. They are more readily observed aloft than at the surface. As the trough moves westward in the easterly circulation, there is general divergence, and subsidence takes place ahead, or west, of the trough line. There dry, fair weather prevails (Fig. 6.25). Just behind the trough line, the temperature inversion lifts and the moist surface layer of air attains great depth. This, in conjunction with widespread convergence and lifting, results in extensive convective showers and thunderstorms, with rain falling from cumulus clouds. Severe line squalls are sometimes associated with

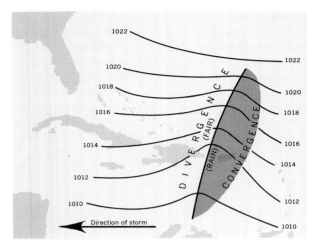

Figure 6.25 A tropical disturbance of a type known as an easterly wave. (*After Riehl.*)

easterly waves in Tropical West Africa. A few easterly waves in the more unstable western parts of the oceanic trades develop into vigorous vortex storms, some of them of the dreaded hurricane variety.

Less important than the three disturbance types just mentioned are others that result when modified polar air, preceded by a polar front, invades the tropics. The result is a trough aloft accompanied by disturbed weather. The whole system moves slowly eastward, or in the opposite direction to that of an easterly wave. On occasion what is called an "induced trough" may develop within the trades, forming in response to a young frontal wave located on the poleward flanks of a subtropical high.

Violent tropical cyclones

Vortex or revolving storms of all grades of intensity exist within the tropics, but relatively few are severe enough to warrant being officially designated a storm. And of these, very few qualify as genuine hurricanes, which must have a minimum wind speed of 75 miles per hour. The number of hurricanes is never large; even in a "good year" no more than 50 develop in the whole Northern Hemisphere. By comparison there may be 20 to 30 middle-latitude cyclones per day during the winter months. But although they are infrequent, and are restricted to a few oceanic areas and adjacent islands and continental margins, the violence and destructiveness of hurricanes cause them to be the subject of great public interest and intensive scientific investigation.

The hurricane (called typhoon in the East Asia sector) is a vertically deep and violent cyclonic whirl of air (Figs. 6.26 and 6.27). There is a spiraling inflow of air at lower levels, a rapid upward movement at intermediate levels, and a spiral outward flow aloft. The whole storm system may extend up to 35,000 or 45,000 ft. Wind speeds of over 100 miles per hour are not uncommon. Within the area of violent winds there are dense nimbostratus clouds which extend upward to more than 20,000 ft. At the very core of the cyclonic whirl of air is a region of serene calm—sultry and nearly cloud-free—where the movement of air is downward. This is called the eye of the storm.

Torrential rains, associated with large-scale rapid uplift, accompany a hurricane. Commonly 5 to 10 in. fall during a single storm, and there is an instance of 46 in. in one day. Isobars in hurricanes are nearly circular, with the diameter of the storm varying from 100 to 400 miles, or roughly one-third that of middle-latitude cyclones. Pressure gradients are excessively steep, and the pressure readings near the center of the storm may be as low as 27.0 in. (914.3 mb). Temperature distribution around the center is relatively similar in all directions, for there are no contrasting air masses or fronts. The source of energy for the maintenance of these violent storms is latent heat of condensation.

Precisely what mechanism is responsible for hurricane formation remains in doubt. Still, there is agreement on some of the requirements which must be satisfied if such a storm is to develop. These include (1) a relatively large ocean area where the sea surface temperatures are high (above about 26.5C or 80.0°F), so that the air above is warm and humid; (2) a sufficiently large value of Coriolis force to cause a vortex circulation of air—a requirement which excludes hurricanes from a belt 5 to 8° wide

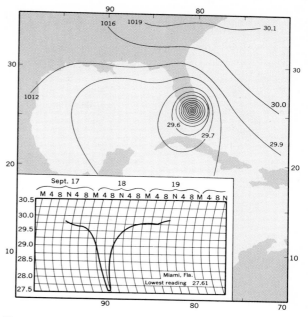

Figure 6.26 A Caribbean hurricane, together with its barograph trace as recorded at Miami, Florida.

on both sides of the equator; (3) weak vertical wind shear in the basic current, which limits hurricane formation to latitudes equatorward of the subtropical jet stream; (4) some preexisting weak tropical disturbance acting as the parent; and (5) upper-level outflow above the surface disturbance. While there is no season entirely free of hurricanes, the four months from July to October have a great preponderance of them (nearly 90 percent of the Pacific hurricanes and 75 percent of those in the Atlantic). It is probably significant that hurricanes are most prevalent in late summer and early fall. Not only is this the time of greatest oceanic warmth, but it is the period when the ITC and unstable equatorial air masses are displaced farthest from the equator, thus becoming subject to appreciable Coriolis force.

As noted above, a hurricane has its beginnings in a preexisting weak disturbance. But it is only an occasional such disturbance that develops into a violent storm, although all would appear to have that potential. A hurricane may exist in the weak-

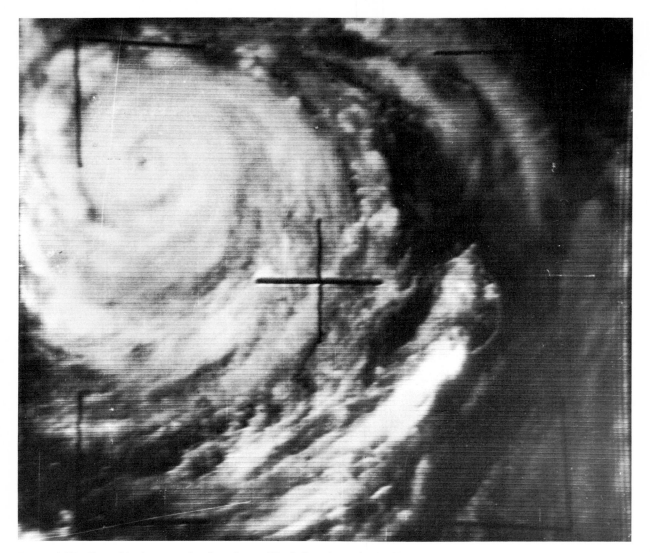

Figure 6.27 Tiros V photograph of typhoon "Ruth," at latitude 31°N, longitude 141°E, August 18, 1962. (*U.S. Weather Bureau.*)

disturbance stage for 2 to 5 days before it begins to intensify markedly, and then another 2 to 6 days may elapse before it reaches its greatest intensity. At first such disturbances move slowly westward in the trade-wind circulation, but toward the western side of the ocean they tend to recurve poleward, and subsequently eastward, as they follow the circu-lation around the western end of a subtropical anti-cyclone. As they move out of the tropics and into the lower middle latitudes, they quickly lose their intensity and become only strong polar-front cyclones. If their course takes them inland, they even more promptly lose many of their hurricane characteristics. It is mainly during the process of recurving from a westerly to an easterly course around the western end of an anticyclonic cell that they wreak their fury upon the continental margins and carry excessive rains even into the middle latitudes. The destructive force of a hurricane comes not from the violence of its winds alone, or from the flooding and washing caused by the deluges of rainfall; probably just as often it is from the sea's high water or storm surge which so frequently accompanies this type of storm as it crosses a coastline.

Regional distribution and frequency of hurricanes. Although all parts of tropical oceans in both hemispheres probably have tropical storms, there is no record of one with hurricane intensity in the South Atlantic Ocean or the eastern South Pacific. Rarely, if ever, has one been observed within 5° of the equator, where Coriolis force is absent. There are at least six general regions of hurricane origin and concentration. These are noted below, together with the special areas and seasons of occurrence and the average number of storms per year. The figure in parentheses is the number of storms reaching hurricane intensity, but for some areas these data are not known.

1. The south, and especially the southwestern, portion of the North Atlantic Ocean: 7.3 (3.5).
   a. Cape Verde Island area—August and September. Possibly these storms do not attain hurricane intensity until they reach the western part of the ocean.
   b. East and north of the West Indies, including Florida and the South Atlantic coast of the United States—June through October.
   c. Northern Caribbean Sea—late May through November.
   d. Southwestern Caribbean Sea—principally June and October.
   e. Gulf of Mexico—June through October.
2. North Pacific Ocean off the west coast of Mexico—June through November: 5.7 (2.2).
3. Southwestern North Pacific Ocean, especially the China Sea, the Philippine Islands, and southern Japan—chiefly May through December: 21.1. Here the tropical cyclones are known as typhoons.
4. North Indian Ocean
   a. Bay of Bengal—April through December: 6.0.
   b. Arabian Sea—April through June, September through December: 1.5.
5. South Indian Ocean
   a. Madagascar eastward to 90°E—November through April: 0.9.
6. South Pacific Ocean eastward from Australia to about 140°W—December through April.

The region of greatest frequency of tropical cyclones, and perhaps the one which has storms of the greatest intensity, is the typhoon area of the Philippines and the China Sea. On the other hand, the best known hurricane region is the Caribbean and western North Atlantic region. Because of numerous inhabited islands and a relatively important ocean traffic in those parts, observational data are abundant.

THUNDERSTORMS

General characteristics and structure. A thunderstorm differs from a convectional shower system only in that lightning and thunder are present. And since lightning appears to be associated with unusually vigorous updrafts, it is fair to say that a thunderstorm is simply a more violent type of instability shower. It represents atmospheric convection of unusual strength.

Fundamentally a thunderstorm is a thermodynamic machine in which the potential energy of latent heat of condensation and fusion in moist conditionally or convectively unstable air is rapidly

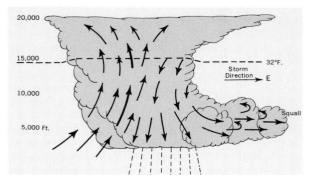

Figure 6.28 A vertical section through a thunderstorm and its cumulonimbus cloud. (*After Byers and others.*)

converted into the kinetic energy of violent vertical air currents. These produce torrential rain, hail, gusty surface squall winds, lightning, and thunder. The intensity of the storm depends upon the supply of latent energy and the rate at which the available energy is expended.

In structure, the typical mature thunderstorm is an agglomeration of convective cells. Each cell, the diameter of which may be 1 to a few miles across, is composed of chimneys of rapidly ascending warm humid air separated by compensating cooler downdrafts with less velocity. The characteristic turbulence within a thunderstorm, associated with its cellular structure, is visibly evident in the dramatic convulsions to be observed in a mass of thunderhead clouds (Fig. 6.28).

Conditions favoring thunderstorm development. Probably the most important condition for thunderstorm development is atmospheric instability, which in turn favors vertical upward motion. Moreover, instability is closely related to vertical temperature and moisture stratification in the atmosphere. The degree of instability mainly determines the intensity of the storm.

As a rule, thunderstorms are most prevalent in the warmer latitudes of the earth, in the warmer seasons of middle latitudes, and in the warmer hours of the diurnal period. This all suggests that high surface temperatures provide a favorable environment for thunderstorm development. Surface

heat is significant because it steepens the lapse rate in the lower air and thus increases its buoyancy, while at the same time it increases the capacity of the air for evaporation and for moisture.

But heat alone is not sufficient for thunderstorm development; there are many intensely hot regions, such as the tropical-subtropical deserts, where few thunderstorms occur. An ample supply of atmospheric moisture is also required in order to supply the necessary energy in the form of latent heat of condensation to maintain the storm. Normally the relative humidity of the warm air must exceed 75 percent. In addition, the lapse rate must be conditionally or convectively unstable.

Further prerequisites for thunderstorm development are some agent to trigger the convection, and a cumulonimbus cloud of great vertical development, usually over 10,000 ft. Instability in itself is not adequate; some trigger is needed to start convection even in conditionally or convectively unstable air. Such trigger action is of several kinds: (1) uplift over hills and mountains; (2) turbulence resulting from friction between the moving air and the earth's surface, or from internal forces within the air itself; (3) convergence in the airflow, with resulting uplift; and (4) upglide along fronts. Possibly intense solar heating of a land surface may sometimes trigger upward motion of the air, although there is increasing skepticism concerning the effectiveness of this method when acting alone. Nevertheless, strong surface heating does markedly raise the atmosphere's buoyancy and instability by steepening the lapse rate.

Since the lapse rate in the lower air over water bodies is normally steepest at night, it is not surprising that thunderstorm frequency over large seas and oceans should show a nocturnal maximum. More unusual is the fact that nighttime thunderstorms also predominate over certain continental regions, such as an extensive area in the middle western United States centering on Nebraska-Kansas and another smaller area in southern Arizona. Various explanations have been suggested for this phenomenon, but none seem entirely satisfactory.

Precipitation in thunderstorms.   The general characteristics of convectional rainfall—including its localism, intensity, and brief duration—were discussed in Chapter 4. One speaks of *thundershowers* rather than *thunder rains.* The rainfall pattern within a thunderstorm is closely associated with the arrangement of the cells and their stages of development. Rain is heaviest under the core of a cell and decreases toward its margins. The duration of the heavy rain from a single cell varies from a few minutes in weak ones to as much as an hour in large and very active cells. In the eastern and southern United States the average duration of a thundershower at an individual station is about 25 min, although this is a mean of a strong variant. Half the total rainfall in the tropics is believed to fall in 10 percent of any time interval—from 24 hr to $\frac{1}{10}$ hr. The unusual localism of thundershower rainfall causes its distribution to be highly patchy and uneven. It is common for one part of a city to suffer a deluge while other parts remain dry.

Hail.   Occasionally hail, locally the most destructive form of precipitation, falls from the cumulonimbus cloud of a thunderstorm. Fortunately it occurs in only a few storms and is confined to narrow belts within any individual storm. Hail is most frequent in spring and early summer and in the afternoon hours of the day—all times when atmospheric instability and vertical updrafts of air are strongest. Hailstones vary in diameter from a fraction of an inch to 2 or 3 in. Large stones consist of onion-like concentric shells around an ice-pellet nucleus. This indicates that they are a product of great turbulence within a vertically deep cumulonimbus cloud in which supercooled water drops, snow, and ice particles are all present. Strong vertical currents carry these precipitation products up and down through repeated cycles of melting and freezing. When hail falls to earth, its impact may be disastrous for growing crops along the narrow path of the hailstorm. Local variations in frequency of hail and amount of hail damage are striking.

The regional distribution of hail frequency is not easy to explain. It should be noted that hail frequency and thunderstorm frequency do not necessarily show the same distribution. For example, hail is rare in the wet tropics, where thunderstorms are most numerous. Within the United States, thunderstorms are most frequent in the subtropical southeastern part, especially in Florida and along the Gulf Coast, but this region has only very infrequent hail. Hail frequency is greatest over the interior of the United States, particularly over a belt about 100 miles wide east of the Rocky Mountain Continental Divide, where newly formed intense showers are relatively frequent.

Lightning and thunder.   Lightning is the flash of light accompanying a discharge of atmospheric electricity. Fortunately, in terms of damage potential, lightning flashes occur more frequently within a cloud, or from one cloud to another, than from cloud to the land surface. The origin of lightning is not clearly understood. It may well be a part of the mechanism by which the standard condition of a negatively charged ground and a positively charged high atmosphere (above 50 miles) is maintained. The normal fair-weather electric current is from atmosphere to ground. A return current from ground to atmosphere is required, and it has been suggested that this occurs largely by means of the electric discharges in the cumulonimbus clouds of the earth's thunderstorms. About 3,000 such storms (averaged over a 24-hr period) are believed to be in progress over the earth at any one time.

For a return electric current to pass from ground to atmosphere, a separation of electric charges must take place within the thunderstorm cloud. Such a separation is known to exist, but how it is produced is not fully understood. However, the presence of both liquid water drops and ice particles within a cumulonimbus cloud appears to provide one favorable condition. Thunder is produced by the violent expansion of the intensely heated air along the path of the lightning flash.

The squall wind.   In the beginning, or cumulus, stage of a thunderstorm the surface winds are mildly convergent. As the cell grows and the downdraft develops, the cold divergent outflow from the downdraft is strong and gusty (Fig. 6.28). The cold dome of outflowing downdraft air is asymetrically developed, extending out much farther on the downwind side, or the front, of the storm. This is the so-called thundersquall. It is associated with the squall cloud, an onrushing dark-gray boiling arch or roll of cloud which is the forward projection of the lower portion of the storm cloud.

The velocity of the squall wind at times attains hurricane violence, so that it may do serious damage to crops and buildings. The force of the squall is due in part to the cool air that has been brought down from aloft with the mass of falling rain. Being denser than the warm surface air, it spreads out in front of the storm, underrunning the warm air. In part its velocity is due also to the onrushing motion of the storm mass itself as it is carried by the upper westerlies, so that forward and outward motions are combined.

Types of thunderstorms

Observations show that a vast majority of thunderstorms occur in more or less distinct patterns; their distribution exhibits an organized arrangement. With the increase in information about thunderstorms, the less they appear to be a consequence of random thermal convection associated with solar surface heating, and the more they seem to occur in conjunction with the convergent circulations of extensive atmospheric disturbances of various kinds. In the Rift Valley of Kenya in East Africa, it was observed that only 4 percent of the annual rainfall there might possibly be attributed to random thermal convection. It is the opinion of one weather observer at Nairobi that for the tropics as a whole, the idea of scattered convectional showers, random in occurrence, is a figment of the imagination. A study of Wisconsin thunderstorms has shown nearly all to be associated with large-scale synoptic patterns. It seems probable, therefore, that so-called heat thunderstorms, local and random in their occurrence and distribution, are infrequent or possibly even nonexistent.

Air-mass thunderstorms. Close to the equator, where thunderstorm frequency is at a strong maximum, general temperature homogeneity almost precludes the existence of density fronts. Here, convective showers and thunderstorms must be largely intra-air-mass in character. Their normal occurrence is within the convergent circulations of a variety of weak tropical disturbances of an extensive character. On the surface weather maps of successive days one is able to observe the shifting patterns of thunderstorm distribution in conjunction with the movements of these widespread disturbances, of which the shower activity is an integral part. Tropical thunderstorms over lands usually show a marked diurnal periodicity, with a maximum in the afternoon and evening hours when surface heating has produced a steepened lapse rate and greater atmospheric buoyancy.

Air-mass thunderstorms in middle latitudes are moderately common when the air within the fairly homogeneous warm sectors of cyclonic storms is unstable. Here the heat and humidity are favoring factors, while the prevailing general convergence may provide the trigger action initiating uplift. Others are produced by warm-air advection at low levels or cold-air advection aloft.

Orographic thunderstorms. In both low and middle latitudes orographic thunderstorms result from the mechanical lifting of a conditionally or convectively unstable air mass by highlands. Such upward deflection furnishes the trigger effect necessary to release large reserves of latent energy of condensation in air masses which are already potentially unstable. It follows, therefore, that thunderstorms are usually more numerous in hilly and mountainous regions than they are where the terrain is flat. In the United States they are common in summer over the mountainous regions of the West and Southwest and over the Appalachians in the East. Because their formation is governed to a great extent by the local terrain, orographic thunderstorms have individual peculiarities of time, occurrence, intensity, and location which make them an easy type to forecast. A unique characteristic of many is their tendency to remain almost stationary, probably because the upward thrusts of air that feed them are distinctly localized. This feature explains their tendency to produce cloudburst rains.

Frontal thunderstorms. These are largely phenomena of the middle latitudes, where they are triggered by a forced ascent of conditionally or convectively unstable air over a wedge of colder, denser air. Although frontal thunderstorms may be scattered within the frontal zones (cold front, warm front, occluded front) of a cyclone, they are definitely organized in their general distribution. Since they move along in conjunction with the storm's fronts and their cloud regions, such thunderstorms may occur at any time of day or night. Still, even this type benefits from daytime solar heating, and so is most numerous in afternoon and evening.

Cold-front thunderstorms, because the upthrust of warm air along the blunter nose of the advancing cold front is more rapid than that along the less steeply inclined warm front, are likely to be more vigorous. Also, much of the awesome thundercloud activity is likely to be visible to an observer at the ground. In addition, cold-front storms are normally followed by clearing weather and cooler temperatures associated with the northwesterly winds on the hind side of the front. When the cold front is also a sharp wind-shift line, as in a V-shaped cyclone, the belt of thunderstorm activity may be several hundred miles long, with individual storms strung out along the wind-shift line (Fig. 6.29). Lines of violent prefrontal squall-line thunderstorms are occasionally developed in the warm sector as much as 150 to 200 miles ahead of the cold front. In the United States, tornadoes occur most frequently in

Figure 6.29 *A.* A V-shaped summer cyclone with a well-developed cold front and associated severe thunderstorms. Regions with polar air are west of the cold front; those covered by *mT* air are between the cold and warm fronts. *B.* Traces of pressure and temperature made at the Naval Air Station, Anacostia, Washington, D.C., during the approach and passage of the cold front shown in *A.* Hours are indicated at top.

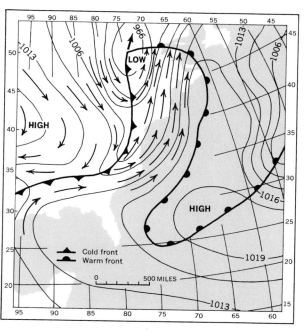

A

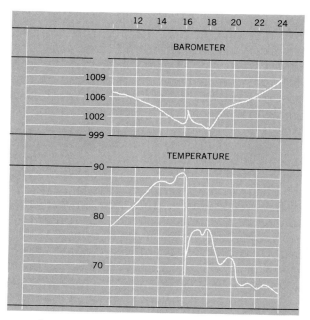

B

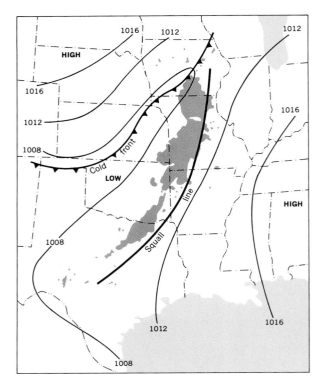

Figure 6.30 Radar echo (shaded) of precipitating clouds associated with a squall line over 1,000 miles long in the interior United States, together with a surface weather map indicating the relation of radar observations to surface fronts and isobars (*After S. G. Bigler.*)

connection with vigorous cold fronts and squall lines (Fig. 6.30).

Warm-front thunderstorms normally are less severe than those along cold fronts and squall lines. They are also less likely to be arranged in linear belts, and they are not associated with a marked wind shift followed by clear, cool weather. To the observer on the ground, their towering thunderhead clouds usually remain hidden by the heavy stratus cloud deck just above the surface of discontinuity.

Distribution of thunderstorms. It has been estimated that as many as 44,000 thunderstorms occur daily over the entire earth. They reach their greatest frequency in equatorial latitudes, where the maxi-

mum is much stronger over the warmer lands than over oceans (Figs. 6.31 and 6.32). The combination of heat, humidity, buoyancy, convergence, and the presence of numerous weak disturbances in the general vicinity of the equatorial convergence creates an ideal environment there for thunderstorm genesis. As pointed out in Chapter 3, the strong concentration of towering cumulonimbus clouds in the vicinity of the equatorial pressure trough plays an important role in the earth's latitudinal energy exchange.

From equatorial latitudes there is a general decrease in their frequency toward the poles, but with some variations (Fig. 6.31). Normally, the latitudes of the subtropical highs, more especially in the land-rich Northern Hemisphere, show up as interruptions in the orderly decrease poleward of thunderstorm frequency. At first thought one might expect to find an even greater decrease in the latitudes of the subtropical highs than actually does occur, because the tropical deserts are concentrated in these latitudes. But it must be recalled that by no means are all longitudes in the subtropics dry. Beyond latitudes 60 or 70°, thunderstorms are very few. The general latitudinal decrease of thunderstorms poleward is evidence of the importance of high surface-air temperatures in the production of strong convection. In extensive areas of the equatorial latitudes, thunderstorms occur on 100 to 180 days, and there are a few places recording more

Figure 6.31 Zonal distribution by latitude belts of days with thunderstorms. (*After C. E. P. Brooks.*)

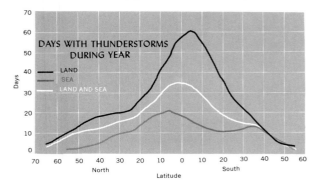

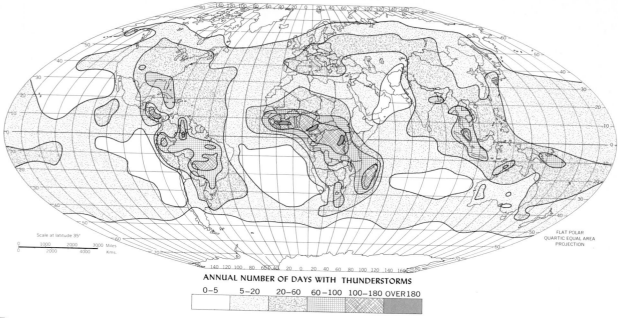

Figure 6.32 Average number of days with thunderstorms for the year. (*After World Meteorological Organization, 1950.*)

than 180. Most parts of low-latitude deserts have only 5 to 20 days with thunderstorms, and some local areas have fewer than 5.

There is a latitudinal displacement of thunderstorm belts following a similar seasonal shift of solar radiation. Since thunderstorm genesis is intimately related to high surface temperatures, it is to be expected that they will reach a maximum in the summer hemisphere. Land areas, with their high summer temperatures, show a greater number of thunderstorms in the warm season than oceans do.

In the United States the fewest days with thunderstorms are found in the Pacific Coast lowlands, which in summer are dominated by cool coastal waters and by stable *mP* air from the eastern end of an anticyclonic circulation (Fig. 6.33). There are two regions of maxima: (1) the southern and southeastern states, especially the Gulf Coast and Florida, and (2) the Rocky Mountain area. The eastern Gulf Coast region of the United States is the most

thundery area outside the tropics, for it has 70 to 90 days a year with thunderstorms. Here heat, humidity, and atmospheric instability combine to produce an environment conducive to strong convectional activity. The number decreases to the

Figure 6.33 Average number of days with thunderstorms in the United States.

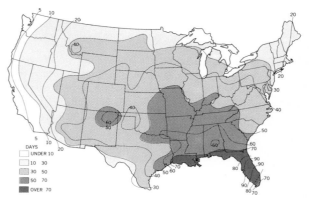

north and west: the central tier of states have 40 to 50 such days, and the northern tier 30±.

Tornadoes. This most violent of the earth's storms consists of a small funnel cloud generally less than ¼ mile in diameter with revolving winds of several hundred miles an hour. The origin of tornadoes is not known, but they usually occur in conjunction with severe thunderstorms along vigorous cold fronts and squall lines. Although they are highly spectacular, tornadoes are a relatively unimportant feature of the earth's climates. In the United States they are apparently confined to areas east of the Continental Divide, where a few hundred are reported each year. The greatest frequency of occurrence is in a belt extending from Iowa through Kansas, Arkansas, and Oklahoma to Mississippi. Iowa has the most: an average of 2.8 tornadoes per year. A few tornadoes each year wreak genuine havoc, but since the average area of a tornado's surface path is about 3 square miles, only a minute part of the earth, or even the United States, feels their terror.

SELECTED REFERENCES FOR FURTHER STUDY OF TOPICS IN CHAPTER SIX

1. **Middle-latitude cyclones and anticyclones.** *Compendium of Meteorology*, pp. 577–629; Byers, *General Meteorology*, 3d ed., pp. 304–324, 302–303; Critchfield, pp. 108–121; Petterssen, *Introduction to Meteorology*, pp. 218–235; Blair and Fite, pp. 152–165, 178–183; Hare, *The Restless Atmosphere*, pp. 70–90; Willett and Sanders, pp. 244–286; Reiter, *Jet-stream Weather*, pp. 324–351; William H. Klein, "Principal Tracks and Mean Frequencies of Cyclones and Anticyclones in the Northern Hemisphere," U.S. Weather Bureau, Research Paper 40, 1957; Riehl, *Introduction to the Atmosphere*, pp. 145–169; *Aviation Weather*, pp. 73–91.
    a. **Broad-scale weather situations and singularities.** *Compendium of Meteorology*, pp. 814–840; Reid A. Bryson and James F. Lahey, *The March of the Seasons*, University of Wisconsin Department of Meteorology, March, 1958; Eberhard Wahl, "A Weather Singularity over the U.S. in October," *Bull. Amer. Meteorol. Soc.*, Vol. 35, pp. 351–356, 1954.

2. **Tropical disturbances.** *Compendium of Meteorology*, pp. 868–878, 887–913; Blair and Fite, pp. 165–167, 186–195; Byers, pp. 368–392; Petterssen, pp. 236–250; Hare, pp. 104–115; E. M. Watts, *Equatorial Weather*, pp. 45–49, 156–177; P. Koteswaram and C. A. George, "On the Formation of Monsoon Depressions in the Bay of Bengal," *Indian J. Meteorol. Geophys.*, Vol. 9, pp. 9–22, 1958; R. H. Eldridge, "A Synoptic Study of West African Disturbance Lines," *Quart. J. Roy. Meteorol. Soc.*, Vol. 83, pp. 312–312, 1957; Willett and Sanders, pp. 208–243. A. G. Forsdyke, *Weather Forecasting in Tropical Regions, Great Britain Meteorological Office*, Geophysical Memoirs, no. 82, pp. 35–44, 1949; Riehl, *Introduction to the Atmosphere*, pp. 171–195.

3. **Thunderstorms.** Byers, pp. 458–479; Blair and Fite, pp. 221–230; Petterssen, pp. 113–131; Riehl, *Introduction to the Atmosphere*, pp. 101–115; *Aviation Weather*, pp. 92–113. *Compendium of Meteorology*, pp. 681–693; Willett and Sanders, pp. 297–303, 308–312.

4. **General references on atmospheric disturbances.** J. S. Arnett, "Principal Tracks of Southern Hemisphere Extratropical Cyclones," *Monthly Weather Rev.*, Vol. 86, pp. 41–44, 1958; Louis J. Battan, *The Nature of Violent Storms*, Doubleday & Co., Inc., Anchor Books, Garden City, N.Y., 1961; C. J. Boyden, "Jet Streams in Relation to Fronts and Flow at Low Levels," *Meteorol. Mag.*, Vol. 92, pp. 319–328, 1963; J. L. Galloway, "The Three-front Model, the Developing Depression and the Occluding Process," *Weather*, Vol. 15, pp. 293–309, 1960; F. K. Hare, "The Westerlies," *Geog. Rev.*, Vol. 50, pp. 345–367, 1960; W. H. Klein, *Principal Tracks and Mean Frequencies of Cyclones and Anticyclones in the Northern Hemisphere*, Res. Paper No. 40, U.S. Weather Bureau, Washington, D.C., 1957; N. E. LeSeur, "Synoptic Models in the Tropics," *Symposium on Tropical Meteorology*, pp. 319–328; J. S. Malkus, "Tropical Weather Disturbances: Why So Few

Become Hurricanes, *Weather*, Vol. 13, pp. 75–89, 1958; Sverre Petterssen, *Weather Analysis and Forecasting*, 2d ed., Vol. II, *Weather and Weather Systems*, McGraw Hill Book Company, New York, 1956; Daniel F. Rex, "Blocking Action on the Middle Troposphere and Its Effect upon Regional Climate," *Tellus*, Vol. II, 1950, pp. 196–211, 275–301; Herbert Riehl, *Tropical Meteorology*, pp. 210–357; Herbert Riehl, "On the Origin and Possible Modification of Hurricanes," *Science*, Vol. 141, pp. 1001–1010, 1963; *Symposium on Monsoons of the World*, Indian Meteorological Department, New Delhi, 1960, pp. 145–174, 237–256; Ivan Ray Tannehill, *Hurricanes*, 7th ed., Princeton University Press, Princeton, N.J., 1950;

Morris Tepper, "Tornadoes," *Scientific American*, Vol. 198, no. 5, pp. 31–37, 1958; "The Thunderstorm," *Report of the Thunderstorm Project*, U.S. Weather Bureau, Washington, D.C., 1949; Glenn T. Trewartha, *The Earth's Problem Climates*, The University of Wisconsin Press, Madison, Wis., 1961, pp. 56, 76–80, 96–97, 100–102, 113–116, 126–129, 140–144, 154–157, 160–164, 172–174, 182–186, 207–211, 219–222, 229–233; J. Vederman, "The Life Cycle of Jet Streams and Extratropical Cyclones," *Bull. Amer. Meteorol. Soc.*, Vol. 35, pp. 239–244, 1954; *World Distribution of Thunderstorm Days*, Tech. Pub. 6, 1953, and Tech. Pub. 21, 1956, World Meteorological Organization, Geneva.

THE WORLD PATTERN OF CLIMATES

CLIMATIC TYPES AND THEIR DISTRIBUTION

CHAPTER SEVEN

CLASSIFICATION OF CLIMATES AND THE WORLD PATTERN

The composite climate of an earth region. Part One of this book analyzed the important individual elements out of which climates are composed, and discussed the controls which affect the distribution of these elements over the earth's surface. Regional variations in the amount, intensity, and seasonal distribution of the elements, as determined by the climatic controls, create the variety of climates which will be described in Part Two.

A composite picture of a climate cannot be assembled apart from its individual climatic elements and the weather types which generate them. This might seem to suggest that the student who has mastered Part One already understands the composite climates of the earth's regions—and in one way this is true. The climate of any locality, however, is not created by a single climatic element but by the distinctive regional combination and interrelation of several elements. Thus in another way, the material of Part One requires focusing and integrating, together with further refinement and detail, before it becomes meaningful in terms of individual regions. This is the function of Part Two, which will describe the earth's climatic types and regions according to a system of classification outlined in the present chapter.

Periodic and nonperiodic climatic elements. Most descriptions of regional climates are made from instrumental data collected by national weather services and organized according to solar controls, fundamentally periodic in nature, whose two basic rhythms are diurnal and seasonal. Published weather bureau records fail to present in its essential integrity that other great control, the nonperiodic weather disturbance, whose beginning, ending, and duration are not readily related to the regular course of radiation geometry, and so to the calendar units of days, weeks, and months. There has been little or no attempt to collect and organize climatic data by this irregular synoptic unit, either ephemeral or quasi-stationary. Ward had this deficiency in mind when he wrote, "Anyone who seriously attempts to study the climatology of the United States should have a series of weather maps in one hand and a set of climatic charts in the other."

Classifying climates. Classification, a process basic to all sciences, consists of recognizing individuals with certain important characteristics in common and grouping them into a few classes or types. By reducing the many to a few, the scientist introduces simplicity and order into bewildering multiplicity and thereby establishes general truths from a myriad of individual instances.

The intent in formulating a classification of climates is scientific, educational, and philosophical. The scientific need is to process the vast amount of climatic data available for the earth so that distribution patterns become apparent. Since the climate of even a very small area is composed of a variety of climatic elements, it is nearly impossible for two places to have identical climates. Accordingly an unaltered and unsynthesized climatic map of the world would be a mosaic composed of thousands of parts. To reduce the thousands to a few (probably under 20) so that the great mass of climatic facts can be comprehended, understood, appraised, and remembered is a feat of no mean educational value. But such a condensing and ordering process, besides exemplifying scientific method, also has the still more important function of facilitating scientific analysis and explanation. Moreover, the process of classifying, generalizing, and mapping presents philosophical problems involving the merits of genetic versus empirical classification methods, the selection of criteria for the grouping of local climates, and the choosing of meaningful climatic boundaries.

A sometimes-heard criticism of all schemes of climatic classification covering the entire world is that if they are generalized enough to be memorized and remembered, they are already too generalized to be valuable. This is by no means the case. Even a simple climatic classification provides a basic framework for a comparative study of climates, and one within which subsequently acquired knowledge can be fitted.

Classification by solar illumination zones. One of the first and simplest climatic classifications, many elements of which are derived from the ancient Greeks, divides the earth into five zones (one tropical, two temperates, two polars) bounded by four astronomically significant parallels. These are the two tropics at $23\frac{1}{2}°$ N and S, and the Arctic and Antarctic Circles at $66\frac{1}{2}°$ in either hemisphere. The former define the extreme seasonal limits of the sun's vertical rays; the latter the limits of the tangent rays.

But although the five zones are astronomically determined, and are based on solar illumination, they obviously have important temperature differences. Between the two tropics, the warm winterless zone, all localities are crossed twice by the vertical rays of the noon sun during the year. Days and nights are nearly equal in length, and the sun's noon rays are never far from the vertical. As a consequence seasonal contrasts in temperature are small, the differences between day and night usually being much larger. By contrast, solar illumination within the summerless polar areas is more a seasonal and less a diurnal affair. Continuous daylight and continuous darkness prevail for long periods, which reach a maximum of 6 months each at the poles. Diurnal variations in temperature are at a minimum, while seasonal variations are strong. Between the $23\frac{1}{2}°$ and the $66\frac{1}{2}°$ parallels, in the intermediate or temperate zones, diurnal and seasonal variations in solar illumination are both strong. Temperature variations are also, with the seasonal usually the larger, so that a cold winter and a warm summer are ordinarily well defined.

Climatic regions and climatic types. The above classification of the earth's climates into tropical, middle-latitude, and polar zones chiefly emphasizes two elements: solar illumination and (indirectly) temperature. It does not take into consideration that other primary element, precipitation. Yet within both the low and the middle latitudes there are moist and dry climates, and some moist and some dry climates have both wet and dry seasons, while others have no dry season or no wet season at all. It is obvious that the geographer-climatologist requires not only a more detailed and refined classification, but also one in which the climatic subdivisions are based upon precipitation as well as directly upon solar illumination, and hence indirectly temperature, characteristics. Such a subdivision of the land areas of the earth into climatic types and climatic regions is presented on the map inside the front cover of this book.

Any portion of the earth's surface over which the important climatic elements, and therefore the broad climatic characteristics, are similar (not necessarily identical) is called a climatic region. Note that not all the subdivisions on the color map inside the front cover differ from one another climatically, for regions with similar climates are found in widely separated parts of the earth, although often in corresponding latitudinal and continental locations. The fact that this is so suggests that there is order and system in the origin and distribution of the climatic elements. It also makes possible the classification of the numerous climatic *regions* into a relatively few great climatic *types* (11). The types in turn can be reduced to still fewer climatic *groups* (6). Thus regions are combined into types and types into groups (Figs. 7.1, 7.2).

If the earth's surface were homogeneous (mainly either land or water) and lacking in terrain irregularities, doubtless winds, temperature, and precipitation would be arranged in zonal or east-west belts resembling the zones of solar illumination. An approach to such a zonal arrangement can be ob-

Figure 7.1 Arrangement of the six great groups of climate on a hypothetical continent. Note that *B*, the dry group (shaded), cuts across four of the five others. *B* is the only group not defined in terms of temperature.

CLIMATIC GROUPS

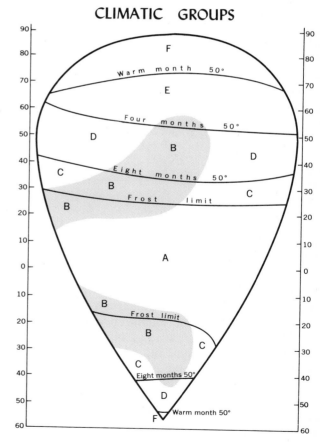

CLIMATIC TYPES

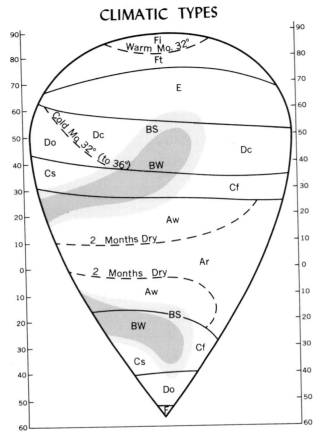

Figure 7.2 Arrangement of the principal types of climate on a hypothetical continent.

tribution for each of the climatic elements. It follows that the climatic regions and types, which are composites of the various climatic elements, also have a patterned arrangement. This means that the operation of the great planetary controls is sufficiently regular and dependable that it tends to produce relatively similar climatic conditions in corresponding latitudinal and continental locations, even though the latter may be separated by great linear distances. Of course it is this repetition of approximate climatic conditions in far-flung regions that makes it possible to classify the many regional, and innumerable local, climates into a relatively few general types. Figures 7.1 and 7.2 show climatic arrangement, or the world pattern of climatic groups and types, as they might appear on a hypothetical continent of relatively low and uniform elevation.

In its climatic arrangement, each of the world's actual continents exhibits features which, in varying degrees, correspond to the general world pattern as determined by the great planetary controls, and as represented in Figs. 7.1 and 7.2. But there are numerous modifications of, and deviations from, any idealized genetic scheme of world climatic distribution. This is one indication that there are important terrestrial controls such as size, shape, and position of land and water bodies, magnitude and directional alignment of terrain features—besides others which are suspected, but whose operations are not entirely understood. One of the most fertile fields of climatic investigation involves a search for explanations of the numerous departures from what is considered to be the normal world pattern. While world pattern is emphasized in this book, the individual climatic regions comprising each of the types, as they are developed on the separate continents, are given brief attention when they depart somewhat from the standard for their type.

Climate and other world patterns. Climate is the most fundamental dynamic force affecting distribution of most natural things on the face of the earth. This is particularly true of natural (wild) vegetation and soil, and to a much smaller degree, of many

served in the more uniform Southern Hemisphere. Because the earth's surface is not uniform but composed of continents and oceans of various sizes, shapes, and latitudinal locations, with land surfaces characterized by many sorts of terrain irregularities, there is a considerable modification of the zonal belts of pressure and winds. This in turn greatly complicates temperature and precipitation distributions.

In spite of the complications imposed by a non-homogeneous earth's surface—and by land masses with great variety in location, dimension, shape, and terrain—Part One of this book has demonstrated that there is a recognizable world pattern of dis-

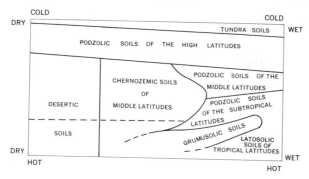

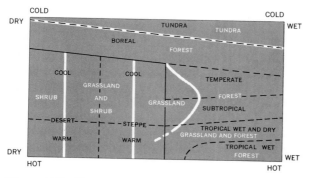

Figure 7.3 The approximate spatial correspondence on the earth between the broad categories of climate, vegetation, and soils is impressive. Note that on these diagrams, annual temperature increases from top to bottom, and annual rainfall from left to right.

gradational terrain features as well. Since all living things require moisture and warmth in varying amounts, their distributional arrangement on the earth is influenced by these climatic elements. In turn, climate and the climatically induced plant formations combine to become important determinants of soil character. Thus when the three great world patterns of climate, natural vegetation, and soil are superimposed, their boundaries roughly coincide (Fig. 7.3). Indeed, the broad coincidence between climate and vegetation is so well recognized that certain vegetation formation types, such as tropical rainforest, tundra, and steppe are employed as names for the approximately corresponding types of climate.

Genetic versus empirical classifications.[1]   There are over a score of different schemes of classification of world climates, most of them possessing genuine merit. Unfortunately a good number distinguish so many types that their value for educational purposes is reduced.

Almost all schemes of climatic classification have subdivisions and boundaries partly based upon temperature and rainfall parameters which are meaningful not in themselves, but in terms of some nonclimatic feature such as wild vegetation, individual cultivated plants, human comfort, or the like. The reason seems to be that such a limited number of statistical values for either temperature or precipitation have an objective existence. In the range of precipitation from 1 to 100 in., there is no particular value that is much more important than any other. The same is true for temperatures from 0° to 100° except one—the value of 32° (0°C)—which is unique because it marks the freezing point of water. However, other temperatures are sometimes employed as climatic boundaries: the average monthly temperature of about 42° as the growth threshold for important domesticated plants, 60 to 65° as the lower limit of human comfort, 50° for the warmest month as broadly coincident with the poleward limit of forest, and 65° (coolest month) as the limit of some tropical crops. These are significant only in terms of their supposed effects on something other than climate (and probably most classifications have been constructed to a greater or lesser degree upon the relation of climate to plant life). If one disregards the distribution of nonclimatic phenomena, it is difficult to provide meaningful temperature-rainfall limits of climatic types. The majority of classifications of climate, therefore, are of an "applied" character.

One basis for grouping the various schemes of climatic classification is to divide them into genetic and empirical types. In genetic classifications an

[1]For a survey of various schemes of climatic classification, see the following: K. Knoch and A. Schulze, "Methoden der Klimaklassifikation," Erg.-Heft 249, *Petermanns Geog. Mitt.* Gotha, 1954: Joachim Blüthgen, *Allgemeine Klimageographie,* Walter de Gruyter and Co., Berlin, 1964, pp. 453–488.

attempt is made to group climates according to the causative factors (e.g., air masses, wind zones) that may be responsible for them. In empirical classifications, origin is discarded as an organizing principle, and observation and experience provide the essential elements for climatic differentiation. The history of classification as a process seems to indicate an increasing dependence upon observed characteristics and a declining use of origin as a basis for the ordering and systematizing of natural phenomena. Among the reasons for this trend toward empirical classification is the fact that ideas regarding origin are likely to change much more as knowledge increases than are carefully observed and described facts about vegetation, animals, soils, or climate. Many descriptive observations of climate made by the classical Greeks and Romans are as valid today as when they were recorded two millenniums or more ago. The same cannot be said for most of their proposed climatic explanations.

It is the author's belief that empirical method is basic to good climatic classification—that a purely genetic approach cannot do justice to the complex climatic patterns of this earth. If the genetic classification is not to be pushed beyond its natural capacity, a genetic map of world climates should be restricted to a relatively few climatic zones. This is not to say that origin should be completely abandoned as an organizing principle in climatic classification, but only that it should be supplemental to careful observation.

The very fact that there is an observable world pattern of climatic distribution underscores the operation of genetic factors. Obviously some of the most universally recognized and extensively developed of the earth's climatic types can be related to well-known features of the atmosphere's general circulation. But it is equally true that within a particular wind system's domain, there are numerous areas whose climates do not fit the expected regime. These should not be unnaturally constrained within the genetic pattern. If empirical and genetic methods can be combined in forming a classification of climates, so much the better. But the empirical should always be paramount, while the genetic should pro-

vide a background which illuminates and rationalizes certain broad zonal and meridional climatic patterns. One such empirical-genetic scheme is the widely used Köppen classification of climates.

Three recent attempts at genetic schemes of climatic classification, accompanied by world maps showing distribution, have been made by Kupfer, Hendl, and Allisow.[2] Flohn also has proposed what is a somewhat more rudimentary scheme involving only seven climatic zones, unaccompanied by a world map, but with the position of the seven zones represented on a hypothetical continent.[3] For each climatic zone he indicates the corresponding pressure and wind belts, as well as the prevailing winds in the extreme seasons. Flohn attempts to relate each of the seven zones to a corresponding climatic type and to suggest the typical vegetation forms. Four of the zones remain perennially within one wind belt; the others undergo alternations of wind belts with the seasons. Flohn refers to the first group as *constant* climatic zones and to the second as *alternating* climatic zones. In the genetic classification he proposes, which is outlined in the following table, precipitation plays the major role in differentiating the seven climatic zones. Flohn considers temperature to be a more general element which changes only gradually.

Köppen classification of climates. Köppen's scheme of classification (used only in greatly modified form in this book) has outstanding merits demonstrated by its widespread acceptance for more than four decades. First published in 1901 and subsequently changed and modified a number

[2] E. Kupfer, "Entwurf einer Klimakarte auf genetischer Grundlage." *Zeitschrift für der Erdkundeunterricht*, Vol. 6, pp. 5–13, 1954; M. Hendl, "Entwurf einer genetischen Klimaklassifikation auf Zirkulationsbasis," *Zeitschrift für Meteorologie*, Vol. 14, pp. 46–50, 1960; B. P. Allisow, O. A. Drosdow, and E. S. Rubenstein, *Lehrbuch der Klimatologie*, VEB Deutscher Verlag der Wissenschaft, Berlin, 1956 (in Russian: Leningrad, 1952).
[3] H. Flohn, "Neue Anschauungen über die allgemeine Zirkulation der Atmosphäre und ihre Klimatische Bedeutung," *Erdkunde*, Vol. 4, pp. 141–162, 1950; Flohn, "Zur Frage der Einteilung der Klimazonen," *Erdkunde*, Vol. 11, pp. 161–175, 1957.

Atmospheric climatic zones of the earth

Zone	Precipitation	Wind belt and pressure belt	Winds*		Typical climate, Köppen	Typical vegetation forms
			Summer	Winter		
1. Inner tropical zone	Constantly wet, mostly heavy rains	Equatorial west-wind zone† ≧8 months	T	T	Af, Am	Tropical rain-forest, monsoon forest
2. Outer tropical zone (tropical margins)	Summer rain (zenithal rain)	Equatorial west-wind zone† <8 months, alternating with trades	T	P	Aw, in part Cw	Savanna with galeria forest, dry forest
3. Subtropical dry zone‡	Prevailingly dry (infrequent downpours)	Trades or subtropical high	P	P	BS, BW	Steppe, desert-steppe, semi-desert, true desert
4. Subtropical winter-rain zone‡	Winter rain, in part equinoctial rain	Summer sub-tropical high, winter middle-latitude westerlies	P	W	Cs	Hardleaf woods
5. Moist temperate zone	Rain in all seasons	Middle-latitude westerlies	W	W	Cf, in part Cw	Broadleaf and mixed forest
6. Boreal zone§	Rainfall pre-dominantly in summer, winter snow cover	Middle-latitude westerlies, in part polar esterlies	W	E	Df, Dw	Coniferous forest and birches
6. Subpolar zone	Meager rain throughout year	Polar easterlies and west winds (subpolar low)	W	E	ET	Tundra
7. High polar zone	Meager snowfall throughout year	Polar easterlies	E	E	EF	Cold desert (ice)

* T = equatorial westerlies and doldrums.
  P = tropical easterlies (trades) and subtropical high.
  W = middle-latitude westerlies.
  E = polar easterlies.
† Or doldrums.
‡ Absent on east side of continent.
§ Only on the continents; absent in Southern Hemisphere.
Source: After Flohn.

of times, the latest world map by the author himself appears in Köppen's book, *Grundriss der Klimakunde*, Berlin, 1931.[4]

The Köppen classification is based upon annual and monthly means of temperature and precipitation. It accepts native vegetation as the best expression of the totality of a climate, so that many of the climatic boundaries are selected with vegetation limits in mind. Köppen recognizes that the effectiveness of precipitation in plant development and growth depends not alone upon amount of precipitation, but also upon the intensity of evaporation and transpiration, by which processes water is lost from soil and plants. The part of the rainfall which is evaporated is of no direct value in vegetation growth. Köppen's method of indicating evaporation intensity, and hence precipitation effectiveness, is to combine precipitation and temperature in a single formula. Thus the same number of inches of rainfall falling in a hot climate, or concentrated in a hot season when evaporation is great, is less effective for plants than the same amount falling in a cooler climate or season. Although useful, this method of measuring precipitation effectiveness is not entirely satisfactory.

A unique feature of the Köppen system is its ingenious symbolic nomenclature for the climatic types. This makes the repetition of cumbersome descriptive terms unnecessary. Each principal type of climate is described by a formula consisting of a pair of letters with precise meanings. Thus the formula *Af* for "tropical wet climate" may be translated as follows: A = constantly hot, average temperature of the coldest month above $18°C$ ($64.4°F$); f = constantly wet, no month of the year having on the average less than 6 cm (2.4 in.) of precipitation.

An additional advantage of the Köppen classification is its adaptability for use at different educational levels. It is at the same time both simple and detailed. It recognizes five great groups of climate, and these, together with the eleven principal types of climate into which the groups are divided, provide the essentials for a rudimentary knowledge of the earth's climatic pattern. But through the use of additional symbols which can be joined to the pair representing the main types, this classification provides an abundance of useful detail for those who wish it.

Another benefit is that Köppen's is a quantitative system. It uses numerical values for defining the boundaries and symbols not only of the great groups and types of climate, but of the lesser subdivisions and characteristics as well. Since specific values of temperature and precipitation are given, it becomes possible for other scientists to question the validity of particular boundaries. Moreover, such boundaries are subject to checking and revision as new data become available. Köppen selected many of the numerical values he employed to establish climatic boundaries because they appear to coincide fairly well with certain significant landscape features, particularly those of native vegetation.

Although climatic boundaries quantitatively defined are a commendable feature in climatic classification, there is a danger in such quantification, for it tends to give a false impression of the degree of accuracy intended by the author. All climatic boundary definitions should be viewed as convenient approximations only. Such a boundary drawn on a map usually represents a broad zone of transition, and it is an exercise in futility to debate the correct classification of stations situated within some scores of miles of the boundary. Indeed, a climatic boundary on a map indicates only the mean position of numerous individual climatic-year boundaries, which typically depart by hundreds of

---

[4] Still more recent and detailed maps, covering only individual continents or parts of continents, are contained in Volumes II to V of the Köppen-Geiger *Handbuch der Klimatologie* (see the reference list at the end of this chapter). Volume I, Part C (W. Köppen, *Das geographischen System der Klimate*, 1936) of the above handbook contains the author's latest analysis of his own classification. See the appendix for a detailed presentation of the Köppen classification, and refer to the world map showing distribution of his climatic types. What is probably the latest Rudolph Geiger edition of a somewhat simplified Köppen map has been published in Joachim Blüthgen, *Allgemeine Klimageographie*. In wall-map form (scale 1:16 mm), under the authorship of Rudolph Geiger and Wolfgang Pohl, it has been issued by Justus Perthes, 1961.

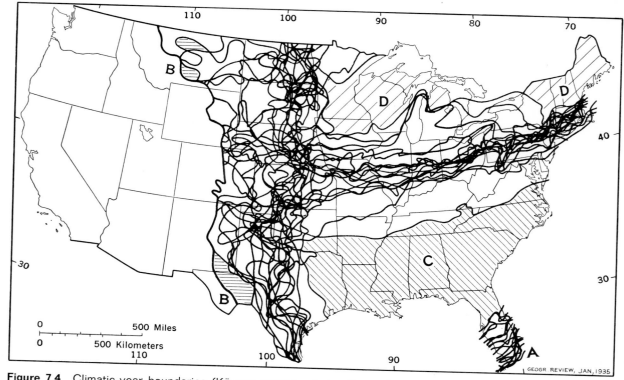

Figure 7.4 Climatic-year boundaries (Köppen system of classification) in the central and eastern United States. Each heavy line represents the climatic boundary between Köppen's dry (*B*) and humid climates, or between his mesothermal (*C*) and microthermal (*D*) climates, for a single year. Clearly, climatic boundaries shift position from one year to the next. Therefore the boundaries drawn on a map showing climatic types and regions (such as the map inside the front cover) are simply the means of different locations for a large number of individual climatic years.

miles from the mean (Fig. 7.4). Because of the inherent pseudoaccuracy of a mean climatic boundary, not much time should be spent in trying to draw it too precisely or in memorizing its exact location after it has been drawn.

Köppen's is an empirical classification in that its climatic types and boundaries are based upon observed features of temperature and precipitation, and have not been constrained to fit a genetic pattern emphasizing the general atmospheric circulation or other controls. Yet a considerable number of the types and subtypes have some degree of areal coincidence with certain broad-scale features of the atmospheric circulation and are partly to be

explained by them. This system therefore has the merit of being an empirical classification with a genetic underpinning. (See the appendix for a detailed presentation of the Köppen classification.)

Climatic classification in this book. The general scheme of climatic classification used here is designed for the first course in a geography of climate, which concentrates on the world pattern of climates. Such a classification should be kept simple, and consequently ought to recognize only a limited number of principal climatic types (usually under 15). Increasing the number tends to obscure the elemental features of the climatic pattern and makes it difficult to memorize the climatic types and their distributions. Where further detail is required, as many second-order and third-order subdivisions can be added within a climatic type or region as are needed.[5] It is believed there is great value in orienting climatic description with respect to the framework of some standard and recognized classification of climates.

Since the needs of geographers provide the essential focus for this book, an empirical type of classification seems to be the most appropriate. People such as geographers, biologists, or agriculturists, who need to understand and use the climatic environment for their own purposes, should have the

[5] For examples of detailed regional studies of climate which have followed the principle suggested above of creating subdivisions within the recognized world types of a standard climatic classification, see R. J. Russell, "Climates of California," *Univ. Calif. Pub. in Geog.*, Vol. 2, no. 4, pp. 73–84, 1926; "Climates of Texas," *Ann. Assoc. Amer. Geographers*, Vol. 35, pp. 37–52, June, 1945. See also John Kesseli, "The Climates of California according to the Köppen Classification," *Geog. Rev.*, Vol. 32, pp. 476–480, 1942.

facts of climate presented realistically, not forced into a preconceived genetic structure or scheme of type location. First and foremost they must be guided by the observed data.

On the other hand, description without reference to genesis or origin soon becomes dull and tiresome—features which unfortunately characterize much of the climatic literature on regions. A recognition of the importance of genesis not only increases interest and adds to the scientific quality of climatic analysis, but also gives an extra dimension of insight to the student's understanding of the description. The very fact that there is a clearly recognizable world pattern of climates suggests the operation of genetic controls. To be sure, there are numerous and important deviations from the standard world pattern, but these departures require explanation just as much. Therefore it seems best to supplement the empirical classification of climates in every way possible by genetic classification.

Earlier editions of this book employed a greatly modified form of the Köppen classification. But in

Figure 7.5 Generalized arrangement of the main climatic groups with respect to temperature and precipitation. Note that temperature increases from left to right, precipitation from bottom to top.

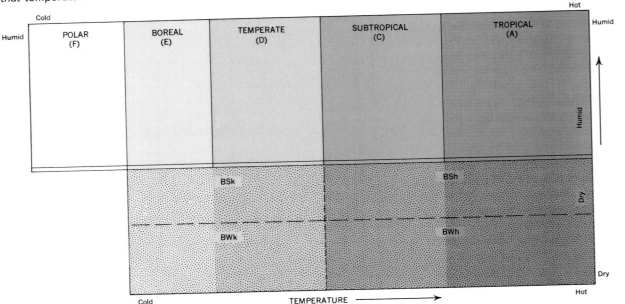

the fourth edition so many further alterations have been made that in fairness to the Köppen system its name probably should no longer appear on the classification. Still, there remain significant vestiges of that distinguished ancestor, and a continuing debt to Köppen is gladly acknowledged. (Again, those who prefer the Köppen system of climatic classification can find it in the appendix and the world map near the end of this book.)

Like most other climatic classifications, the one used here recognizes temperature and precipitation as the two elements of paramount importance. Whether they should be coequal or of different rank is debatable. In three fairly recent climatic classifications and climatic maps by European geographers, the primary groupings are thermic, while rainfall differences create the subdivisions within the great temperature zones.[6] There is no primary group of dry climates. Thornthwaite's well-known classification recognizes five grades of precipitation effectiveness, six of temperature efficiency, and four of seasonal concentration of precipitation. Of his eight main climatic groups, five give primary emphasis to precipitation and three to temperature. The Köppen classification identifies five main groups of climate, all but one—the dry group—being thermally defined. They are as follows:

A. Tropical rainy climates
B. Dry climates
C. Warm temperate rainy climates
D. Boreal climates
E. Snow climates

Unfortunately, the one which is nonthermally defined (*B*, dry climates) is sandwiched in with the other four, rather than being set apart in a position at the end or the beginning of the list to emphasize its unlike basis of classification.

The classification used in this book has six main climatic groups (Figs. 7.1 and 7.5). Five are based on five great thermic zones; the sixth is the dry group, which cuts across four of the thermic zones,

[6]H. von Wissman (in Blüthgen, *Allgemeine Klimalehre*), Carl Troll (with C. Paffen) and N. Creutzburg (see reference list at end of this chapter).

the polar alone excepted. Accordingly, temperature differentiates five of the main climatic subdivisions, and precipitation one. As might be expected, the five thermic groups have a strong zonal orientation. This is much less true of the dry group.

The 6 great climatic groups and their poleward boundaries[7]

Based on temperature criteria

A. Tropical: Frost limit in continental locations; in marine areas 65° (18°C) for the coolest month (after Wissmann)
C. Subtropical: 8 months 50° (10°C) or above[8]
D. Temperate: 4 months 50° or above
E. Boreal: 1 (warmest) month 50° or above
F. Polar: All months below 50°

Based on precipitation criteria

B. Dry: Outer limits, where potential evaporation equals precipitation

In the low latitudes astride the equator (Fig. 7.1), there is a winterless belt with constantly high temperatures and adequate annual rainfall (*A*). This is the zone of the humid tropics. It has no frost; in marine areas the temperature for the coolest month

[7]Because of the widespread use of, and familiarity with, the Köppen classification among American geographers, it was decided to continue the use of the letter *B* to designate dry climates even though that procedure fails to adequately set the dry group apart from all the others. To emphasize the uniqueness of the *B* group, it has been put at the end of the list.

[8]For more than three-quarters of a century climatologists have been employing the average monthly temperature of 50° as a significant climatic boundary in terms of plant growth and human comfort. Thus 1 month (warmest) with a temperature of 50° is thought to coincide approximately with the poleward limits of forest in continental locations. A period of 4 months with 50° or more is commonly used as the boundary separating cool from warm climates. Climates having fewer than 4 months with average temperatures of 50° are considered to be lacking in significant agricultural potential. While the temperature threshold for growth varies considerably between different plants, it is probably true that an average monthly temperature of about 50° seems to mark the threshold of active growth for a number of them. Conrad suggests 50° as the lower limit of human comfort and relates that temperature to the length of the active season at many resorts.

is 65° or above. Within this group are two main types of climate (Fig. 7.2): the tropical wet type (Ar)[9] with a wet season 10 to 12 months in length, and the tropical wet-and-dry type (Aw), which has predominantly zenithal or summer rains and a low-sun dry season of over 2 months (see table following).[10] The tropical wet type coincides fairly well with the ITC, doldrums, and equatorial westerlies, where general wind convergence is characteristic. This is also the realm of the tropical rainforest. The tropical wet-and-dry type is alternately under the influence of the ITC–equatorial westerlies during high sun, and of the dry trades and subtropical anticyclone during low sun. Native vegetation varies from semideciduous or deciduous forest to savannas, or combinations of trees and grass.

Within the humid middle latitudes are three climatic groups: subtropical (C), temperate (D), and boreal (E) (Fig. 7.1). The subtropical group has occasional freezing temperatures in its continental parts, but in its frostless marine areas the equatorward boundary, is the isotherm of 65° for the coolest month. In this group 8 months or more are above 50° and therefore classed as warm. Two principal climatic types are included within the subtropics (Fig. 7.2): the subtropical humid $(Cf [w])$ and the subtropical dry-summer (Cs). The subtropical humid type is characteristically positioned on the eastern side of a continent. There it is influenced by the unstable or neutral air in the western end of a subtropical anticyclone in summer, and by the westerlies and their cyclonic storms in the cooler months, so that it may have no distinctly dry season (see the following table). Summer is the wettest season more often than not, and when a monsoonal circulation is well developed, as it is in parts of eastern Asia, winters may be so moisture-deficient that the summer emphasis is strong. The subtropical dry-summer type with its winter rainfall and summer drought is alternately influenced by middle-latitude westerlies and their wave disturbances in

winter, and the stable eastern end of an anticyclonic cell in summer. Unlike its more humid counterpart, this dry-summer type is typically located on the subtropical western side of a continent.

Within the temperate group (D), 4 to 7 months have average temperatures of 50° or more. Two divisions are recognized (Fig. 7.2): the oceanic or marine (Do), and the continental (Dc). The boundary separating the two is usually the isotherm of 32° (0°C) for the coolest month,[11] so that mild winters are characteristic of the oceanic type, and more severe winters of the continental type (refer to the table of climatic groups and types). Typically the oceanic type is located on the windward western side of a continent, or on islands. The continental type is situated either farther inland on the western side of a continent than Do, or else on the leeward eastern side.

The boreal climate (E) of the higher middle latitudes is supercontinental in its temperature features, with a short cool summer, a long cold winter, and a very short freeze-free season. It has only 1 to 3 months with an average temperature of 50° or over. Coniferous forests dominate. The boreal climate is so severe that it discourages human settlement, and population is sparse.

Finally, in the high latitudes there are the summerless polar climates (F), where the average temperatures of all months are below 50°. Two subdivisions are recognized: the tundra type (Ft), where the warmest month, while less than 50°, is above 32° (0°C); and the icecap type (Fi), where all months are below 32°.

The dry climates (B) form the only group not defined by temperature criteria. Here a deficiency of precipitation (i.e., evaporation exceeds precipitation) is the dominant climatic characteristic. In terms of temperature, dry climates may be tropical, subtropical, temperate, or boreal. They are divided into an arid or desert type (BW) and a semiarid or steppe type (BS). A further subdivision separates the *hot* tropical-subtropical deserts and steppes $(BWh,$

[9] Letter symbols are defined in a later section.

[10] Only rarely in the tropics is the wet-and-dry type characterized by zenithal drought and winter rains (As), so that this is not recognized as a significant climatic type.

[11] Or up to 36° (2°C) in some locations inland. An annual temperature range of 36° has also been suggested for the Do/Dc boundary.

Scheme of climatic groups and climatic types

| Groups of climate | Types of climate | Pressure system and wind belt | | Precipitation |
		Summer	Winter	
A. Tropical humid	Ar, tropical wet	ITC, doldrums, equatorial westerlies	ITC, doldrums, equatorial westerlies	Not over two dry months
	Aw, tropical wet-and-dry	ITC, doldrums, equatorial westerlies	Drier trades	High-sun wet (zenithal rains), low-sun dry
C. Subtropical	Cs, subtropical dry-summer	Subtropical high (stable east side)	Westerlies	Summer drought, winter rain
	Cf, subtropical humid	Subtropical high (unstable west side)	Westerlies	Rain in all seasons
D. Temperate	Do, oceanic	Westerlies	Westerlies	Rain in all seasons
	Dc, continental	Westerlies	Westerlies and winter anticyclone	Rain in all seasons, accent on summer; winter snow cover
E. Boreal	E, boreal	Westerlies	Winter anticyclone and polar winds	Meager precipitation throughout year
F. Polar	Ft, tundra	Polar winds	Polar winds	Meager precipitation throughout year
	Fi, icecap	Polar winds	Polar winds	Meager precipitation throughout year
B. Dry	BS, semiarid (steppe)			
	BSh (hot), tropical-subtropical	Subtropical high and dry trades	Subtropical high and dry trades	Short moist season
	BSk (cold), temperate-boreal		Continental winter anticyclone	Meager rainfall, most in summer
	BW, arid (desert)			
	BWh (hot), tropical-subtropical	Subtropical high and dry trades	Subtropical high and dry trades	Constantly dry
	BWk (cold), temperate-boreal		Continental winter anticyclone	Constantly dry

Undifferentiated highlands

BSh) from the *cold* temperate-boreal deserts and steppes (*BWk, BSk*) of middle latitudes. The *h/k* boundary separating hot from cold dry climates is the isotherm of 8 months with a temperature of 50°F (10°C) or above—the same that separates the subtropical and temperate climatic groups (Fig. 7.1). Hot tropical-subtropical dry climates more or less coincide with the more stable parts of subtropical anticyclones and trades, where subsidence and divergence are characteristic. The cold dry climates, which are a poleward extension of the hot variety, normally have type locations leeward of high mountains or in the interiors of large continents.

Definitions of climatic symbols and boundaries

A = killing frost absent; in marine areas, cold month over 65°F (18.3°C)

    r (rainy) = 10 to 12 months wet; 0–2 months dry

    w = winter (low-sun period) dry; more than 2 months dry

    s = summer (high-sun period) dry; rare in *A* climates

B = evaporation exceeds precipitation. Boundary,

$$R = \tfrac{1}{2}T - \tfrac{1}{4}PW$$

where R = rainfall, in.

T = temperature, °F

PW = % annual rainfall in winter half year[12]

Desert/steppe boundary is

$$R = \frac{\tfrac{1}{2}T - \tfrac{1}{4}PW}{2}$$

or half the amount of the steppe/humid boundary

W (German *Wüste*) = desert or arid

S = steppe or semiarid

h = hot; 8 months or more with average temperature over 50°

k (German *kalt*) = cold; fewer than 8 months average temperature above 50°

s = summer dry

w = winter dry; *n* = frequent fog

C = 8 to 12 months over 50°; coolest month below 65°F (18.3°C)

    a = hot summer; warmest months over 72°F (22.2°C)

    b = cool summer; warmest month below 72°

    f = no dry season; difference between driest and wettest month less than required for *s* and *w*; driest month of summer more than 1.2 in. (3 cm)

    s = summer dry; at least three times as much rain in winter half year as in summer half year; driest summer month less than 1.2 in. (3 cm); annual total under 35 in. (88.9 cm)

    w = winter dry; at least ten times as much rain in summer half year as in winter half year

D = 4 to 7 months inclusive over 50°F (10°C)

    o = oceanic or marine; cold month over 32°F (0°C) [to 36° (2°C) in some locations inland]

    c = continental; cold month under 32° (to 36°)

    a = same as in *C*

    b = same as in *C*

    f = same as in *C*

    s = same as in *C*

    w = same as in *C*

E = 1 to 3 months inclusive over 50°F (10°C)

F = all months below 50°F (10°C)

    t = tundra; warmest month between 32°F (0°C) and 50°F (10°C)

    i = icecap; all months below 32°

A/C boundary = equatorial limits of freeze; in marine locations the isotherm of 65° (18°C), for the coolest month

C/D boundary = 8 months 50° (10°C)

[12] This single formula represents Patton's approximation of the more elaborate Köppen formulas, three in number, for the boundary of dry climates.

D/E boundary = 4 months 50°

E/F boundary = 50° for warmest month

t/i boundary in F climates = 32° (0°C) for warmest month

B/A, B/C, B/D, B/E boundary = evaporation equals precipitation

BS/BW boundary = one-half the B/A, B/C, B/D, B/E boundary

h/k boundary in dry climates = same as C/D

Do/Dc boundary = 32° (0°C) [to 36°F (2°C)] for coolest month

Owing to the scarcity of climatic data for many highland areas and also to the great complexity of climates within highlands, no attempt has been made to differentiate climatic types within the highland areas on the climate map inside the front cover.

Distribution of climates on a hypothetical continent. In the following chapters on climate types, keep referring to the color map at the front of the book, which illustrates distribution of the types over the earth's land areas, and Figs. 7.1 and 7.2, which diagram characteristic arrangements of climatic groups and types on a hypothetical continent. This continent's shape roughly corresponds to that of a composite of the world's actual land masses, so that it illustrates the positions of the climatic types as they probably would be if there were no modifications and complications resulting from varying shapes, sizes, positions, and elevations of land masses. The strong resemblances between Fig. 7.2 and the map of actual climates inside the front cover are obvious.

GENERAL REFERENCES FOR PART TWO

Ackerman, Edward A., "The Köppen Classification of Climates in North America," *Geog. Rev.,* Vol. 31, pp. 105–111, 1941.

Alissow, B. P., et al., *Lehrbuch der Klimatologie,* VEB Deutscher Verlag der Wissenschaft, Berlin, 1956.

Atlas of American Agriculture, Part 2, Climate, 3 sections: "Frost and the Growing Season"; "Temperature, Sunshine and Wind"; "Precipitation and Humidity"; U.S. Government Printing Office, Washington, D.C. Contains excellent and detailed maps of the climatic elements.

Atlas of Climatic Types in the United States, 1900–1939, Misc. Publ. 421, U.S. Department of Agriculture, 1941.

Blair, Thomas A., *Climatology: General and Regional,* Prentice-Hall, Inc., Englewood Cliffs, N.J., 1942.

Blüthgen, Joachim, *Allgemeine Klimageographie,* Walter de Gruyter and Co., Berlin, 1964, pp. 453–488.

Brooks, Charles F., A. J. Connor, et al., *Climatic Maps of North America,* Harvard University Press, Cambridge, Mass., 1936.

Chang, Jen-Hu, "Comparative Climatology of the Tropical Western Margins of the Northern Oceans," *Ann. Assoc. Amer. Geographers,* Vol. 52, pp. 221–227, 1962.

Climate and Man, Yearbook of Agriculture, 1941, U.S. Department of Agriculture, Washington, D.C.

Critchfield, Howard J., *General Climatology,* 2d ed., Prentice-Hall, Inc., Englewood Cliffs, N.J., 1966.

Creutzburg, N., "Klima, Klimatypen und Klimakarten," *Petermanns Geograph. Mitt.,* Vol. 94, 57–69, 1950.

Flohn, H., "Neue Anschauungen über die allgemeine Zirkulation der Atmosphäre und ihre Klimatische Bedeutung," *Erdkunde,* Vol. 4, pp. 141–162, 1950.

———, "Zur Frage der Einteilung der Klimazonen," *Erdkunde,* Vol. 11, pp. 161–175, 1957.

Garbell, Maurice A., *Tropical and Equatorial Meteorology,* Pitman Publishing Corporation, New York, 1947.

Hann, Julius, *Handbook of Climatology,* Part I, English translation by Robert De C. Ward, The Macmillan Company, New York, 1903. 4th rev. German ed. by Karl Knoch, J. Engelhorn's Nachfolger, Stuttgart, 1932. Parts II and III (untranslated) deal with regional climates.

Hare, F. Kenneth, "Climatic Classification," Chap. VII in L. Dudley Stamp and S. W. Woolridge (eds.). *London Essays in Geography,* Harvard University Press, Cambridge, Mass. 1951.

————, *The Restless Atmosphere,* Harper & Row, Publishers, Incorporated, New York, 1963, pp. 116–182.

Haurwitz, Bernard, and James M. Austin, *Climatology,* McGraw-Hill Book Company, New York, 1944.

Hendl, Manfred, "Einführung in die physikalische Klimatologie," Band II, *Systematische Klimatologie,* VEB Deutscher Verlag der Wissenschaften, Berlin, 1963.

Heyer, Ernst, *Witterung und Klima,* B. G. Teubner Verlagsgesellschaft, Leipzig, 1963.

Kendrew, W. G., *Climates of the Continents,* 4th ed. Oxford University Press, Fair Lawn, N.J.,

————, *Climatology* (3d ed. of *Climate*), Oxford University Press, Fair Lawn, N.J., 1949.

Knoch, K., and A. Schulze, "Methoden der Klimaklassifikation," Erganzungsheft N. 249, *Petermanns Geog. Mitt.,* 1952.

Köppen, W., *Grundriss der Klimakunde,* Walter de Gruyter and Co., Berlin, 1931. Contains a relatively complete analysis of the Köppen scheme of climatic classification.

———— and R. Geiger, *Handbuch der Klimatologie,* Gebrüder Borntraeger, Berlin, 1930 and later, 5 vols, not completed. Vol. 1 covers the field of general climatology; the other four are on regions. Those parts dealing with the United States, Mexico, West Indies, Central America, Australia, New Zealand, and parts of eastern Africa are in English; the other parts are in German.

> Knoch, K., "Klimakunde von Südamerika," Vol. 2, part G, 1930.
> Sapper, K., "Klimakunde von Mittelamerika," 1932.
> Ward, Robert De C., and Charles F. Brooks, "Climatology of the West Indies," Vol. 2, part I, 1934.
> Birkeland, B. J., and N. J. Föyn, "Klima von Nordwesteuropa und den Inseln von Island bis Franz-Josef-Land," Vol. 3, part L, 1932.
> Alt, E., "Klimakunde von Mittel- und Südeuropa," Vol. 3, part M, 1932.

> Braak, C., "Klimakunde von Hinterindien und Insulinde," Vol. 4, part R, 1931.
> Taylor, G., "Climatology of Australia," and Kidson, E., "Climatology of New Zealand," Vol. 4, part S, 1932.
> Robertson, C. L., and N. P. Sellick, "The Climate of Rhodesia, Nyasaland, and Mozambique Colony," Vol. 5, part X, 1933.
> Ward, Robert De C., and Charles F. Brooks, "The Climates of North America: The United States, Mexico, Alaska," Vol. 2, part J, 1936.
> Köppen, W., "Das geographische System der Klimate." Vol. I, part C, 1936.
> Sverdrup, H. U., Helge Petersen and Fritz Loewe, "Klima des kanadischen Archipels and Grönlands," Vol. 2, part K, 1935.
> Connor, A. J., "The Climates of North America: Canada," Vol. 2, part J, 1938.
> Schott, G., "Klimakunde der Südsee-Inseln," Vol. 4, part T, 1938.
> Meinardus, W., "Klimakunde der Antarktis," Vol. 4, part U, 1938.
> Köppen, W., "Klimakunde von Russland," Vol. 3, part N, 1939.

Kupfer, E., "Entwurf einer Klimakarte auf genetischer Grundlage," *Zeitschrift für den Erdkundeunterricht,* Vol. 6, pp. 5–13, 1954.

Lauer, W., "Humide und aride Jahreszeiten in Afrika und Südamerika und ihre Beziehung zu den Vegetationsgürteln," *Bonner Geog. Abh.,* Vol. 9, pp. 15–98, 1952.

de Martonne, Emmanuel, *Traité de géographie physique,* 7th ed., Librairie Armand Colin, Paris, 1948.

Miller, A. Austin, "Air Mass Climatology," *Geography,* Vol. 38, pp. 55–67, 1953.

————, *Climatology,* 3d ed. E. P. Dutton & Co., Inc., New York, 1953.

Riehl, Herbert, *Introduction to the Atmosphere,* McGraw-Hill Book Company, New York, 1965, pp. 215–244.

————, *Tropical Meteorology,* McGraw-Hill Book Company, New York, 1954.

Schulze, A., "Weg und Ziel der Klimaklassifikation," *Geographischen Taschenbuch,* pp. 429–433, 1956–1957.

Shear, James A, "The Polar Marine Climate," *Ann. Assoc. Amer. Geographers,* Vol. 54, pp. 310–317, 1964.

Thornthwaite, C. Warren, "The Climates of North America According to a New Classification," *Geog. Rev.,* Vol. 21, pp. 633–655, 1931.

———, "The Climates of the Earth," *Geog. Rev.,* Vol. 23, pp. 433–440, 1933.

———, "Problems in the Classification of Climates," *Geog. Rev.,* Vol. 33, pp. 232–255, 1943.

———, "An Approach toward a Rational Classification of Climate," *Geog. Rev.,* Vol. 38, pp. 55–94, 1948.

Trewartha, Glenn T., *The Earth's Problem Climates,* University of Wisconsin Press, Madison, Wis., 1961.

Troll, C., "Climatic Seasons and Climatic Classification," *Oriental Geographer,* Vol. 2, pp. 141–165, 1958.

———, "Karte der Jahreszeiten-Klimate der Erde," *Erdkunde,* Vol. 18, pp. 5–28, 1964 (colored map).

Visher, S. S., *Climatic Atlas of the United States,* Harvard University Press, Cambridge, Mass., 1954.

Ward, Robert De C., *Climates of the United States,* Ginn and Company, Boston, 1925.

Watts, I. E. M., *Equatorial Weather,* University of London Press, Ltd., London, 1955.

Sources of World Climatic Data:

Tables of Temperature, Relative Humidity, and Precipitation for the World, Parts I, II, III, IV, V, and VI, 1958, Great Britain Meteorological Office.

World Weather Records. Issued 1921–1930, 1931–1940 as Smithsonian Institution Miscellaneous Collections; 1941–1950 by the U.S. Weather Bureau, 1959.

CHAPTER EIGHT

TROPICAL HUMID CLIMATES (A)

Type location and boundaries. The humid tropics form a somewhat interrupted and irregular belt 20 to 40° wide around the earth astride the equator (see Fig. 7.1 and climate map inside the front cover). This belt is distinguished from all other humid regions of the earth by the fact that it is constantly warm; in other words, it lacks a winter. Throughout the tropical humid climates the temperature difference between day and night, while modest, exceeds by several times the difference between the warmest and coolest months of the year (Fig. 8.1).

Heat is so constant and so all-pervading in the humid tropics that their poleward limits are set by low temperatures (except where dry climates establish the poleward boundaries). As noted in Chapter 7, the temperature limits are either the equatorial limits of freeze, or in marine locations that are without freezing temperatures, the isotherm of 65° for the coolest month. These temperatures were selected because they generally coincide with the poleward limit of certain plants which grow only in the warmest regions and cannot tolerate marked seasonal changes in temperature. The chief interruptions of the belt of humid tropical climates over the continents are caused by highlands, where altitude reduces temperatures below the permissible limits. In the vicinity of the

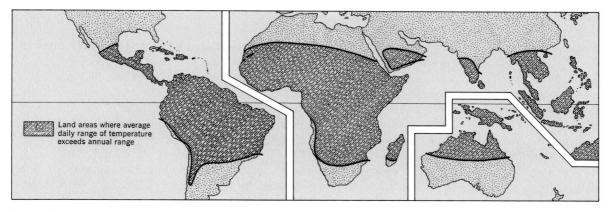

Figure 8.1 A distinctive feature of most tropical climates is that average daily range of temperature exceeds average annual range. (*After Troll.*)

equator, *A* climates usually prevail up to elevations of 3,000 to 4,000 ft.

A climates are by far the most widespread of any of the great climatic groups. According to Hermann Wagner, they occupy about 36 percent of the entire earth's surface—nearly 20 percent of the land surface, and approximately 43 percent of the ocean surface.

Normally the tropical humid group has its greatest latitudinal spread toward the eastern or windward side of a continent (Fig. 7.1). It is typically more restricted toward the center and west. On the eastern side, *A* climates are usually bounded on their poleward margins by subtropical humid regions (*Cf*), one of the milder middle-latitude climates (Fig. 7.2). Since both *A* and *C* climates are humid, the *A/C* boundary must be one of reduced temperature, as explained above. Toward the center and western side, on the other hand, the *A* group is terminated poleward by dry climates, *B* (Fig. 7.1). Instead of low temperatures in winter, the limiting factor in these sectors is increasing aridity. This arid boundary usually shows a maximum equatorward thrust along the western oceanic margins.

The somewhat asymmetrical development of the *A* group, and the contrasting climates and boundaries which limit it in its eastern and western parts, are largely attributable to certain features of the atmospheric circulation. The western margins of the oceanic subtropical anticyclones, which affect the eastern sides of tropical continents, are characterized by neutral or unstable maritime air masses. Here subsidence is weak or nonexistent and the trade-wind inversion is either absent altogether or positioned at a high altitude, so that deep convectional overturning is possible. This unstable air favors the development of various kinds of rain-bringing atmospheric disturbances. In addition, the eastern side is predominantly the windward side of the continent, so that much of the time there is an onshore movement of humid air coming off a warm ocean and warm ocean currents. Any orographic lifting of this air is likely to produce heavy precipitation.

Western sides of tropical continents, except for those close to the equator, are more likely than the eastern sides to be affected by the stable eastern margins of the oceanic subtropical anticyclones, where subsidence is strong and the inversion layer low. In some continents, especially western South America and Africa south of the equator, the drying effects of the anticyclones cause the *B* climates to thrust within about 5° of the equator (see climate map). Anticyclonic subsidence, with perhaps some aid from the prevailing cool ocean currents, produces a stable atmosphere which is opposed to the development of convective overturning and rainmaking disturbances.

The intensity of sunlight, both direct and reflected, is another noteworthy feature of *A* climates. Maud D. Haviland writes:

Except in the very deepest jungles, some sunlight filters to the forest floor through chinks in the foliage canopy overhead. These shafts of light strike as bright and hot as the rays from a burning glass; and as they move over the leaves, small invertebrate life creeps out of their way. I once watched such a sunbeam searchlight overtake a party of ants who were devouring the decomposing body of a caterpillar. Within half a minute they fled helterskelter and the remains of their meal shrivelled up as if before a furnace.[1]

Precipitation. Rainfall in the humid tropics is relatively abundant—rarely lower than 30 to 35 in., and usually well over that amount. Its origin is multiple. Most of it falls as heavy intermittent convective showers accompanied by thunder and lightning. However, this shower activity is mainly of an organized type, occurring, as mentioned above, in conjunction with the convergent circulations of weak atmospheric disturbances.

To be sure, much remains unknown about the dynamics of tropical convection. It is observed, however, that tropical cumuli are organized into extensive systems, which appear to be controlled by feeble but large-scale disturbance circulations. And in turn the development, growth, and dissipation of tropical cumuli alter the delicate balance of heat and humidity sufficiently to react upon the wind systems. Cloud systems are such a dominant feature in tropical weather—a fact recently made evident by satellite photographs—to suggest that descriptions of the cloud systems may provide the best means of characterizing this weather.

Unlike the uniform temperature conditions, rainfall in *A* climates is more variable in amount and in both seasonal and areal distribution. Accordingly the two principal climatic types within the humid tropics are distinguished from each other mainly on the basis of their seasonal distribution of precipitation. Throughout both types, however, much of the rainfall is associated with the ITC and its unstable air masses, so that precipitation reaches a maximum at the time of high sun when the ITC prevails. In the tropics, rainfall emphatically follows the sun.

Diurnal variations of rainfall in the tropics present a complex situation, difficult to describe in simple terms and even more difficult to explain. Many, but by no means all, inland stations show a primary afternoon rainfall maximum, probably induced by solar heating (Fig. 8.2). On the other hand, over tropical oceans, including small islands, and many coasts, it has been fairly well established that a night rainfall maximum exists. But complicated diurnal rainfall curves are numerous in both continental and marine locations.

Based upon seasonal rainfall contrasts, two subdivisions of tropical humid climates are recognized—tropical wet (*Ar*), and tropical wet-and-dry (*Aw*) (Fig. 7.2). In tropical wet, 10 or more months of the year are classed as wet; if there is any dry season at all, it must be brief (Fig. 8.3). In tropical wet-

Figure 8.2 Diurnal march of rainfall at Kuala Lumpur on the west side of Malaya. Here there is a striking rainfall peak in the warmer afternoon hours. Such a daily pattern is fairly common over tropical lands, including large islands; but there are many complications. A night rainfall maximum is more frequent over tropical oceans. (*After Ooi Jin-bee in Journal of Tropical Geography, Vol. 12, March, 1959.*)

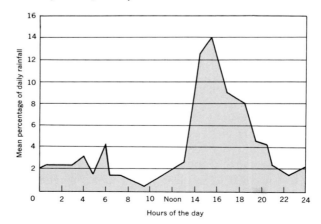

[1] Maud D. Haviland, *Forest, Steppe, and Tundra,* Harvard University Press, Cambridge, Mass., 1926, pp. 37–38.

and-dry, the dry season is longer than 2 months, and ordinarily the drought is more severe as well. Total annual rainfall is usually less in *Aw* than in

Ar. A dense broadleaf evergreen forest is characteristic of *Ar* climate; lighter deciduous forest, thorn forest, and tall grasses are found in *Aw*.

TROPICAL WET CLIMATE (*Ar*)[2]
(Tropical rainforest)

Type location. Uniformly high temperatures and heavy annual precipitation distributed throughout the year, so that there is either no dry season or at most 2 dry months, are the two distinguishing characteristics of tropical wet climate. Typically this

climate is found astride the equator and extending out 5 or 10° on either side. (Fig. 7.2). The latitudinal spread may be increased to 15 or even 20° along the eastern or windward margins of a continent, where the subtropical anticyclone is relatively weak and its air neutral or unstable. *Ar* is associated with the equatorial trough of low pressure and with the ITC and equatorial westerlies, which prevail in that general low-pressure zone (Fig. 8.3). Tropical wet regions may be encroached upon by the dry trade winds at the time of the trades' maximum equatorial migration, but never for long enough to produce more than a brief slowing up of the rains. Along

[2] In the symbol *Ar*, *r* = rainy for 10 to 12 months, or only 0 to 2 months dry. Definition of a dry month is derived from Clyde Patton's simplification of Köppen's three formulas for the dry-humid boundary (see page 250). A dry month will, therefore, be $\frac{1}{12}$ of the annual value derived from the Patton formula. In equatorial lowlands, where the average annual temperature is close to 80°, a dry month will usually have about 2.3 in. of average rainfall.

Figure 8.3 Hypothetical arrangement of tropical wet (*Ar*), tropical wet-and-dry (*Aw*), and dry (*B*) climates with respect to latitude, ITC, and wind and pressure belts at the time of the times of the extreme seasons.

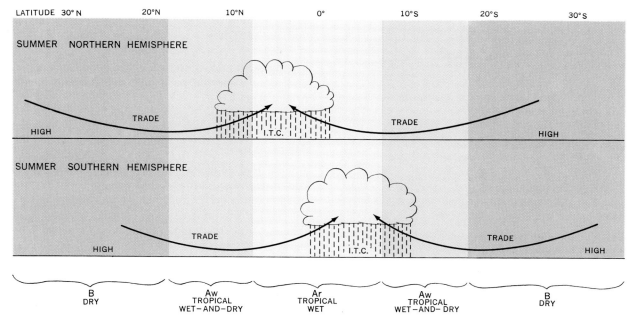

most of its poleward margins *Ar* climate gradually passes over into *Aw*, the boundary between them being the isoline of 2 dry months. Along the wetter eastern margins of a continent, however, *Ar* climates may extend far enough poleward that they adjoin subtropical humid climates (*Cf*) of middle latitudes (Fig. 7.2 and climate map). This boundary, as mentioned earlier, is set by low winter temperature instead of by length of the dry season.

Geographical location. The Amazon Basin of South America; the Congo Basin of Africa and oceanic margins of parts of Tropical West Africa; and the insular-peninsular area of Southeast Asia comprise the most extensive regions with tropical wet climate. Eastern Central America and parts of the Caribbean islands, western Colombia, the Guianas, the Atlantic margins of Brazil, eastern

Figure 8.4 Average monthly temperatures and precipitation for a representative station with tropical wet climate. Monthly temperatures are much more uniform than monthly amounts of precipitation. Note that Colombo has no dry month.

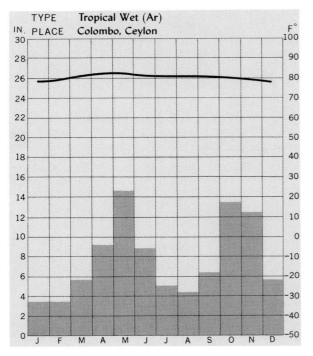

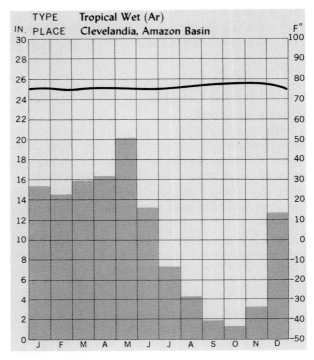

Figure 8.5 At Clevelandia in the Amazon Valley, monthly temperatures are very similar to those at Colombo. But here there are 2 months that may be classed as dry. Extensive areas of *Ar* climate have 1, or at most 2, dry months.

Madagascar, and southern Ceylon are other smaller and more dispersed parts (see climate map inside front cover).

Temperature

Annual and seasonal temperatures. Lying, as it commonly does, athwart the equator and consequently in the belt of near-maximum solar radiation, *Ar* climate has monthly and annual temperatures that are uniformly high. Yearly averages usually lie between 77 and 80° +. Since the sun's noon rays are never far from a vertical position, and days and nights vary little in length from one part of the year to another, not only are annual averages high, but temperature hardly changes from month to month (Figs. 8.4, 8.5, and 8.6). On continents, the annual

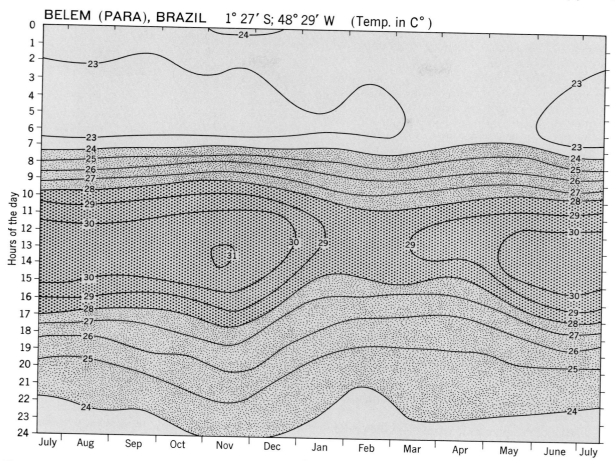

Figure 8.6 Thermoisopleths for Belém, an *Ar* station in the Amazon Valley. Note how much greater the diurnal range is than the annual range. (*After Troll.*)

temperature range, or difference between the warmest and coolest months, is usually less than 5°. Ranges over the oceans in these low latitudes are even smaller: Jaluit in the Marshall Islands in mid-Pacific records only 0.8° difference between the extreme months.

It is not excessively high monthly averages but rather the uniformity and monotony of this constant succession of hot months, with no relief, that characterize the tropical rainforest climate. Actually, the average July temperatures of many subtropical American cities—such as Charleston with 82°, Galveston with 83°, and Montgomery with 82°—

may equal, or even exceed by a few degrees, those of the hottest months at stations near the equator. In the tropical rainforest the abundance of cloud and heavy forest cover act to prevent excessively high temperatures such as occur in tropical deserts. The hottest month at Belém (Amazon Basin) is only 80°, and at most stations in the Congo Basin it is slightly under 80°. The minor temperature contrasts between the warmest and coolest months are determined not so much by the position of the sun as by the amount of cloudiness, for the highest temperature usually occurs during the seasons of least rain and clearest skies.

Daily temperatures. The daily, or diurnal, range of temperature is usually 10 to 25°, or several times greater than the annual range. For example, at Bolobo in the Congo, the average daily range is 16° while the annual range is only 2°. This range differential is a typical feature of the wet tropics. During afternoons the thermometer ordinarily rises to temperatures varying from about 85 to 93° and at night sinks to 68 to 75° (Figs. 8.7 and 8.8). It is often said that night is the winter of the tropics. Even the daily extremes of temperature are never very great, however; the average of the daily maxima at Belém is only 91°, that of the daily minima, 68°. The highest temperature ever recorded at Santarém (Amazon Basin) is 96°, while the lowest is only 65°. This absolute maximum of 96° may be compared with 105° for Chicago and 112° for St. Louis, located in the middle latitudes.

Although daytime temperatures may not be excessively high in *Ar* climate, the heat, together with slight air movement, intense light, and high relative and absolute humidity, produces an atmospheric condition with low cooling power. It is oppressive and sultry, so that one's vitality and energy are sapped. Sensible temperatures are therefore excessively high, even though the thermometer readings may not indicate abnormal heat.

Even the nights give little relief from the oppressive warmth. Yet to the poorly clad natives, who are sensitive to even the slightest drop in temperature, the humid night air may appear chilly. Fires are often lighted at night if the temperature drops

Figure 8.7 Daily maximum and minimum temperatures for the extreme months in one year at a representative station with tropical wet climate (*Ar*). Diurnal solar control is almost complete, as shown by the regular daily rise and fall of temperature.

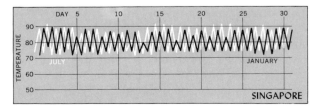

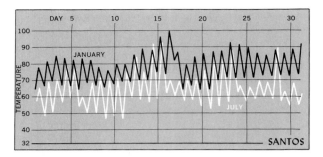

Figure 8.8 Daily maximum and minimum temperatures for the extreme months in one year at a station with tropical wet climate located on the east coast of Brazil at 20°S. Note the somewhat greater nonperiodic temperature changes caused by advection associated with atmospheric disturbances from the middle latitudes. Because of the station's considerable distance from the equator, low-sun seasonal temperatures are somewhat lower than normal for *Ar*.

below 70°. Some Negro groups in central Africa use heated sleeping quarters during the coolest season, when the average temperature may drop into the upper 60s. Rapid nocturnal cooling is not to be expected in regions of such excessive humidity and abundant cloudiness. The nighttime temperature drop is often sufficient, nevertheless, to cause surface condensation in the near-saturated air, so that lowland radiation fog and heavy dew are common. The periods of least rainfall and clearest skies have the lowest night temperatures, which on occasions fall below 60°.

Daily march of temperature. Figures 8.7 and 8.8 show the daily march of temperature for the extreme months at representative stations within the tropical wet (*Ar*) climate. These graphs illustrate a temperature regime in which sun is almost completely in control. There is a striking diurnal regularity about the changes: Temperatures rise to about the same height each day and fall to about the same level each night, so that one 24-hr period almost duplicates every other. Irregular invasions of warm and cold air masses, of the type so common in the middle latitudes, are practically unknown. This absence of nonperiodic and erratic temperature variations results from (1) the great width of the tropical zone, which makes invasions of cold air relatively impossible, and (2) the general lack of vigorous, moving cold anticyclones, which are the instigators of many of the great nonperiodic temperature changes of middle latitudes. What small temperature variations do exist from day to day are largely a result of variations in cloudiness.

Precipitation

Amount. Annual rainfall in the wet tropics is heavy—commonly 70 to over 100 in., with less than average over the continents and more over the oceans (Figs. 8.4 and 8.5; also see data on page 262). Taken as a whole, *Ar* climate coincides with the earth's belt of heaviest precipitation. Because of the abundant precipitation, surface ocean waters in the doldrums are less salty than in the trades. The largest measured amount of rainfall in *Ar* climates, and perhaps the greatest for any place on the earth, is on the windward side of a low mountain on the island of Kauai in the Hawaiian group, where the average annual amount is over 450 in. Near the foot of Cameroon Mountain in equatorial Africa, the annual rainfall averages about 400 in. Cherrapunji, in the Khasi Hills of northeastern India, receives an average of 426 in. In all these stations, however, the excessive annual precipitation is associated with orographic effects. Such amounts of rain are unknown over extensive lowlands.

Conditions close to the equator are ideal for rain formation. A main reason is the directional convergence of tropical air masses, with associated lifting of the air. Supplementing this convergence are the high temperature and abundant humidity, which produce a nearly saturated condition throughout a deep atmospheric layer. The reservoir of atmospheric moisture for precipitation is therefore unusually great. In this warm, humid, and highly unstable air, only a slight amount of original lifting is required to trigger strong convective overturning with associated heavy showers. Weak atmospheric disturbances are numerous.

Cloudiness, much of it cumulus in character, is fairly abundant in *Ar* climates, averaging about 60 percent. At Manaus, in the Amazon Valley, cloudiness varies between $6/10$ and $7/10$ for each month. At Belém it is $4/10$ in the driest month and $8/10$ in the rainiest. Completely clear days are almost unknown. Considering the fact that *Ar* climate is the rainiest of the earth, it may seem strange that the percentage of cloudiness is not as great near the equator as it is in the higher middle latitudes, where rainfall is much less. This is related to the fact that a larger proportion of the cloud in equatorial latitudes is of the cumulus type, which is capable of producing a maximum of rainfall with a minimal duration of cloudiness. A dull gray solid overcast is atypical. Cloudiness in continental *Ar* climates has a distinct diurnal periodicity, with the maximum usually in the warmer hours of afternoon or evening when convectional overturning is at a peak. Nights and early morning hours are usually clearer.

Diurnal march of cloudiness at Belém (In tenths of sky covered)

6 A.M.	3.6	2 P.M.	5.7
8 A.M.	3.8	4 P.M.	6.0
10 A.M.	5.5	10 P.M.	5.2

Sky conditions are lively and animated as the upward-surging air currents produce clouds full of motion and variety in form and shadings. The predominantly convectional rain, reflecting the unstable character of the atmosphere, falls in hard showers from towering cumulonimbus clouds. Maximum precipitation over lands usually occurs during the warmer hours of the day, when convectional ascent is at a maximum. Early mornings may be relatively clear, but as the sun climbs toward the zenith and temperature increases, cumulus clouds begin to appear. These grow in number, diameter, and height with the heat of the day, until by afternoon they become giant ominous thunderheads if the circulation is convergent, as it is in the presence of a weak disturbance. Several thunderstorms, accompanied by thunder and lightning, in a single afternoon are not unusual. The rain may continue on into the evening, although there is a tendency for skies to become clearer as the heat wanes. The cloud cover and downpour of rain temporarily cool the air, but with the passing of the storm and the sun's appearance again, the usual oppressive conditions return. As described in Chapter 3, the few thousand cumulus "hot towers" active at one time in the equatorial pressure trough play a key role in maintaining the tropical circulation. Vast amounts of heat of condensation are released in them.

Within the equatorial-trough region, thunderstorms reach their maximum development for any latitude of the earth; on the average there are 75 to 150 days with such storms during the course of the year. Tropical thunderstorms, although localized, are extremely intense. The great thickness of the heated and humidified air in equatorial latitudes provides an abundance of storm energy. These paroxysms of nature, with their fierce lightning, crashing thunder, and deluges of rainfall, are awesome spectacles. One traveler writes:

The force of the downpour is another factor in the oecology of the forest. In the wet season thunderstorms of great violence are frequent, and the rain descends with a suddenness and volume unknown outside the tropics. The sun is shining, the forest glitters with a million lights, birds are on the move, and insects hum and dance from leaf to leaf. All at once a shadow is drawn over the sun, and all activity of bird and beast ceases as the sound of rushing rain rapidly approaches. An avalanche of water then crashes down, blotting out surrounding objects and, as it seems, sweeping the very breath from the nostrils, bewildering and benumbing the senses. Every twig and leaf is bent and battered, and in a few seconds streams pour down the paths and the world seems changed into a thundering cataract. Then, as suddenly as it came, the storm passes, and the sun blazes out again before the roar of the storm sweeping over the treetops has died away in the distance. Even before the leaves have ceased to drip, or the land-crabs, tempted forth by the teeming water, have scuttled to cover again, the life of the forest is resumed. It is almost incredible how some fragile forms escape destruction under such terrific bombardments. . . .[3]

While solar heating unquestionably produces a strong diurnal periodicity in thunderstorm, cloud,

[3] Haviland, *op. cit.*, p. 39.

Climatic data for representative stations with tropical wet climate (*Ar*)

	J	F	M	A	M	J	J	A	S	O	N	D	Yr	Range
Belém, Amazon Valley, 1°18′S														
Temp., °F	77.4	77.0	77.2	77.9	78.4	78.4	78.4	78.6	78.4	79.0	79.3	78.6	78.3	2.3
Precip., in.	13.4	16.0	17.2	13.5	11.3	6.9	5.7	5.0	4.7	3.6	3.4	6.9	108.0	
Stanleyville, Congo Basin, 32′N														
Temp., °F	77.2	77.0	77.4	77.5	76.6	75.7	74.3	74.3	75.2	75.6	75.7	75.7	76.0	3.1
Precip., in.	3.3	4.1	6.9	5.6	6.1	3.5	4.4	8.9	7.5	9.8	6.7	2.8	69.4	
Singapore, 1°18′N, 103°52′E														
Temp., °F	79.5	80.5	81.5	81.5	82.0	81.5	81.5	81.0	81.0	80.5	80.5	80.5	80.5	2.5
Precip., in.	9.9	6.8	7.6	7.4	6.8	6.8	6.7	7.7	7.0	8.2	10.0	10.1	95.0	
Belize, British Honduras, 17°31′N														
Temp., °F	74.8	76.8	79.2	79.2	81.9	82.4	82.6	82.6	82.0	79.3	76.1	73.6	79.3	9
Precip., in.	5.1	2.6	1.6	1.5	4.1	9.1	9.6	8.5	9.4	11.0	10.2	6.3	79.0	
Manaus, Amazon Valley, 3°1′S														
Temp., °F	79.2	79.2	79.5	79.2	79.3	79.9	80.2	81.5	82.2	82.0	81.7	80.2	80.4	3
Precip., in.	10.5	9.7	10.6	10.5	7.6	4.0	2.5	1.5	2.3	4.9	6.0	8.5	78.6	

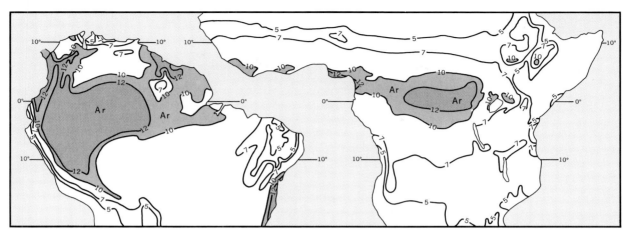

Figure 8.9 Differentiating those parts of tropical South America and Africa where 12 months are classed as wet (none dry), 10 or 11 months are wet (1 or 2 dry); and 9 or fewer months are wet (over 2 dry). (*After W. Lauer.*)

and shower occurrence at many land stations, there are significant departures from this regular pattern. Usually the course of the weather over a period of several weeks combines features of periodic solar control with some of nonperiodic control. The nonperiodic weather element originates in the frequent weak but extensive atmospheric disturbances whose passage brings spells of bad weather involving increases in cloud, shower activity, and the number of thunderstorms. In the irregular intervals between such disturbances skies are clearer and there is less, or even no, rainfall. But even during a cloudy, rainy spell accompanying a weak disturbance, the diurnal pattern ordinarily is not obliterated, for daytime solar heating adds its effects to the instability associated with general convergence within the disturbance. The violent hurricane type of storm does not occur in those *Ar* climates situated close to the equator where Coriolis force is absent. Only littoral and insular locations poleward of about 10° may be afflicted with hurricanes.

Seasonal distribution, or annual march.   It is untrue that in *Ar* climate as a type, abundant rainfall is so well distributed throughout the year that no months are dry (Figs. 8.4 and 8.5). There are extensive regions, to be sure, where such perfection in

wet equatorial climate exists: for example, the western Amazon Basin and a core region in the innermost Congo Basin. But more extensively developed in equatorial latitudes is a slightly different rainfall regime which, while annual rainfall may total equally as much, is characterized by a brief dry season of 1, or at most 2, months (Fig. 8.9). In both types a rainforest vegetation prevails. Almost certainly more than half the total area with tropical wet climate seems to have this short dry season; it represents more the standard for *Ar*, therefore, than the constantly wet regime.

Heavy equatorial rainfall with a brief dry season not only is typical of *Ar* climate in parts of the Asiatic sector, where a monsoon system of winds prevails. It probably also predominates in equatorial Africa, and it is extensively developed in the middle and lower Amazon Basin. According to Lauer, the single largest area of *Ar* with no dry season is the upper three-fifths of the Amazon Basin.

It would seem logical if the constantly wet phase of *Ar* everywhere went through a gradual change to tropical wet-and-dry climate, or *Aw*, by way of a transitional zone of *Ar* with a brief dry season. But actually, in many equatorial parts the brief dry season appears to be typical of core areas, so that it cannot be considered only as a transition phase.

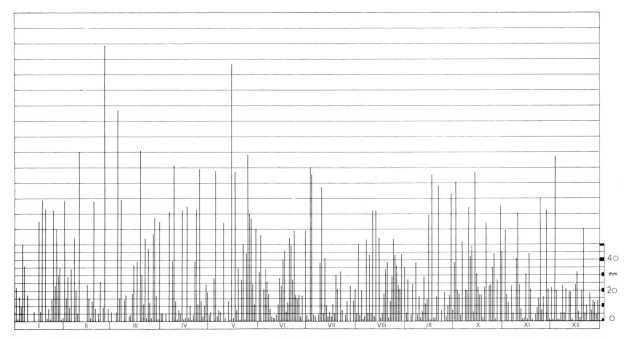

Figure 8.10 Daily rainfall amounts for one year (1928) at Andagoya, an *Ar* station in western Colombia inland from the Pacific coast. Here the average annual rainfall is exceedingly heavy (7,089 mm, 282 in.), and no month is dry. In 1928 there were only 48 days on which no rain fell. (*After M. Hendl, Einführung in die physikalische Klimatologie, Band II.*)

Even in places where all 12 months are wet, however, it should not be inferred that the year's rainfall is evenly distributed. In comparison with the rainiest months, there are others that are much less wet, although scarcely dry.

Theoretically it might be supposed that stations in equatorial latitudes would have two periods of maximum and two periods of minimum rainfall during the course of a year. The two maxima logically would coincide with, or closely follow, the zenithal positions of the sun, and hence the ITC. The two periods of less heavy rain should coincide with or follow the solstices, when the drier trades advance closest to the equator. Actually this ideal equatorial regime with two symmetrical maxima and two minima, although it is found in some areas, certainly is not a widespread feature. The seasonal rainfall mechanism is not that perfectly geared to the astronomical controls.

Rain-days.   Not only is rainfall abundant in the wet tropics, but the number of days with rain is large (Fig. 8.10). Belém in the Amazon Valley has 94 in. of rain and 243 rainy days. During the rainier season or seasons, precipitation falls on a large majority of the days, although there are always some without any. During the drier periods, not only are rain-days fewer, but the amount per day is also less. Thus at Belém, March, which is the wettest month, has 6 to 7 times as much rain as November, which is the driest (see the table on page 265); but even in November there are 10+ rainy days with a total of more than 2 in. of precipitation. On the other hand, March has nearly 14 in. of rain, falling on 28+ days. Thus 91 percent of the days in the wettest month are rainy, and only 34 percent of those in the driest month. The fact that not all days have rain in spite of constant heat and humidity is another indication that thermal convec-

Average precipitation and rain-days at Belém, Amazon Valley

Month	Rainfall, in.	Rain-days
January	12.68	26.8
February	13.90	26.1
March	13.94	28.2
April	13.07	26.4
May	9.45	22.6
June	5.87	20.5
July	5.24	18.5
August	4.72	16.2
September	3.70	15.4
October	3.35	12.6
November	2.13	10.3
December	5.98	19.5
Year	94.03	243.1

tion due to solar heating cannot entirely explain the temporal distribution of rainfall.

It is clear that in tropical wet climate there are rainy spells in the drier seasons and short dry spells even in the rainy seasons. This comes about partly as a result of fluctuations in the positions of the intertropical convergence—fluctuations which not only are regular and seasonal, but also occur irregularly within short periods of time. The more frequent shower activity of the wet season indicates the proximity of the ITC and equatorial westerlies, while rainy spells even in the drier months probably are signs of a temporary reestablishment of these controls. The less rainy seasons and the drier spells denote a greater prevalence of dry trade-wind air masses. A second feature controlling the frequency of spells of showery weather is the passage of traveling weak disturbances, whose occurrence is nonperiodic.

Rainfall variability. In spite of the heavy rainfall characteristic of this climate, there may be considerable variations from year to year in total amount, although usually they are not large enough to have serious consequences. Certainly annual fluctuations of rainfall are much greater than those of temperature. Variability in precipitation is a definite feature of subhumid and dry climates, but strange as it may seem, droughts sometimes also occur in the humid climates near the equator. Occasionally they may

Figure 8.11 Daily cloudiness (in tenths of sky covered) and precipitation amounts (in millimeters) for one year (1914) at Batavia (now Djakarta), Java. Note the single dry season, June–October, which is also the period of least cloud. In this particular year Batavia was *Aw*, not *Ar*. (*After J. Blüthgen.*)

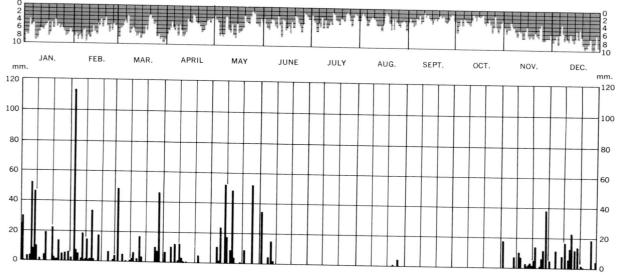

even jeopardize crops, though not necessarily because the rainfall is meager in an absolute sense, but because it is relatively so. In these low latitudes where temperatures are constantly high, evaporation is excessive. Moreover, the crops grown are of a kind that require abundant precipitation. Consequently a total that might be judged more than adequate in the middle latitudes could represent a deficiency and be injurious in the wet tropics.

Rainfall variability at two tropical stations (Inches)

Station	Driest year	Wettest year	Average
Colombo, Ceylon	51.6	139.7	83.8
Singapore	32.7	158.7	92

Winds

The feeble temperature gradients create weak pressure gradients, so that air movement is commonly slight, and also fickle. The whole region is poorly ventilated, and this, in conjunction with high temperatures and excessive humidity, causes physical discomfort. Temporary relief may be brought by the strong but brief squall winds associated with thunderstorms. Occasionally, especially in the drier periods, the trades advance far enough equatorward to bring brief spells of desiccating weather. Such a wind, known as the harmattan along the Atlantic coast of Tropical West Africa, is usually described as cool, particularly at night. This is probably owing to its great evaporation.

Along coasts in the low latitudes, sea breezes are important climatic phenomena. The importation of cooler air from the sea during the heat of the day is a great advantage to residents along the littoral, causing tropical coasts to be much more livable than the interiors. Ordinarily the effects of the sea breeze reach inland only 30 to 60 miles.

Weather in *Ar* climate

Preceding sections have emphasized the diurnal regularity found in individual elements of *Ar* climates. This is especially true of temperature, for deviation from a sun-controlled regime is very slight.

Rarely does advection significantly modify the diurnal monotony. Cloud may be somewhat more effective, since the seasons or periods of greatest cloudiness usually show appreciably lower daytime temperatures, as well as smaller diurnal ranges.

Cloud and precipitation do not have the same high degree of diurnal regularity as the temperature element. This reflects the extensive traveling disturbances, nonperiodic in occurrence, which stimulate convective shower activity of an organized character. However, solar daytime heating also acts to intensify the shower activity during the warmest hours.

Regional features[4]

Although this book is concerned mainly with the world pattern of climates as exemplified by the climatic types and their distribution, that does not necessarily exclude giving some attention to separate climatic regions on the various continents, providing that such a region has individuality within the type.

Latin America. The Amazon Basin in South America is distinguished both by the extensiveness of *Ar* development and by the great overall quantity of its annual precipitation. No other large region equals it in this combination of features. Not only is the basin freely open to equatorial air from the Atlantic, but as Fig. 8.12 shows, during the Southern Hemisphere summer it is flooded by a humid, unstable northwesterly current, probably a segment of the equatorial westerlies (see Chapter 3). Still, the Amazon lowland is by no means a region of rainfall uniformity, either in annual amounts or in annual march. The belt of heavy precipitation (2,000–3,000 mm; 80–120 in.) lies in the deepest interior of the middle and upper Amazon Basin. This very wet interior is partly the result of its invasion by humid unstable northwesterly air in the Southern Hemisphere summer. The westerly flow appears to provide a bountiful advected moisture supply. But in

[4]Much of this section is taken from the author's book, *The Earth's Problem Climates,* University of Wisconsin Press, Madison, Wis., 1961.

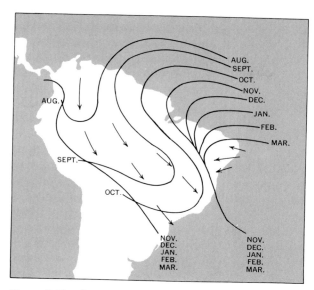

Figure 8.12 Stages in the advance south and east over Brazil in spring and summer (Southern Hemisphere) of the humid, unstable equatorial westerlies and the ITC. Note that the rate of southeastward advance of this northwesterly air is much more retarded along the north coast than it is in the interior, and that it normally does not succeed in overrunning extreme eastern Brazil. (*After Coyle. From Trewartha, The Earth's Problem Climates.*)

the downstream part of the basin between the interior and coastal regions of heavy rainfall, precipitation drops as low as 70 in. (1,750 mm), for reasons which are obscure.

The heavy annual rainfall over most of the Amazon Basin also shows a distinct periodicity. All 12 months are wet only in the middle and upper parts upriver from Manaus; most of the remainder has 1 or 2 dry months. But in that less wet part of the lower basin where rainfall drops as low as 70 in., there appear to be spots, or even a northwest-southeast corridor, of *Aw* climate with more than 2 dry months. Over nearly the whole basin, however, including the wet upper part, winter and spring of the Southern Hemisphere have lower rainfalls than the other seasons (Fig. 8.13). This is because of the deep westward penetration of the South Atlantic subtropical anticyclone and its more stable easterly flow during the low-sun period.

Little is known about rainbringing disturbances in the Amazon country, though one in particular has often been described. This is a wind surge representing a speed convergence in the humid unstable flow from the northwest which invades the basin from August to March. Such surges travel southeastward in the northwesterly flow. They appear to resemble the speed convergences described for the Indian summer monsoon, and like them they generate spells of bad weather with cloud and instability showers.

Much smaller regions of *Ar* climate in Latin America occur in (1) the Guianas, which is a northwestward continuation of the Amazonian *Ar*, (2) the coastal strip and slopes in Brazil bordering the South Atlantic, (3) the Pacific lowland and slopes of Colombia, and (4) the Caribbean lowlands of Central America, together with certain more elevated islands and parts of islands in the West Indies.

Ar along the Atlantic margins of Brazil south of Cape St. Roque is distinguished by the fact that the northern part from the Cape to about 13°S receives some 70 to 80 percent of the year's rainfall in low-sun autumn-winter. Spring-summer is relatively dry (Fig. 8.14). Such a seasonal distribution, which is unusual in the tropics, appears to be a consequence of a strong buildup in spring-summer of

Figure 8.13 Annual march of rainfall at three stations in the Amazon Basin. Uaupés is located in the extreme western part, where annual rainfall is both heavy and well distributed throughout the year, with no dry months. Belém, on the oceanic eastern margins, likewise has heavy annual rainfall, but there is greater seasonal variation. Obidos, situated between the other two stations, has much less total annual rainfall, and even a short dry season. The Southern Hemisphere spring is normally the driest season.

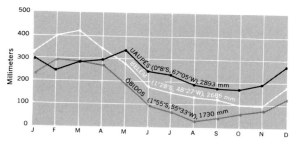

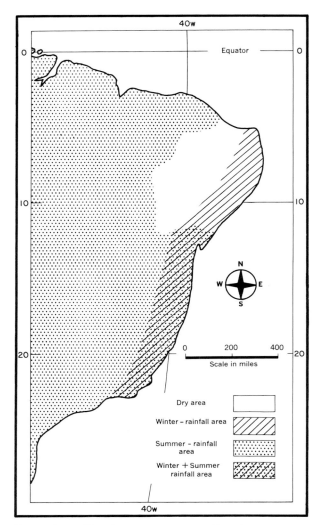

Figure 8.14 The dry area in eastern Brazil occupies an intermediate position between the winter-rainfall region to the east, and the strong summer-rainfall region to the interior and west. It normally escapes the full effects of the disturbances which produce these contrasting seasonal rainfalls. (*From Trewartha, The Earth's Problem Climates.*)

the subtropical anticyclone over the "hump" of easternmost Brazil. This high holds at bay the northwesterly flow of equatorial air which is such a potent rainbringer to the Amazon. It also blocks the normal migration of the ITC south of the equator along the coast (Fig. 8.12).

Ar along the Pacific lowlands and slopes of Colombia is notable for its excessively large annual amounts of precipitation, usually between 150 and 400 in. Such abnormal rainfall reflects the prevalence of westerly-southwesterly flow of unstable equatorial air which is blocked by high mountains not far inland. Supplementing these orographic effects is the influence of an ITC positioned almost constantly a few degrees north of the equator. Another unusual feature is the fact that the coastal strip has a strong predominance of night rains, usually concentrated between midnight and 4 A.M.

A widespread feature of *Ar* in the Caribbean lowlands and slopes of Central America, and many islands in the West Indies as well, is the two summer maxima (May-June and September-October-November), with a secondary minimum in mid- and late-summer (July-August) called the *veranillo*. The summer "dent" in the top of the annual rainfall curve occurs at the time of a slight increase in atmospheric pressure associated with an extended subtropical anticyclone in the westernmost Atlantic. Hurricane rainfall is an important part of the second, or autumn, maximum.

Africa. In the Congo Basin of Africa, annual rainfall is distinctly less than in the Amazon lowland. Although it is generally over 60 in., it exceeds 80 in. only in restricted areas. In part this reduced Congo rainfall reflects the influence of dry Saharan air coming from the north. No such deeply humidified mass of unstable *mT* air invades the Congo as that which floods the Amazon from the northwest for much of the year. Entrance of Indian Ocean air is made difficult by the East Africa highlands. Also, much of the *mT*–South Atlantic air which enters from the west originates in the northern and eastern limb of a subtropical cell and has traveled over a cool current, so that its vertical structure is fairly stable. Why the interior and lowest part of the Congo Basin, the Cuvette Central, should have the most extensive area of greatest rainfall remains unclear.

At least two types of rainbringing disturbance in the Congo have been described: surges in the southwesterly flow, and line squalls that commonly move

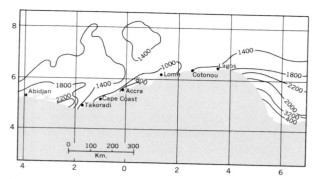

Figure 8.15 Annual rainfall amounts along the West African coast from the Ivory Coast to Nigeria. The 1,000 mm (40 in.) isohyet (line of equal rainfall) approximately defines the Accra dry belt. (*From Climatological Atlas of Africa, 1961.*)

from east to west being carried by the easterly flow aloft. Many of these squalls seem to develop along the convergence zone between easterlies and equatorial westerlies.

Along the Atlantic coast, in Liberia and the Ivory Coast in Tropical West Africa, is a somewhat smaller region of *Ar* climate. Here a well-developed south-

west summer monsoon, or equatorial westerlies, results in a large rainfall total strongly concentrated in the high-sun period. "Winter," when the dry easterlies reach nearly to the coast, is distinctly drier, so that ordinarily 2 months are classed as dry. Separating the Congo and West Africa *Ar* regions is a distinctly drier (30 to 40 in.) *Aw-BS* area in Ghana-Togo-Dahomey (Fig. 8.15). Its origin is obscure, although since the driest part (under 30 in.) is along the coast, it seems likely that the drought is partly associated with winds whose direction is parallel to the coastline or slightly offshore.[5] Madagascar shows a rainy east (windward) side in trades, so that the location of *Ar* climate there resembles the situations in eastern Brazil and in eastern Central America.

Southeast Asia and others. Southeast Asia, a much-fragmented insular and peninsular region with highly diverse directional alignments of coasts and terrain features, naturally exhibits numerous

[5] See Trewartha, *The Earth's Problem Climates*, pp. 106–110.

Figure 8.16 A mercator projection satellite photograph showing the equatorial region centered on the central and eastern Pacific Ocean. A striking feature is a pair of cloud bands, coincident with heavy rainfall, which parallel the equator, each band situated a few degrees north and south of 0°. Between the two cloud bands is a narrow belt with little cloud (shown as a dark band bisected by the equator) which extends westward from South America to beyond 160°E. This is the equatorial Pacific dry belt. The photograph is not a single exposure, but rather is a multiple-exposure "average" from ESSA III and V computer-produced mosaics, March 16–31, 1967. (*Courtesy of J. Kornfield and A. F. Hasler.*)

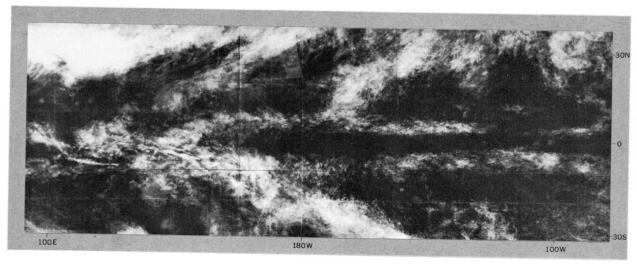

regional and local modifications of *Ar* climate. During the course of a year much of the region is influenced both by the southwest monsoon, or equatorial westerlies, and by the tropical easterlies, or trades. Since the monsoon is usually the heavier rainbringer, a high-sun precipitation maximum is the rule over the greater part of the area, and west sides of islands and peninsuals have the heaviest annual rainfalls. But at least three fairly extensive areas, all located along eastern littorals backed by highlands, exhibit a trade-wind low-sun winter maximum: the northeastern Philippines, the east side of the Indochina Peninsula, and the eastern margins of the Malay Peninsula. Stations with 1 or 2 dry months are very common.

One of the earth's most striking climatic anomalies is the rainfall deficiency in the equatorial eastern Pacific ocean (Fig. 8.16). In the form of a long thin wedge tapering toward the west, its apex in the vicinity of 160° to 170°E, a narrow zone of greatly diminished rainfall follows the equator for a distance of 7,000 miles or more westward from the Ecuador-Peru coast. Rainfall increases westward—from 5 to 10 in. in the extreme east to 30+ in the far western part. Belts of heavy rainfall flank this zone to the north and south. The dry belt is remarkably persistent in location; its latitudinal shift during the year is slight. What the origin of this longitudinally extensive equatorial drought may be remains obscure. Some would attribute it to the narrow band of equatorial cool water that thrusts westward from the South American coast. Others attach importance to the lack of equatorial westerlies, and the prevalence of an easterly flow, probably associated with divergence and subsidence. Additional published solutions are known to be imminent.[6]

TROPICAL WET-AND-DRY CLIMATE (*Aw*)

Aw climate differs in two principal respects from *Ar*: It usually has a smaller total precipitation, and the annual rainfall is less well distributed throughout the year. The wet season is shorter and the dry season longer, with the drought more severe. As a result the dense *Ar* rainforest gives way in *Aw* regions to lighter, more deciduous forest and tree-studded grasslands. It is this widespread occurrence of tall, coarse grass (called *savanna*) which has led to *Aw* being often referred to as savanna climate. This name may not be so appropriate, however, for many students of climate doubt whether tropical grasslands are climatically induced. Moreover, pure savannas, without trees, seem to be the exception rather than the rule.

Type location. Typically *Aw* is bounded by *Ar* on its equatorward side and by dry climates (*B*) or subtropical humid climate (*Cf*) on its poleward side (see Fig. 7.2 and map of climate types inside front cover). These boundaries were precisely defined earlier in the chapter in the general discussion of the *A* group of climates. The reason why *Aw* usually grades into a dry climate on the western side of a continent and into a humid climate on the eastern side lies in the contrasting nature of the oceanic subtropical anticyclones in its eastern and western parts. This too has been discussed earlier.

The typical latitudinal location of *Aw* is from about 5 or 10° out to 15 or even 20° (Figs. 7.2 and 8.3). On the generalized latitudinal profile of sea-level pressure, *Aw* is therefore located between the equatorial low-pressure trough and the subtropical highs. Thus on a mean wind chart it will be found somewhat toward the more humid equatorial margins of the tropical easterlies, or trades. This places *Aw* in an intermediate, or transitional, position between the humid unstable air masses associated with the ITC and equatorial westerlies on one side, and the stable subsiding air masses of the subtropical anticyclones on the other (Fig. 8.3).

During the course of a year, with the north-south

[6]*Ibid.*, pp. 86–88. Heinz H. Lettau, *Physical Coupling between the Dry Belt of the Lower Atmosphere and the Cromwell Current of the Upper Ocean, along the Pacific Equator*, ATS—1 Volume, University of Wisconsin Press, Madison, Wis., 1968. J. Kornfield and K. J. Hanson, *Seasonal Displacement of the ITC*, ATS—1 Volume, University of Wisconsin Press, Madison, Wis., 1968.

shifting of the solar radiation belts and the consequent pressure and wind migrations, *Aw* latitudes are alternately encroached upon by the wet ITC and equatorial westerlies (convergent) at the time of high sun, and by the drier parts of the trades and subtropical anticyclones (divergent) in the low-sun period. In some *Aw* regions this seasonal wind reversal is called a monsoon. The result is a rainy "summer" and a dry "winter." In a genetic sense, *Aw* may be classed as one of the "alternating" climates mentioned in Chapter 7, since its most distinctive characteristic of a wet and dry season is closely associated with its dominance by contrasting elements of the general circulation in the opposite seasons.

Because *Aw* lies between dry climates on its poleward side and wet climates on its equatorward side, it is to be expected that within an *Aw* region there will be a gradual change in climatic character from the wetter equatorial margins toward the drier poleward margins. Naturally the dry season becomes increasingly longer and more severe toward the poleward side. These latitudinal contrasts are less marked toward the eastern side of a continent.

Geographical location. Study of the climate map inside the front cover shows that many, if not most, large *Aw* areas do have the type locations described above and shown graphically in Figs. 7.2 and 8.3. This is true of the two extensive *Aw* areas in South America, one to the north of the Amazonian *Ar*, in Venezuela, Colombia, and the Guianas, and the other to the south of the *Ar*, in south central Brazil and adjacent parts of Bolivia and Paraguay. It is equally true in Africa, where the extensive Sudan *Aw* lies north of the equator and the large veld *Aw* to the south. The *Aw* in equatorial eastern Africa is less typically located, however, and a large part of the *Aw* in both eastern and southern Africa is somewhat atypical because of the many plateaus whose altitude often reduces the temperature below that which is normal for *Aw*. Highlands have the same effect on *Aw* climate in parts of southeastern Brazil. The *Aw* of northern Australia and that of southern and southeastern Asia are relatively typical in location.

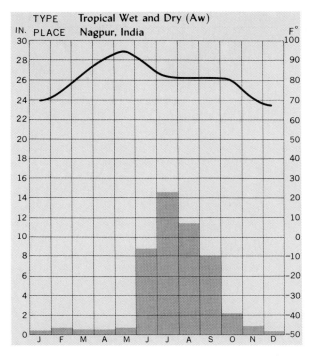

Figure 8.17 Average monthly temperatures and precipitation amounts for a representative station with tropical wet-and-dry climate (*Aw*) in India. Note that the highest temperatures are in May preceding the rains.

Temperature

Temperature characteristics of *Aw* and *Ar* climates are fairly similar. Constantly high temperatures prevail in both, for the noon sun is never far from a vertical position, and days and nights change little in length from one part of the year to another. As a rule the annual range of temperature in *Aw*, while small, is somewhat greater than in *Ar*—over 5° but seldom exceeding 15° (Figs. 8.17 and 8.18). The modest diurnal range during the rainy season in *Aw* resembles that in *Ar*, but in the dry season *Aw*'s range is likely to be greater. Like *Ar* climates, diurnal range in *Aw* is usually consistently larger than annual range.

Many times the hottest month or months do not coincide with the time of highest sun but somewhat precede it (Figs. 8.17 and 8.18). This is because the greatest heat occurs before the height of the rainy

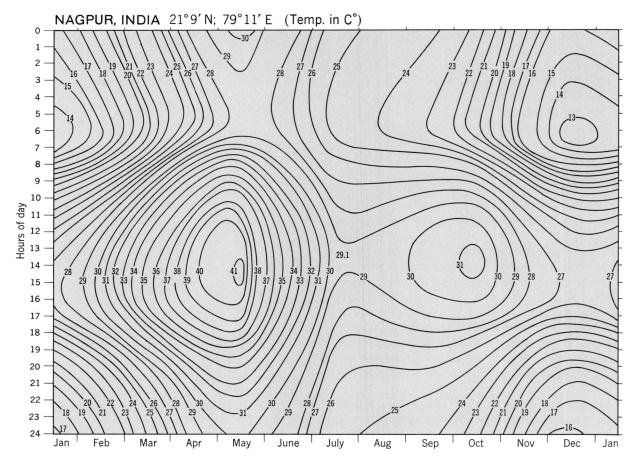

Figure 8.18 Thermoisopleths for an *Aw* station. Note that the annual range of temperature is considerably greater than in the *Ar* climate (Fig. 8.6), for it equals or exceeds the diurnal range in some months. The hottest period is in late April and May, preceding the time of highest sun. There is a secondary peak in October following the rainy season. (*After Troll.*)

period, at which time the more persistent cloud cover and heavier precipitation tend to lower the temperature. Thus in the Northern Hemisphere March, April, and possibly May are likely to be hotter than June or July, which are the rainiest periods for *Aw*. In some *Aw* regions the inhabitants recognize three temperature periods: the *cool dry season* at the time of low sun, the *hot dry season* just preceding the rains, and the *hot wet season* during the rains. Temperatures may rise again very slightly just after the rainy period as a result of the clearer skies and drier atmosphere (see the table on

page 273 for data for representative stations).

During the so-called cool dry season (period of lowest sun), day temperatures are still high, with afternoon maxima between 80 and 90° and occasionally above 90°. The humidity is low, however, so that the heat is not oppressive (Figs. 8.19 and 8.20). Nights at this season are inclined to be pleasantly mild—usually below 70° and often below 60°—for the dry air and clear skies are conducive to rapid terrestrial radiation.

During the hot dry season, which usually begins about the time of the spring equinox, increased

Climatic data for representative stations with tropical wet-and-dry climate (*Aw*)

	J	F	M	A	M	J	J	A	S	O	N	D	Yr	Range
					Navrongo, Ghana, 10° 53′N									
Temp., °F	81	85	89	90	87	82	80	79	79	82	82	80	83	10
Precip., in.	0.0	0.2	0.6	1.9	4.4	5.7	7.9	10.4	9.0	2.7	0.2	0.1	43.1	
					Timbo, Guinea, 10° 40′N									
Temp., °F	72	76	81	80	77	73	72	72	72	73	72	71	74	9.7
Precip., in.	0.0	0.0	1.0	2.4	6.4	9.0	12.4	14.7	10.2	6.7	1.3	0.0	64.1	
					Calcutta, India									
Temp., °F	65	70	79	85	86	85	83	82	83	80	72	65	78	21
Precip., in.	0.4	1.1	1.4	2.0	5.0	11.2	12.1	11.5	9.0	4.3	0.5	0.2	58.8	
					Cuiabá, Brazil, 15° 30′S									
Temp., °F	81	81	81	80	78	75	76	78	82	82	82	81	80	6.6
Precip., in.	9.8	8.3	8.3	4.0	2.1	0.3	0.2	1.1	2.0	4.5	5.9	8.1	54.6	
					Normanton, Australia, 17° 39′S									
Temp., °F	86	85	85	82	78	73	72	75	80	85	88	87	81	15
Precip., in.	10.9	10.0	6.1	1.5	0.3	0.4	0.2	0.1	0.1	0.4	1.8	5.6	37.5	

intensity and duration of solar radiation cause the daily maxima to rise well above 90° and often over 100° (Figs. 8.19 and 8.20). With the beginning of the rains, however, temperature conditions very much resemble those of the tropical wet climate. The diurnal range becomes somewhat smaller, and while the heat is not so intense as in the hot dry season, the higher humidity causes the sensible temperature to be much more oppressive and sultry.

Figure 8.19 Daily maximum and minimum temperatures for the extreme months in one year at a station with tropical wet-and-dry climate (*Aw*) in Brazil. Note the dominance of periodic or solar control and the relatively large diurnal range of temperature at this interior station.

Precipitation

Annual amount. Since temperatures are not very different within the tropics, rainfall becomes the critical element in setting apart the several climatic types of the low latitudes. Characteristically, the total annual amount of rainfall of *Aw* (40 to 60 in.)

Figure 8.20 Daily maximum and minimum temperatures for the extreme months at Havana, Cuba, at about 23°N. Its coastal location causes the daily range to be smaller than at Porto Nacional (Fig. 8.19). Located on the margins of the tropics, Havana also has a somewhat larger temperature difference between the warmest and coolest months than Porto Nacional. In addition, there is evidence of a very weak nonperiodic air-mass control in January.

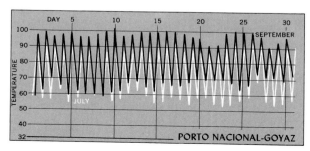

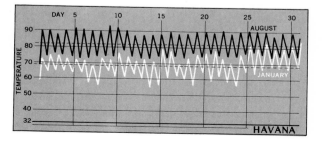

is less than that of *Ar* climate (see data). The smaller annual rainfall in *Aw* reflects the transitional nature of this type, located as it usually is between very wet ITC climates equatorward, and dry climates of the subtropical anticyclones poleward. As mentioned earlier, rainfall within *Aw* declines with increasing latitude.

Rainfall regime. It is seasonal distribution rather than amount of precipitation, however, which chiefly distinguishes the two climates of the humid tropics. The *Aw* type has a marked dry season of more than 2 months' duration, which characteristically comes at the time of low sun. In that season desert-type weather prevails, for usually the drought is intense. This contrast between the two tropical types is mainly due to their latitudinal locations—*Ar* occurs fairly consistently in the equatorial convergence zone, while *Aw* is intermediate between the ITC and dry trades (Fig. 8.3). In the wet season *Aw* is like *Ar*; in the dry season it resembles the desert.

The Sudan *Aw* of northern Africa is a good example of this rainfall regime. As the sun's vertical ray moves northward from the equator after the spring equinox, pressure and wind belts shift in the same direction, although lagging a month or two behind the sun. The convergent ITC belt of heavy rains gradually creeps northward, and thunderstorms begin to appear in March or April over the Sudan. Rainfall continues to increase until July or even August, when the ITC reaches its maximum northward migration. With the southward retreat of the ITC following the sun, the rains decline. By October or November the dry, subsiding trades again prevail over the Sudan, and drought grips the land. The length of the wet and the dry seasons is variable, depending upon distance from the equator.

There is no abrupt boundary between tropical wet *Ar* and tropical wet-and-dry *Aw* climates—only a very gradual transition from one to the other. On the equatorward margins of the *Aw*, the rainy season persists for 8 to 10 months. In such locations there may be even a slight depression at the crest of the annual precipitation curve occurring in the short interval between the northward and southward shifts of the ITC (Fig. 4.31, Zungeru). Usually the farther poleward one travels in the *Aw*, the shorter the period of ITC control and the longer that of the subsiding trade-wind air masses, so that the dry season increases in length while the wet period shrinks. Emphatically, rainfall follows the sun. This rule holds for either hemisphere, although of course it should be kept in mind that when a Northern Hemisphere *Aw* is having its rainy season, a similar region south of the equator is experiencing drought, and vice versa.

Rainfall reliability. Not only is *Aw* rainfall lower in total amount and more seasonal in distribution throughout the year than *Ar*, but it is likewise less reliable, with wider fluctuations in quantity from year to year (Fig. 8.21). One year may bring such an abundance of rain as to flood the fields, rot the crops, and increase the depredations of injurious insects and plant diseases; in the following year there may be even more severe losses from drought. In northern Australia the average rainfall variation from the normal is as much as 25 percent.

Seasonal weather

During its rainy season, the weather of *Aw* climate closely resembles that of *Ar* at its worst. This period usually is ushered in and out by violent thunderstorms and severe squall winds. In these transition periods the weather is very trying, with violent short deluges of rain and intensely hot sunshine alternating. As the rainy season advances, sunny days become rarer, showers more frequent.

During the height of the rains, violent thunderstorms appear to be less prevalent than they are in the transition periods, while convective showers, organized in distribution, reach maximum frequency. These showers, like those of *Ar* climates, probably originate in extensive atmospheric disturbances of the wave and weak-cyclone types. According to the Australian Weather Bureau, most of the mid-rainy-season showers of northern Australia

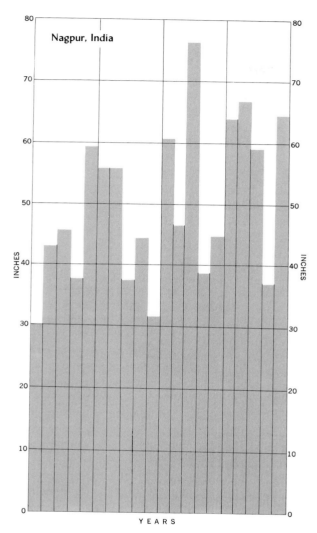

Figure 8.21 Variations in amounts of annual rainfall over a 20-year period at Nagpur, India, a station with tropical wet-and-dry climate. Large annual variations in precipitation are characteristic of this type of climate.

accompany weak tropical lows. Similarly, much of the *Aw* rain of India occurring during the height of the monsoon is received from weak depressions which form over the head of the Bay of Bengal and move slowly in a northwesterly direction across India. Other disturbances in the nature of speed

convergences are also numerous in *Aw* climates.

In the low-sun, or dry, season the weather is like that of the deserts. In spite of the aridity, the dry season is welcomed after the humid, oppressive heat of the rainy periods. Occasional showers may occur during the months of drought; the number depends upon which margin of the *Aw* is being considered. On the dry margin the period of absolute drought may be of several months' duration, while on the rainy margin, where *Aw* makes contact with *Ar* climate, there may be no month absolutely without rain. Remember that none of these climatic boundaries is sharp; there are very gradual transitions from one type to another. During the dry season the savanna landscape is parched and brown. The trees lose their leaves, the rivers become low, the soil cracks, and all nature appears dormant. Dust and smoke from grass fires fill the air, so that visibility is usually low.

Following is a description of the seasonal weather and related landscape changes in an *Aw* region *south* of the equator in Africa.

The winter months, or dry season, extend, with a slight variation, from April to November. They are, as I have said, pleasant and healthy in the extreme. Now the traveller and hunter of big game make their appearance; the deciduous trees are leafless; the grasses dry, yellow, and ready for the chance spark or deliberate act which, with the aid of a steady breeze, will turn vast expanses of golden grasslands into so many hideous, bare deserts of heat-tremulous black. All nature seems to be at a standstill, hibernating. The rivers are low. Where, but a few short months since, wide, watery expanses rushed headlong toward the sea . . . there now remain but tranquil, placid channels, flowing smilingly at the bottom of steep, cliff-like banks. . . .

With October the heat becomes very great. Vast belts of electrically charged, yellowish clouds, with cumulus, rounded extremities, begin to gather and at the close of day are seen to be flickering in their murky centres with a menacing tremor of constant lightning. This may go on for a week or more, and then Nature arises like a strong man in anger and looses the long pent-up voice of the thunder and the irresistible torrents of the early rains. The first manifestation may come at evening and is a soul-moving display of natural force. . . .

After such a disturbance as the one I have just described, rain is fairly continuous for some time, and the effect of this copious irrigation makes itself felt in every branch of animal and vegetable life. Within a few days the change is startling; the paths and roadways choke themselves with a rich clothing of newly sprung grasses, whilst the trees, the extremities of whose twigs and branches have been visibly swelling, now leap into leaf and blossom. The mosses, which for months past have looked like dry, bedraggled, colourless rags, regain once more their vivid, tender green. Now the forest throws off its puritanical greyness and, with an activity and rapidity beyond belief, decks itself in flowers of a thousand gorgeous shades of colour, from chrome-yellow and purple to grateful mauve.

The birds now put on their finest feathers, the animals appear in their brightest hues. Colour and warmth run riot in the brilliantly clear air now washed clean from the mist and smoke which for so many months have obscured it. The clear verdant green of rapid-springing grasses and opening fronds clothes the landscape, and the distant peaks of the mountains lose their pale, bluey-grey haziness and stand boldly out in the light of the sun. The months succeed each other, bringing with them new and strange beauties, for summer is now at its height, and trees and flowers at their most perfect period. . . . April comes, and suddenly Nature holds her hand. The swollen rivers and inundated plains shake themselves free from the redundant waters. The grasses have now reached a formidable height. The rains now cease, and the land begins to dry up. Rich greens turn to copper, and brown, and yellow, and little by little, with the advent of May, the winter returns with its sober greyness.[7]

Upland *Aw*

In tropical latitudes on several continents, particularly in Africa and South America, there are extensive uplands a few thousand feet in elevation which

[7] R. C. F. Maugham, *Zambezia*, John Murray, London, 1910, pp. 383–388.

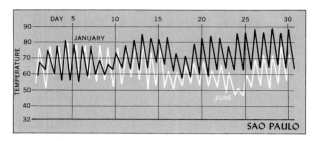

Figure 8.22 Daily maximum and minimum temperatures for a station with tropical wet-and-dry climate on the Brazilian upland at nearly 24°S. Note the lower temperatures imposed both by altitude and by the latitude. While solar control is dominant, nonperiodic air-mass control is also evident here on the margins of the humid tropics.

have most of the normal *Aw* characteristics but differ chiefly in lower temperatures (Fig. 8.22). Depending on their height, some of these upland stations are likely to show cool-month temperatures that are below the minimum for the tropics, and so represent *C* climates. Others qualify as *A*. But in most instances, whether they are tropical (*A*) or subtropical (*C*) in temperature, their seasonal distribution of rainfall remains that of the *Aw* type. Consequently a tropical upland often shows an intermingling of local *Aw* and *C* climates. On the climate map in the front of the book these slightly atypical tropical upland climates continue to be represented as *Aw*, even though some parts certainly are subtropical (*C*) rather than tropical. Their very general distribution is indicated on the climate map by a light overshading. (For climatic modifications imposed by altitude, see Chapter 11.)

Regional features

Latin America. Interior Brazil south of the equator, called the Campos, represents one of the earth's most extensive regions of *Aw* climate. Since it is also relatively uncomplicated in its main climatic lineaments, it constitutes almost a textbook example of *Aw* characteristics. The annual rainfall of 40 to 60+ in. is strongly concentrated in the high-sun period, when the Campos is dominated by a northwesterly

Climatic data for an upland *Aw-Cw* **station: Zomba, Northern Rhodesia** (Lat. 15°22′S, long 38°18′E, elevation 3,042 ft)

	J	F	M	A	M	J	J	A	S	O	N	D	Yr
Temp., °F	72	72	72	70	67	64	63	65	70	74	75	73	69.7
Precip., in.	13.9	8.9	8.9	36	0.5	0.6	0.3	0.4	0.2	1.0	4.5	10.4	53.3

flow of moist, unstable equatorial air from the Amazon forests. In the eastern parts, which have uplands a few thousand feet high, areas of *Aw* climate are intermingled with those that show *C* temperature characteristics.

The smaller Venezuela-Colombia *Aw*, or Llanos, is the Northern Hemisphere counterpart of the Brazilian Campos. Along its Caribbean littoral, there is an anomalous situation where a dry climate—mostly steppe, but possibly with restricted areas of desert—prevails in a location which seemingly should be wet. Parts of the coast have an annual rainfall of less than 500 mm (20 in.); in a few restricted areas it may be as low as 300 to 400 mm (12 to 16 in.). In part the precipitation deficiency is related to a downwind acceleration of the resultant winds, with vertical subsidence aloft. In part also, it is caused by coastal divergence in a surface airflow which nearly parallels the east-west stretches of coast. Significantly, along short stretches of north-south coast where the tendency of onshore winds to pile up is accentuated, rainfall is much heavier. In addition to its dryness, a further peculiarity of

Figure 8.23 Composite annual rainfall profile of 17 stations in the Netherlands West Indies. Average annual rainfall is only 569 mm, or 22.4 in. (*After Lahey. From Trewartha, The Earth's Problem Climates.*)

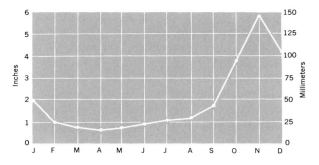

this littoral is that its modest precipitation is concentrated at the time of low sun (Fig. 8.23).[8]

Other *Aw* areas in Latin America include a small one in Pacific Ecuador close to the equator, which typically would be located some 10° farther south. Its northward displacement results from the unusual intensity of the desert-making mechanisms farther south along the coast of Peru. *Aw* is also characteristic of the leeward Pacific side of Central America and southern Mexico. In addition, it prevails on the low-elevation islands of the West Indies and the lowland Yucatán Peninsula.

Africa. In Africa north of the equator, the huge east-west Sudan *Aw* is typically located between the Congo *Ar* and the Sahara *B* climates. It is relatively narrow in a north-south direction, owing to the far southward thrust of Saharan aridity. During midsummer the superheated Sahara has the effect of displacing the surface ITC unusually far poleward (20–21°N). South of this convergence are the equatorial southwesterlies, while above them is dry Saharan air from the northeast. Thus little rain falls along the ITC, and even for several hundred miles south of it, because the surface westerly flow is so shallow, and the rising arid air aloft has too little moisture to permit much precipitation. Still farther south, over the Sudan belt where the southwesterly air is deeper, summer rainfall is much more abundant.

Two weather types which are important rain generators for the Sudan *Aw* are the disturbance line and the surge or speed convergence in the monsoon-like southwesterly flow. The disturbance line superficially resembles a cold front, being accompanied by thunderstorms and turbulent squall winds. In reality, however, these storms develop not

[8] Trewartha, *The Earth's Problem Climates*, pp. 57–64.

along a front, but within the deep easterlies of the Sudan, underlain by the humid equatorial southwesterlies. They move in a westerly direction, carried by the easterly flow aloft; but the accompanying cloud and rain come from the moisture contained in the underlying southwesterlies. The second Sudan weather type, the surge, is a feature of the surface equatorial southwesterly current during the high-sun period. As the surge moves onshore from the Atlantic Ocean and the Gulf of Guinea, it produces cloudy, rainy weather, but without the accompanying turbulence, wind shifts, and squall winds associated with the more vigorous disturbance line. It brings what West African meteorologists call the "monsoon rains."[9]

A special weather type characteristic of the dry season in the Sudan (also found in the Sahara) is a strong southward push of fresh, dry, northeasterly trade-wind air. This is known as the *harmattan.* It is very dry, hot by day and cool by night, and laden with dust picked up during its trajectory over the Sahara wastelands. A dense and widespread dust haze is characteristic.

A great deal of the *Aw* in equatorial East Africa and in southern Africa is of the cooler upland variety, with many areas showing *C* instead of *A* temperatures. Equatorial East Africa with its moderately high plateaus is scarcely typical of *Aw* in either temperature or seasonal rainfall distribution. There two wet and two dry seasons are common, reflecting the dual passage of the ITC. As the maps inside the front and back covers show, equatorial East Africa is drier than normal for the eastern side of a continent close to the equator. There *A(C)w* and *B* climates prevail instead of *Ar* (compare Africa with South America and Asia).

Asia.   Like most *Aw* areas, the Indian subcontinent has a marked seasonal wind shift, the so-called Indian monsoon, between low sun and high sun. But in the usual, or genetic, sense, India's seasonal wind reversal is no more a monsoon than those of most other *Aw* regions, but rather a latitudinal migration of planetary wind belts following the annual course of the sun. During the dry season of winter, the weak northeasterly flow that prevails over most of the subcontinent is essentially a trade wind. Owing to a bifurcation of the upper westerlies by the Tibetan highlands, one branch of the westerlies and its southerly jet stream move southward around the highland mass and are positioned over the northernmost part of the subcontinent. Weak cyclonic storms steered by the jet provide modest cool-season precipitation in northerly and northwesterly parts. As a result, the annual rainfall curve there, in addition to a strong primary summer maximum, has a small secondary winter one as well.[10]

The excessively hot and dry premonsoon period (March–May), with its anticyclonic control, is terminated by the "burst of the monsoon," as the southwesterly flow of unstable equatorial air surges northward over the subcontinent accompanied by turbulent weather. The southwesterly current consists of two branches. A westerly current from the Arabian Sea meets the elevated west coast of India at almost right angles, resulting in heavy orographic rainfall. In modified form it continues northeastward across the broad peninsula. The second, or Bay of Bengal, branch has an exclusively oceanic trajectory. Toward the head of the bay the two branches merge, and a portion of this current appears to recurve westward around the eastern end of a pressure trough positioned over northern India. It flows westward up the Ganges Valley on the northern side of the pressure trough.

India's summer rainfall, which is modest in amount and variable over much of the subcontinent, is a product both of the equatorial southwesterly circulation and of the several types of extensive atmospheric disturbances embedded in it. It is these disturbances that bring the spells of weather with organized cloud and shower areas. One such disturbance has been given the name "monsoon depression." Most monsoon depressions are relatively weak cyclonic circulations, but a few mushroom

[9] *Ibid.*, pp. 97–102.

[10] *Ibid.*, pp. 153–157.

into violent hurricanes. Many seem to originate over the Bay of Bengal, from which they take a course toward the northwest. This concentrates their rainfall effects mainly over the northern, and especially the northeastern, sector of India.

Other spells of India's disturbed rainy weather in summer are generated by pulsations or surges in the southwesterly current, and they move downstream in that monsoonal flow. These surges appear to resemble similar speed convergences mentioned earlier as bringing rain to the Amazon Basin and the Campos in Brazil, and to the Congo Basin and the Sudan in Africa. Significantly, a steady flow, even in a humid unstable circulation, is conducive to fair weather. It is when the steady flow is disturbed that cloud and rain result.

Aw in Southeast Asia is chiefly confined to the more inland sections of the Indochina Peninsula. The seaward margins are mainly *Ar*. The *Aw* region of northern Australia, like those of eastern India and the Caribbean, is occasionally subject to hurricanes.

SELECTED REFERENCES FOR FURTHER STUDY OF TOPICS IN CHAPTER EIGHT

Beckinsale, R. P., "The Nature of Tropical Rainfall," *Tropical Agriculture*, Vol. 34, pp. 76–98, 1957.

Chang, Jen-Hu., "Comparative Climatology of the Tropical Western Margins of the Northern Oceans," *Ann. Assoc. Amer. Geographers*, Vol. 52, pp. 221–227, 1962.

Fosberg, F. R., B. J. Garnier, and A. W. Küchler, "Delimitation of the Humid Tropics," *Geog. Rev.*, Vol. 51, pp. 333–347, 1961.

Garbell, Maurice A., *Tropical and Equatorial Meteorology*, Pitman Publishing Corporation, New York, 1947.

Lauer, Wilhelm, "Humide und aride Jahreszeiten in Afrika und Südamerika und ihre Beziehung zu den Vegetationsgürteln," in Lauer et al., *Studien zur Klima- und-Vegetationskunde der Tropen, Bonner Geog. Abhand.*, Geog. Inst. der Univ. Bonn, 1952.

Lockwood, J. G., "The Indian Monsoon: A Review," *Weather*, Vol. 20, pp. 2–8, 1965.

Palmer, C. E. "Tropical Meteorology," in Thomas F. Malone (ed.), *Compendium of Meteorology*, American Meteorological Society, Boston, pp. 859–880, 1951.

Portig, W. H., "Central American Rainfall," *Geog. Rev.*, Vol. 55, pp. 68–90, 1965.

Ramage, C. S., "Diurnal Variation of Summer Rainfall in Malaya," *J. Tropical Geog.*, Vol. 19, pp. 62–68, 1964.

Riehl, Herbert, *Tropical Meteorology*, McGraw-Hill Book Company, New York, 1954.

Symposium on Monsoons of the World, Indian Meteorological Department, New Delhi, 1960.

Symposium on Tropical Meteorology, J. W. Hutchings (ed.), New Zealand Meteorological Service, Wellington, 1964.

Thompson, B. W., "An Essay on the General Circulation of the Atmosphere over Southeast Asia and the West Pacific," *Quart. J. Roy. Meteorol. Soc.*, Vol. LXXVII, pp. 569–597, 1951.

Thornthwaite, C. W., "The Water Balance in Tropical Climates," *Bull. Amer. Meteorol. Soc.*, Vol. 32, pp. 166–173, 1951.

Trewartha, Glenn T., "Climate as Related to the Jet Stream in the Orient," *Erdkunde*, Vol. XII, pp. 205–214, 1958.

Trewartha, Glenn T., *The Earth's Problem Climates*, The University of Wisconsin Press, Madison, Wis., 1961.

Watts, I. E. M., *Equatorial Weather*, University of London Press, Ltd., London, 1955.

Yin, M. T., "A Synoptic-Aerologic Study of the Onset of the Summer Monsoon over India and Burma," *J. Meteorol.*, Vol. 6, pp. 393–400, 1949.

CHAPTER NINE

SUBTROPICAL CLIMATES (C)

Although the subtropics are transitional in climatic character between tropics and middle latitudes, the classification system used here places the humid-subhumid parts of subtropical climates in the middle latitudes. Lacking the constant heat of the tropics and the constant cold of the polar icecaps, middle-latitude climates are characterized by a strong seasonal rhythm in temperatures. The largest annual temperature ranges on earth are found in the Northern Hemisphere middle latitudes with their vast continents. Thus temperature becomes coequal with rainfall in determining the various climatic types of middle latitudes. In the tropics seasons are distinguished as wet and dry, since seasonal temperature contrasts are small; in the middle latitudes they are called winter and summer, and the dormant season for plant growth usually is one of low temperatures rather than drought.

In the middle latitudes the changeableness of the weather is proverbial, for they are the realm of conflict between air masses expelled from polar and tropical source regions. Within the humid parts of these latitudes, three large groups of climates are recognized: subtropical, temperate, and boreal.

Type location. The subtropical group, the warmest and mildest of the three, occupies the equatorward margins of middle latitudes (Fig. 9.1). On its poleward frontier it joins the temperate group; on its low-latitude side it borders the tropics. Many areas of the subtropics have occasional snow, but it does not remain on the ground to form a durable snow cover. Most of this group also is subject to some freezes, although in marine locations they may be rare or even absent. The tropical boundary chosen for subtropical climates is the equatorward limits of freeze, or in marine locations, the 65° isotherm for the coolest month. The poleward boundary consists of 8 months with an average temperature of at least 50°.

Within the subtropical belt two types of humid climates are distinguished: subtropical dry-summer, usually situated on the western side of a continent, and subtropical humid on the eastern side (Fig. 9.1).

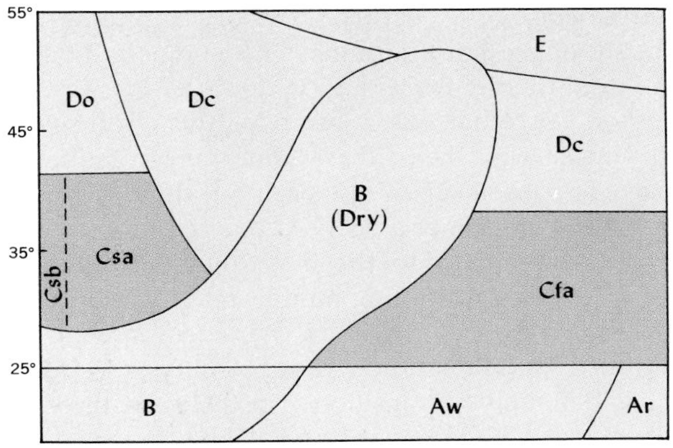

Figure 9.1 Type location of the two subtropical climates (*Cs* and *Cfa*) with respect to latitude, position on continents, and bordering types of climates.

SUBTROPICAL DRY-SUMMER CLIMATE (*Cs*)[1]
(Mediterranean)

In its simplest form this climate has three main features: (1) a concentration of the year's modest amount of precipitation in the winter season, while summers are nearly or completely dry; (2) warm-to-hot summers and unusually mild winters; and (3) abundant sunshine and meager cloudiness, especially in summer. Quite deservedly this climate, with its bright sunny weather, blue skies, few rainy days, and mild winters—and its usually plentiful fruit, flowers, and winter vegetables—has acquired a glamorous reputation.

The *Cs* type has marked climatic characteristics which are duplicated with notable similarity in the

five regions where it occurs: the borderlands of the Mediterranean Sea, central and coastal southern California, central Chile, the southern tip of South Africa, and parts of southernmost Australia. According to Köppen, *Cs* occupies only 1.7 percent of the earth's land area; yet in spite of its limited extent it is one of the most distinctive and best-known types. It is the only extensive humid climate that has drought in summer and a strong rainfall maximum in winter.

Type location. Typically, subtropical dry-summer climate is located on the tropical margins of the middle latitudes (30 to 40°) along the western sides of continents (Figs. 7.2 and 9.1). It is the western sides that are affected by the stable eastern end of an oceanic subtropical high. Lying thus on the poleward slope of a subtropical high-pressure cell's

[1]*s* = Dry season in summer of the respective hemisphere. At least three times as much rain in the winter half year as in the summer half year. Driest summer month receives less than 3 cm or 1.2 in. Total annual precipitation usually less than 89 cm or 35 in.

stable side, *Cs* is between the dry subsiding air masses of the horse latitudes on the one hand and the rainbringing fronts and cyclones of the westerlies on the other. As a result of the north-south shifting of wind belts, the Mediterranean latitudes are encroached upon by each of these systems during the course of a year. At one season they are joined climatically to the dry tropics and at the opposite season to the humid middle latitudes. Tropical constancy therefore characterizes them in summer, middle-latitude changeability in winter. *Cs* is definitely a transition type between tropical dry climates equatorward and temperate climates, usually temperate oceanic (*Do*), poleward (see Fig. 7.2 and the climate map inside the front cover).

In both central Chile and California, mountains terminate the dry-summer type abruptly on the land side, and dry climates prevail on the other side of the mountains. In South Africa and southwestern Australia, the farthest poleward extent of these continents carries them barely into Mediterranean latitudes, so that subtropical dry-summer climate occupies southern and southwestern extremities rather than distinctly west-coast locations. Only in the region of the Mediterranean Basin, which is an important cool-season convergence zone and a route of winter cyclones, is this type of climate found far inland, and there it extends for 2,500 miles or more. It is the relative warmth of the Mediterranean Sea in winter, and the resulting low-pressure trough coincident with it, that makes the Mediterranean Basin a region of air-mass convergence in the cool months, with a resulting development of fronts and cyclones.

Continental interiors and eastern margins of continents, with their tendencies toward monsoon wind systems and summer maxima of rainfall, are governed by conditions opposed to the development of *Cs* climate, especially its characteristic rainfall regime. And while the eastern sides of subtropical continents come under the influence of a subtropical anticyclone in summer, they are affected by the unstable air of the western parts of the cell and so are likely to have abundant summer rainfall.

Temperature

Because of its subtropical latitude and its west-side position on the continents, *Cs* climate has very little cold weather. Average temperatures of the winter months are usually between 40 and 50° and of the summer months between 70 and 80°, so that mean annual ranges of 20 to 30° + are common. These are relatively small for the middle latitudes but are larger than those of the tropics, except possibly for some tropical dry climates.

Subtropical dry-summer climates may be classified in two subdivisions based mainly on the degree of summer heat. This in turn is largely a function of location. First and far more extensive is a warm-summer subtype (*Csa*), situated either inland from the coast or on a coast bordered by warm water. The second, a cool-summer subtype (*Csb*),[2] occupies limited areas mainly where coasts are washed by cool currents fed by upwelling, such as the California and Chile littorals. Some elevated sites may also be *Csb*. Since *Csa* is so much more widespread that it represents the standard condition for *Cs*, the cool-water subtype is treated as only a minor variant.

Summer.   Except along cool-water coasts and on uplands, *Cs* summer temperatures are hot, resembling those of tropical-subtropical dry climates (Fig. 9.2). In July and August the dry-summer subtropics blaze under a pitiless sky. Red Bluff, California, located inland in the Sacramento Valley behind the coastal ranges, has an average July temperature of about 82°. Such temperatures are about 16° higher than the July average at Santa Monica, California, located farther south but on a cool-water coast. Of course, the summer climate of the interior valleys of California represents a somewhat extreme variety of dry-summer subtropical. Lowland stations along the borderlands of the Mediterranean Sea usually

---

[2] In the symbols *Csa* and *Csb* the letter *a* indicates a hot summer in which the average temperature of the warmest month is 22°C (71.6°F) or above. The letter *b* indicates a cool summer in which the average temperature of the warmest month is below 22°C.

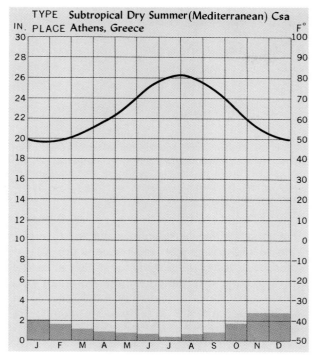

TYPE Subtropical Dry Summer(Mediterranean) Csa
PLACE Athens, Greece

Figure 9.2 Annual march of temperature and rainfall for a subtropical dry-summer station with a hot summer (*Csa*).

have July averages between 75° and 80°+. *Cs* stations in the Southern Hemisphere are not so warm; Perth and Adelaide in Australia record January temperatures of 74°. But while summer days are hot in most regions with this climate, it is not sultry heat, except along a coast. Dry heat like that of the desert is typical of inland California.

Averages of the daily maxima in the Sacramento Valley usually are close to 90° (Fig. 9.3). In a year selected at random, Sacramento had 27 days in July and 16 in August with maximum temperatures of 90° or above. Similar figures for Red Bluff were 30 and 30. Sacramento has recorded a temperature as high as 114°. Clear skies, dry air, and a nearly vertical sun provide ideal conditions for strong daytime heating in *Csa* climates.

These same conditions are also conducive to

rapid nocturnal cooling. Consequently there is a marked contrast between day and night, especially in the drier Mediterranean climates. At Sacramento in the Great Valley, hot, clear, summer days with afternoon temperatures of 85 to 100° are followed by nights when the thermometer sinks between 55 and 60°. The daily range for this city in a recent July was 36.6°—a figure characteristic of deserts. Following the hot, glaring days, the relatively cool nights, in which a light topcoat may be comfortable, are much appreciated by the inhabitants of inland Mediterranean climates. One 24-hr period is much like another in summer, for sun is in control (Fig. 9.3).

Cool-summer coastal subtype (Csb). Compared with areas inland, any coastal station is likely to have somewhat moderated summer heat. Perth, Australia has a hot-month temperature of 74°, Lisbon 72°, Casablanca (Morocco) 74°, and Istanbul 74° —all of which are 5 to 8° lower than those of interior California. Still, these are not cool by the standards of cool-water coasts (see Fig. 9.4 and the data for Santa Monica, page 286). The warm-month temperature of Santa Monica on the California coast is only 66°, San Francisco 59°, and Valparaiso in coastal Chile 66°. In summer such locations fre-

Figure 9.3 Daily maximum and minimum temperatures for the extreme months in one year at a subtropical dry-summer station inland in California (*Csa*). Note the hot summer and the large diurnal range of temperature. Solar control is dominant in summer, but irregular, nonperiodic air-mass control is conspicuous in winter.

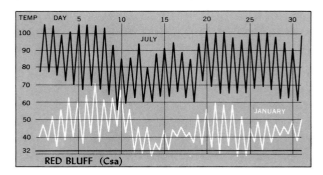

RED BLUFF (Csa)

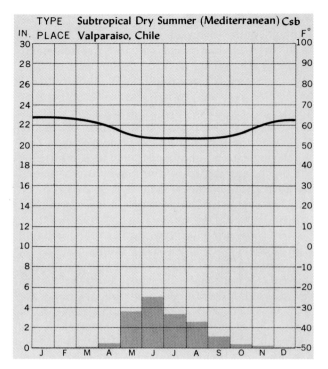

Figure 9.4 (TYPE Subtropical Dry Summer (Mediterranean) Csb; PLACE Valparaiso, Chile)

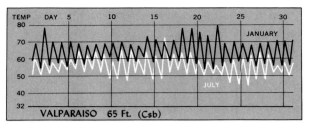

VALPARAISO 65 Ft. (Csb)

Figure 9.5 Daily maximum and minimum temperatures of the extreme months for the same station as in Fig. 9.4. Note the small diurnal range of temperature. Although sun control dominates, there is evidence of some weak nonperiodic control associated with disturbances and their advective effects.

Figure 9.4 Annual march of temperature and rainfall for a Southern Hemisphere subtropical dry-summer station, located along a cool-water coast in Chile (*Csb*). Note the cool summer.

quently have marine fog and low stratus fog, which are related to the cool water and the above-surface inversion of temperature. In many of its features—low temperature, small temperature range, fog, and aridity—the summer climate of coastal subtropical

Chile and California resembles that of the cool, foggy desert coasts (*Bn*) farther equatorward. As would be expected winters too are milder in coastal locations than in the interiors, and frosts are rare. Annual ranges are consequently small (only 9° at San Francisco and 11° at Valparaiso), and so are daily ranges (Fig. 9.5).

Winter. It is for their mild, bright winters, characterized by congenial days and crisp nights, that subtropical dry-summer climates are justly famed. People from the higher latitudes seek them out as winter playgrounds and health resorts. In northern latitudes the word winter has a congealing sound, but winter in the Mediterranean borderlands transforms the barren hills with a glow of color. Even inland locations have average cold-month temperatures 10 to 15° above freezing. Thus inland Sacra-

Maximum and minimum temperatures for some *Cs* stations

Station	Average annual minimum	Absolute minimum	Average annual maximum	Absolute maximum
Valencia	31	20	96	107
Naples	33	24	94	101
Athens	29	20	100	109
Fresno	26	17	109	115
San Diego	36	25	91	110
Perth	38	34	105	112
Cape Town	32	28	97	103

mento's average January temperature is 46°. But along coasts middle winter is even milder—55° at Perth, 53° at Santa Monica, and 48° at Naples. In southern California in January, midday temperatures rise to between 50 and 60° and at night drop to 40°± (e.g., 39° in Sacramento at night, 45.5° in Los Angeles).

Freeze and the growing season. The growing season in *Cs* climates is not quite the whole year, for there are occasional freezes during the 3 winter months. But to say that the growing season is 9± months does not adequately describe the situation, for while freezing temperatures do occur during midwinter months, they come on only a few nights and are rarely severe. During a period of 41 years at Los Angeles, there were 28 in which no freeze occurred and the growing season was 12 months long. In southern California, freezes may be expected at intervals of about 10 to 15 years. During a recent year at Red Bluff there were 10 nights when the temperature dropped below 32°; at Sacramento there were 7. The lowest temperature ever recorded at Los Angeles is 28°, at Naples 24°, and at Sacramento 17°. Even on the occasional nights when temperatures do slip a few degrees below freezing, they rise well above 32° again the following day. Never does the thermometer stay below the freezing point for an entire day.

What freezes do occur are usually the result of radiation cooling following an importation of cold polar air. The subfreezing temperatures are confined to a shallow layer of surface atmosphere, particularly to depressions in which the cool, dense air has collected. For this reason such sensitive crops as citrus are commonly planted on slopes. Occasionally fires must be lighted among the citrus trees in order to prevent serious damage from freezing. At first it may seem odd that in Mediterranean climates, where freezes are neither frequent nor severe, unusual losses should result occasionally from low temperatures. But it is this infrequency and lack of severity that make frost so treacherous, since the mild winters tempt farmers to grow types of crops which are particularly sensitive to cold, such as out-of-season vegetables and citrus.

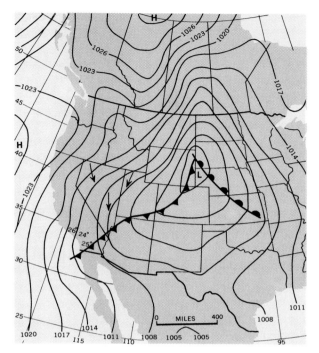

Figure 9.6 Surface weather map (January 2, 1949, 4:30 A.M.), showing a synoptic condition which brought freezing temperatures to southern California. (*After Durrenberger.*)

Weather controls associated with freeze. Typical weather controls that produce occasional killing frosts in California are illustrated by the weather map in Fig. 9.6, which shows atmospheric conditions in the American Southwest for a frost night in January. A well-developed low traveling on a southerly track, with steep gradients on its poleward side, was followed by an invasion of cold polar continental air associated with a large anticyclone which spread southward over the whole Southwest. The clear skies and dry air of the anticyclone permitted strong nocturnal cooling in the already chilly air. It was advection processes bringing fresh polar Canadian air from northerly latitudes that set the stage for the subsequent freeze. Radiation processes were effective in dropping the surface temperatures of the advected polar air below freezing during the relative calm on the night of January 2.

Climatic data for representative subtropical dry-summer stations (*Cs*)

	J	F	M	A	M	J	J	A	S	O	N	D	Yr	Range
Red Bluff, California (*Csa*, interior)														
Temp., °F	45	50	54	59	67	75	82	80	73	64	54	46	62.3	36.3
Precip., in.	4.6	3.9	3.2	1.7	1.1	0.5	0.0	0.1	0.8	1.3	2.9	4.3	24.3	
Santa Monica, California (*Csb*, cool-water coast)														
Temp., °F	53	53	55	58	60	63	66	66	65	62	58	55	59.5	13.6
Precip., in.	3.5	3.0	2.9	0.5	0.5	0.0	0.0	0.0	0.1	0.6	1.4	2.3	14.8	
Perth, Australia (*Csa*, coast)														
Temp., °F	74	74	71	67	61	57	55	56	58	61	66	71	64	19
Precip., in.	0.3	0.5	0.7	1.6	4.9	6.9	6.5	5.7	3.3	2.1	0.8	0.6	33.9	
Naples, Italy (*Csa*, coast)														
Temp., °F	48	49	53	59	65	72	77	77	72	64	56	51	62	29
Precip., in.	4.8	3.5	1.7	1.8	2.2	0.7	0.6	1.3	4.3	4.6	4.1	4.7	34.3	
Haifa, Israel (*Csa*, coast)														
Temp., °F	57	58	62	67	74	78	82	83	81	76	69	60	71	26
Precip., in.	7.1	5.7	0.9	0.7	0.1	0.0	0.0	0.0	0.0	0.5	2.7	6.7	24.4	

Precipitation

Annual amount. As a general rule dry-summer regions have too little annual rainfall rather than too much. The climate is not only subtropical but also subhumid, 15 to 25 or 30 in. being a fair annual average. If this modest amount were concentrated in the hot Mediterranean summer when evaporation is great, conditions would be semiarid rather than subhumid (Fig. 9.7). The name *subtropical dry-summer* is useful, therefore, in distinguishing this climate from its wetter counterpart, *subtropical humid* climate, situated on the eastern sides of continents in similar latitudes.

Lying as it typically does between dry climates (*B*) equatorward and moist temperate oceanic climates (*Do*) poleward, *Cs* climate usually shows a gradual increase in rainfall from its equatorward to its poleward margins. This is well illustrated by three California cities: San Diego, the farthest south, has only 10 in. of rain; Los Angeles has 16 in.; and San Francisco to the north has 23 in. Precipitation also tends to increase from the interiors toward the coasts, except where terrain features modify the pattern.

Seasonal distribution. More distinctive than the annual amount of rain, however, is its distribution over the 12-month period. Characteristically *Cs* climate has a pronounced rainfall concentration in the cooler months, while summer is relatively, and in some places absolutely, dry (Figs. 9.2 and 9.4). Since most rain comes in the cool season, less moisture is lost through evaporation, and more is available for plant and human use. Mild winter temperatures that permit plant growth also give maximum effectiveness to the modest amount of precipitation.

At Los Angeles 78 percent of the precipitation falls from December through March, and less than 2 percent from June through September. Rainfall in the European rimlands of the Mediterranean Basin does not show quite such strong seasonal contrasts: comparable amounts for Seville, Spain, are 50 and 7 percent; for Genoa, Italy, 40 and 18. The less dry summers of Mediterranean Europe are partly a result of higher latitude, which leads to development of some cyclonic rainfall even in the warm season.

Typically the rainfall regime of *Cs* regions is alternately that of the dry climates in summer and the

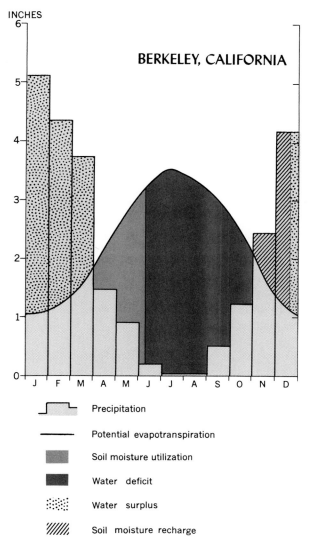

Figure 9.7 The average annual water balance at Berkeley, California, based on mean monthly values. Note that in this dry-summer climate, evapotranspiration exceeds precipitation for 7 of the warm months of the year, and there is an actual water deficit for nearly 5 months. (*Data from Thornthwaite and Mather.*)

As pointed out earlier, they move poleward in summer, bringing Mediterranean latitudes under the influence of the stable eastern margins of a subtropical anticyclone. As they move equatorward in winter, the same latitudes fall within the zone of middle-latitude fronts and cyclones. Thus rainfall in the dry-summer subtropics is predominantly of frontal or cyclonic origin. In Pacific North America in summer, stable anticyclonic air prevails along the coast, extending beyond southernmost California (Fig. 9.8). In addition, its basal layers have been chilled over the waters of the North Pacific and over

Figure 9.8 Pressure, winds, and ocean temperatures along the Pacific Coast of the United States in mid-summer. Coastal climates with cool, dry summers are a consequence of these controls. (*After Patton.*)

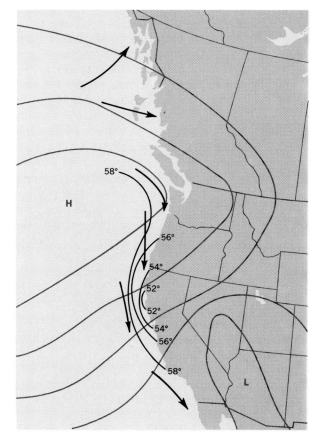

cyclonic westerlies in winter, when rain is relatively abundant. This seasonal alternation of summer drought and winter rain results from a latitudinal shifting of wind and rain belts following the sun.

the cool coastal currents, so that inversions, both at the surface and aloft, are common and the whole air-mass structure is opposed to summer precipitation. In winter, Mediterranean California's rains are associated with an overrunning of colder air masses by *mT* air, or with *mT* or *mP* air being forced up over highland barriers.

Snowfall.   At low elevations *Cs* climate has no durable snow cover, and even snowfall is rare. Over central and southern California (excluding the mountains), annual snowfall averages less than 1 in., and there is none at all along the coast from San Luis Obispo southward. In all Mediterranean lowland regions, snow is so rare that it attracts much attention and comment when it does fall.

But where highlands are present in Mediterranean latitudes, as they commonly are, they usually have an abundance of snow. Their total precipitation is greater than that of the lowlands, and the lower temperatures due to altitude cause a larger share of it to fall as snow. Because of the reduced temperatures these highlands are probably more commonly *D* climate than *C*. Some of the heaviest snowfall anywhere in the United States is on the windward western slopes of the Sierra Nevada mountains of California at elevations of 7,000 to 8,000 ft. At Summit, California (7,017 ft), the average annual snowfall is over 400 in., while 679 in., or more than 60 ft, have been recorded in a single season. Tamarack, California (about 8,000 ft.), has more than 500 in. of snowfall annually, and 884 in. have been recorded in a year. Snow falls in the Atlas region of North Africa from November to April, and in protected places the snow cover lasts well into the summer. In the southern and middle parts of Greece at elevations of 2,000 to 2,500 ft, there is relatively abundant snowfall that remains on the ground for several weeks. Heavy snowfall is also characteristic of the highlands of northern Iran and Afghanistan. This snowfall of Mediterranean highlands provides an invaluable source of water for irrigation in the adjacent lowlands.

Winter rainfall and cloudiness.   Although winter is the rainy season, it is by no means dismal and gloomy, as are the temperate oceanic west-coast regions farther poleward. Since Mediterranean latitudes usually lie on the equatorward sides of the main storm tracks and centers, and are far removed from most of them, they have a less persistent cloud cover and more abundant sunshine. To be sure, winters are considerably cloudier than summers, but they are still relatively bright and sunny.

Perhaps fewer than one-tenth of the storms that cross the northern Pacific Coast region of the United States have any effect upon the weather of San Diego in extreme southern California. Occasionally a cyclone enters the country from the southwest, or one of the northern storms moves southward so that its center approaches close to Mediterranean latitudes. Under these conditions general rains may occur throughout the region. Since main cyclonic tracks coincides with the Mediterranean Basin in winter (Fig. 9.13), cool-season cloudiness is greater in southern Europe than in California.

In interior subtropical California, midsummer months have over 90 percent of the possible sunshine. In winter this is reduced to 50 percent or less, although well south (in the vicinity of Los Angeles) it reaches 60 to 70± percent. Dull, gray days with persistent rain do occur, but showery conditions with a broken sky are more common. After the rain the sun seems to shine more brilliantly than ever in the washed and dust-free atmosphere. These winter showers often are fairly heavy—more so than in regions farther poleward which lie nearer the storm centers. On the average, rain falls on only 7 days in the San Bernardino–Los Angeles region in January, the rainiest month. At Red Bluff farther north and inland, there are 11 rainy days in December, 12 in January, and 10 in both February and March. Bellingham, Washington, and Los Angeles, California, have almost identical amounts of January precipitation (3.4 and 3.3 in., respectively); yet it falls on 17 days at Bellingham and on only 7 at Los Angeles. In other words, precipitation falls harder but less frequently, and with a less persistent cloud cover, in the Mediterranean latitudes (Fig. 9.9). Even the rainiest months have few days when people are kept indoors by precipitation. At the same time the relatively abundant sunshine and

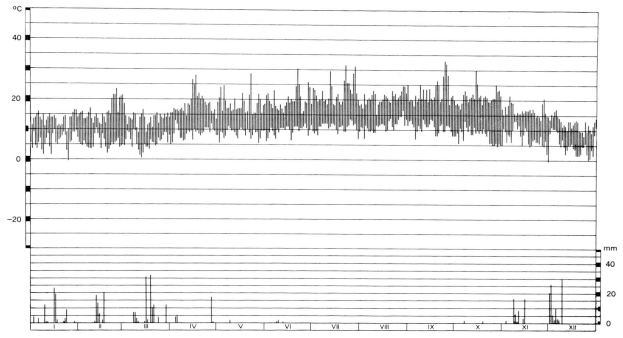

Figure 9.9 Temperature (in °C) and rainfall (in mm) at San Francisco for each day of one year (1954). The cool-season rainfall is concentrated in a number of episodes associated with the passage of cyclonic storms and their fronts. But rainless days are numerous even in the wetter months. The march of temperature shows a modest nonperiodic control. There were only two nights when the minimum temperatures dipped below freezing. The occasional hot days in summer occur when the usual flow of air from the cool ocean temporarily slackens. (*From M. Hendl, Einführung in die physikalische Klimatologie, Band II.*)

mild temperatures, make these regions ideal winter resorts.

Summer drought and sunshine. Summers in most of the dry-summer subtropics are periods of brilliant sunshine, extremely meager precipitation, nearly cloudless skies, and desertlike humidity, except along the coast. Sacramento usually has no rain at all in July and August, and in those months 95 and 96 percent, respectively, of the possible sunshine is received. Afternoon relative humidity is only about 30 to 40 percent. On the average, Los Angeles has only 1 rainy day during the 3 summer months; San Bernardino has 2, and Red Bluff 3. Mediterranean Europe's summers have more cloud and rainfall. The low rainfall, dry heat, abundant sunshine, and excessive evaporation characteristic of interior Mediterranean summers are well suited

for out-of-doors drying of fruits on a large scale.

In spite of the summer heat, thunderstorms are rare, except possibly in the mountains or hills. In southern California 2 to 4 a year are the usual number. The dry, subsiding summer air of these regions, in which inversions are common, is opposed to the formation of deep cumulus clouds.

Coastal regions that have strong anticyclonic subsidence, and are paralleled by cool ocean currents fed by upwelled water, are characterized by high relative humidity and much fog and low stratus cloud in summer. It rarely remains foggy over the coastlands for the entire day, since the mists are usually burned off by the sun after 9 to 10 A.M. Nights, however, may be damp and unpleasant. The California coast is one of the foggiest areas in the United States: parts of it have 40 days per year with dense fog. Coastal Chile has even more. The coastal redwood forests of central and northern California derive a portion of their summer moisture from fogs, while summer crops of beans and certain other vegetables on the coastal terraces of southern California also partly depend on moisture from this source. Another area with frequent summer fogs is the Atlantic coast of Morocco, where Cape Spartel has an average of 28 summer days with fog. All these regions are characterized by temperature inversions, usually close to and well above the surface.

Dependability of precipitation.   Like most subhumid climates, subtropical dry-summer regions have unreliable rainfall, although it fluctuates less than a summer-maximum regime with the same yearly amount (Fig. 4.35). At San Bernardino, California, where the annual rainfall averages 16 in., it has been as low as 5.5 and as high as 37 in. during a 48-year period. In 5 of the 48 years, it dropped below 10 in. At Santiago, Chile, rainfall is below normal in 7 out of 10 years, with a few excessively wet years making up the deficiency. The somewhat precarious, as well as subhumid, character of the precipitation compels farmers to depend upon irrigation to a relatively large degree.

Seasonal weather

Daily weather is less fickle in the subtropical climates than it is farther poleward, where it is increasingly dominated by the conflict of air masses and the development of fronts and cyclones. Because the subtropical climates border the low latitudes, where sun control prevails, they possess many of the tropics' weather characteristics. This is particularly true at the time of high sun, or *summer,* when they are essentially a part of the low latitudes. Thus a typical summer day in the California *Csa* type is almost a duplicate of one in a low-latitude desert. Moreover, one day is much like another. Drought, brilliant sunshine, low relative humidity, high daytime temperatures, and marked nocturnal cooling are repeated day after day with only minor varia-

tions. Diurnal regularity is the keynote of weather at that season. (Here again, more northerly Mediterranean Europe is something of an exception.) Regions with cool-summer *Csb* climates are famous for their well-developed sea breezes, since the cool water offshore and the excessive heating of the dry land under intense solar radiation provide ideal conditions for strong daytime indrafts of air. As a consequence the desert heat is greatly meliorated.

In *autumn,* winds become less regular and uniform. As the cyclone belt creeps equatorward following the sun's rays, an occasional low appears, with associated cloud cover and rain. The dry and dusty land begins to assume new life under the increasing precipitation. Temperatures are still relatively high, and these, together with the increased humidity, create some oppressive and sultry days. As sun control loses the almost total dominance it had in summer, diurnal regularity becomes less prevalent and "spells" of weather more frequent. Importations from the lower latitudes arriving on the front of an advancing cyclone may temporarily reestablish summer heat, as well as bring clouds and showers. This warm importation is followed by advection of cooler air from higher latitudes as the storm center retreats.

In *winter* the frequency and strength of cyclones increases, and it is then that irregular, nonperiodic weather changes are most marked. Rainy days, brought by lows whose centers are often well poleward of subtropical latitudes, are sandwiched between delightfully sunny ones in which temperatures are comfortably mild, even though the nights may be chilly with occasional frosts. The relatively abundant rainfall of this period, together with the mild temperatures, produces a green winter landscape which is uncommon throughout most of the middle latitudes, where winter is a dormant season.

Spring is a delightful season of the Mediterranean year—fresh and yet warm. On the whole, it is cooler than autumn. This is the harvest period for many grains. Passing cyclones gradually become fewer as summer approaches, although nonperiodic weather changes are still significant. The cyclonic centers are preceded by advection of hot, parching air (called *sirocco* in the Mediterranean Basin) from the already superheated low-latitude deserts. These sirocco winds, with temperatures of 100°± and relative humidities of 10 to 20 percent, blow for a day or two without halting and may do serious damage to crops. Fine, choking dust accompanies the temperature importation and at times almost obscures the sun. Following the storm center, the hot sirocco winds are replaced by chilly importations from higher latitudes. These northerly winds have the name *mistral* in southern France and *bora* in the eastern Adriatic. Such cyclonic winds are especially prominent in lands bordering the Mediterranean Sea, since traveling lows are relatively numerous there, and since an extraordinarily large and hot desert lies to the south while cold lands are located behind the bordering highlands to the north. Cool-season temperature gradients along the northern margins of the Mediterranean Basin are extraordinarily steep.

Regional features of subtropical dry-summer climates

The California subdivision of *Cs* is almost a textbook model of subtropical dry-summer climate. Its interior valleys have very hot desertlike summers, while the coast is strikingly cool, with summer fog and stratus. Total annual rainfall is modest, and there is a single winter maximum and a practically rainless summer (Fig. 9.10).

By contrast, the Mediterranean Basin, representing by far the most extensive development of *Cs* climate, and the one that has given the type its regional name, shows much more variation (Fig. 9.11). Chiefly it is the Asiatic and North African sectors of the basin's rimlands that exhibit *Cs* climate in its purest development. The three great peninsulas of southern Europe, on the other hand, display forms of *Cs* climate which are less true to type and more transitional in nature. This is scarcely surprising, since much of Mediterranean Europe lies poleward of 40° and so extends beyond subtropical latitudes and their usual weather controls. As described in the previous section, other modifications

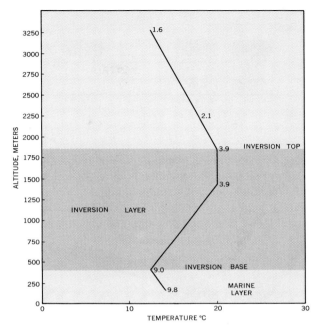

Figure 9.10 Typical lapse rate along the southern California coast in summer. The striking feature is the well-developed inversion of temperature some 1,450 m thick. Below the inversion is a shallow marine layer, where there is a normal decrease in temperature with height, and where humidity is high. The numerals to the right of the lapse rate show mixing ratios in parts per thousand. Relative humidity decreases sharply from the bottom to the top of the inversion layer. The tops of the clouds formed in the moist marine layer practically coincide with the inversion base. (*After Nieberger, Johnson, and Chien.*)

result from the fact that the European sector lies between a cold winter continent to the north, and an unusually extensive and warm inland sea and the Sahara to the south. Consequently advection plays an exceptionally important role in weather processes. Specially named local hot and cold winds, such as sirocco, bora, and mistral, have already been mentioned.

The deep inland penetration of *Cs* in the region of the Mediterranean Basin is found nowhere else. Such an extensive development of the type there appears to be a result of the greater warmth of the

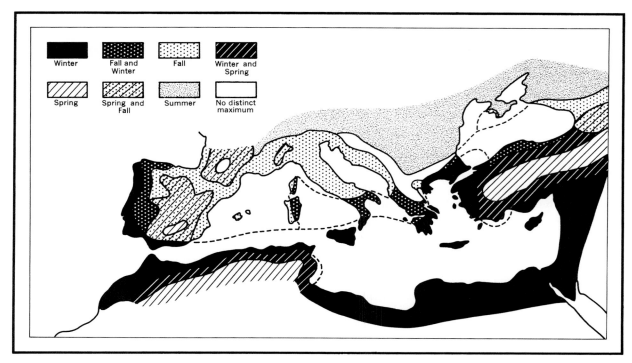

Figure 9.11 Seasons of maximum rainfall in the Mediterranean Basin borderlands. Much of the northern basin does not have a simple single winter maximum typical of the subtropical dry-summer climate. (*After Huttary. From Trewartha, The Earth's Problem Climates.*)

Mediterranean Sea in winter compared with that of the surrounding lands, especially Eurasia to the north. In the cooler months the effect of the warm sea surface is to make the cool air masses invading from higher latitudes unstable and showery. It also

provides an environment distinctly favorable to cyclogenesis, so that the Mediterranean Basin is one of the earth's main tracks of winter cyclones and of air-mass convergence (Figs. 9.12 to 9.14). These winter storms move eastward along the basin for

Figure 9.12 Atmospheric streamlines in the Mediterranean Basin in January and in July. Note that in winter the Mediterranean is a region of atmospheric convergence, a condition favoring cyclogenesis. In summer, divergence and drought are more prevalent. (*After Conrad.*)

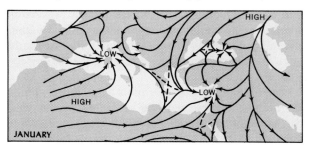
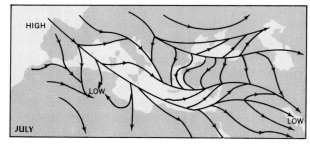

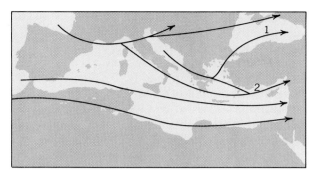

Figure 9.13 Main routes followed by cool-season cyclones in the Mediterranean Basin. Cyclones are relatively infrequent here in summer. Note how far inland the disturbances travel; it is this feature which causes the subtropical dry-summer climate to penetrate so deeply. (*After Weickmann and van Bebber.*)

several thousand miles, carrying with them a potential for creating a cool-season maximum of cloud and precipitation. It can be observed that many Mediterranean cyclonic storms develop in conjunction with an invasion of the subtropical sea by cold northerly air on the rear of an eastward-moving upper trough situated over western and central Europe. The most concentrated region of cyclogenesis is over the Ligurian-Tyrrhenian Sea just west of Italy, where surges of cold air are channeled into the Mediterranean Basin from the north through the Rhone and Carcassonne gaps in southern France.[3]

Throughout Iberia, southern France, and much of Italy, the seasonal march of rainfall becomes more complicated than the simple *Cs* winter maximum–summer minimum regime. A rainfall profile with two maxima and two minima is common, often with one of the minima in summer and the other in winter. Doubtless these rainfall complications arise because the southern European peninsulas lie on the poleward margins of the subtropics or even beyond, so that their rainfall regimes reflect continental as well as subtropical Mediterranean controls.

Subtropical Chile in South America resembles

[3] Glenn T. Trewartha, *The Earth's Problem Climates,* The University of Wisconsin Press, Madison, Wis., 1961, pp. 224–231.

California in that it has a cool coastal *Cs*, with abundant summer fog and stratus, where the warmest-month temperature is only about 64°. It is like California too in that it has a great interior north-south lowland. But unlike the Sacramento lowland, that of Chile is not a region of intense summer heat (*Csa*), for it has a moderate elevation. Santiago at 1,700 ft in the Vale of Chile has an average warm-month temperature of only 69°, which is barely 5° warmer than Valparaiso at sea level on a cool-water coast (Fig. 9.15). Winters are also more extreme in the Vale, however, with the result that interior Santiago has an annual range of 23° and coastal Valparaiso a highly marine 11°.

Cs development in southern Africa is very limited,

Figure 9.14 Seasonal frequency (percent) of cyclones on tracks 1 and 2 as shown in Fig. 9.13. (*After Conrad.*)

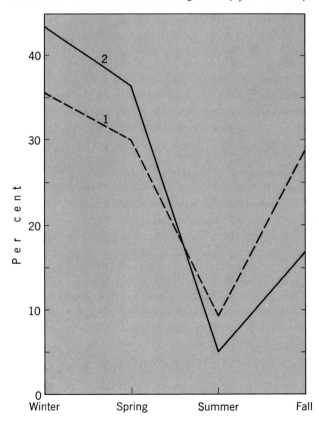

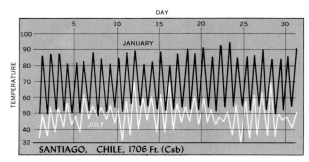

Figure 9.15 Daily maximum and minimum temperatures for the extreme months of one year at Santiago, Chile. This station, although located inland, has a cool summer (*Csb*) because of its altitude of about 1,700 ft. Compare with Fig. 9.5.

being confined to the southwesternmost extremity around Cape Town. Although the region is strongly marine in its temperature characteristics (warm month only 71°), it lacks the chilling effects of a well-developed cold current and upwelling. While summer is much drier than winter, it is not rainless, as in southern California; the 3 summer months have a total rainfall of about 2 in.

Australia's *Cs* is in two far-separated locations. While both are marine in character, with warm-month temperatures below 75°, neither of them is a genuinely cool-water coastal type. Perth in the southwest, with about 35 in. of annual rainfall, is wet by *Cs* standards. Adelaide on the west side of southeastern Australia, with only 21 in., is closer to the type's normal, although Adelaide's 3 summer months are far from rainless, receiving about 2.5 in. The annual range of rainfall is much greater at Perth than at Adelaide.

SUBTROPICAL HUMID CLIMATE (*Cf*)

Subtropical humid climate usually differs from the dry-summer variety in at least three important respects: (1) it is located on the eastern, rather than the western, side of a continent, (2) it has more precipitation, and (3) the precipitation is distributed throughout the year, although ordinarily with some degree of concentration in the warm months. Summer drought, a hallmark of Mediterranean climates, does not occur in *Cf* regions.

Type location. The two subtropical climates are similar in their latitudinal positions in that both are on the equatorward margins of the intermediate zones, although the subtropical humid type may be shifted a few degrees farther equatorward. Characteristically its spread is from about 25° to 35 or 40° (see Figs. 7.2, 9.1, and the map of climate types inside the front cover).

In their continental locations, however, the two subtropical types are dissimilar, for subtropical humid climate ordinarily is situated on the eastern side of a land mass and hence to the east of the dry interior. In summer this east-side location is dominated by the unstable or neutral *mT* air masses within the western parts of an oceanic subtropical anticyclone, where subsidence is weak. It is the stable structure of the eastern limb of an anticyclone which mainly accounts for the summer drought in California; the unstable western limb of such a cell favors abundant summer rainfall in the southeastern United States. A supplementary cause of the east-side location of *Cf* is the tendency for monsoon circulations to develop there, at least on the large continents of Asia and North America. Such a wind system favors wet summers and drier winters—the very antithesis of the rainfall regime in *Cs* climate. In addition, the warm ocean currents which parallel subtropical east coasts may act directly or indirectly to increase total rainfall, and make the summer season wetter.

The two subtropical climates, situated in similar latitudinal but unlike continental locations, are usually flanked by contrasting types on their equatorward and poleward frontiers (Fig. 9.1). While *Cs*

Rainfall regimes for four pairs of subtropical stations, *Cs* on the west side of a continent, *Cf* on the east (Inches)

Station	Latitude	Winter half-year	Summer half-year	Annual
San Luis Obispo (Cs)	35° N	18.7	2.7	21.4
Hatteras (Cf)	35° N	26.3	28.7	55.0
Mogador (Cs)	32° N	10.6	2.6	13.2
Chungking (Cf)	30° N	10.4	31.5	41.9
Perth (Cs)	32° S	28.9	5.0	33.9
Port Macquarie (Cf)	31° S	28.3	33.1	61.4
Valparaiso (Cs)	33° S	18.3	1.4	19.7
Rosario (Cf)	33° S	10.5	24.4	34.9

commonly adjoins hot dry climates on its low-latitude side, *Cf* borders tropical humid climates there. This difference has a great effect upon the nature of thermal and humidity advection in the two climates: parching dry heat with dust is imported into *Cs*, humid sultry heat into *Cf*. On their poleward sides these types are likewise bordered by contrasting climates. Subtropical dry-summer type generally merges into a mild marine climate, while subtropical humid is bounded by a severe continental climate. This contrast is especially strong in the large continents of Eurasia and North America, whose *Cf*

climates are more susceptible to advection of severe cold than their dry-summer subtropics are. Dry climates ordinarily flank *Cs* on its eastern side, but they adjoin *Cf* on its western margin.

Temperature

In temperature features the two subtropical types are relatively similar, although subtropical humid usually shows less contrast between coastal and interior locations (Figs. 9.16 and 9.17). This general similarity in temperature between *Cs* and *Cf* is not

Climatic data for representative subtropical humid stations (*Cf*)

	J	F	M	A	M	J	J	A	S	O	N	D	Yr	Range
					Charleston, South Carolina									
Temp., °F	50	52	58	65	73	79	82	81	77	68	58	51	66.1	31.4
Precip., in.	3.0	3.1	3.3	2.4	3.3	5.1	6.2	6.5	5.2	3.7	2.5	3.2	47.3	
					Chungking, China									
Temp., °F	48	50	58	68	74	80	83	86	77	68	59	50	67	38
Precip., in.	0.7	0.9	1.3	4.0	5.3	6.7	5.3	4.4	5.8	4.6	2.0	0.9	41.9	
					Memphis, Tennessee									
Temp., °F	41	44	53	62	70	78	81	80	74	63	52	43	62	30
Precip., in.	4.9	4.3	5.4	5.1	4.2	3.2	3.4	3.4	2.8	2.9	4.2	4.3	48.3	
					Port Macquarie, Australia									
Temp., °F	71	72	69	65	60	56	54	56	59	63	67	70	63	18
Precip., in.	5.9	7.5	6.5	5.9	5.6	4.6	4.5	3.8	3.9	3.2	4.1	5.9	61.5	
					Dallas, Texas									
Temp., °F	45	47	56	65	73	81	84	83	77	66	55	47	65	35
Precip., in.	2.2	2.0	2.7	3.9	4.4	3.1	2.5	2.1	2.9	2.8	2.5	2.4	33.6	

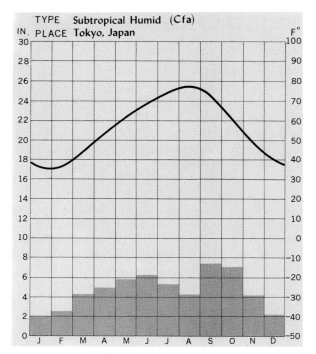

Figure 9.16 Compare with Figs. 9.2 and 9.4. The annual march of rainfall is quite different in the two subtropical climates.

surprising, since the two correspond fairly well in latitude. But because *Cf* is not subject to the influence of cool ocean currents, it has no distinctly cool-summer coastal subtype with cool, foggy stations like San Francisco. With very few exceptions, its summer is sufficiently warm to warrant adding the letter *a* to the *Cf* symbol, so that it becomes predominantly *Cfa*.

Summer. Summers are distinctly hot; average hot-month temperatures are usually around 75 to 80° (81° at Charleston, South Carolina; 80° at Shanghai, China; 77° at Brisbane, Australia, and Durban, South Africa; and 74° at Buenos Aires). Along the immediate coasts of the smaller Southern Hemisphere land masses, warm-month temperatures are as often slightly below 75° as they are above (Figs. 9.16 to 9.19).

Not only is the air temperature high, but absolute and relative humidity are too. In the American South Atlantic and Gulf states, where *mT* air almost completely dominates in summer, the average July relative humidity (8 P.M. reading) is 70 to 80 percent. High humidity in conjunction with high temperature produces sultry, oppressive weather with low cooling power. Summer heat in most *Csa* climates is drier, so that sensible temperatures are commonly higher in the humid subtropics than in *Cs* regions even when the thermometer registers the same. Summer heat in the American Gulf states closely resembles that of tropical wet climates. At New Orleans during June, July, and August, the average temperatures are 2 to 3° higher than they are at Belém in the Amazon Valley, while the amount of rainfall is nearly the same. The average

Figure 9.17 Annual march of temperature and rainfall at Memphis on the poleward margins of the humid subtropics in the United States. Note the cool-season maximum of precipitation.

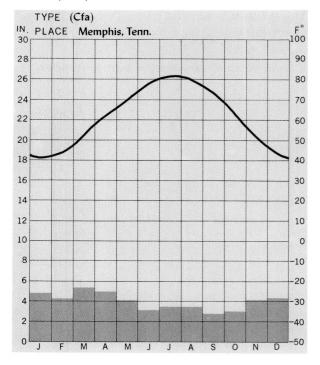

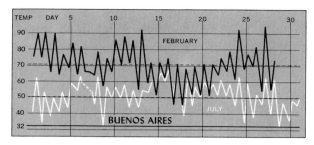

Figure 9.18 Daily maximum and minimum temperatures for the extreme months of one year at a subtropical humid station in the Southern Hemisphere. Nonperiodic air-mass control is conspicuous in both winter and summer. Winters are mild.

of the daily maxima in July throughout most of the American Cotton Belt is between 90 and 100°, while the highest temperatures ever recorded are usually between 100 and 110° (Fig. 9.19). Thus Montgomery, Alabama, has had a temperature of 106° in July and 107° in August and Savannah, Georgia, 105° in July.

Night temperatures. Not only are the days hot and sultry, but nights are oppressive as well, resembling those in the humid tropics. The humid atmosphere, with more cloud than in *Csa* climates, pre-

Figure 9.19 Daily maximum and minimum temperatures for the extreme months of one year at a subtropical humid station in the United States. Note the strong periodic, or solar, control in summer. By contrast, winter shows strong nonperiodic, or air-mass, control. Both winter and summer temperatures are more extreme at Montgomery than at Buenos Aires (Fig. 9.18), owing to the effects of the broad North American continent.

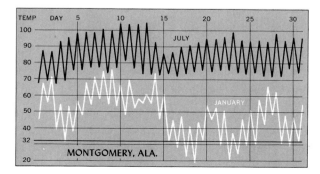

vents the same rapid loss of heat that takes place in their drier air and clearer skies. The slower night cooling results in relatively small diurnal ranges, usually only one-half to two-thirds as large as those in the inland dry-summer subtropics (Figs. 9.2, 9.3, and 9.19). For example, the averages of the July maxima for Sacramento and Montgomery are 89 and 91°, respectively, and their average July minima are 58 and 72°. This gives a diurnal range of 31° for Sacramento and 19° for Montgomery, the difference being due chiefly to Sacramento's cooler nights. Since the sun is very much in control of the daily weather, one day in summer is relatively like another in the humid subtropics. However, there are some nonperiodic variations (Fig. 9.19), which are often the result of variations in the amount of cloud.

Winter. Winters are, of course, relatively mild in subtropical latitudes, with cool-month temperatures that usually average between 40 and 55° (Figs. 9.16 and 9.17). Montgomery, Alabama, has an average cool-month temperature of 49°; Shanghai, China, 38°; Buenos Aires, Argentina, 50°; and Sydney, Australia, 52°. Even in winter, the southeastern United States receives much warm, humid, tropical air from the Gulf of Mexico and the tropical Atlantic, but the normal seaward pressure gradient of that season makes the progress of *mT* air inland more sporadic than in summer. Usually the winter indrafts of tropical air are associated with passing cyclones. In eastern Asia, the strong monsoonal outpouring of cold *cP* air in winter from the large continent to the north and west results in the lowest *average* winter temperatures in those latitudes for any comparable part of the world (Fig. 9.20). Mobile, Alabama and Shanghai, which are approximately in the same latitude, have average January temperatures of 52 and 38°, respectively.

After subtropical China, the southeastern United States has the lowest average winter temperatures of any of the subtropical humid regions. Here, as in eastern Asia, there is a well-developed, although less strong, winter-monsoon tendency. The prevailing winter winds come from the north and north-

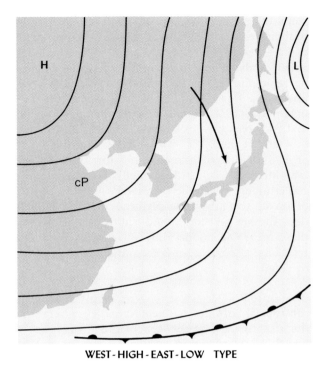

WEST-HIGH-EAST-LOW TYPE

Figure 9.20 A polar-outbreak high in eastern Asia flooding the subtropics of that region with cold air and carrying freezing temperatures far into southern China and over all of Japan. (*From Trewartha, Japan: A Geography, University of Wisconsin Press.*)

west and consequently from the cold interior and higher latitudes. In the eastern United States as compared with eastern Asia, the more numerous and better-developed moving cyclones tend to disrupt the monsoonal winds. The humid subtropical regions of the Southern Hemisphere, which have no severe continental climates on their poleward margins because of the size or the latitudinal positions of their respecitve land masses, receive no such advections of severe cP air and are consequently milder.

Midday temperatures in winter are likely to be pleasantly warm in the humid subtropics, with the thermometer usually rising to 55 or 60° (Figs. 9.18 and 9.19). On winter nights temperatures fall to about 35 to 45°. These certainly are not low, but combined with Cf's high humidity they are likely to produce a sensible temperature which is distinctly chilly and uncomfortable. Summer is emphatically the dominant season, so that little effort is made to provide efficient heating systems in many regions with Cfa climate. Consequently people often feel more uncomfortable indoors than they do in colder lands farther poleward where winter heating is adequate.

Annual range of temperature. In Cf climates the annual range is usually modest, although there is considerable variation depending upon the size of the continent and the latitude of the particular station (Figs. 9.16 and 9.17). At Buenos Aires the annual range is only 23° and at Sydney only 19°, but at Montgomery it is 33°, and at Shanghai 43°. Apparently, the larger the land mass and the better the development of a continental system of monsoon winds, the colder the winters and the larger the annual range.

Minimum temperatures and frost. As would be expected, the growing season, or period between killing frosts, is long in subtropical humid climates— usually at least 7 months, and at most nearly, if not quite, the entire year. Even though freezing temperatures may be *expected* during a period of several months, they actually occur on only a relatively few nights of the winter season. In Cf, like Cs, climates, the long growing season and infrequent freezes create good conditions for sensitive crops and for those requiring a long maturing period.

There is scarcely any part of the continental humid subtropics which has not had freezing temperatures, except for some areas on the tropical oceanic margins of the subtropics. In sections of the Southern Hemisphere Cf, frost does not occur every winter, and usually it is light when it does come. Thus the average lowest winter temperature at Brisbane, Australia, is 37°, and at Sydney 39°. The lowest temperature ever recorded at Buenos Aires is 23°; at Montevideo, Uruguay, 20°; and at Brisbane, Australia, 32°.

In the humid subtropics of both the United States and China, winter minima are considerably lower than those of the Southern Hemisphere continents. Shanghai has recorded a temperature as low as 10°, and Montgomery, Alabama, −5°. New Orleans, with a normal January temperature of 55°, has the extremely low absolute minimum of 7°. Even daily maximum temperatures below freezing are not uncommon in poleward parts of the North American and Asiatic sectors.

One of the distinguishing features of the South Atlantic and Gulf states of the United States, a region where the average winter temperatures are relatively high, is the unusually low winter minima— even lower than those of China. Thus while southeastern China has lower *average* winter temperatures, the American humid subtropics have severer *cold spells* and consequently lower minima. Severe

freezes are annual in occurrence, and temperatures as low as 10° have been recorded along the ocean margins of all the Gulf states. No other part of the world near sea level in these latitudes has such low winter minima. Serious damage to sensitive vegetation results from these cold spells.

The occasional severe invasions of *cP* air which characterize the winter weather of the seaboard states of the American Gulf and South Atlantic can be attributed to the openness of the North American continent east of the Rocky Mountains. The lack of terrain barriers permits the surges of cold *cP* air to move rapidly southward into subtropical latitudes with only moderate modifications (Fig. 9.21). In subtropical California, on the other hand,

Figure 9.21   A synoptic type resulting in a severe freeze in the North American humid subtropics. Numerals on the map represent night minimum temperatures.

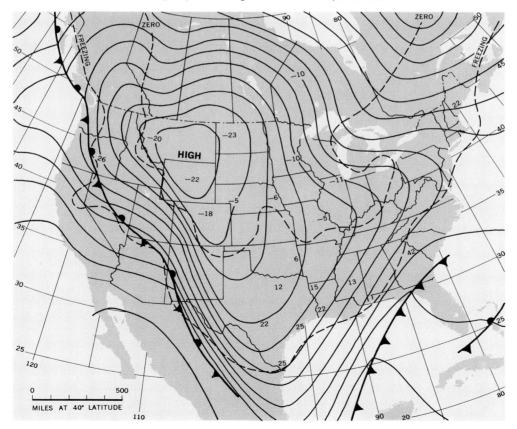

mountains tend to retard such severe invasions of cold air, so that there the absolute minima are much higher. Thus while commercial citrus production extends north to about 38° in California, it is confined to regions south of latitude 30 or 31° in the southeastern United States. In China, the more hilly and mountainous surface configuration prevents such unrestricted advection of cold *cP* air.

Precipitation

Annual amount and seasonal distribution.   Rainfall is relatively abundant within the humid subtropics, 30 to 65 in. representing the normal spread. Still, there are considerable differences between the several continental regions. On the landward frontiers of this type, where it adjoins steppe climate, rainfall is the lowest.

In contrast to the dry-summer subtropics, there is no season of drought over much the greater part of the earth's humid subtropics. Usually summer has more precipitation than winter, although winter is far from dry (Figs. 9.16 and 9.17). Summer is never a dry season. Thus as a rule, the annual march of rainfall is opposite in the two subtropical types. In the humid subtropics it is mainly where a strong monsoon circulation prevails that winter is dry, as it is in parts of interior southern China (which is therefore classed as *Cw* instead of *Cf*; see the definitions given on page 250). In the Southern Hemisphere, seasonal contrasts in rainfall are not marked. It is a fair generalization about the humid subtropics on all continents that such seasonal contrasts tend to be accentuated toward the drier interior parts (Fig. 9.22).

Summer.   Most of the summer rainfall on lowlands is in the form of convectional showers which originate in *mT* air made increasingly unstable by passage over a warmed land surface. Clouds are chiefly of the cumulus type, and the showers are frequently accompanied by lightning and thunder. In fact, the American humid subtropics is the most thundery region of the United States, for a large part of it has over 60 electrical storms a year, while a part of Florida has over 90. A great majority of the

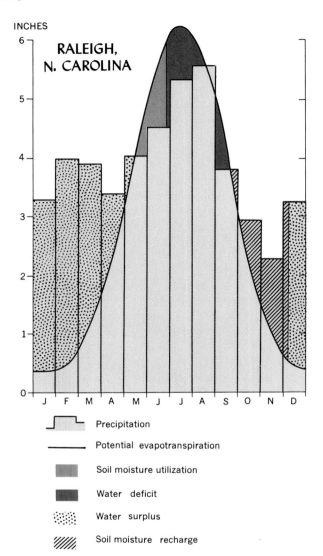

Figure 9.22   The average annual water balance at Raleigh, North Carolina, based on mean monthly values. Note that even this humid station, with a rainfall maximum in the warm months, has a period in summer when evapotranspiration exceeds precipitation, and there is a short period of actual water deficit. (*Data from Thornthwaite, Mather, and Carter.*)

showers and thunderstorms occur in association with extensive but weak atmospheric disturbances in which convergence lifts the warm humid air of tropical origin and makes it unstable (Fig. 9.25).

Consequently the showers are organized, not random, in their distribution.

Hurricane rainfall is largely confined to the American and Asiatic humid subtropics, where it contributes to the late-summer and fall rainfall maximum characteristic of parts of these regions. Not only do the heavy late-summer and early-autumn hurricane rains greatly damage ripening crops and cause serious floods, but their violent winds may play havoc with coastal shipping and port cities. In one typhoon affecting the China coast, 40,000 people are estimated to have perished, chiefly by drowning.

In spite of the plentiful summer rainfall of the humid subtropics, sunshine is relatively abundant, even though it is much less so than in *Csa* summer climates. Montgomery, Alabama, receives 73 percent of the possible sunshine in June and 62 percent in July.

Winter. In winter the land surface in the humid subtropics is likely to be colder than maritime air arriving from tropical source regions. As a result the poleward movement of such an air mass over the cooler land chills its base layers and thereby increases stability. Consequently, convectional overturning promoted by surface heating, so prevalent in summer, is uncommon in winter. Only when the stabilized tropical maritime air masses are forced to rise over relief barriers, or when they become involved in convergences along fronts and in cyclonic systems, does precipitation usually occur. Winter rainfall over lowlands, therefore, is chiefly frontal or cyclonic in origin.

As a consequence, winter rain is usually accompanied by a general and persistent cloud cover extending over wide areas, from which precipitation may fall steadily for hours on end. Typically it is less intense but more prolonged than the showery rain in summer thunderstorms. Because there are more cyclones in winter, skies are cloudier than in summer. At Shanghai in an average January, only 2 in. of rain falls, but there are 12 rain-days; yet in August, the 6 in. of precipitation falls on only 11 days. Each rainy day in August, therefore, produces three times as much precipitation as a rainy day in January. Montgomery, which has 73 percent of the possible sunshine in June, receives only 49 percent in January and 44 percent in December. Gray, overcast days with rain may be unpleasantly chilly.

Snow falls occasionally when a vigorous winter cyclone swings well equatorward, but it rarely stays on the ground for more than a few days. In the southernmost parts of the United States' humid subtropics snow does not even fall on an average of one day a year, and the snow cover likewise does not last for a day. On the northern margins, however, 5 to 15 days have snow, and the ground may be snow-covered for an equally long period.

Seasonal weather

A majority of the cyclonic storms in the eastern United States follow a track which approximately coincides with the Great Lakes and the St. Lawrence Valley, so that they do not strongly influence the Gulf states. It is when cyclones and anticyclones move on a southern circuit that the Cotton Belt is most affected by them.

In *summer,* when the cyclone belt is farthest poleward and the sun is largely in control, irregular weather changes are at a minimum (Fig. 9.19). Weak cyclones resembling those of the wet tropics may bring spells of showery, cloudy weather, but no marked temperature changes. Humid, sultry days, each much like the others, are the rule. The thermometer rises to about the same height each day and sinks to similar minima each night.

Late summer and fall is the season of the dreaded hurricanes, whose severity more than makes up for their infrequency. Otherwise, sunny autumn days provide delightful weather, although the equatorward-advancing cyclone belt gradually produces more gray, cloudy days and unseasonable temperature importations as winter closes in.

In *winter,* including early spring, the storm belt is farthest equatorward, so that irregular weather changes are most frequent and extreme then (Fig. 9.19). Advection of tropical air on the front of an advancing cyclone may push daytime temperatures well above 60 or even 70°, while the following *cP* air masses may reduce the temperature as much as 30° within 24 hr, occasionally bringing severe freezes. Bright, sunny winter days are pleasant and exhilarating outdoors. In *spring* the cyclonic belt retreats and regular diurnal sun control is gradually reestablished (Fig. 9.19).

Regional features of subtropical humid climates

In a number of ways the humid subtropical southeastern United States is climatically exceptional. Perhaps the outstanding anomaly is that over an extensive part of this region, characterized by leeward location on a great continent and situated mainly inland, winter rainfall exceeds that of summer (see Figs. 9.17, 9.23, and the data for Memphis). Summer is also wet, of course, but it is less so. Both to the north and west and to the east and south of the winter-rain region, the usual summer maximum prevails. More specifically, the oval-shaped region of cool-season maximum reaches from eastern Texas on the southwest to eastern Kentucky and westernmost North Carolina on the northeast. Its long and short dimensions are about 1,000 by

350 miles. Although within this large region winter's rainfall exceeds summer's, winter is not everywhere the season of maximum, for spring equals or even exceeds it in parts. There is a core, nevertheless, where winter is the wettest season (Fig. 9.23). Much of the core region has the heaviest winter precipitation anywhere in the United States east of the Rocky Mountains.

The heavy rainfall in winter and early spring is associated with a high frequency of cyclonic storms, which appear first in the Texas region and subsequently follow a northeasterly course toward the Atlantic Seaboard south of Cape Hatteras. This concentration of winter depressions and their heavy rains generally coincides with the southernmost position of the upper-troposphere jet streams during the period December through March. In summer-fall, which has the least rain, though it is still far from dry, jets, fronts, and storm tracks retreat northward. In addition, during this drier period the North Atlantic subtropical anticyclone extends westward over the southeastern United States.[4]

But while a large part of this country's subtropical interior shows a cool-month maximum, the reverse is true of the Gulf and Atlantic margins, where a strong summer maximum prevails. Here the weather in summer is conditioned by a deep current of humid tropical air aloft which is associated with an upper trough positioned over the area. An additional factor may be the directional or speed convergence associated with the sea breeze.

A second distinctive feature of the American humid subtropics is its abnormally low winter minimum temperatures. This characteristic was described in the earlier section on temperature.

Two distinguishing properties of the Asiatic humid subtropics—unusually low average winter temperatures, and the strikingly dry winters of China well inland—were also discussed earlier. An additional feature is the bimodal summer rainfall maximum which occurs in large areas of southern China and southwestern Japan (Fig. 9.24). The first of these maxima is in May-June, while the second

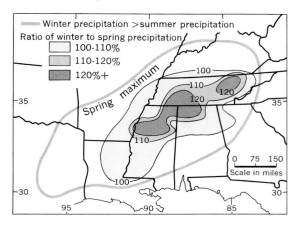

Figure 9.23 In an extensive region of the subtropical southeastern United States, winter precipitation exceeds that of summer. Winter and spring are the wettest seasons, and winter is wetter than spring in the central part of the region. (*From Trewartha, The Earth's Problem Climates.*)

[4] Trewartha, *The Earth's Problem Climates*, pp. 293–299.

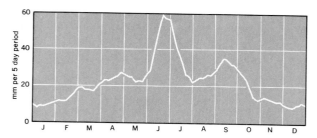

Figure 9.24 Composite annual profile of 5-day means of rainfall (overlapping method) for five stations in subtropical Japan. The primary maximum, known as *Baiu*, is in June–July. In September there is a second and less conspicuous maximum, known as *Shurin*, which originates mainly from typhoon rainfall. (*From Trewartha, The Earth's Problem Climates.*)

comes in late summer and early fall. Separating the two is a midsummer secondary minimum.

The June maximum has been linked with the summer circulation pattern over eastern Asia. During winter, the southern branch of the zonal westerlies and of the jet stream follow a course to the south of Tibet and across northern India-Pakistan, northern Southeast Asia, and southern China. In early summer, the jet stream and zonal westerlies abruptly shift to a position north of the highlands. The equatorial westerlies, or Indian monsoon, then surge northward and invade not only tropical southern Asia but also China and Japan, bringing those regions an air-mass environment in which weak disturbances can induce abundant rainfall (Fig. 9.25). The midsummer secondary minimum has been attributed to the intensification and northward and westward displacement of the North Pacific subtropical anticyclone as it follows the sun, with an associated decline in vapor pressure at the midtroposphere level. The second rainfall peak coincides with a withdrawal of the anticyclone and a restrengthening of the southerly monsoon circulation. Significantly, atmospheric pressure is lowest in August. An important part of the August-September rains is generated by tropical storms, including typhoons, which are most numerous and intense in late summer and fall.[5]

[5] *Ibid.*, pp. 191–199.

Another special variation of *Cf* climate in Asia is the heavy cool-season precipitation along the northwest side of Japan. The cold, dry winds from Asia are warmed, humidified, and made unstable over the Sea of Japan, so that they yield abundant precipitation when forced upward over western Japan's highlands. Many stations even show a winter maximum. Much of the winter precipitation falls as snow showers, resulting in a deep snow cover.

All three Southern Hemisphere humid subtropics are distinguished by unusually mild winters and small annual ranges of temperature, for none of them have a broad land source immediately poleward to provide cold advection. In both South Africa

Figure 9.25 A weak low of the early summer period in subtropical eastern Asia. Such a synoptic pattern is known in Japan as the *Baiu* type. It is disturbances of this type which produce the earlier of the two summer maxima of rainfall characteristic of much of the Asiatic humid subtropics. (*From Trewartha, Japan: A Geography, University of Wisconsin Press.*)

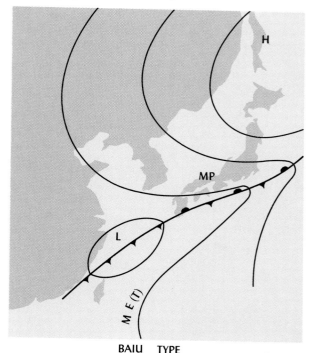

BAIU   TYPE

and Australia, subtropical humid climates are confined to fairly narrow coastal strips, so that Durban has a cold-month temperature of 63° and a range of only 13°, while the comparable figures for Brisbane are 59° and 18°, and for Sydney, 53° and 19°. In the South American sector—where the humid subtropics extend somewhat farther inland, and a land area, even though narrow, flanks it to the south—winters are a trifle more severe and ranges are slightly greater (Buenos Aires has a cold-month average of 49°, with a range of 25°).

SELECTED REFERENCES FOR FURTHER STUDY OF TOPICS IN CHAPTER NINE

Byers, H. R., "Summer Sea Fogs of the Central California Coast," *Univ. Calif. Publ. Geog.*, no. 3, pp. 291–338, 1930.

Huttary, Josef, "Die Verteilung der Niederschläge auf die Jahreszeiten im Mittelmeergebiet," *Meteorol. Rundschau,* Vol. 3, pp. 111–119, 1950.

Institute of Geophysics and Meteorology, Academia Sinica, Peking, staff members of the Section of Synoptic and Dynamic Meteorology, "On the General Circulation over Eastern Asia," *Tellus,* Vol. 9, pp. 432–446, 1957.

Lu, A., "The Winter Frontology of China," *Bull. Amer. Meteorol. Soc.,* Vol. 26, pp. 309–314, 1945.

Miller, James E., and T. Mantis Homer, "Extratropical Cyclogenesis in the Pacific Coastal Regions of Asia," *J. Meteorology,* Vol. 4, pp. 29–34, 1947.

Patton, Clyde P., "Climatology of Summer Fogs in the San Francisco Bay Area," *Univ. Calif. Publ. Geog.*, Vol. 10, no. 3, 121–125, 1956.

Ramage, C. S., "Variation of Rainfall over South China through the Wet Season," *Bull. Amer. Meteorol. Soc.,* Vol. 33, pp. 308–311, 1952.

Reichel, Eberhard, "Die Niederschlagshäufigkeit im Mittelmeergebiet," *Meteorol. Rundschau,* Vol. 2, pp. 129–142, 1949.

Trewartha, Glenn T., *The Earth's Problem Climates,* The University of Wisconsin Press, Madison, Wis., 1961, pp. 180–199, 223–247, 293–304.

Yao, C. S., "On the Origin of the Depressions of Southern China," *Bull. Amer. Meteorol. Soc.,* Vol. 21, pp. 351–355, 1940.

CHAPTER TEN

TEMPERATE CLIMATES (D)

This group of humid climates occupies a medial position within the middle latitudes—usually between subtropical climates equatorward and boreal climates poleward. Its equatorward and poleward boundaries are 8 months and 4 months, respectively, with average temperatures of 50° or above. Actually the name "temperate" is not wholly suited to all climates within this group, for the more severe continental type is scarcely moderate in its seasonal temperatures. But for want of a better title, temperate is used here to designate the climates which are intermediate in temperature between boreal cold and subtropical heat.

Two types comprise the temperate group: a milder climate called oceanic or marine (Do), and a more severe one that is continental (Dc). The boundary adopted for separating the two types is the 32°F (0°C) [to 36° (2°C) in some locations] isotherm for the coldest month. In the oceanic type, all months have average temperatures of 32°F or above.

TEMPERATE OCEANIC CLIMATE (*Do*)

Type location. As this climate's name suggests, its weather is like that of the adjacent ocean. Typically it occupies a position on the western or windward side of middle-latitude continents, poleward of about 40°. There the onshore westerly winds bring marine climate conditions to the land (refer to Fig. 7.2 and the climate map).

Temperate oceanic climate may not be completely limited to the western side in cases where land areas are relatively narrow, as they are in islands, such as Tasmania, New Zealand, and Great Britain, or where the continent extends for only a short distance into the belt of westerlies, as Australia and Africa do. But an extensive development of this type along east coasts of large middle-latitude continents is unlikely (in spite of the proximity of oceans), because there severe seasonal temperatures result from leeward location and the monsoon wind systems.

On its equatorward margins, *Do* climate characteristically borders the subtropical dry-summer type. Because of its higher latitude, ordinarily it is not strongly influenced by the stable air in the eastern parts of an oceanic subtropical anticyclone, so that *Do*, unlike *Cs*, usually does not have a pronounced dry summer. On its poleward side the temperate oceanic climate extends far into the higher middle latitudes, where it is eventually terminated by either the boreal or the tundra type. The far poleward extension of this mild climate is the result of oceanic control offsetting normal latitudinal influence. Warm ocean currents, which parallel the west coasts of continents in middle latitudes poleward from the subtropics, tend to accentuate the normal tempering effects of the ocean itself.

The depth to which oceanic climate extends into the interior of a continent is determined largely by terrain. Where mountains closely parallel the west coast, as in North and South America and Scandinavia, marine conditions are restricted to relatively narrow belts of littoral seaward from the highlands. But where extensive lowlands reach to the coast,

Climatic data for representative temperate oceanic stations in climate (*Do*)

	J	F	M	A	M	J	J	A	S	O	N	D	Yr	Range
Valentia, Ireland														
Temp., °F	44	44	45	48	52	57	59	59	57	52	48	45	50.8	15
Precip., in.	5.5	5.2	4.5	3.7	3.2'	3.2	3.8	4.8	4.1	5.6	5.5	6.6	55.6	
Seattle, Washington														
Temp., °F	40	42	45	50	55	60	64	64	59	52	46	42	51.4	24
Precip., in.	4.9	3.8	3.1	2.4	1.8	1.3	0.6	0.7	1.7	2.8	4.8	5.5	33.4	
Paris, France														
Temp., °F	37	39	43	51	56	62	66	64	59	51	43	37	50.5	27
Precip., in.	1.5	1.2	1.6	1.7	2.1	2.3	2.2	2.2	2.0	2.3	1.8	1.7	22.6	
Hokitika, New Zealand														
Temp., °F	60	61	59	55	49	47	45	46	49	53	55	58	53.0	16
Precip., in.	9.8	7.3	9.7	9.2	9.8	9.7	9.0	9.4	9.2	11.8	10.6	10.6	116.1	
Portland, Oregon														
Temp., °F	39	42	46	51	57	61	67	66	61	54	46	41	53.0	28
Precip., in.	6.7	5.5	4.8	3.1	2.3	1.6	0.6	0.6	1.9	3.3	6.5	6.9	43.8	

as in most of western Europe, the effects of the sea are carried well inland, and *Do* climate prevails over a greater east-west area. On its land side, temperate oceanic climate is characteristically bordered by severe continental types, either dry or humid. In North America, where highlands closely parallel the west coast, *Do* climate is replaced by dry climate to the east of the mountains. In Europe, on the other hand, the westerlies blowing off the ocean have freer entrance, and *Do* climate gradually passes over to temperate continental climate (Fig. 10.13 and the climate map at the front of the book).

Temperature

Summer. Although the word "temperate" is a poor one to apply to thermal conditions in middle-

Figure 10.1 Average monthly temperature and rainfall for a temperate oceanic station in lowland western Europe. Note the small annual range of temperature and the modest amount of precipitation well distributed throughout the year.

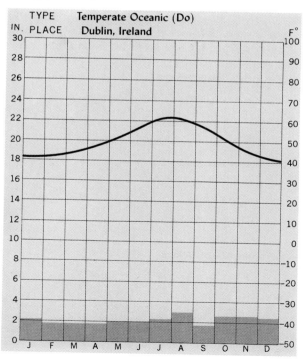

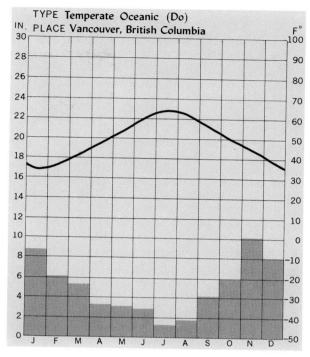

Figure 10.2 Monthly temperatures and rainfall for a temperate oceanic station on the southern Pacific coast of Canada, which is backed by highlands. Here the annual precipitation (60 in.) is greater than that for most lowland stations, and there is a pronounced maximum in winter. A strong winter maximum normally is characteristic of those parts of *Do* climate which are located closest to the subtropical dry-summer type (*Cs*).

latitude continental climates, it is relatively appropriate for the *Do* type (Figs. 10.1, 10.2, and 10.5). Summers are moderate: several degrees below the average temperature for the latitude. While they are more or less ideal for human efficiency and comfort, they are somewhat too cool for the best growth and the maturing of a number of cereal crops. Seattle, Washington, has a mean July temperature of only 63°; Dublin, Ireland, 60°; and Paris, France, 66°. Night cooling is not rapid in these humid, cloudy, marine climates, as it is in Mediterranean regions in the summer. The average of the daily minima in July is only 55° at Seattle and 51° at Bellingham, Washington, while the daily

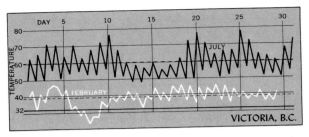

Figure 10.3 Maximum and minimum temperatures for the extreme months of one year at a station with a marine location in Pacific Canada. Note the small diurnal range, especially in winter, when cloudy skies prevail.

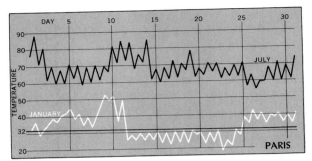

Figure 10.4 A station in western Europe inland from the coast. Here the temperature range between the warmest and coldest months is much greater than at Victoria (Fig. 10.3). Nonperiodic air-mass control of temperature is fairly conspicuous.

maxima are 73 and 72°, so that the normal diurnal range is only around 20° (Figs. 10.3 to 10.5). The average *annual* maximum temperature at Cambridge, England, is only 87°.

Occasional hot days may occur when a passing disturbance temporarily halts the invasion of cool sea air and substitutes air of land or tropical origin. Under such conditions both Seattle and Bellingham have had a temperature as high as 96°, and Paris, 100°. But severe and prolonged hot waves are rare in the temperate oceanic climate (see temperature data, page 306). A great majority of the cyclonic disturbances which affect these west coasts are strongly occluded and therefore lack extensive warm sectors that would tend to produce frequent spells of warm weather and raise average summer temperatures.

Winter. *Do* winters are much milder for the latitude than the summers are cool. This is particularly true in western Europe, where a great mass of relatively warm water, the North Atlantic Drift, lies offshore. Thus the coastal parts of western Europe are 20 to 40° too warm for their latitudes in January, while western North America, with its less extensive and less warm ocean current, is 10°+ too mild. The

Features related to freezing temperatures

	J	F	M	A	M	J	J	A	S	O	N	D	Yr
Ice days (maximum temperature below freezing)													
Seattle	1.4	0.4								0.1	0.8		2.7
Birmingham, England	0.9	1.5	0.2							0.1	0.9		3.6
Bergen, Norway	4.3	3.6	1.2								1.1	3.6	13.8
Days with freeze (minimum temperature 32° or below)													
Birmingham, England	8.6	9.5	7.4	3.2	0.1					0.7	6.8	6.7	42.6
Falmouth, England	3.2	2.9	3.2	0.6						0.1	1.9	2.8	14.8
Paris	15.8	13.3	10.7	2.5	0.2					3.0	6.5	13.5	65.5
Seattle	7.6	5.5	1.7	0.1						0.1	0.9	5.6	21.3

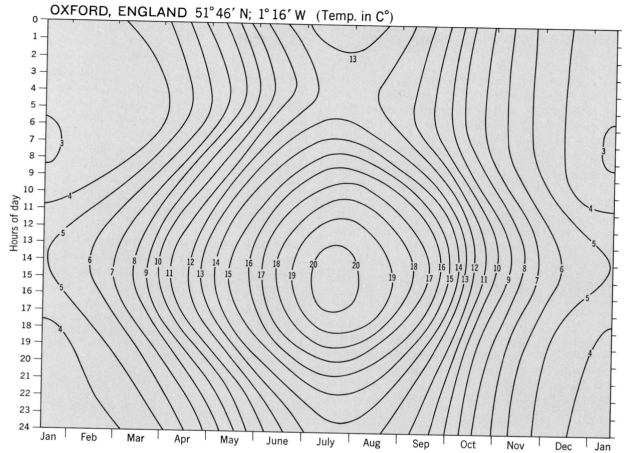

OXFORD, ENGLAND 51°46′ N; 1°16′ W (Temp. in C°)

Figure 10.5 Thermoisopleths for a *Do* station in western Europe. Diurnal range, small at all seasons, is somewhat greater in summer than in winter. Annual range is also small. (*After Troll.*)

dominance of oceanic control is indicated by the fact that isotherms tend to parallel these coasts in winter rather than to follow the lines of latitude. Consequently the decrease in temperature is much more rapid from the coast toward the interior than it is in a north-south direction. Paris is 7° colder in January than Brest, which is 310 miles nearer the ocean. In Norway the heads of some of the longest fiords are 10° colder than the open coasts. And Hammerfest on the coast of Norway at 71°N is an ice-free port, while icebreakers are required to keep open the river harbor of Hamburg 17° farther south. January averages of 35 to 50° in west-

ern Europe contrast with averages of 0 to −40° in the continental climates of interior Asia in similar latitudes.

Winter minima and frosts. The average cold-month temperature at London is 39°; Seattle, 40°; Valentia, Ireland, 45°; and Valdivia, Chile, 46°. Annual ranges are small: 23° at London, 24° at Seattle, 15° at Valentia, and 13.5° at Valdivia. For Seattle the average of the January daily minima is 35°, so that on a majority of nights freeze is absent. At Paris freezing temperatures occur on nearly half the nights in the 3 winter months. In London the thermometer remains above freezing on more January nights than it goes below. It is unusual for London to have a temperature below 15°, while 4° is the lowest minimum ever recorded. At Seattle the thermometer has fallen as low as 3°. The predominantly cloudy skies and humid atmosphere in winter tend to decrease daytime heating and to retard nighttime cooling, so that a flat temperature curve with small diurnal temperature variation is characteristic. Clear days show greater ranges (Figs. 10.3 and 10.4).

Freezes are more frequent, as well as more severe, than in subtropical dry-summer climates, and the freeze-free season is shorter. Nevertheless, the growing season is unusually long for the latitude: 180 to 210 days in the American North Pacific Coast region, for example. Seattle has only 4 months when temperatures below freezing are to be expected, and an average of only 21 days a year with freeze. Birmingham, England, averages 43 days with freeze, and Paris 65.

Winter is usually severe enough to produce a dormant season for most plant life, which is not true in subtropical dry-summer climates. Killing frosts occasionally occur in *Cs* regions, but freeze is almost entirely confined to night hours; temperatures rarely remain below freezing during the entire 24 hours. In *Do* climates, on the other hand, temperatures during abnormally cold spells may stay constantly below freezing for a period of several days. On the average Seattle has 2 to 3 ice days each year (i.e., days when the maximum temperature is below freezing). The comparable figure for Birmingham, England, is 3 to 4 days, and for Bergen, Norway, 13 to 14 (see the table on page 308). Midday temperatures of normal winter days are relatively high, however; the average of the daily maxima for January at Seattle is 45°, and the daily range is less than 10°. On the whole, day-to-day temperature changes are much less diurnal and regular in winter than in summer, for winter is more completely controlled by the succession of cyclones and anticyclones.

Cold spells. Severe cold spells in these marine climates are usually caused by advection of frigid polar continental air whose source is a cold anticyclone located interior and poleward. A well-de-

veloped winter high stationed over the Columbia Plateau region may force an approaching low to move southward along the American North Pacific Coast. The low may be followed by an outpouring of cold air from the interior, which descends the western slopes of the Cascades into western Washington and Oregon. Northeasterly, and not northwesterly, winds bring the coldest weather to the American North Pacific Coast. The same is true for western Europe, where both unusually cold spells and abnormally cold winters are caused by a substitution of cold northeasterly *cP* air for the normal southwesterly winds from the sea (Fig. 10.6).

During one unusual winter cold spell in Europe, the influence of the continental anticyclone and its polar air masses persisted for several weeks. As reported by Kendrew, temperatures in eastern Kent, England, remained continuously below freezing for

Figure 10.6   A synoptic pattern favoring unseasonably low winter temperatures in western Europe. A cold anticyclone to the north and east, and a trough of low pressure over the eastern Atlantic, result in advection of cold *cP* air. (*After Kendrew.*)

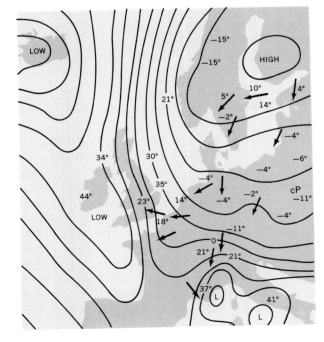

226 hr, the Thames was frozen over in many parts, and practically the whole of the British Isles was frost-bound for 5 weeks. On the continent at this time, German coastal cities had temperatures below zero, while the Rhine was frozen throughout almost its entire course. But such invasions of cP air are infrequent in temperate oceanic climates, for, coming as they do from the northeast, they are opposed to the general westerly air movement of the middle latitudes. The American North Pacific Coast is further protected against cP invasions by mountain barriers.

Annual march of temperature.   In the most marine parts of these west-coast climates the march of the seasons is commonly retarded, so that autumn is relatively warmer than spring. This is particularly true in western Europe, which is more subject to invasions of cold air from the continent in spring than in autumn. At Valentia, Ireland, April has a mean temperature of 48°, while October registers 51°; at Bordeaux, France, the means for these months are 53° and 55°. The seasons of maximum and minimum temperatures are also somewhat retarded, so that at Valentia February is a trifle colder than January, and August is warmer than July.

Precipitation

Annual amount.   This is a humid climate with adequate rainfall at all seasons (Fig. 10.1). Precipitation has a high degree of reliability, and droughts are uncommon. The total amount of rainfall, however, varies greatly from region to region, depending in large measure upon the character of the relief. Where lowlands predominate, as they do in parts of western Europe, rainfall is moderate—usually 20 to 35 in. But where west coasts are elevated and bordered by mountain ranges, as in Norway, Chile, and Pacific North America, precipitation may be excessive—even reaching 100 in. or more on windward mountain slopes (see the climatic data).

Moreover, where lowlands exist, moderate rainfall prevails well into the interior of the continent, but where coastal mountains intercept the rain-bearing winds, precipitation is strongly concentrated close to the littoral. East of the mountains drought conditions are likely to prevail. Of course, an extensive distribution of moderate rains is economically more desirable than the concentration of large and unusable quantities along a mountainous coast. Unfortunately, of the three continents extending well into the westerlies only Europe is freely open to the entrance of rain-bearing winds.

Seasonal distribution.   Temperate oceanic climates are characterized by *adequate rainfall at all seasons,* rather than by a particular season of excess or deficiency (see Fig. 10.1 and the climatic data). This does not necessarily mean that all months have much the same amount, but only that in most parts of *Do* there is no marked seasonal emphasis, such as exists in subtropical dry-summer and tropical wet-and-dry climates, or in regions with strong monsoons. In other words, it is not a seasonal rainfall deficiency which causes the dormant period for plant growth.

In very marine and mountainous locations it is not uncommon to find that the cooler months have most precipitation, with the summers somewhat drier. At Brest, 59 percent of the year's rain falls in the winter half-year, and 41 percent in the summer half. Valentia has approximately the same seasonal distribution. But this winter maximum occurs mainly on exposed seaboard locations and ceases to be characteristic a short distance inland (Fig. 10.7). Thus while Brest shows a slight cool-season maximum, Paris, 310 miles farther inland but still within the oceanic climate, has 55 percent of the year's total in the summer half-year. It has been suggested that this coastal winter maximum may be partly the result of lifting caused by the frictional effects of the coast upon onshore highly saturated maritime air masses. Throughout much of Europe's *Do* climate, fall is slightly the wettest season and spring the driest.

As a general rule, *Do* regions that lie nearest to *Cs*—which usually means the equatorward parts of *Do*—have a somewhat drier summer than is true

Regions with more rain in winter
half year than in summer half year

Figure 10.7 The stippled areas have more rain in the winter half-year than in the summer half-year. In western Europe it is mainly the more maritime sections that fall in this category. (*After Kendrew.*)

for the type as a whole. This is not surprising, for the same strengthened and poleward-displaced subtropical anticyclone which produces the summer drought in *Cs* is likely to cause reduced summer rainfall in adjacent equatorward parts of *Do*. Thus *Do* climate on the North American Pacific Coast south of 50° has a striking summer minimum of precipitation (Fig. 10.2).

Snowfall. Although winter is characteristically a wet season, snowfall is not abundant on temperate oceanic lowlands, for temperatures are too high. Snow is infrequent enough in most of northwest Europe to be a prime topic of conversation when it remains on the ground more than a few days. Paris has an average of 15 snow-days during the year. In the Puget Sound lowland there are some

10 to 15 snow-days, and the snow cover lasts for approximately the same length of time. In the northeastern part of the British Isles the lowlands have about 25 days with snow, but the southwest coast has only 4. The snow that falls is wet and heavy because of the relatively high temperatures. Upon the ground it quickly turns to slush and creates unpleasant conditions underfoot.

Where mountains border these west coasts, however, the lower temperatures and more abundant winter precipitation cause snowfall to be extremely heavy. On the western slopes of the Cascade Range, an average of 300 to 400 in. of snow falls each year. Snowfall is also heavy on the western slopes of the British Columbia Coast Ranges, the Scandinavian highlands, the mountains of southern New Zealand, and the southern Andes. In each of these regions the mountain snowfields have in the past created numerous valley glaciers, which in turn have carved the characteristically irregular, fiorded coasts.

Origin and nature of precipitation. Over lowlands precipitation is chiefly frontal or cyclonic, and a significant part of it falls as steady, long-continued light rain from a gray, leaden sky. Since it is in fall-winter that these storms reach their maximum development, it is then that cloudy, rainy days are most numerous. Cyclones are weaker and less numerous in the warm seasons; yet even so, the absolute humidity is higher at those periods, and the entrance of lows into the continents is facilitated by lower pressures. Consequently summer rain may nearly, if not quite, equal winter totals, although it falls in heavier showers on fewer days. At London, July has 13 rainy days with 2.4 in. of rain, while in January the respective figures are 15 and 1.9. Summers, therefore, are usually brighter and sunnier than winters.

The relatively cool summers and the prevalence of stable maritime air masses at that season are not conducive to the development of thunderstorms and their convective systems. Coastal stations of the American North Pacific states record only two to four a year, although thunderstorms are more numerous at higher elevations and farther inland, where turbulence and overturning are induced by the rugged land surface. At Oxford, England, only 28 percent of the rain in June, and 32 percent in July, falls on days in which thunder is heard. Still, showery, intermittent convective rain is not unusual. Oddly enough, however, this type is associated with the coldest maritime air. In Europe it is cool *mP* air from the north or northwest, which has been humidified and made unstable over the ocean, that is most likely to develop cumulus clouds and showery weather. By contrast, warmer *mP* air from the west and southwest is more stable and more likely to produce a gray overcast and light rain.

A distinguishing feature of the precipitation in *Do* climates is the relatively small amount of rain that actually falls, considering the large number of cloudy, rainy days. Although Paris has only 22.6 in. of precipitation, this is spread over 160 rain-days (an average of 0.14 in. for each rain-day). Seattle, with 33.4 in. of precipitation, has 151 rain-days; London has 24.5 in. and 164 rain-days; Sumburgh Head on the Shetland Islands has 36.7 in. spread out over 260 rain-days. London has had 72 rain-days in succession. The prevalence of strongly occluded cyclones is an important factor causing the many days in lowlands that have dull, lowering skies with low-hanging clouds and light rain or none (Fig. 10.8).

Where coasts are precipitous, abundant rains of orographic origin supplement those from cyclones and from the few convectional storms. Not only is the total precipitation greater there, but the rate of fall is heavier as well. Connor[1] has pointed out that the winter precipitation of the North American *Do* climate is frontal as well as orographic. During winter, the prevailing winds in the interior valleys and even along the coast are easterly. Hence some sort of persistent discontinuity, or front, exists between the locally generated cool easterlies of the mainland and coast, and the warmer Pacific air masses moving in from the west. No doubt a considerable

[1] A. J. Connor, "The Climates of North America: Canada," in W. Köppen and R. Geiger, *Handbuch der Klimatologie*, Vol. 2, part J, 1938, p. 345.

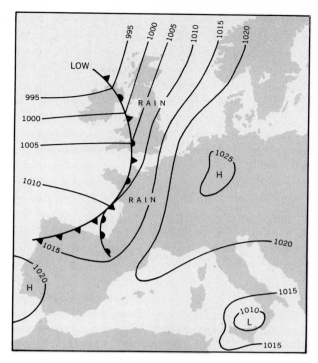

Figure 10.8 A strongly occluded storm in western Europe, producing light but steady and widespread rainfall, a low cloud ceiling, and low visibility. Most of the cyclones which affect western Europe are in an advanced stage of occlusion. Such storms are inclined to produce much cloud, but only a modest amount of precipitation on lowlands.

part of the so-called orographic rainfall is produced by the trigger effect of highlands upon conditionally or convectively unstable maritime air.

Cloudiness and sunshine. Temperate oceanic climate is one of the cloudiest of the whole earth. Relative humidity is almost always high, particularly in winter. On the American North Pacific Coast the mean annual cloudiness is 60 to 70 percent, making it the region with the most cloudiness and least sunshine of any part of the United States (Fig. 10.9). Over wide areas of western Europe cloudiness is greater than 70 percent, and the sun is sometimes hidden for several weeks in succession. Winter and fall, the seasons with most cyclones, are

Precipitation occurrence

	J	F	M	A	M	J	J	A	S	O	N	D	Yr
Days with precipitation													
Paris	14	14	14	13	14	12	12	12	11	15	15	15	160
Oxford, England	15	15	14	13	13	12	14	14	11	16	15	17	168
Bergen, Norway	20	20	18	16	16	15	16	20	19	19	20	21	219
Vancouver, Canada	21	16	17	14	12	10	6	7	11	17	21	22	173
Days with snowfall													
Oxford	4.0	3.9	4.1	0.9	0.1					0.1	1.2	2.6	16.8
Valentia	1.1	1.4	1.3	0.3	0.1					0.1	0.7	1.7	10.8
Bergen	9.0	9.5	8.3	4.2	0.6					0.7	4.5	5.8	42.3
Paris	3.9	3.4	3.3	0.9							0.9	2.3	14.8
Vancouver	4.9	2.2	1.4	0.1							0.6	2.7	11.9
Seattle	3.5	2.6	1.6	0.5							0.5	1.6	10.2

much darker and gloomier than spring and summer. Seattle, which has only 22 percent of the possible sunshine in November and 21 percent in December, receives 65 percent in July and 60 percent in August, so that summers there are relatively bright and pleasant. Valentia, Ireland, has only 17 percent of the possible sunshine in December, but 43 percent in May. Even though summers are sunnier than

Sunshine and cloud

	J	F	M	A	M	J	J	A	S	O	N	D	Yr
Duration of sunshine (percent)													
Paris	21	31	36	39	49	47	50	50	42	33	23	20	37
Number of overcast days													
Kew, England	17.4	14.8	13.0	10.2	8.6	9.5	11.9	10.0	7.8	10.5	12.6	15.0	141.3
Bergen	16.8	17.5	14.9	13.5	14.3	13.3	14.9	16.6	15.8	15.7	16.4	17.4	187.1
Number of clear days (cloudiness 0 to 20 percent)													
Victoria, Canada	1.8	4.2	8.0	6.2	6.0	7.2	13.6	12.6	6.8	4.4	0.8	1.6	73.2
Kew	1.7	2.2	1.3	2.9	3.8	2.8	1.6	2.1	3.7	2.6	2.1	1.1	27.7
Bergen	4.2	2.5	3.1	3.6	3.1	3.8	2.1	1.5	2.3	3.0	2.8	3.0	35.1
Cloudiness (in tenths of sky covered)													
Dublin	6.4	6.4	5.9	5.8	6.0	6.2	6.8	6.3	5.8	6.0	6.2	6.3	6.2
Bergen	7.1	6.9	6.8	6.1	6.2	5.9	6.8	7.1	7.0	6.9	7.1	7.2	6.8
Paris	7.0	6.5	5.8	5.7	5.7	5.5	5.2	4.9	5.1	5.9	6.9	7.2	6.0
Vancouver	7.6	6.9	6.3	6.0	6.1	5.6	4.1	4.3	5.2	6.8	7.7	7.7	6.2

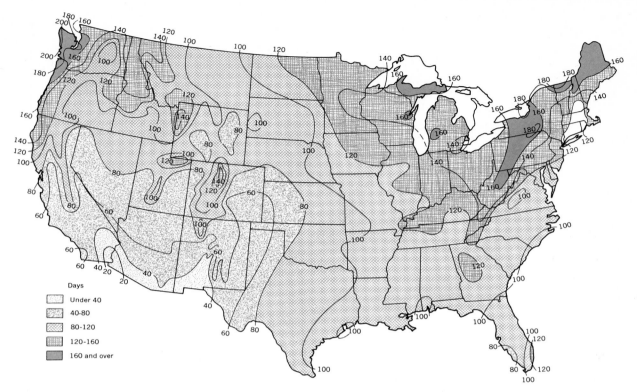

Figure 10.9   Average number of cloudy days in the United States. Note that the Northwest and the Northeast, both regions of cyclone concentration, have the most cloudiness. (*U.S. Department of Agriculture.*)

winters, they are still much cloudier than those of Mediterranean climates. Fog and mist are characteristic weather elements of the marine climate. The American North Pacific Coast has over 40 days with dense fog during the year; Bergen, Norway, has 37.7; and Fanö, Denmark, 53.6.

Seasonal weather

The nonperiodic or cyclonic element.   Because *Do* regions have so many cyclonic storms, the weather is dominated by these disturbances and their accompanying nonperiodic temperature and precipitation changes. The diurnal element, or sun control, is correspondingly weak.

But in spite of cyclonic control of weather, temperature changes associated with the approach and passage of disturbances are not nearly so striking

as they are in continental climates—the North American Middle West, for example. Temperature changes are moderate because the air masses involved in the cyclonic circulations affecting *Do* climates are usually maritime, so that temperature contrasts along fronts are not great. In addition, since most *Do* cyclones are strongly occluded, the majority do not have an extensive warm sector of *mT* air; they are essentially converging systems of relatively cool maritime air (Fig. 10.8). Air temperatures in these storms show nothing comparable to the contrasts between winter *cP* and *mT*-Gulf present in interior North America. As previously noted, isotherms tend to parallel the coast, especially in the cooler seasons, so that advected temperatures from either north or south are not severe. It is only from the eastern continental interior that severe cold can be derived, and such air movement is opposed to the general west-to-east circulation in the middle latitudes.

Traveling disturbances in temperate oceanic climates also produce great nonperiodic variability in the precipitation element. As explained in the previous section, overcast days with light precipitation of frontal origin are numerous. When the cyclonic disturbances slow up or stagnate, as they commonly do, a succession of gray, dripping days results.

Winter, in spite of its mild temperatures, is a stormy period. The westerlies themselves are strongest at that season, and at frequent intervals the pressure gradients are made steeper by passing cyclones. In coastal locations, gales are numerous as one storm follows another in rapid succession. The high seas generated by winter winds are strong enough to make navigation difficult, and unusually severe storms may do serious damage to shipping. The fog and mist create poor visibility and add to the difficulties of navigation.

Winter precipitation, most of it rain rather than snow, is relatively abundant and very frequent. Long periods of dark, gloomy, dripping weather make winters depressing and hard to endure. Not only is frontal cloud abundant, but it also persists in the strong flow of *mP* air following the cold front. If this air has had a northerly or northwesterly trajectory it is unstable, producing cumulonimbus clouds and showery precipitation over the sea and along the coast. Inland there is some clearing at night, but showers predominate by day. If the airflow is westerly and southwesterly, it is likely to be less chilly but more stable, often resulting in gray skies with stratocumulus cloud and light rain.

Between the frequent cyclones there are occasional sunny winter days with crisper weather, but these are the exception rather than the rule. Night freezes are not unusual, especially when skies are clear, but ordinarily they are not severe. Now and then *cP* air masses push westward and create a succession of clear days in which temperatures may remain continuously below freezing.

As the days lengthen with the advance of *spring,* cyclones become less frequent and sunshine more abundant. The air is still cool but the sun is warm, and in western Europe spring is acclaimed the most delightful season.

Summer temperatures are pleasant and conducive to physical well-being. Where sunny days are numerous, as they are in the United States Pacific Northwest, a more charming summer climate would be hard to find (Fig. 9.8). In the higher middle latitudes, or in very exposed marine locations, there are still many chilly, gray, overcast days even in summer. Rain is also relatively abundant, but it falls on fewer days than in winter.

In *autumn* the equatorward swing of the storm belt brings an increase in cloudiness, and perhaps in precipitation. On the whole, autumn is cloudier than spring. The season remains mild, however; September and October (in the Northern Hemisphere) are usually 2 or 3° warmer than May and April.

Regional characteristics

North America. There are two unusual features of this climatic region, both of them in the southern part. The first is the modest total annual rainfall along the margins of northern Puget Sound and the straits of Georgia and Juan de Fuca. Here, over a considerable area which includes southeastern Vancouver Island and parts of the state of Washington, annual rainfall is below 30 in., and there is a core area where it is only 15 to 20 in. (Fig. 10.10). Coupeville, Washington, has only 17 in., Sequim 16 in., and Port Townsend 18 in. Authorities usually attribute these abnormally low totals to the rain-shadow effects of the Olympic Mountains and Vancouver Island. However, it seems likely also that local airstreams moving through restricted oceanic channels, together with peculiarities of coastal alignment with respect to these local air currents, may be an auxiliary factor in reducing rainfall along littorals.

The second abnormal feature of the United State's *Do* is the dry summers. A number of stations show July and August rainfalls of only $\frac{1}{2}$ to 1 in. (Fig. 10.11). Victoria has 3 summer months each with less than 1 in., Seattle 2, and Portland 1. Such summer highest pressure in summer appears to be as far north as 50°. A temperature inversion exists about 20 percent of the time at Tatoosh Island, 48°N. The northward-displaced anticyclone is not only effective in stabilizing the air, but also acts to block most summer cyclones from penetrating to the coast.[2]

Europe. The European sector, the only extensive lowland region with temperate oceanic climate, is also by far the largest sector with abnormally high winter temperatures. That winters are exceptionally mild for the latitude over larger areas in Europe than elsewhere is related both to the openness of the continent to Atlantic air and to the unusual

[2] Glenn T. Trewartha, *The Earth's Problem Climates,* The University of Wisconsin Press, Madison, Wis., 1961, pp. 270–272.

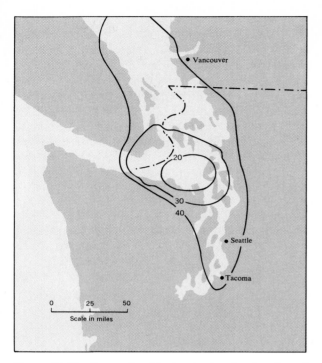

Figure 10.10 Generalized isohyets showing location of areas of modest annual precipitation (in inches) in the northern Puget Sound region of the United States and Canada. (*From Trewartha, The Earth's Problem Climates.*)

drought is more typical of subtropical dry-summer climates than of *Do.* To be sure the total annual rainfall in most parts is large for *Cs.* Still, this is not the case in the drier sections mentioned in the paragraph above. These are saved from being classified as *Cs* only because they have too many months in which the average temperature is below 50°.

This abnormally low summer precipitation in a west-coast position so far poleward is caused by the unusual northward displacement along the coast of an arm of the North Pacific high (Fig. 9.8). The cool coastal waters generated by the cell's circulation may be an auxiliary factor. While the center of the anticyclone over the ocean is positioned at about 38°N, the isobars on its land side bulge in a northeasterly direction, so that along the coast the

Figure 10.11 Annual march of rainfall at a station in the Puget Sound region of modest rainfall. The small annual total, together with the relatively dry summer, makes this station resemble subtropical dry-summer climate, except that its temperatures are not subtropical.

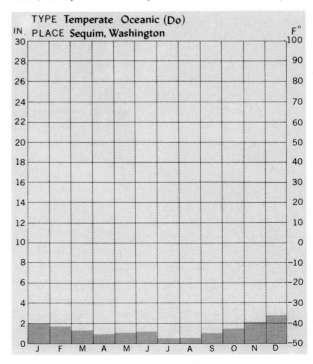

warmth of the eastern North Atlantic Ocean. A January air temperature 50° too warm for the latitude (termed a positive or plus anomaly of 50°) is characteristic of the ocean space 200 miles off the coast of Norway at about 65 to 70°N. The 20° January isanomal follows the coast of France and West Germany and crosses southernmost Sweden.

Another temperature peculiarity of the maritime sections of northwestern Europe is the rapid decrease with altitude in the length of the growing season. The unusual coolness of uplands in Britain creates a low tree line; landscapes in the Pennines at elevations as low as 1,700 ft resemble tundra. The rapid shortening of the growing season with elevation is chiefly a consequence of the steep lapse rate in the predominately cool maritime air masses.

Noteworthy features of annual rainfall on low-lands in western and central Europe are its modest amount (commonly 20 to 35 in.) and its slow decline inland. A few areas have less than 20 in. Some of the lowest annual totals are to be found in the lee of higher land, or along flat coasts and on low islands where winds are strong. The generally mod-est rainfall of these lowlands, freely open to Atlantic

Figure 10.12   Annual frequency of cyclones with central pressure less than 1,013 mb. High frequency is characteristic of the Mediterranean–Black Sea borderlands in the south, and likewise of the higher middle latitudes in the vicinity of ±60°N. Low fre-quency is typical of western and central Europe in the general proximity of ±50°N. (*After Schedler. From Trewartha, The Earth's Problem Climates.*)

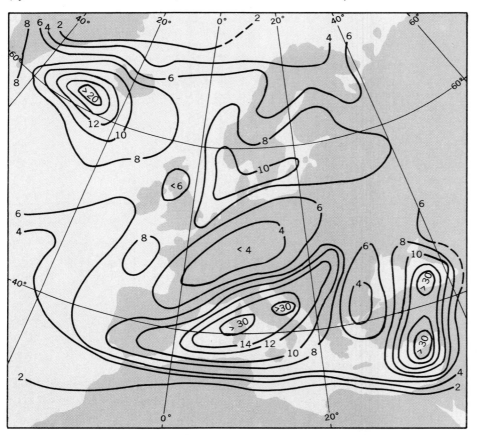

air, is related to the relative infrequency of rain-bringing disturbances in latitudes 45 to 55°N, the approximate position of Europe's lowlands (Fig. 10.12). Cyclones are more frequent to both the north and the south. This minimal frequency of cyclones in a zone astride the 50°N parallel results at least partly from the frequent blocking of Atlantic storms by stagnant warm highs centered in the easternmost Atlantic at about latitudes 45 to 60°N. During such periods of anticyclonic blocking, not only are depressions prevented from passing directly inland, but subsidence in the highs themselves also promotes stability.[3]

Other sectors.   The Chilean *Do* is a mountainous, cloudy, rain-drenched region swept by strong winds. Valdivia, situated close to the region's subtropical margins, has an annual precipitation of 3,700 mm (150 in.). Of this 70 percent is concentrated in the winter half-year, which suggests that some anticyclonic influence is present in summer. Southward the anticyclonic effects diminish rapidly, and so also does any seasonal accent of rainfall.

Restricted areas of temperate oceanic climate are found in southernmost eastern Australia, Tasmania, and South Island, New Zealand. In South Island, precipitation distribution resembles that in South America poleward of 40°. Abnormal amounts of rain for *Do* climate (150 to 200 in.) fall on the windward mountain slopes of Westland, New Zealand, while the subhumid Canterbury Plain east of the highlands receives only 20 to 30 in. Any seasonal concentration is slight.

TEMPERATE CONTINENTAL CLIMATE (*Dc*)

Type location.   Although it is scarcely "temperate," at least in its winter weather, this fairly severe climate is found in some of the earth's most populous and economically well-developed countries in all three continental sectors—Anglo-America, Europe, and Asia. In this respect it is unlike that still more severe group of continental climates, the boreal type located farther poleward, in which population is indeed very sparse.

In temperate continental climate a genuine winter, with a durable snow mantle, is combined with an authentic summer season to produce the annual climatic cycle. Fall and spring, the transition seasons, not only are brief, but are also chiefly composites of winter and summer weather types. Colder and snowier winters, shorter frost-free seasons, and larger annual ranges of temperature differentiate the more severe continental climate from the temperate oceanic type.

This greater severity is caused primarily by locational differences, for continental climate occupies interior and leeward areas on the great land masses rather than the west coasts dominated by the oceanic type. Emphatically, continental climate is land-controlled (Figs. 7.2 and 10.13). And because it is associated with large continents in middle latitudes, this type is confined to the Northern Hemisphere. Only Eurasia and North America are able to generate it. Of the Southern Hemisphere continents, South America alone extends sufficiently far poleward to have rigorous climates, but its narrowness south of latitude 35° obviates genuinely severe conditions in spite of the latitude. Temperate continental climate, which does not occur on western, or windward, coasts because of their dominance by maritime air masses, occupies the interiors of land masses and commonly extends down to tidewater on their leeward, or eastern, sides. On east coasts, in spite of their proximity to the sea, modified continental conditions usually prevail.

The latitudinal spread of temperate continental climate is mostly about 10 to 20°, but the particular latitude belts occupied are not identical in all three continental sectors. Where, as in North America and Asia, the continental type is on the lee side of the land mass, it extends from about 35 or 40° to 50°±. In Europe, on the other hand, since the continental type is located windward of the Eurasian

[3] *Ibid.*, pp. 207–215.

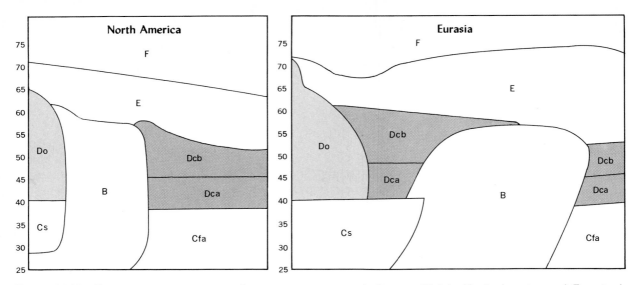

Figure 10.13 Contrasting arrangement of temperate continental climates (*Dc*) in North America and Eurasia. In Eurasia they are found both to the east and to the west of the dry interior; in North America they are located almost exclusively to the east of the dry climates.

land-mass center, temperatures are less severe and the latitudinal spread is from 40° + to 60° or more. Thus the European sector is positioned farther poleward.

The previous section on temperate oceanic climate described the contrasting depths of penetration of marine influence in North America and Eurasia. There are also contrasting arrangements of temperate continental climate on the two continents (compare Fig. 7.2 with the climate map inside the front cover). In North America, because marine conditions close to the west coast are terminated abruptly by mountain ranges, extensive dry climates intervene between temperate oceanic and temperate continental types (Fig. 10.13). Leeward from the west-coast mountains, temperatures suddenly change character from marine to continental, and simultaneously precipitation drops to low levels. Thus a humid oceanic climate passes abruptly to a dry continental one with no humid continental condition between. In Eurasia, on the other hand, where lowlands stretching inland from the west coast permit deep entry of oceanic air into the con-

tinent from the west, temperate continental climate lies not only to the east of the dry interior but also to the west. Consequently this type is to be found both in central and eastern Europe and in eastern Asia. As can be seen on the climate map, the North American and Asiatic temperate continental regions are similar in location; the European is different.

Typically, temperate continental climate is bounded by boreal climate on its poleward side. There the accepted limits are set by the isoline of 4 months with an average temperature of 50° or above. On the south, where its limit is 8 months with a temperature of 50° or over, the *Dc* type adjoins subtropical humid climate in North America and Asia. But in Europe its southern border is mainly set by dry climates or the subtropical dry-summer type.

It may seem unusual that this relatively severe land-controlled climate should extend eastward to the ocean margins. But since the eastern side is the leeward side, the west-to-east general circulation acts to hinder deep entrance of maritime easterly air from the adjacent ocean. In addition, the tendency toward monsoon circulation on the eastern or lee sides of middle-latitude continents accentuates the seasonal temperature extremes, making for cold winters and warm summers.

Temperature

A feature of the annual temperature curve in continental climates is its approximate symmetry, for corresponding spring and fall months have about the same temperatures. In marine climates, by contrast, seasonal lags in temperature distort this symmetry.

Because they have a wide latitudinal spread, *Dc* climates show marked temperature contrasts between their poleward and equatorward parts. An indication of these contrasts is the great difference in values of the isolines marking *Dc*'s northern and southern limits—4 months and 8 months with an average temperature of 50° or above. Still, for any particular latitude this climate is sure to have relatively severe winters and summers, so that annual ranges are large (Fig. 10.14). Of the two extreme

Figure 10.14 Thermoisopleths of a station with only a moderately severe temperate continental climate. (*After Troll.*)

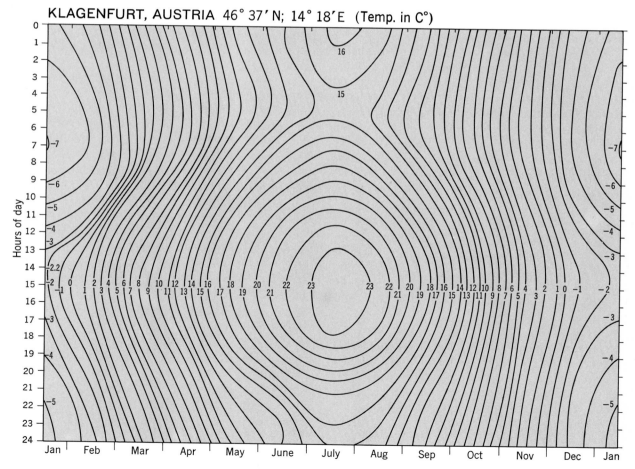

KLAGENFURT, AUSTRIA 46° 37' N; 14° 18' E (Temp. in C°)

seasons, it is the winter cold, rather than the summer heat, which is most pronounced. Nevertheless, summers are warm for the latitude.

Not only are the seasons extreme, but temperatures are variable from one year to another. In marine climates one winter is usually much like another, but wide departures from the normal seasonal temperature are characteristic of continental climates—in extreme instances, as much as 30°.

Effects of a snow cover upon air temperature. Only in continental, boreal, polar, and some highland climates does the snow cover last long enough to have a marked effect upon cool-season air temperatures. Once a region is overlain by a snow mantle, the soil surface itself ceases to influence air temperature. As Chapter 1 explained, light falling upon snow is largely reflected, so that little of it is effective in heating the ground or the atmosphere. Moreover, while a snow surface loses energy very rapidly by earth radiation, the low conductivity of snow acts to retard the upward flow of heat from the ground to replace that which is being lost. Observations made at Leningrad after a fall of 20 in. of loose, dry snow showed a temperature of −39° at the top of the snow surface, while the soil surface underneath was only 27°—a difference of 66°. *cP* air masses moving southward over this cold snow surface are only very slowly modified.

Obviously, the effect of a snow cover is to reduce winter temperatures markedly. As spring advances, it also retards the warming of the air, for much of the solar energy is expended in melting the snow and ice. On the other hand, the snow cover tends to keep the ground warmer and so prevents deep freezing.

Seasons severe. Warm to hot summers and cold winters are characteristic of temperate continental climate. Any monsoon-like winds, advecting temperatures from lower latitudes in summer and from higher latitudes in winter, function to accentuate the normal seasonal severity. In general, the climate becomes more rigorous from south to north and from the coast toward the interior.

Although westerly winds and winter monsoons tend to carry continental air masses down to the eastern littorals, there is some onshore wind of cyclonic origin which meliorates conditions slightly, with the result that east coasts have *modified* continental climates. For example, at New York City and Omaha, Nebraska, which are in similar latitudes although New York is on the Atlantic seaboard and Omaha lies deep in the interior, July temperatures are 74° and 77° respectively, while January temperatures are 31° and 22°. Consequently the annual range is 43.0° at New York and 55.0° at Omaha. However, the higher atmospheric humidity of the air along the seaboard makes summer heat more oppressive and sultry, and winter cold more raw and penetrating, than the drier extremes of the interior. The degree of marine modification is greatest where coasts are deeply indented—as they are in extreme eastern Canada, for example.

Seasonal gradients. Summer and winter in temperate continental climate present marked contrasts in latitudinal temperature gradients. (Compare these with seasonal solar radiation gradients as described in Chapter 1.) In the warm season the few isotherms that cross the central and eastern United States are spaced far apart, so that temperature changes very gradually from north to south: about 1° for every degree of latitude, or approximately every 70 miles (Fig. 10.15). Similar weak summer gradients are found in eastern Asia. In winter, however, temperature changes abruptly from north to south: in the eastern United States, about 2.5° for each degree of latitude, or two to three times more rapidly than in summer. Between Harbin (Haerhpin) and Wuhan in China, there is only 13° difference in July, but 42° in January. Between St. Louis and Winnipeg, the January contrast amounts to 34°, the July contrast to only 13°. Obviously, there is much more reason for northerners to go south to escape winter cold than for southerners to go north to escape summer heat.

The steeper temperature gradients of winter lead to much sharper advected temperature variations

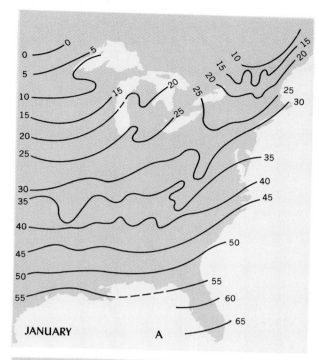

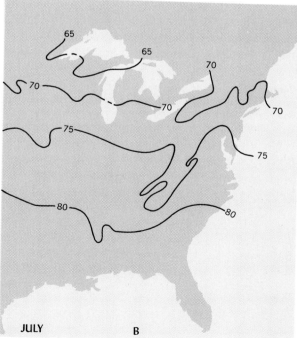

accompanying changes in wind direction. The growing season varies greatly in length from north to south in the continental climates, approaching 200 days on the low-latitude margins and decreasing to 100± days on the subarctic side.

Subtypes. Based largely on temperature contrasts, two subtypes of temperate continental climate will be recognized here: one with a warm summer (*Dca*) and the other with a cool one (*Dcb*). The boundary between them is the July isotherm of 72° (22°C).

Warm-summer subtype. Because this is the warmer and less severe phase of temperate continental climate, it typically occupies the more southerly parts of the general type. As the climate map shows, in the United States it is found in a tier of states extending from central Kansas and Nebraska on the west to the Atlantic seaboard and including Iowa, northern and central Missouri, Illinois, Indiana, Ohio, Pennsylvania, and parts of adjacent states both to the north and to the south. The Corn Belt lies within its borders. In Europe the *Dca* type is found chiefly in the lower Danube Valley—and it is on the plains of the Danube that much of Europe's maize crop is grown. The third principal *Dca* region lies in eastern Asia and includes much of North China, most of Korea, and northern Honshu in Japan.

Summer in this less severe subtype is characteristically long, warm, and humid, owing to the general prevalence of conditionally unstable *mT* air masses (Fig. 10.16). The North American and Asiatic *Dca* regions have summer temperatures that are almost subtropical in character—July at Des Moines averages 77°; New York, 74°; Tientsin, 81°; Mukden (Shenyang), Manchuria, 77°. On the whole, the European *Dca* has less intense summer heat, with July averages more often under than over 75° (e.g., Bucharest is 73°). Typical American Corn Belt cities have average July temperatures in the neighbor-

Figure 10.15 Surface-temperature gradients in the temperate continental climates of the central and eastern United States are much steeper in winter than in summer.

Climatic data for representative *Dca* stations

	J	F	M	A	M	J	J	A	S	O	N	D	Yr	Range
						Peoria, Illinois								
Temp., °F	24	28	40	51	62	71	75	73	65	53	39	28	50.8	51.6
Precip., in.	1.8	2.0	2.7	3.3	3.9	3.8	3.8	3.2	3.8	2.4	2.4	2.0	34.9	
						New York City								
Temp., °F	31	31	38	49	59	69	74	72	66	56	44	34	52	43.0
Precip., in.	3.3	3.3	3.4	3.3	3.4	3.4	4.1	4.3	3.4	3.4	3.4	3.3	42.0	
						Bucharest, Rumania								
Temp., °F	26	29	40	52	61	68	73	71	64	54	41	30	50.7	47.5
Precip., in.	1.2	1.1	1.7	2.0	2.5	3.3	2.8	1.9	1.5	1.5	1.9	1.7	23.0	
						Peking, China (marginal)								
Temp., °F	24	29	41	57	68	76	79	77	68	55	39	27	53	55
Precip., in.	0.1	0.2	0.2	0.6	1.4	3.0	9.4	6.3	2.6	0.6	0.3	0.1	24.9	

Figure 10.16 A station representing the warm-summer subtype of temperate continental climate. Large annual range of temperature, and precipitation concentrated to a moderate degree in the warm season, are characteristic.

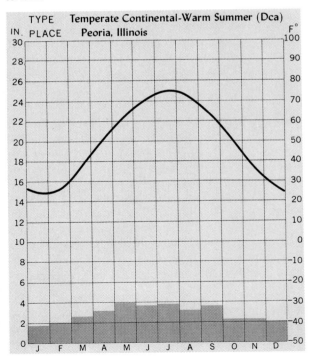

hood of 77°. Not only are summers warm, but they are inclined to be relatively humid as well, so that a summer month contains a large number of sultry, oppressive days which are very uncomfortable for the inhabitants. At Urbana, Illinois, in the heart of the Corn Belt, where the average July temperature is 75°, the mean of the daily maxima is 86°, and the highest temperature ever recorded is 105° (Fig. 10.17). Not only are the days hot, but because of

Figure 10.17 Daily maximum and minimum temperatures for the extreme months of one year at a station with temperate continental climate, warm-summer subtype. Nonperiodic air-mass control is conspicuous, especially in winter.

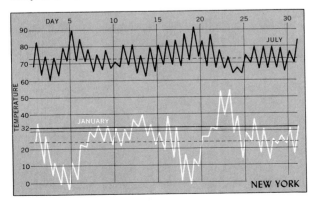

Features related to freezing temperatures

	J	F	M	A	M	J	J	A	S	O	N	D	Yr
Days with freeze (minimum temperature 32° or below)													
Berlin	19.3	17.4	11.3	2.6	0.1					1.2	8.7	16.4	77
Warsaw	23.8	22.6	20.1	5.2	0.3				0.2	3.0	14.8	22.5	112.5
Bucharest	28.3	20.8	13.4	2.0					0.1	1.2	12.4	22.0	100.4
Dubuque, Iowa	29.1	28.0	20.8	6.5	0.3				0.1	3.6	17.5	27.2	132.5
Duluth, Minnesota	31	28	24	16	3					7	25	30	164
Ice days (maximum temperature below freezing)													
Boston	9.4	9.9	2.6								0.8	6.7	29.0
Dubuque	17.4	16.1	5.2	0.2						0.2	3.6	14.1	56.2
Duluth	26.2	25.6	15.3	2.1						1.2	10.5	24.1	104.3

the high absolute humidity, nocturnal cooling is slow, so that nights too are often uncomfortably warm. Urbana has a July mean daily minimum temperature of 64°.

Summer weather, with its prevalence of tropical air masses, is much less variable than that of winter. Occasional summertime invasions of polar air following cold fronts and their associated thunderstorms interrupt the standard succession of hot days (Fig. 10.17). In addition to being hot, summers are relatively long: the period between killing frosts is in the neighborhood of 150 to 200 days. As a result, this is a highly productive climate agriculturally.

Winters are likely to be relatively cold. Spells of overcast, raw, sometimes foggy, and generally disagreeable weather associated with tropical air masses are sandwiched between periods of crisp cold. The average January temperature at Urbana is 26°, and the average of the daily minima is only 18°, although 25° below zero has been known to occur (Figs. 10.16 and 10.17). Peoria, Illinois, has an average January temperature of 24°; Bucharest, Rumania, 26°; and Peking, China, 24°. Monthly averages, however, are not really appropriate for describing winter temperatures, since these are composed of wide variants. Because *Dca* climate has such different temperature conditions on its northern and southern frontiers, either subpolar or

subtropical temperatures can be advected there in the space of a few hours.

Cool-summer subtype.   This more severe of the two subtypes occupies the northerly parts of the temperate continental realm. It is sometimes called the "spring wheat" climate, since that important commercial crop reaches its most specialized development in the subhumid parts of *Dcb*. In North America the subtype is found in general east of the 100th meridian and includes the northern tier of states in the United States and portions of southern Canada (see the climate map). In Eurasia it includes most of Poland, eastern Germany, southern Sweden, and a large part of the central Russian plain between latitudes 50 and 60°+. Beyond the Urals it extends on into Siberia as a narrowing belt in the vicinity of latitude 55°. In much of the U.S.S.R. it is terminated on the south by steppe climate. The third large representative area is in northeastern Asia—especially central and northern Manchuria, southeastern Siberia, and northern Japan.

Because of its characteristic location in higher latitudes, temperatures are generally lower than in the *Dca* subtype farther south (Figs. 10.18 and 10.19). The differential is much larger in winter than in summer, for while the hot months are only 5 to 10° cooler than those in *Dca*, average winter temperatures are 10 to 30° lower. It is the severer

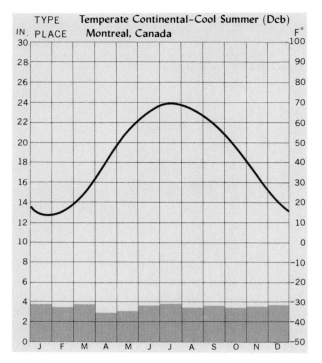

Figure 10.18 Temperature and rainfall conditions at a station with temperate continental climate, cool-summer subtype. Note the large annual range of temperature. At this station there is no seasonal concentration of precipitation—a feature characteristic of the northeastern United States and adjacent parts of Canada, where winter cyclones are numerous.

winters, then, that chiefly account for the larger annual ranges.

Summers are usually moderately warm for a few months: average July temperatures are about 65 to 70° (Grand Forks, North Dakota, 69°; Montreal, 69°; Moscow, 66°; Barnaul, Siberia, 67°; Sapporo, Japan, 69°). But the climate is handicapped by the short duration of summer. Thus while Indianapolis (*Dca*) has 7 months with average temperatures of 50° or over, Fargo and Winnipeg (*Dcb*) have only 5. Freeze comes early and remains late, so that the growing season is only 3+ to 5 months long—too brief for a number of crops. Somewhat offsetting the disadvantages of shorter and cooler summers are the longer summer days in the higher latitudes.

At the time of the Northern Hemisphere summer solstice, Winnipeg has a daily period of solar energy which is more than 1 hr longer than that of St. Louis.

Summer midday temperatures in July are likely to be warm to hot, especially when the sun is shining. Overcast days, on the other hand, are apt to be cool. Days with temperatures of 90° and above are not unusual (Fig. 10.19). In a recent summer, Bismarck, North Dakota, had 20 days when the thermometer reached 90° or above and 4 when it touched or exceeded 100°. The next year there were 14 days with 90° or above, but none that reached 100°. *Dcb*'s hot waves, which are similar in origin to those in regions farther south, are usually not quite so severe or so prolonged. On the other hand, there are more spells of cool weather associated with polar air masses than in the *Dca* type, for (in North America, at least) *Dcb* is closer to the tracks of summer cyclones. It is the more frequent invasions of cool polar air that pull down the general summer temperature average.

Winter is characteristically the dominant season. At Barnaul, Siberia, the January average is −2°;

Figure 10.19 Temperate continental climate, cool-summer subtype. Note the very large and irregular temperature changes indicating strong air-mass control associated with cyclones and anticyclones.

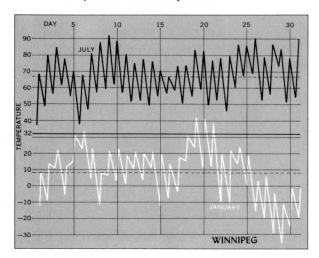

Climatic data for representative *Dcb* stations

	J	F	M	A	M	J	J	A	S	O	N	D	Yr	Range
Montreal, Canada														
Temp., °F	13	15	25	41	55	65	69	67	59	47	33	19	42	56
Precip., in.	3.7	3.2	3.7	2.4	3.1	3.5	3.8	3.4	3.5	3.3	3.4	3.7	40.7	
Moscow, U.S.S.R.														
Temp., °F	12	15	23	38	53	62	66	63	52	40	28	17	39	54
Precip., in.	1.1	1.0	1.2	1.5	1.9	2.0	2.8	2.9	2.2	1.4	1.6	1.5	21.1	
Duluth, Minnesota														
Temp., °F	9	13	24	38	48	58	65	64	56	45	29	16	39	56
Precip., in.	1.0	0.9	1.5	1.9	2.9	3.7	3.6	3.0	3.3	2.2	1.5	1.0	26.5	
Harbin (Haerhpin), China														
Temp., °F	−2	5	24	42	56	66	71	69	58	40	21	3	38	74
Precip., in.	0.1	0.2	0.4	0.9	1.7	3.8	4.5	4.1	1.8	1.3	0.3	0.2	19.3	

Bismarck, North Dakota, 8°; Sapporo, Japan, 21°; Uppsala, Sweden, 24°. But these averages are composed of very unlike temperature elements, since the succession of fronts and air masses brings much subzero weather as well as some that is considerably above freezing (Fig. 10.18). At Bismarck, although the average January temperature is 8°, the average of the daily minima is −2.8°. Even the highest temperatures on most January days are well below freezing, for the mean of the daily maxima for that month is only 18.0. At Winnipeg the mean of the daily maxima for January is 6.2°, and that of the daily minima −14.7° (Fig. 10.19). The comparable figures for milder Quebec are 17.7° and 1.6°. In a recent January, Bismarck had 5 days in which the temperature did not rise to zero and 13 in which it was constantly above zero. There were 14 days in which the minimum temperatures went below −10°. The lowest temperature for that particular year was about −37°, although −45° has been recorded. But winters differ greatly from year to year in the continental climates; the following January, Bismarck had only 1 day with a temperature below −10°, as compared with 14 the previous year. As in the *Dca* type, large nonperiodic air-mass-controlled temperature fluctuations within a short period of time are also characteristic. Changes of 40° within 24 hr are relatively common at Bismarck.

On one March day, a rise from −15° to nearly 50° was recorded.

In the western portions of Europe's *Dcb* (Germany, Poland, southern Sweden), although summers are cool so that the region must be classified as *b*, the winters are not correspondingly severe. The mean January temperatures of Vienna (29°) and Warsaw (26°), for example, are higher than those of many *Dca* regions in eastern Asia and the eastern United States.

Precipitation

Annual amount. Although temperate continental is classed as a humid climate, more of its total area has too little annual precipitation than too much. Since probably more than half the area with *Dc* climate receives an average of 30 in. or less, large parts must be judged subhumid. Apparently this is to be expected in land-controlled climates.

Rainfall decreases from the seaward margins toward the interiors, and usually from the lower toward the higher latitudes as well (see the climatic data). Thus along its interior margins temperate continental climate borders dry climate, and it is these interior parts which are especially subhumid. The prairies, in interior parts of both North America and Eurasia, represent this drier subtype.

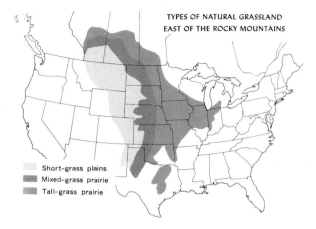

Figure 10.20   Extensive areas in the western parts of the North American temperate continental region are included within what is known as the "prairie wedge." Here the original vegetation was tall-grass prairie, which thrust eastward like a wedge into the humid and forested east central United States. (*After J. Richard Carpenter.*)

In North America the prairie of the temperate continental climates extended eastward in the form of a wedge into humid Illinois and Indiana (Fig. 10.20). This eastward projection of the grasslands into the forest has been interpreted various ways. Some authorities maintain that the humid prairie region was climatically capable of supporting forest, but was prevented from doing so by fires (set both by lightning and by man) which killed the woody plants but not the grass, and by the injurious effects of grazing animals. Others believe that the moist prairie had a climatic origin.[4]

The regional distinctiveness of the "prairie wedge" lies basically in its precipitation characteristics: (1) Low snowfall and low rainfall are typical of winter. (2) In the prairie there is a greater risk of a large rainfall deficit in summer than exists in the forest regions to the north and south. (3) The prairie has fewer days with precipitation, less cloud, and lower relative humidity in July and August

[4] John Borchert, "The Climate of the Central North American Grassland," *Ann. Assoc. Amer. Geographers,* Vol. 40, pp. 1–39, 1950.

than the forested region to the north. (4) Dry summers in the prairie wedge are characterized by large positive departures from average temperatures and by frequent hot winds. These climatic characteristics of the prairie wedge are related to its position in the general circulation pattern over North America, for this region approximately coincides with the greatest mean transport of dry westerly continental air from the eastern base of the Rocky Mountains (Fig. 10.21).

Seasonal distribution.   Although winters are by no means dry, summer is predominantly the season of maximum rainfall (Fig. 10.16). This seasonal distribution is related to the following conditions:

Figure 10.21   Average number of months per year with a mean transport of dry continental air eastward from the eastern base of the Rocky Mountains. One of several explanations offered for the prairie wedge is this greater frequency of dry continental air, resulting in a unique combination of climatic characteristics. (*After Borchert.*)

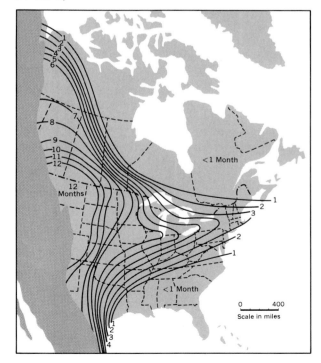

(1) The specific humidity, or reservoir of water vapor in the atmosphere, is much lower over the continents in the winter cold than in the summer heat.

(2) During winter the subsiding air in the continental seasonal anticyclone is likewise conducive to low specific humidity. This subsidence tends to develop a strong temperature inversion and thus to increase the stability of the air mass—a condition which is opposed to precipitation.

(3) Continental anticyclones, which develop over the colder parts of land masses in winter, tend to repel or divert cyclonic storms. The diverging surface winds of these high-pressure cells are opposed to the formation of fronts. Cyclones in summer, although they may be weaker and less frequent, can nevertheless penetrate deeper into the continents.

(4) Convection is at a maximum during the summer months, for the warm land surface has a tendency to make the air masses flowing over it unstable. In winter, on the other hand, the cold snow surface increases low-level stability.

(5) Owing to the seasonal extremes of temperatures, and hence of pressure, seasonal monsoon circulations tend to develop over the large continents. These lead to an inflow of warm, humid tropical maritime air with high rainfall potentialities in summer, and to an outflow of dry, cold polar continental air in winter.

Because of its effect on agricultural productivity, the characteristic summer maximum of rainfall in continental climates is one of the most important facts about the whole climate type. This is especially true where the total amount of precipitation is relatively modest, as it is over extensive areas within *Dc*. In tropical climates, where the *temperature* growing season is 12 months in length, it does not matter much when the maximum precipitation comes, for at any time of year the rainfall will be effective for plant growth. Even in the subtropical parts of middle latitudes—California and Florida, for example—the winters, although they have some frosts, still are so mild that a completely dormant

season for plants is not imposed. Consequently winter rainfall has direct value for crops.

In continental climates, on the other hand, relatively long and severe winters create a completely dormant season for vegetation. There it is essential that the periods of sufficient heat and adequate rainfall coincide, for winter precipitation, much of it in the form of snow, is of no immediate benefit in supplying moisture to plants. A rainfall regime with a winter maximum and summer drought probably is the most efficient one for subhumid subtropical California, but for the upper Mississippi Valley it would be disastrous. A general principle of climatology is that in the middle latitudes, the *distribution of rainfall throughout the year*, or its *regime*, is just as important as the annual *quantity* of rain that falls.

It is especially the deep continental interiors and the regions of marked monsoonal tendencies whose summers are emphatically rainier than winters. At Peking in North China, a station typical of regions with well-developed monsoons, December and January each have only 0.1 in. of precipitation, while July and August have 9 and 6 in., respectively. Throughout North China winters are exceedingly dry (see the climatic data). Omaha, Nebraska, which is typical of an interior regime in North America, has 0.7 in. in January and 4.7 in June. Over much of the United States *east* of the Mississippi, however, the discrepancy between winter and summer precipitation is not so marked (Fig. 10.18). The same is true in most of continental Europe. New York City, which receives 3.3 in. in each of the three winter months, has only slightly more, 4.1 and 4.3 in., in July and August. Its total for the year, however, is 42.5 in. Leningrad, U.S.S.R., has 2.5 and 2.8 in. in July and August but 1.0 and 0.9 in. in January and February.

In spite of a prevailing summer maximum, however, midsummer in extensive areas with continental climate is a period of water deficiency (Fig. 10.22). This is a result of the large evaporation at the time of greatest heat. During July and August pastures and lawns often become sere, and crops suffer from the water deficiency.

Early-summer maximum. In the more subhumid

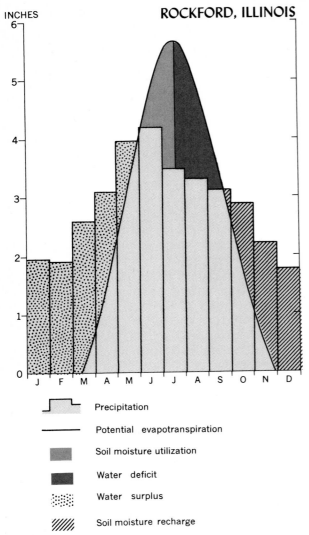

ROCKFORD, ILLINOIS

Precipitation

Potential evapotranspiration

Soil moisture utilization

Water deficit

Water surplus

Soil moisture recharge

Figure 10.22 The average water balance, based on mean monthly values, at Rockford, Illinois, a temperate continental station with warm summers. Note that for about 4 months in summer, water losses through evapotranspiration exceed the water gained through precipitation. An actual water deficit prevails in late summer. As a consequence crops, pastures, and lawns may suffer. (*Data from Thornthwaite, Mather, and Carter.*)

interior locations the period of maximum rainfall, more often than not, is in early summer or even late spring rather than at the time of greatest heat

(Fig. 10.23). This is the case in the Danube Basin and in the western part of the prairie region of the United States. At Belgrade, Yugoslavia, June is the wettest month, and May has more precipitation than July. At Omaha the maximum amount is also received in June. The fact that rainfall in these areas drops off in July, even though heat and moisture are at a maximum then, appears to be related to the decrease in disturbed weather conditions later in the summer compared with June. In frequency of fronts and cyclonic storms, June resembles spring more than summer. Between June and July the circulation features, including the main storm tracks, shift northward considerably.

Agricultural production benefits from having the year's rainfall relatively concentrated in the early part of the growing season. It is in their tillering period, which is spring and early summer, that cereal crops are developing stems and basal leaves and therefore require the greatest amounts of moisture. The maturing and harvesting season for small grains preferably should be warm, bright, and dry.

Winter precipitation. Winter precipitation is largely frontal and cyclonic in origin. In North America, *mT* air masses moving poleward up the Mississippi Valley are continuously being chilled at their base, so that stability is increased and convectional overturning is unlikely. But usually this air does not advance very far northward before coming into conflict with denser polar air masses, over which it is forced to ascend. Sometimes the front develops even before *mT* reaches the northern

Figure 10.23 An early summer maximum (June) of rainfall characterizes extensive subhumid interior parts of the temperate continental regions.

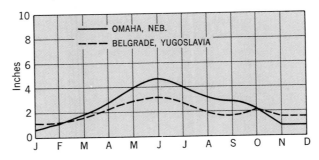

Gulf Coast, and only occasionally does the tropical air advance at the surface farther north than the Great Lakes region.

Even in winter, mT air is sufficiently humid that when it is forced up over the colder air it yields moderate amounts of precipitation, which usually falls from a dull, gray overcast that blankets the sky from horizon to horizon. The lively skies of summer with their tumultuous cumulus clouds are largely lacking in winter. In northeastern Asia, with its stronger winter monsoon, the cP surges are so frequent that surface mT air is practically absent in northern China in the cool months. Winter precipitation in these regions is very meager.

Snow. Cool-season precipitation in continental climates is partly in the form of snow, and a permanent snow cover, varying from a few weeks to several months in duration, is typical. Because it takes 5 to 15 in. of snow to equal an inch of rain, and because snow tends to remain on the ground while rain does not, the smaller total precipitation of winter may be more impressive than summer's greater amount. Also, the cyclonic winter precipitation is continuous over longer periods of time than the briefer convectional showers of summer.

In those parts of the northeastern United States and Canada where winter cyclones are particularly numerous and well developed (Great Lakes region, St. Lawrence Valley, New England, and the Canadian Maritime Provinces), snow becomes exceedingly deep. Northern New England and New York have more than 7 ft of snowfall during an average winter, and the snow cover remains on the ground for more than 4 months (Fig. 10.24). In parts of the Adirondack Mountains 150 in. or more of snow falls annually. Over the American Great Plains, on the other hand, the total is only 20 to 30 in.

Summer precipitation. A large proportion of the total summer rain in temperate continental climates is in the form of convectional showers falling from cumulonimbus clouds. These summer showers are less extensive than the rains and snows of winter, are shorter in duration, and are likely to be more intense. Daytime solar heating of the land surface unquestionably encourages convective overturning and hence shower activity. But the majority of the

Figure 10.24 Except in highlands, the snow cover lasts longest in the north central and northeastern parts of the United States. It is almost nonexistent along the Gulf and in the extreme southwest.

summer showers are not haphazardly distributed, and of course they are not a feature of all warm afternoons. Rather, they are organized in their pattern and occur in conjunction with passing atmospheric disturbances of an extensive character.

Unfortunately, these summer showers have certain characteristics that make the warm-season rainfall less effective than its total amount might indicate. Many of them come as pelting rain, so that too much of the water runs off rather than soaking in. Furthermore, they are local and spotty; crops in some localities may be suffering from a water deficiency, while in others they are being injured by an excess. Feast or famine, too much or too little, is a feature all too common. If the $4\pm$ in. of rainfall typical of the individual midsummer months in the upper Mississippi Valley could be depended upon, and if this amount were spaced so that about $\frac{1}{2}$ in. fell twice a week as widespread soaking rains, the actual amounts would be adequate for field crops and lawns. But that is not the way nature provides it.

In spite of the fact that the warm season has more total precipitation than winter, it has a smaller proportion of cloud. Thunder and lightning frequently accompany the summer showers. In Hungary, 61 percent of the rain in June comes on days with thunderstorms. Steady cyclonic rains falling from gray, overcast skies are not absent but this type of weather is less common in summer than in the cooler seasons when the polar front is farther equatorward. On the whole, cyclonic weather is most typical of *Dc*'s poleward margins.

Seasonal weather

Strong nonperiodic weather element. No other type of climate has such rapid and marked non-periodic weather changes in winter as the temperate continental, for it is there that the conflict between polar and tropical air masses reaches its maximum. Actually, spells of weather associated with the passage of cyclonic and anticyclonic storms are equally, if not more, numerous in temperate oceanic climates, but in those locations the accompanying temperature changes are not nearly so severe. In the central and eastern United States, which is freely open to air masses from both north and south, storm control is especially powerful. The weather element is less pronounced in eastern Asia, where an unusually strong winter anticyclone and its monsoon wind system hinders cyclonic development. Greater surface relief in eastern Asia likewise obstructs the free latitudinal movement of air masses. The net result is that the principal tracks of winter cyclones are located over the ocean to the east of Asia, and there is less storm activity over the mainland itself than in North America.

It is in the cold season, when the sun, and with it the storm belt, has retreated farthest south that nonperiodic control of weather is strongest in the continental climates. At that season the diurnal sun control is usually subordinate; weather is dominated by moving cyclones and anticyclones associated with rapidly shifting polar and tropical air masses and the fronts that develop along their boundaries. Many times the daily rise and fall of temperatures with the course of the sun are obscured by larger nonperiodic oscillations resulting from temperature advections by polar and tropical air masses (Figs. 10.17 and 10.19). These sharp temperature changes are made possible by the steep thermal gradients of winter, because of which a north wind brings

frigid cold, while a south wind advects unseasonable warmth.

Both cyclones and anticyclones tend to move faster—and to be larger, more frequent, and more severe—in winter than in the warmer seasons. As a consequence, changes in temperature, wind, and weather in general are more severe and more frequent in winter. Since there is a distinct concentration of winter storm tracks over the northeastern United States and adjacent parts of Canada, it is this region which has the most weather changes. With increasing distance from the northeastern region of concentration, both to the south and to the west, storm control weakens and weather changes diminish.

Winter weather types. The usual cycle of weather changes that occur with the passage of a winter cyclone, followed by an anticyclone, has been described in Chapter 6. But there can be an almost infinite number of weather variations in these cycles, depending on the season, the intensity and extensiveness of the storm, the types and characteristics of the air masses and fronts involved, the tracks taken by the storm, and the patterns of high-level atmospheric circulation. As a consequence, so far it has been impossible to work out a satisfactory classification of weather types. However, even the layman is conscious of certain weather types: among the examples mentioned in Chapter 6 were cold waves, warm waves, January thaw, and Indian Summer. But there are many more recognizable although less distinctive ones without names. A number of the commoner weather types are illustrated by the sketches of synoptic conditions in Figs. 10.25 to 10.34. These are worth careful study, for continental climates cannot be understood apart from the weather types which dominate them.

Among the weather types of winter is the *blizzard,* which is so spectacular that it has gained a reputation out of proportion to its climatic importance and frequency. The genuine blizzard is more than just a blinding snowstorm, for it involves a gale wind, below-zero cold, and drifting powdery snow. Actually there may be no precipitation falling at the time; yet the air is filled to a height of several hundred feet with swirling masses of dry, finely pulverized snow whipped up from the freshly fallen snow cover. Sometimes the sun can be dimly seen shining wanly through the shroud of flakes. It is the combination of wind, cold, and blinding snow that makes these storms dangerous to both men and animals caught out in them.

In one notorious blizzard over the northern American Great Plains winds of 50 miles an hour were recorded, accompanied by a temperature of 20° below zero. In this storm between two and three hundred persons are reported to have lost their lives, and thousands of cattle perished. The genuine blizzard is extremely rare in the eastern states. It occurs occasionally in the Middle West but is most truly representative of the western Prairie and Plains states. The blizzard is also found in the prairie lands of Russia and Siberia, where it has the name of *buran.*

On the synoptic chart this weather type is associated with unusually steep pressure gradients in the rapidly advancing forward part of a polar-outbreak high, and hence behind a cold front. The frontal precipitation in the cyclone lays down the necessary new cover of loose, soft snow. The violent northwest winds and rapidly falling temperatures on the rear of the cold front provide the remaining prerequisites.

More frequent and more widespread than the genuine blizzard is the *cold wave,* a typical feature of continental winters, particularly in the North American sector. Not every sharp drop in temperature is a cold wave; the thermometer must fall a certain number of degrees within 24 hr, and it must drop below a certain fixed minimum. In the northernmost United States, the temperature in an authentic winter cold wave must fall at least 20° within 24 hr and must reach zero or below.

This weather type is associated with a rapidly southward-moving polar-outbreak high (Fig. 6.15). It is usually preceded by mild weather with rain or snow—features of the front side of a cyclone. The sharp drop in temperature, which *is* the cold wave, occurs when the wind shifts from a southerly

to a northerly direction, or in other words, when the cold front separating tropical and *cP* air passes. However, the lowest temperatures may not be reached during the period when the strong northwest wind is blowing, but a day or so later, in the calm air near the center of the anticyclone where radiation produces the maximum surface cooling. Anticyclones which take the form of intercyclonic entities and are composed of modified *mP* air may bring fine, clear, crisp weather, but without subzero temperatures (Fig. 10.25).

A deep cyclonic storm, especially if it originates in the general Texas region and takes a course northeastward across the country, will probably bring widespread, and perhaps heavy, snowfall to the temperate continental regions of the Mississippi Valley and eastward (Fig. 10.26). If the cyclone travels a more northerly route, temperatures are likely to be milder and the rain (instead of snow) areas more extensive (Fig. 10.27). A weak low that follows a route north of the Great Lakes may bring

Figure 10.25 Fine, clear winter weather over the central United States associated with a high-pressure system composed mainly of modified *mP* air from the Pacific. Temperatures are for 1:30 A.M.

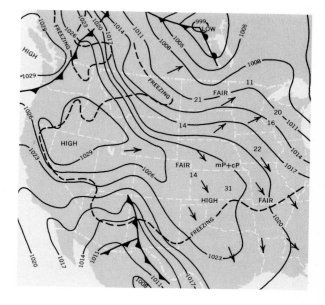

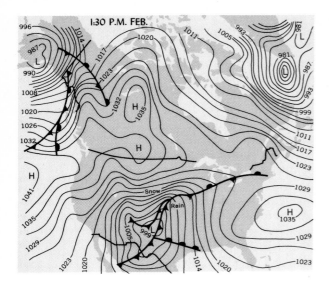

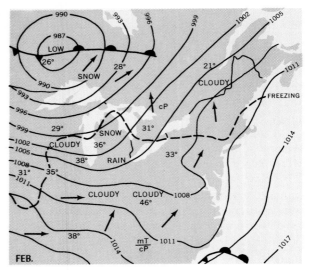

Figure 10.26 A well-developed winter storm originating in the Texas area and moving northeastward across the United States. Such storms are likely to bring heavy precipitation, much of it in the form of snow on the northern side of the storm.

Figure 10.27 A deep and extensive winter storm centered over the upper Great Lakes region. It was accompanied by widespread precipitation, snow to the north and west, and rain to the east and south in the warm sector. Temperatures are for 1:30 A.M.

Figure 10.28 A common winter weather type. Here a February cyclone traveling on a northern track is producing cloudy, mild weather and light precipitation over extensive areas of the north central United States. Temperatures shown are for 1:30 A.M.

much gray, overcast, moderate weather, but only modest amounts of rain or snow (Fig. 10.28).

Summer weather types. Compared with the stronger nonperiodic storm control of winter, summer has greater diurnal regularity of weather (Figs. 10.16 and 10.18). Since the polar-front storm belt swings northward with the sun, the humid continental regions lie on its southern margins. As a result weather becomes more stable and temperatures are likely to be fairly similar over wide areas. Not only are cyclones and anticyclones fewer and weaker, but the normal seasonal temperature gradients are also weak, so that advections from north or south cannot bring such striking changes. Several days in succession with south winds, however, may force maximum temperatures to 90°+ and occasionally even to 100° and over. Diurnal variations in wind velocity and temperature are characteristic. Clouds of the cumulus type and showery rainfall tend to replace winter's sullen gray skies and long-continued spells of precipitation. Summer has some of the aspects of tropical weather.

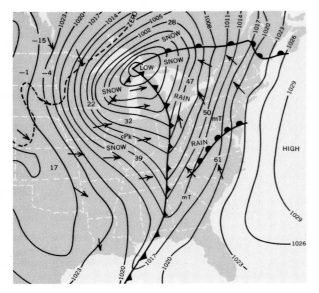

Cyclones and anticyclones, while both less frequent and weaker in summer, nevertheless serve to break the regularity of sun-controlled weather. Hot waves, frontal thunderstorms, summer cool waves, and drought spells are a few of the more common weather types of the hot season. *Hot waves* or spells of unusually hot weather, with maximum temperatures of over 90° on several successive days, as a rule are caused by long-continued advection of tropical air masses by southerly winds. The arrangement of storm areas on the weather map for such an invasion of tropical air is nearly opposite to that for cold waves. Hot waves occur with northward-sloping barometric gradients, when a relatively stagnant warm high is over the south and southwest, and a weak, slowly moving low, with little cloud, lies to the north (Fig. 10.29).

Cold-front thunderstorms are associated with elongated V-shaped lows in which there is a distinct wind-shift line (Fig. 6.30)). Preceded by days with south winds and high temperatures, the heat wave is broken when the eastward-advancing cold front,

Figure 10.29 A summer weather type in the form of a July heat wave over the central and eastern parts of the country. Temperatures shown are the maxima for the 12 hr preceding. Tropical air from the western end of a warm anticyclone centered over the Gulf of Mexico and the western Caribbean controls the weather.

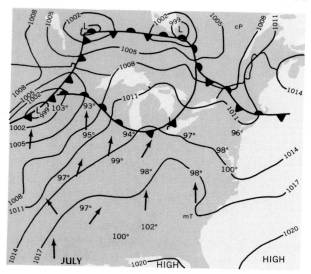

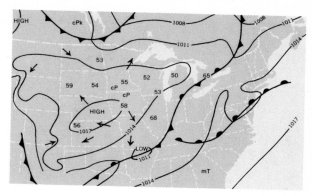

Figure 10.30 A spell of fine cool summer weather (August) produced by a high-pressure system composed of Canadian air. Temperatures are for 1:30 A.M.

with its general turbulence and thunderstorms, is reached, and the wind shifts to northwest. The delightfully cool polar air on the rear of such a storm brings the *summer cool waves* (Fig. 10.30). Many of the summer *drought spells* in the humid continental regions of the United States appear to be associated with the development of a warm anticyclone over the Gulf states. Such a high blocks the normal northward progress of *mT* air and substitutes its own dry subsiding air.

Spring and fall weather.   Spring and fall, the transition seasons, exhibit a more even struggle between storm and sun control. At times the one and then the other is in the ascendance, so that there is something of an oscillation between summer and winter conditions. Mild, warm days in April and early May, with regular diurnal rises and falls of the thermometer resembling summer, may be followed by a reestablishment of winter conditions as a passing cyclone lays down a snow cover, and the succeeding polar air mass drops temperatures to an unseasonable freeze. Continental climates are famous for the capriciousness of their spring weather (Figs. 10.31 and 10.32).

Autumn brings some of the loveliest days of the entire year but likewise some of the rawest and gloomiest. Bright, clear weather with warm midday temperatures and crisp, frosty nights comes with polar air-mass control. A reestablishment of hot-

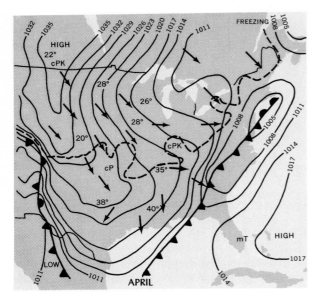

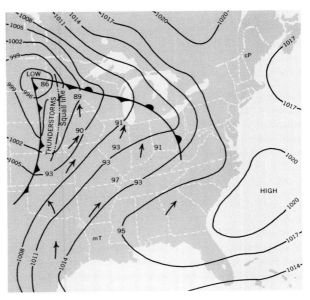

Figure 10.31 A spring weather type. Here a cold anticyclone advancing southward as a mass of cold *cP* air with northwest winds carries low temperatures deep into the subtropics, causing a severe spring (April) freeze in the north central states. Temperatures shown are for 1:30 A.M.

Figure 10.32 A spring heat wave over the central United States resulting from a northward flow of air from the western flanks of the Bermuda High. This air was drawn into a cyclone centered over the eastern Dakotas. Note the squall line, with thunderstorms, in advance of the cold front. Temperatures are daytime maxima for the preceding 12 hours.

Figure 10.33 Fine, clear, crisp weather in October associated with a cold anticyclone from Canada, composed of *cP* air. Temperatures are for 1:30 A.M. Isotherm of 32° shown by dashed lines.

Figure 10.34 A deep and extensive autumn cyclone, with steep pressure gradients, accompanied by strong winds and extensive precipitation in the form of rain and snow. Temperatures are for 1:30 A.M. Isotherm of 32° is shown by dashed line.

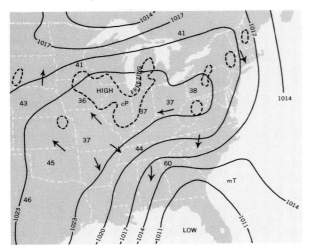

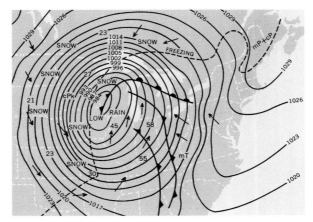

wave gradients in October and November after severe frost, and perhaps even snow, causes a temporary return of summer conditions. This is the warm weather with hazy, smoky atmosphere known as Indian Summer. But well-developed cyclonic storms may also bring raw gray days with chilly rain, and occasionally a temporary snowy winter landscape may be produced as early as October (Figs. 10.33, 10.34).

Regional characteristics

The relative severity of a climate is sometimes expressed by the index of continentality, which denotes the annual amplitude of temperature and is mainly a function of the size of a land mass. It is slightly different also for each latitude, increasing poleward as the amplitude of annual variation of solar radiation increases. The following formula is one of several that have been proposed for computing the index of continentality (C).

$$C = \frac{1.6A}{\sin \phi} - 14$$

where

A = annual temperature amplitude in °C
ϕ = latitude

From the table below it may be noted that Europe has the lowest index of continentality at all three latitudes, and Asia the highest. The continental differentials vary at different latitudes.

Some indexes of continentality

Continent	Latitude		
	68°	50°	40°
Europe	40	38	40
North America	59	53	51
Asia	72	71	60

North America. Within an extensive area in the upper Mississippi Valley–Great Lakes region, the typical *Dc* annual march of rainfall is significantly modified. Instead of a single summer maximum, the crest of the annual profile is caved in, so that there is a mid- or late-summer secondary minimum. This results in a twinned summer maximum, with one peak usually in June and the other in September. July and August are drier. The core of the region with a twinned summer maximum is in southern Wisconsin, northern Illinois, and eastern Iowa. In modified form it extends out for several hundred miles around the core.

What requires explanation is not so much the two peaks of rainfall as the secondary minimum in July-August, which produces the two peaks. At least a partial answer is found in the summer circulation aloft. An anticyclonic ridge and its dry northerly tongue of air are positioned over the region in summer, and they reach their maximum development in July-August. The total effect is to depress the rain-making processes in their vicinity. It is the dry current aloft which appears to cause a falling off in the showery precipitation in mid- and late-summer.

If a rainfall profile based on 7-day means is constructed for the period May to October, the simple twinned maximum based on monthly means becomes much more complicated, with numerous ridges and hollows. In this profile there appears to be a moderately good positive correlation between summer wet periods and the combined frequency of warm, stationary, and occluded fronts, and warm sectors. There is also a general inverse correlation between the frequency of highs and the peaks in the 7-day rainfall profile.[5]

Local climatic effects of the Great Lakes, although not so striking as might be expected considering the size of the water surfaces, can be seen in the locally heavy snowfalls along the windward shores. The northern coast of upper Michigan facing Lake Superior, the western shore of lower Michigan fronting on Lake Michigan, peninsular Ontario, and those parts of New York State and Pennsylvania adjacent to Lakes Erie and Ontario, all show an excess of snowfall over that received by surrounding

---

[5] Trewartha, *The Earth's Problem Climates,* pp. 284–289.

areas farther removed from the lakes. This excess snow along downwind coasts comes in the form of heavy snow showers during periods of rapidly falling temperature and strong northwest winds. At these times a polar-outbreak high is advancing rapidly to the south and east. The cold dry air, warmed and humidified over the open water, is lifted by orographic effects and frictional turbulence when it comes onshore, so that it produces bursts of snow. Normally the lake snow is confined to a narrow strip not more than 20 miles wide inland from a lake.

Annual rainfall profiles within the American temperate continental region change shape from west to east. In the deep interior, rainfall is modest in amount and strongly concentrated in the warmer months, winter being relatively dry. Thus at Omaha, Nebraska, with 29 in. of precipitation, rainfall in the 3 summer months is nearly five times that of the 3 winter months. Of Winnipeg's 20.2 in., the ratio is 3.4 to 1. Eastward there is a gradual increase in the total precipitation and simultaneously winters become relatively wetter, so that the ratio of summer to winter precipitation decreases (Fig. 10.18). Thus the annual rainfall profile gradually flattens toward the east, until along the Atlantic seaboard in New England and Canada some stations actually show a slight winter maximum. There is not much change in total warm-season rainfall; it is the cool-season precipitation that increases eastward, both relatively and absolutely. This rise in winter precipitation toward the Atlantic is related to the increasing humidity caused by proximity to the ocean and also to the fact that winter storm tracks tend to converge in the Lower Great Lakes–New England area.

Europe.   The European sector of temperate continental climate is the exception in location, for it is found to the west, or windward, of the dry interior. On the west it borders the temperate oceanic type, while the Asiatic and North American sectors on that side are terminated by severe dry climates. As noted earlier, Europe's Dc is also located 5 to 10°

farther poleward than those of the other two continents.

Because of its location windward from the continent's core, the European sector has temperature features somewhat unlike those in Asia and North America. In fact, it is difficult to find stations in Europe whose summer and winter temperatures both match those of stations in the other two continents. If they are similar in either the cold season or the warm season, they are different in the other, for the amplitude of the annual temperature curve is smaller in Europe. For example, Berlin and Duluth, Minnesota, have similar July temperatures (64° and 65°), but their January means (Berlin 31° and Duluth 9°) are far apart.

Temperate continental Europe also has a relatively modest amount of annual precipitation: nearly all of it under 25 in., and considerable areas under 20. This feature was discussed in an earlier section on temperate oceanic climate in western Europe. In addition to the reasons given there, it may be noted that much of the Dc in Soviet Russia is bordered on its south side by dry climate, so that drought-making controls are always close at hand.

Asia.   Asia's leeward location on the largest continent and its remarkably well-developed monsoon circulation provide the necessary controls to produce unusually severe seasonal temperatures, large annual ranges, and a modest precipitation strongly concentrated in summer. Winters are very dry. Insular Japan and peninsular Korea do not fit this description so well, although their temperatures are still continental. For any selected latitude, Japan's seasonal temperatures closely match those of the eastern United States, yet they are less extreme than those in mainland Asia. Also, rainfall is much heavier in Japan and is less summer-concentrated. Most of mainland China's temperate continental climate on lowlands is so deficient in annual rainfall that a large part of the region is marginal between humid and dry climates. Dry climate may actually reach down to tidewater on the great Huang Plain.

SELECTED REFERENCES FOR FURTHER STUDY OF TOPICS IN CHAPTER TEN

Bailey, Harry P., "Toward a Unified Concept of the Temperate Climate," *Geog. Rev.*, Vol. 54, pp. 516–545, 1964.

Barry, R. G., "Aspects of the Synoptic Climatology of Central South England," *Meteorol. Mag.*, Vol. 92, pp. 300–308, 1963.

Borchert, John, "The Climate of the Central North American Grassland," *Ann. Assoc. Amer. Geographers*, Vol. 40, pp. 1–39, 1950.

Brooks, C. E. P., *The English Climate*, English Universities Press, London, 1954.

Brunnschweiler, Dieter H., "The Geographic Distribution of Air Masses in North America," *Vierteljahrschr. Naturforch. Ges.*, Zürich, Vol. 97, pp. 42–49, 1952.

Bryson, Reid A., and James F. Lahey, *The March of the Seasons*, Department of Meteorology, The University of Wisconsin, Madison, Wis., 1958.

Gregory, Stanley, "Climatic Classification and Climatic Change," *Erdkunde*, Vol. 8, pp. 246–252, 1954.

Horn, Lyle H., Reid A. Bryson, and William P. Lowry, *An Objective Precipitation Climatology of the United States*, Sci. Repts. 6, The University of Wisconsin, Department of Meteorology, Madison, Wis., 1957.

Kendall, Henry M., "Notes on Climatic Boundaries in the Eastern United States," *Geog. Rev.*, Vol. 25, no. 1, pp. 117–124, 1935.

Kimble, George H. T., *Our American Weather*, McGraw-Hill Book Company, New York, 1955.

Lamb, H. H., "British Weather around the Year," *Weather*, Vol. 8, pp. 131–136, 176–182, 1953.

Manley, Gordon, "The Effective Rate of Altitudinal Temperature Change in Temperate Atlantic Climates," *Geog. Rev.*, Vol. 34, pp. 408–417, 1945.

———, *Climate and the British Scene*, Collins, London, 1952.

Rex, Daniel F., "The Effect of Atlantic Blocking Action upon European Climate," *Tellus*, Vol. 3, pp. 100–111, 1951.

Schroeder, Mark J., "Maritime Air Invasion of the Pacific Coast: A Problem Analysis." *Bull. Amer. Meteorol. Soc.*, Vol. 48, pp. 802–808, 1967.

Trewartha, Glenn T., *The Earth's Problem Climates*, The University of Wisconsin Press, Madison, Wis., 1961, pp. 203–222, 267–306.

Villmow, Jack Richard, "Daily Weather Maps as Illustrations of Weather Types," *Ohio J. Sci.*, Vol. 58, no. 6, pp. 335–342, 1958.

Wahl, Eberhard W., "The January Thaw in New England: An Example of a Weather Singularity," *Bull. Amer. Meteorol. Soc.*, Vol. 33, no. 9, pp. 380–386, 1952.

Wallén, C. C., "Climate," in Axel Somme (ed.), *The Geography of Norden*, J.W. Cappelens Forlag, Oslo, 1960, pp. 41–53.

CHAPTER ELEVEN

BOREAL (E), POLAR (F), AND HIGHLAND CLIMATES (H)

This chapter treats three different climates whose common feature is the prevalence of cold. In all three, cold is the main obstacle to large-scale settlement and agricultural development. In them all, especially in boreal and polar climates but in many high-altitude regions as well, human population is exceedingly sparse or even completely absent in large areas.

BOREAL CLIMATE (*E*)
(Subarctic)

Type location. Boreal climate has many features in common with temperate continental, except that it is characterized by greater severity because of its location farther poleward. Still, the differences are of degree rather than kind, for both are continental in character. Boreal represents the archetype of continentality, for no other climate has such large annual ranges of temperature. Indeed, climatic rigor is so great that cereal crops cannot be grown except in a few favored locations.

Boreal, or subarctic, climate is found only in the higher middle latitudes (50 or 55° to 65 or 70°) on the two great Northern Hemisphere continents. On its poleward side it ordinarily grades into tundra, one of the polar climates (compare Fig. 7.2 with the climate map inside the front cover). This northern boundary is approximately the isotherm of 50° for the warmest month (usually July), which coincides fairly well with the poleward limit of tree growth. In the tundra beyond the July isotherm of 50°, lowly vegetation forms, such as mosses, lichens, and bushes, predominate.

On its southern margins boreal climate usually adjoins the cool-summer phase of temperate continental climate, where the boundary is the isoline of 4 months with an average temperature of 50° or above. In some parts, middle-latitude dry climate lies adjacent to boreal on the south. The Russians have given the name *taiga* to the subarctic lands of Eurasia with their extensive coniferous forests, and this term has come to be applied not only to the comparable region in North America, but also to the type of climate that both of them possess. Thus boreal and taiga are to a large degree regionally synonymous.

The Eurasian subarctic area extends from Sweden and Finland in Europe across the whole of the continent to the coast of Siberia. It widens toward the Pacific, or leeward, side as continentality increases. In North America the boreal region stretches from Alaska on the Pacific across Canada to Labrador and Newfoundland on the Atlantic. The boreal regions of both North America and Eurasia are portions of the general high-latitude source regions for polar air masses which so greatly influence weather conditions farther south.

Temperature

Summer. Long, bitterly cold winters, very short and cool summers, and brief springs and autumns are characteristic (Figs. 11.1 and 11.2). Since the isotherm of 50° for the warmest month has been accepted as the poleward boundary of this type of climate, at least 1 month must have an average temperature of 50° or above. Yakutsk, Siberia, which

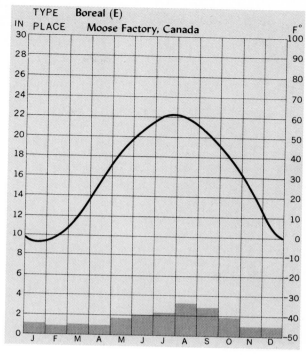

Figure 11.1 Cool summers, severe winters, large annual ranges of temperature, and modest precipitation, usually concentrated in summer, are characteristic of boreal (subarctic) climate. A marginal station.

at nearly 62°N represents the extreme in subarctic climates, has an average temperature in July, the warmest month, as high as 66°. This is 2 or 3° higher than the same month at London or Berlin and 9° higher than July at San Francisco, a subtropical station. Midsummer daily maxima of 80° are common at Yakutsk, and the thermometer occasionally reaches 90°. The absolute maximum is 102°. But at the same station, June and August have mean temperatures of only 59°, so summer is distinctly on the cool side.

Summer's briefness is even more striking than its coolness, however. At Yakutsk, there are only 3 months in which the mean temperatures exceed 50°, for both May and September have averages in the low forties. Another representative boreal station, Fort Vermilion, at 58°27'N in Canada, has a cooler July than Yakutsk's, with an average tem-

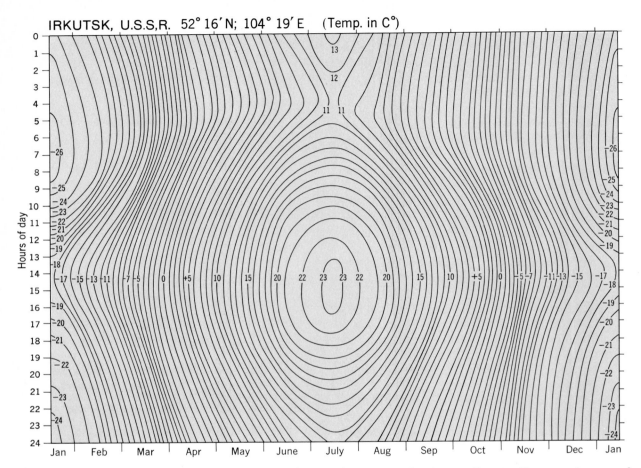

Figure 11.2 Thermoisopleths for a boreal station deep in the interior of subarctic Eurasia. The annual range of temperature is remarkably large; the diurnal range is much smaller. The pattern is that of a strongly continental climate. (*After Troll.*)

perature of only 60°. June and August average 55 and 59°, respectively. The mean of the daily maxima in July is 74°, and of the minima, 46°. Temperatures over 90° have been recorded at Fort Vermilion in both June and July. If they are sunny, subarctic summer days may be pleasantly warm and occasionally even hot (Fig. 11.3).

Another striking temperature feature of boreal climate is the small daily range in summer (and winter as well) compared with the annual range. This just reverses the situation in the humid tropics.

At Irkutsk in the U.S.S.R., where the annual range is 70°, the daily range in midsummer is only 21° (average July maximum is 73°, minimum is 52°). Such a small range is not surprising, since the hours of darkness are so few (Fig. 11.2).

Long summer days. Thus the unusually long summer days in these higher latitudes compensate somewhat for the brief and cool summers. Although the intensity of sunlight is not so great, the large number of hours during which the sun shines is an offsetting factor. Moreover, the short nights do not

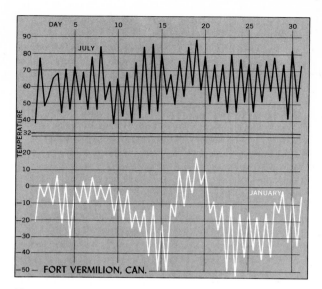

Figure 11.3 A boreal station in Canada. Note the unusually strong nonperiodic air-mass control of temperature changes in winter. Summer shows somewhat greater diurnal regularity.

permit a long period of cooling. At latitude 55°N, June days average 17.3 hr of possible sunshine; at latitude 60°N, 18.8 hr; and at latitude 65°N, 22.1 hr. Moreover, since there is still twilight when the sun is as much as 18° below the horizon, it is evident that in summer the hours of genuine darkness are few indeed. At the time of the summer solstice in the more northerly portions of the subarctic lands, one can read a paper outdoors even at midnight. Because of the long daylight period at this solstice, average daily possible receipts of solar radiation at latitude 60°N are almost equal to those at the equator.

Growing season. Unfortunately, the boreal lands have very short periods that are entirely without freeze. The growing season in the Mackenzie Valley of Canada varies from about 50 to 75 days, and many stations must expect freezing temperatures in July and August in at least half the years. A shift of wind to the north at any time brings with it the chill of the ice-laden Arctic. Thus although it is the occasional midwinter freeze which is dan-

gerous in subtropical climates, it is the midsummer freeze which damages crops in this boreal type. The coolness, shortness, and precariousness of the growing season in subarctic lands are the most serious handicaps to their agricultural development. Indeed, these are the chief disadvantages that have retarded permanent settlement. At present much of the subarctic either is totally unoccupied, or has a meager sprinkling of frontier farmers and people engaged in exploiting mineral, forest, and wild-animal resources.

Winter. Winter follows summer with only a fleeting autumn season. Frosts may arrive in late August; ice begins to form on pools in September. By the middle of October, navigation of small craft has become difficult on the subarctic lakes of Canada. At Verkhoyansk, Siberia, the mean temperature drops 39° from October to November. Subarctic Siberia holds the world records for minimum temperatures at low elevations, for they are even lower than those of polar climates. Oimekon in the northeastern part has an average January temperature of −60°, and −90° has been recorded. This same station has an average January minimum of −85° and an average maximum of −26°. In July, the warmest month, the average minimum is only 24°, which indicates that there are some freezing temperatures; the average maximum is 90.5°. This, of course, represents an extreme case. At Yakutsk, however, where July has an average temperature of 66°, the January mean drops to approximately −46°, producing an annual range of 112°. For 7 months at Yakutsk the average temperatures are below freezing, and during 5 months they are below zero. No other type of climate can show such contrasts between summer and winter temperatures. Still, notice that it is the excessively cold winters, rather than any temperature peculiarity of the summers, which cause these much greater mean annual ranges of temperature in the taiga, as compared with those in temperate continental climates. If annual temperature curves for stations in continental and boreal climates are superimposed, the July portions are not markedly different; it is the bottoms,

Climatic data for representative boreal stations (E)

	J	F	M	A	M	J	J	A	S	O	N	D	Yr	Range
Fort Vermilion, Alberta, Canada (58°27′N)														
Temp., °F	−14	−6	8	30	47	55	60	57	46	32	10	−4	26.7	74.3
Precip., in.	0.6	0.3	0.5	0.7	1.0	1.9	2.1	2.1	1.4	0.7	0.5	0.4	12.3	
Moose Factory, Ontario, Canada (51°16′N) Marginal														
Temp., °F	−4	−2	10	28	42	54	61	59	51	39	22	5	30.4	65.6
Precip., in.	1.3	0.9	1.1	1.0	1.8	2.2	2.4	3.3	2.9	1.8	1.1	1.2	21.0	
Yakutsk, Siberia, U.S.S.R. (62°13′N)														
Temp., °F	−46	−35	−10	16	41	59	66	60	42	16	−21	−41	12	112
Precip., in.	0.9	0.2	0.4	0.6	1.1	2.1	1.7	2.6	1.2	1.4	0.6	0.9	13.7	
Archangel, U.S.S.R. (54°25′N)														
Temp., °F	8	9	18	30	41	53	60	56	46	34	22	12	33	52
Precip., in.	0.9	0.7	0.8	0.7	1.2	1.8	2.4	2.4	2.2	1.6	1.2	0.9	16.8	

or winter parts, of the curves that are much farther apart because of the colder boreal winters. Nevertheless, the smaller differences in summer temperatures between temperate continental and boreal climates are much more significant as far as agricultural land use is concerned. A July average of 60° as compared with 75° has far greater economic consequences than January averages that differ two or three times as much—provided, of course, the January averages are below freezing anyway.

Concerning the boreal Siberian winter, the early German climatologist Julius Hann wrote:

It is not possible to describe the terrible cold one has to endure; one has to experience it to appreciate it. The quicksilver freezes solid and can be cut and hammered like lead; iron becomes brittle, and the hatchet breaks like glass; wood, depending upon the degree of moisture in it, becomes harder than iron and withstands the ax so that only completely dry wood can be split. Every step in the dry snow can be heard for long distances; the bursting of the ice cover and the frozen ground sound like the cannonading of distant batteries.

The winter temperatures of northeastern Siberia, though bitter, are not impossible to bear, for air movement is usually slight, skies are clear, and, since absolute humidity is very low, the air is physiologically dry. For example, when −20° air with a relative humidity of 80 percent comes in contact with the human body and is heated to 60°, the relative humidity falls to nearly 2 percent.

Boreal winters in North America are not quite so severe as those in Siberia. This is partly because Asia is a broader land mass. In addition, however, the mountains of eastern Siberia tend to dam up the outflow of cold winter air toward the Pacific. This contributes to the excessive accumulation of cP air over Asia, an accumulation which in turn produces the great continental anticyclone. No such blocking effect is possible in less mountainous eastern Canada, so that such representative stations as Moose Factory, Dawson, and Fort Good Hope show average January temperatures of −4, −22, and −32°, respectively. Annual ranges at these stations are 66, 82, and 93°. At Dawson in the Yukon at 64°3′N, the thermometer on an average January night falls to approximately −29° and rises to nearly −16° during the warmest hours of the day (Fig. 11.3). During the 3 winter months at this station, the average of the daily maxima does not reach zero. The lowest temperature ever recorded is −64°.

The extreme and long-continued cold of boreal winters causes large parts of these regions to be permanently frozen down to great depths (Fig. 11.4). Over extensive areas of the subarctic lands only the upper few feet thaw out during the short summers.

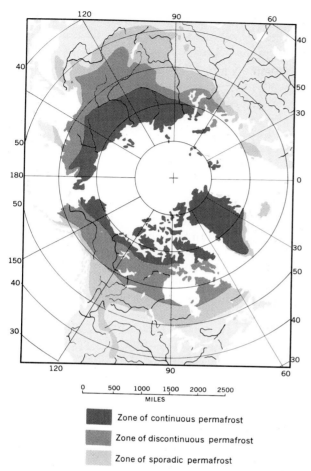

0   500   1000   1500   2000   2500
MILES

▓ Zone of continuous permafrost

▒ Zone of discontinuous permafrost

░ Zone of sporadic permafrost

Figure 11.4 Distribution of permafrost in the Northern Hemisphere. Continuous permafrost characterizes large parts of the Eurasian subarctic; it is discontinuous in the North American sector. (*After Robert Black.*)

Just as long days are characteristic of boreal summers, so long nights are a feature of the winters. For example, on December 21 all places on the 60°N parallel can receive a maximum of only 5.7 hr of sunshine, while at latitude 65°N the maximum is only 3.3 hr. These long daily periods of darkness are not only dismal and wearisome, but they are in considerable measure responsible for the low winter temperatures.

As in midsummer, so in midwinter, the daily range of temperature is small. At Irkutsk, where the

annual range is 70°, the January diurnal range is only 15°. This reflects the long daily period of darkness and very short day, with a weak sun positioned not far above the horizon.

Spring, like autumn, is a short and inconspicuous season. At Yakutsk there is a difference of 25° between the mean temperatures of April and May, and 18° between May and June. The average April temperature at Yakutsk is like that of Madison, Wisconsin, in January, while May is only 4 to 5° lower than April at Madison.

Precipitation

Annual amount.   Annual precipitation in boreal climates is usually meager (see Fig. 11.1 and the rainfall map inside the back cover). Over much of the Siberian taiga it is no more than 15 in., while most of subarctic Canada receives less than 20, and parts of it less than 15, in. In both Eurasia and North America the oceanic margins are the principal areas where rainfall exceeds 20 in. In most middle-latitude climates these small amounts would cause the region to be classed as semiarid, but where such low temperatures and therefore low evaporation rates prevail, and where the ground is frozen so much of the year, the precipitation is sufficient for forest growth. The relatively modest annual precipitation is related to the prevailing low temperatures and correspondingly low specific humidity, a strong continental winter anticyclone with settling air and diverging winds, and the great breadth of the land masses in these subarctic latitudes.

Seasonal distribution.   Precipitation is concentrated in the warmer months, a feature typical of strong continentality (Fig. 11.1). Thus although the total amounts are small, they are adequate in most years for whatever meager cropping is carried out. At Yakutsk, where the total annual rainfall is 13.7 in., August is the wettest month with 2.6 in., and February the driest with 0.2 in. At Dawson in the Yukon the total is 12.5 in., with 1.5 in July and 0.8 in January (0.7 in. in February and 0.5 in March). Winter—with low temperatures, low specific humidity, strong anticyclone with settling air, and seaward-moving winds—presents a situation that is antagonistic to precipitation. Over east central Siberia winters are especially dry; the 3 winter months have only 10 percent of the annual precipitation, while the 3 summer months have 58 percent. This is the region of most intense cold and strongest anticyclonic control.

Because of the prevalence of anticyclones, winters over the more interior and continental portions of boreal lands are likely to have much bright, clear weather. East central Siberia is especially cloudless. Summers there are more cloudy than winters, the average being less than $^3/_{10}$ in winter but almost $^7/_{10}$ in August. In its prevailing clear skies northeastern Siberia resembles Italy; in winter it has almost as little cloud as some deserts. Hann writes, "The air in these high latitudes is remarkably clear, the color of the sky is a blue-violet, like the dress of the Sistine Madonna; sounds can be heard for long distances, so that one can hear the bark of a dog 20 km away." The meager cloud cover during long winter nights is an additional factor contributing to low temperatures.

Over eastern interior Canada in subarctic latitudes, cloudiness is not far from $^5/_{10}$ in both winter and summer. Winter days often are cloudy even when there is no precipitation, for considerable low stratus cloud and ground fog occur during that season.

Winter precipitation. Over lowlands the winter precipitation, practically all of it in the form of snow, is frontal in origin. The prevalence of the strong continental anticyclone tends to suppress cyclogenesis and to repel moving cyclones from outside, so that snow-bearing storms are usually forced to follow the southern margins of boreal regions rather than traveling directly across their centers. Over the center of North America, for example, the principal winter storm tracks loop well southward.

The fact that principal winter cyclonic tracks tend to avoid the centers of boreal regions helps to explain the meager winter precipitation. But the few fronts that do cross these areas commonly yield enough relatively dry, hard snow to form a perma-

nent snow cover lasting 5 to 7 months. Because of the shelter provided by the forest, little melting or evaporation occurs, so that winter snows accumulate to a depth of 2 to 3 ft in the taiga. The forest protection also retards melting of the snow cover in spring. In parts of Siberia, however, winter precipitation is so meager that sleighing is sometimes difficult. Strong winter winds blowing the snow and thus producing blizzard-like weather are not so characteristic of the forested taiga as they are of the open prairies to the south and the treeless tundra to the north.

Summer precipitation. Summer, the season of maximum surface heating, steepest vertical temperature gradients, and highest specific humidity, provides conditions that are relatively more favorable for rainfall than those of winter. Summer rain, like winter snow, is largely frontal in origin. Thunderstorms are not common; the total number in the Mackenzie Valley of Canada, for example, is only about 5 to 10 a year. Fort Vermilion in the Mackenzie Valley has an average of 5.3 rain-days in June, 9.1 in July, and 7.5 in August. Comparable data for Dawson in the Yukon are 11.7, 10.3, and 10.9.

Seasonal weather

The nonperiodic weather element is generally less marked in boreal lands than in the temperate continental climates farther south, where the alternation of well-developed cyclones and anticyclones is more frequent. Still, weather in the taiga varies greatly between different parts. As a rule, it is in regions closest to the sea that cyclonic control is strongest.

Winter in the taiga has much anticyclonic weather with settling air and light-to-moderate winds. The greater the continentality and the stronger the winter anticyclone, the clearer, drier, and colder the weather. Occasional cyclones bring gray skies with hard, dry, blowing snow. Cyclones are often more numerous in the short transition seasons than in the dead of winter, for in winter the subarctic lands are included within the source regions of those great masses of cold anticyclonic air which periodically surge equatorward, affecting even the lower middle latitudes. *Summer* weather is also rather settled, although cyclones make themselves felt in occasional rains and in the advection of warm and cold air masses. A strong invasion of *cP* air from higher latitudes can cause freezing temperatures even in midsummer. Hot waves are rare but not unknown.

Regional characteristics

North America.   As far as weather is concerned, the American boreal region may be divided into a somewhat quieter and drier west and an appreciably more stormy east. Of the two main routes of cyclones, one from the Pacific and the other from the interior of the continent farther south, only the first group affects the region west of the 100th or 105th meridian. Moreover, the cyclones from the Pacific which have crossed the western mountains are weak and relatively inefficient as rain producers.

The somewhat stormier and wetter east is subject to both the regenerated Pacific cyclones and those from the interior. Moreover, these disturbances are operating in air that has a higher humidity content. As a result, winter weather in eastern subarctic Canada is more disturbed, with gales, heavier snowfall, and high wind chill. All the eastern half has over 10 in. of annual precipitation, and the part east of Hudson Bay has 20 to 40 in.

A common feature of boreal localities situated along the windward side of open water is a fall maximum of precipitation in the cold months. This is the case along the western or windward side of the Labrador Peninsula facing Hudson Bay. There subzero arctic air from the northwest, accompanying polar-outbreak anticyclones, is warmed and humidified over the open water of the bay in fall and early winter. Instability snow showers from dense cumulus clouds drop heavy snows on the windward shores. These effects disappear later in the winter, when the areas of open water shrink greatly or even disappear.

Eurasia.   The Eurasian boreal climate somewhat reverses the weather and rainfall patterns of the North American sector. It is the western parts of Eurasia that are more humid and less severe in temperature, for since there are no high mountains the influence of Atlantic air masses is significant. It is the center and east that are most severe as well as drier, particularly in winter. The Pacific coast, however, is distinguished by its cloudy, cool summer, with much fog and drizzle. Winters there are less cold than west of the mountains. A noteworthy feature of the Eurasian boreal sector, as can be seen on the climate map inside the front cover, is its increasing latitudinal spread toward the east, a spread which parallels a similar increase in continentality.

POLAR CLIMATES (*F*)

Just as the tropics are characterized by the absence of a cool season, so the polar regions lack a period of warmth. Monotonous heat prevails in the low latitudes; enduring cold typifies the high. The polar areas cannot be made to appear warm by noting that there have been occasional days with temperatures over 80° beyond the Arctic Circle. The rare warm day does not determine the general climatic character of a region.

Phenomena of light and darkness.   For 6 months at the poles, the sun is out of sight entirely. For another 6 it is constantly above the horizon, although it is never very high, so that solar radiation is weak. At the Arctic and Antarctic Circles, which lie near the equatorward margins of polar climates, the daily period of sunlight varies from 24 hr at the time of the summer solstice to zero hr at the winter solstice. At points between the poles and the 66½° parallels, the lengths of the periods of sunlight, and absence of sunlight, are intermediate between the two extremes.

These peculiarities of daylight and darkness introduce new conceptions of certain climatic phenomena. Diurnal range is bound to be small in a region that may have continuous day or continuous night, so that this measure loses some of its significance. The annual march of temperature also has a different meaning in the altered context. At the North Pole, for instance, there is no incoming direct solar radiation for a period of 6 months. Loss of heat by terrestrial radiation goes on continuously, however, so that a minimum temperature is usually reached a month or so before the spring equinox.

Type locations and boundaries. Polar climates are mostly confined to the high latitudes of the earth. Somewhat similar conditions can be found at high altitudes in a great variety of latitudes, but usually these regions are very isolated and fragmentary, and in this book they are included within the group designated as highland climates (see Fig. 7.2 and the climate map).

In the classification of climates used here, the boundary of polar climates is the same as the one chosen by Köppen: the mean temperature of 50° for the warmest month. Thus in polar climates the average temperature of all months is below 50°. The 50° warmest-month isotherm is considered to correspond fairly well with the poleward limit of tree growth in all but a few highly marine locations. Scientists have not yet determined whether this temperature value is the active climatic control limiting tree growth, or simply a coincidence. It is significant that while a cool-month isotherm is often used to bound the tropics, a warm-month isotherm serves the same purpose for polar climates. This suggests that while a period of coolness is of great importance in influencing the type of plant life in the low latitudes, a period of warmth is likely to be the crucial element in high latitudes.

In the Northern Hemisphere, the July isotherm of 50° swings well poleward of the Arctic Circle over most of Asia and Alaska, coincides reasonably well with it over lowland Europe, and lies to the south of it over much of eastern North America and Greenland. Large inland bodies of water, the cool Labrador Current, and the Greenland icecap cause summers in eastern North America and Greenland to be distinctly cooler than in Asia. In the Southern Hemisphere, the only extensive land area with polar climates is ice-covered Antarctica. North of the equator, it is the Arctic Sea borderlands of Eurasia and North America, extensive island groups north of both continents, and ice-covered Greenland which are classified as *F* regions. In both hemispheres the lands with polar climates are important source regions for the polar air masses which so powerfully affect the weather of middle latitudes.

The Arctic and Antarctic serve the same thermodynamic function of maintaining the earth's latitudinal heat balance. Vast quantities of heat are transported by atmospheric and oceanic circulations from low latitudes to the polar regions, where it is dissipated to space in the form of long-wave radiation. In polar regions where the water vapor content is low, as it is over Antarctica, which has only about one-tenth the atmospheric moisture that the middle latitudes do, the escape of long-wave earth radiation is very rapid. It is less rapid in the Arctic, where the air's moisture content is somewhat higher.

The Antarctic is essentially a seagirt land which takes the form of a high ice plateau. By contrast, the Arctic is almost a landlocked frozen sea, whose marginal lands are mainly lowlands in Eurasia and North America, though they also include the high ice plateau of Greenland. Naturally the different configurations of the two polar regions result in important climatic differences between them. Because its single land mass is centered near the pole and surrounded on all sides by extensive oceans with uniform temperatures, the Antarctic shows much greater uniformity and simplicity in its climate than the more complex Arctic. Pressure and wind systems are symmetrically developed about the South Pole, so that Antarctica is dominated by a strongly zonal circulation, and there is only modest change in these elements throughout the year. The Arctic has far less symmetry, and greater seasonal variation, in pressure and winds, and also a stronger meridional component in the circulation. It is characterized by more frequent sporadic outbreaks of polar air toward lower latitudes.

Mean Arctic sea-level pressure in January is dominated by four great cells: two continental highs (Siberia and northwestern North America), and two oceanic lows (Iceland and Aleutian). It is noteworthy that there is no Arctic high as such on the mean winter map. In July, a continuous ring of low pressure encircles the ice-covered polar sea. Over the pack ice, pressure is nearly flat. Like winter, summer has no Arctic high. Because of the asymmetry and eccentricity of the sea-level pressure field, polar easterlies are not well defined. Moving disturbances, both cyclones and anticyclones, are numerous at all seasons and penetrate all parts of the Arctic. Many cyclones enter from outside. Because of the great synoptic activity, mean pressure and circulation patterns are not very useful in a description of the climate; irregular weather occurrences are the rule. As a consequence there is a high degree of interdiurnal temperature variability. Along the margins of open water, such variability may be associated with advection; over the inner-core frozen area, it is largely due to changes in wind and cloud and their effects on the prevailing surface temperature inversion.

Temperature and precipitation. Polar climates have the distinction of producing the lowest *mean annual,* as well as the lowest *summer,* temperatures for any part of the earth (Figs. 11.5 and 11.6). Temperatures remain low in spite of the long days of summer, for the sun's rays are too oblique to be genuinely effective. Moreover, much of the solar energy is either reflected by the snow and ice or consumed in melting the snow cover and evaporating the meltwater, so that neither the land surface nor the air adjacent to it becomes warm. Winters are bitterly cold, rivaling the severest in boreal regions. In spite of the cool summers, winter cold is so extreme that annual ranges are large. One of the most conspicuous features of polar climates is the intense low-level temperature inversions. In the

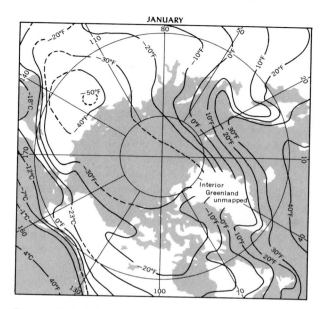

Figure 11.5 Mean daily surface temperature of the Arctic region for January. (*After Hare and Orvig.*)

Figure 11.6 Mean daily surface temperature of the Arctic region for July. (*After Hare and Orvig.*)

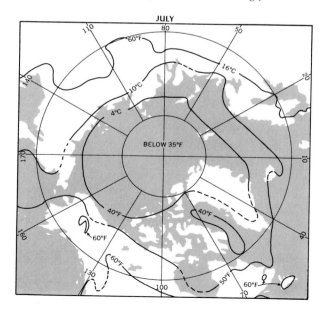

Arctic these are less common in summer than in winter.

Precipitation is meager throughout the very high latitudes: less than 10 in. over large parts of the land areas. But in spite of its meagerness, the low evaporation permits some runoff, part of it in the form of glaciers. Because of the low evaporation and the small amount of melting, great permanent snow and ice fields several thousand feet thick have accumulated on Greenland and the Antarctic continent. The scantiness of polar precipitation is not surprising in view of the low specific humidity which must accompany the low temperatures. The reservoir of water vapor is small at all times. Moreover, in these latitudes there is a general settling of the cold upper-air masses, a process which stabilizes the atmosphere and creates conditions unfavorable to condensation. Precipitation is usually heavier in the warmer months, when the air's moisture content is most abundant.

Tundra and icecap climates.   Polar climates are subdivided here into two types, with the *warmest-month* isotherm of 32°F (0°C) serving as the boundary between them. Where the average temperatures of all months are below freezing, the growth of vegetation is impossible, and a permanent snow-and-ice cover prevails. These are the *icecap climates.* Where one or more of the warm-season months has an average temperature above 32° (but not over 50°), so that the ground is free from snow for a short period and a sparse and lowly vegetation cover is possible, the climate is designated as *tundra.*

TUNDRA CLIMATE (*Ft*)

Type location.   Tundra climate is transitional between the icecap, or region of perpetual snow and ice, on the one hand, and middle-latitude climates, usually boreal, on the other (see Fig. 7.2 and the climate map). Its accepted equatorward and poleward boundaries are the warmest-month isotherms of 50 and 32°, respectively—which, as mentioned above, approximately coincide with important vegetation boundaries.

The isotherm of 50° for the warmest month seems a fairly satisfactory boundary for tundra climate in continental regions. However, in a few strongly marine areas (the Aleutian Islands, part of Iceland, and the southernmost tip of South America), the consistently cool (below 50°) but mild year-round climate permits tree growth in locations that are protected from strong winds. If it is desired to eliminate such minor areas, the definition can be expanded to include the further requirement that the coldest winter month must have a mean temperature of 20° or lower, so that the mean annual range will be over 30°.[1]

Tundra climate on land areas is almost exclusively confined to the Northern Hemisphere. In the Antarctic, open ocean occupies the latitudes where tundra would normally develop. Only some of the most northerly fringes of the Antarctic continent, the southernmost tip of South America, and certain small Antarctic islands might be included. The most extensive tundra areas are the Arctic Sea rimlands of both North America and Eurasia. Most of North America's Arctic archipelago, as well as the coastal fringe of Greenland, is likewise included. According to definition, the ice-covered Arctic Sea must also be classed as having a tundra climate, since the inner pack ice is uniformly cold, with a surface temperature of about 34°.

Temperature

Summer.   Since the land area occupied by marine tundra climate is negligible, it can be said that continentality in temperature characteristics is a distinguishing feature of almost all tundra regions. Long, bitterly cold winters and short, cool summers are the rule (Figs. 11.7 and 11.8). By the definition given

[1] James A. Shear, "The Polar Marine Climate," *Ann. Assoc. Amer. Geographers,* Vol. 54, pp. 310–317, September, 1964.

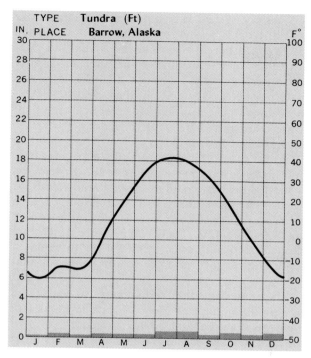

Figure 11.7 A tundra station of the severe continental type. Large annual range of temperature and meager annual precipitation are characteristic.

for tundra climate boundaries, the average temperature of the warmest month can be no lower than 32° and no higher than 50°.

Raw and chilly, the warmest months of the tundra resemble March and April in southern Wisconsin and January in the American Cotton Belt. Usually only 2 to 4 months have average temperatures above freezing, and killing frost is likely to occur at any time. Summer cool waves in middle latitudes are associated with air masses moving southward from the tundra and boreal source regions. Along coasts, where water, ice, and land are in close proximity, fogs are very common. These fogs may last for days at a time and are extraordinarily depressing.

Under the influence of unusually long summer days, the snow cover begins to disappear in May, and the lakes are usually free of their ice cover in June. Because of the permanently frozen subsoil,

called permafrost, subsurface drainage is deficient, and bog and swamp are prevalent (Fig. 11.4). Myriads of mosquitoes and blackflies make life almost unbearable for both men and animals during the summer period of wet earth.

At Ponds Inlet, Canada, a tundra station at 72°43′N, the average July temperature is 42°, with the thermometer rising to about 49° during the warmest hours of the day and sinking to an average of 35 or 36° at "night." A striking feature of tundra climate is the small diurnal temperature range even though there is a large annual range, a feature typical of continental climates (Fig. 11.9). At Sagastyr in Siberia, where the annual range is 85°, the daily range in July is only about 6°. This is because of the continuous daylight beyond the Arctic Circle in summer. No frost occurs on most July "nights" in

Figure 11.8 Annual march of temperature at a marine tundra station located on an island in the high-latitude Atlantic Ocean. A cool summer, and a remarkably mild winter considering the latitude, result in an unusually small annual range of temperature.

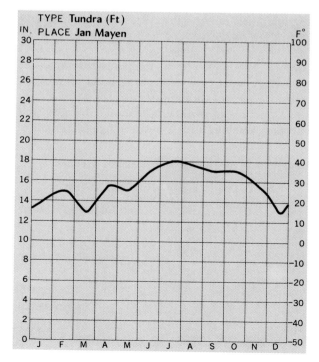

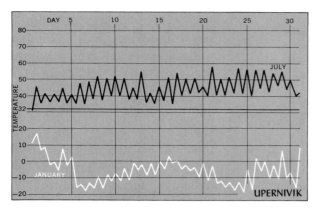

Figure 11.9 Small diurnal temperature ranges, in both winter and summer, are typical of this tundra station in Greenland. Nonperiodic temperature changes are relatively large in winter.

tundra climates, although it is not an unusual event when the thermometer slips a few degrees below freezing (Fig. 11.9). Warm days occur now and then. On at least one occasion, Ponds Inlet has recorded a temperature of 77°, and Chesterfield Inlet, another tundra station in Canada, has had a temperature of 84°. However, such high temperatures are a rare exception and not the rule.

Winter. While summer temperatures are not much different from one tundra region to another, there are greater variations in the winters. Along the Arctic coasts of Siberia, average January and February temperatures are in the neighborhood of −35 or −40°, and it is appreciably colder farther inland. At this season winds are mostly from the bitterly cold boreal region to the south, and their advections of cold air further intensify the severity of temperatures in the tundra. The tundra of northeastern Siberia, where the winter-month averages are below −40°, represents the extreme condition; winters are not quite so severe along the Arctic borderlands of North America. A coastal station in Labrador has a January mean of −8°; Ponds Inlet in Canada records an average of −28° for January, −30° for February, and even −24° for March. At Ponds Inlet 5 months—November through March —have average temperatures below zero, while 9 are below freezing. Daily temperature ranges are exceedingly small: even smaller than in summer, for in midwinter most of the tundra lies in constant darkness.

The annual march of temperature is represented by a profile whose crest and base have much weaker gradients than the flanks. Thus at the times of both constant daylight and constant night the temporal change in temperature is not large. It increases in the intermediate seasons, when there are rapidly changing proportions of light and darkness (Figs. 11.7, 11.8, and 11.10).

Precipitation

Over most of the tundra lands, precipitation probably does not exceed 10 or 12 in. (Fig. 11.1). In portions of eastern arctic Canada (particularly the Lab-

Climatic data for representative tundra stations (*Ft*)

	J	F	M	A	M	J	J	A	S	O	N	D	Yr	Range
Sagastyr, Siberia, U.S.S.R. (73° N, 124° E)														
Temp., °F	−34	−36	−30	−7	15	32	41	38	33	6	−16	−28	1	77
Precip., in.	0.1	0.1	0.0	0.0	0.2	0.4	0.3	1.4	0.4	0.1	0.1	0.2	3.3	
Upernivik, Western Greenland (73° N, 56° W)														
Temp., °F	−7	−10	−6	6	25	35	41	41	33	25	14	1	16.5	50.6
Precip., in.	0.4	0.4	0.6	0.6	0.6	0.6	0.1	1.1	1.0	1.1	1.1	0.5	9.2	
Spitsbergen Island (78° N, 14° E)														
Temp., °F	4	−2	−2	8	23	35	42	40	32	22	11	6	18	44
Precip., in.	1.4	1.3	1.1	0.9	0.5	0.4	0.6	0.9	1.0	1.2	1.0	1.5	11.8	

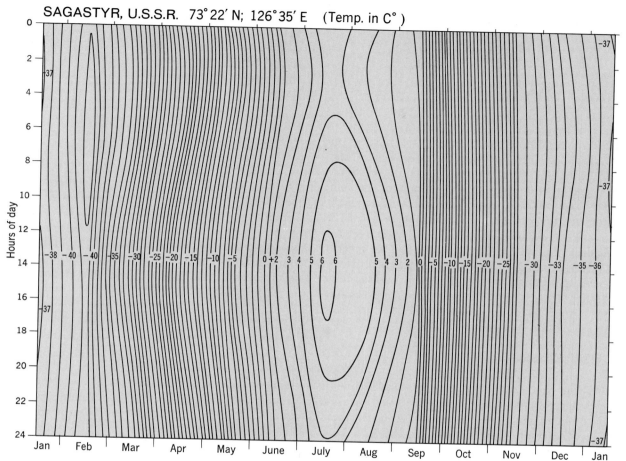

Figure 11.10 Thermoisopleths for a continental tundra station in northern Siberia. Exceedingly small diurnal ranges of temperature and large annual ranges are conspicuous features. Note the rapid change in temperature during the transition seasons. (*After Troll.*)

rador Peninsula) and a few other widely distributed areas, it is somewhat greater. Low temperatures and winter anticyclonic conditions are not conducive to much precipitation. In addition, convective turbulence induced by surface heating is weak. Summer and autumn, the warmest seasons, are the periods of maximum precipitation throughout the tundra as a whole, although in the more marine locations fall and winter may show larger totals than summer.

Precipitation is predominantly cyclonic, or frontal, in origin. In the warm season much of it falls as rain, with occasional wet snows. The meager winter snowfall is usually dry and powdery, so that it forms a very compact cover. It is only this very compact snow, a few inches of which may equal an inch of rain, that the Eskimos use in constructing their igloos. The actual amount of dry, sandlike snow that falls is not easy to measure, since it is often accompanied by strong winds which heap it up in depressions and on the lee sides of hills while they sweep all exposed surfaces bare. There are no forests, as in the taiga, to break the force of the wind and hold the snow cover. Stefansson estimates that 75 to 90 percent of the surface of Arctic lands is nearly free of snow at all seasons. As a result both of the small amount of snow and of its strong tendency to drift, sledging may be difficult.

Seasonal weather

There are conflicting reports concerning the character of *winter* weather in the tundra, undoubtedly because conditions differ from one part of the Arctic to another and because successive winters can be unlike. Certainly the variability of winters is much greater than that of summers. In some years clear, cold, anticyclonic conditions appear to dominate the weather, while in others, at least in some areas, cyclones accompanied by blizzards follow each other in fairly rapid succession. In general, however, winters in both the Eurasian and North American sectors probably have much anticyclonic weather. The dryness of the air causes rapid evapo-

ration from the human body, and explorers report the prevalence of torturing thirst. Blizzards on the treeless tundra are awesome weather phenomena, their low temperatures, gale winds, and driving snow combining to force all living creatures to seek shelter.

The transition seasons of *spring and fall* are somewhat different; autumn is a quiet dying away of nature, spring a more sudden awakening. Both are likely to be stormier than winter, however. Autumn is one of the worst seasons for severe storms, since the seas are still unfrozen, and their surface temperatures are markedly higher than those of the rapidly cooling lands.

Summer is a relatively quiet and uniformly cool season (Fig. 11.9). The wet, swampy ground makes the air damp and raw. Fogs are common, the sky is often gray and cheerless, and it is not unusual to have wet snowfalls.

ICECAP CLIMATE (*Fi*)

This least well-known among the world's climatic types is characteristically developed over the great durable continental ice plateaus of Antarctica and Greenland. Only fragmentary climatic data have been obtained from these deserts of snow and ice, where the average temperature of no month rises above freezing.

Circulation patterns

As noted earlier, polar circulation patterns are complex, and owing probably to contrasting surface features, they are not the same in the two hemispheres. Also, obviously they are not the same on and around the high ice plateaus of Antarctica and Greenland as they are at or near sea level (Fig. 11.11). Over Antarctica at stratospheric levels there exists a deep cyclonic vortex with high-velocity westerly winds. In summer this pattern weakens, permitting a greater influx of warmer air from lower latitudes. The Arctic cyclonic vortex circulation

at high levels is less persistent and is disrupted more frequently, in winter as well as in summer, so that great outbreaks of cold air pour southward over Canada and Eurasia.

Close to the ice surface of interior Antarctica and Greenland, persistent shallow anticyclones of thermal origin prevail. Within them steep temperature inversions extend upward from the ice surface for several hundred meters. Radiating outward

Figure 11.11 Hypothetical distribution of zonal pressure and wind in the South Pole region. (*After Meinardus, Mintz, and Dean.*)

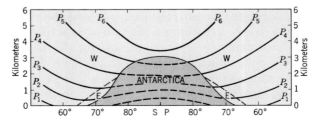

from the center of the anticyclone, and from the highest part of the convex ice plateau, a shallow layer of gravity winds drains downslope over the ice surface toward the lower marginal areas. Such downward-moving winds, called katabatic winds, dominate the circulation over the icecaps. At Eismitte in central Greenland near the crest of the icecap, the recorded wind velocity is only 10 miles an hour. But along the steep margins of the ice plateaus, winds may reach gale force in local areas. Subsiding air replaces this constant outflow, thereby probably strengthening the shallow surface anticyclone's circulation. Without the constant downslope drainage of cold air, a very cold and deep anticyclone would probably build up over the ice, with frequent surges of cold air from the interior toward lower latitudes. A steady downslope drainage, which eliminates the necessity for strong surges, may partly account for the fact that the Southern Hemisphere does not have such great polar outbreaks of cold air as the northern latitudes.

Temperature

Conditions on the icecaps all favor low temperatures. During a considerable part of the year the sun is completely below the horizon, and even during the period of continuous sunlight the solar rays are very oblique and deliver little energy per unit area. Because of the excellent reflecting properties of the snow and ice surface, probably 80 percent or more of the solar radiation reaching the icecap surface is sent immediately back into the atmosphere. A part of the remaining 20 percent is consumed in evaporating the ice and snow. On the other hand, the relatively dry air permits rapid losses of heat by earth radiation whenever the sky is clear. The result is a nearly permanent surface temperature inversion. Such an inversion prevails on over 80 percent of the days at the South Pole.

Both Antarctica and Greenland are ice plateaus with convex surfaces whose high interior parts rise to about 10,000 ft in Greenland and 13,000 ft in Antarctica. Hence any recorded air temperatures for the elevated core areas presumably would be

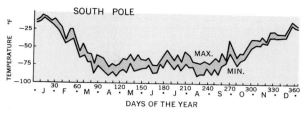

Figure 11.12 Five-day means of maximum and minimum surface temperatures, 1958, South Pole, Antarctica. (*After Sabbagh.*)

some 30° + higher if reduced to sea level. At Eismitte in high interior Greenland, the coldest month has an average temperature of about −53° and July has one of +12°, producing an annual range of 65°. The daily range in July, with constant daylight, is only 12 or 13°. In January there is no consistent daily range. At South Pole Station in interior Antarctica, the average monthly temperatures in 1958 remained below −70°F for the 6 months April to September (Fig. 11.12). The warmest month of the year averaged −10° and the coldest −75°. Certainly this is the lowest summer temperature on the earth, while the winter temperatures rival, or are somewhat lower than, those of northeastern Siberia. The annual range of 65° is continental in its magnitude. The lowest world temperature (−126.9°) and the lowest monthly mean temperature (−96.8°) were both recorded at Vostok in Antarctica.

Some, but not all, icecap stations have an annual march of temperature in which the profile based on month-to-month averages shows a broad, flattish winter minimum (Fig 11.13). At the South Pole, for example, monthly averages over the period April to September remain below −70°, with only minor fluctuations. Nonperiodic short-time temperature fluctuations over Antarctica are not uncommon, but their magnitude and frequency vary from station to station. Such fluctuations are associated with invasions of the inner polar areas by atmospheric disturbances in the form of frontal or nonfrontal low-pressure systems. Much of the temperature changes in these disturbances are related to their winds, which destroy the temperature inversion,

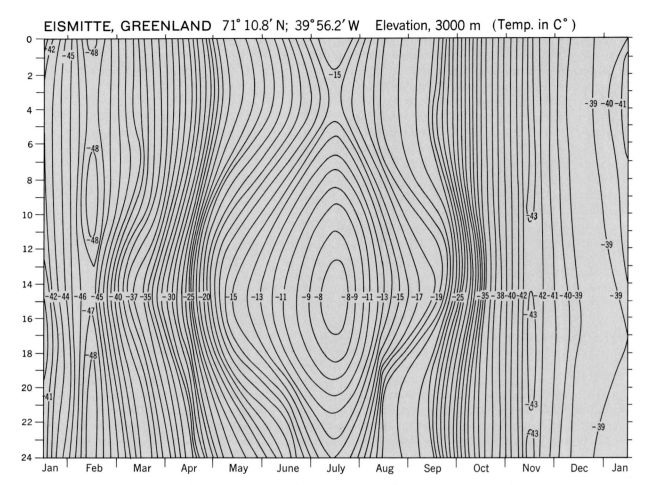

EISMITTE, GREENLAND 71° 10.8' N; 39° 56.2' W   Elevation, 3000 m   (Temp. in C°)

Figure 11.13 Thermoisopleths for a station on the Greenland icecap. Small diurnal ranges of temperature and large annual ranges are conspicuous features. So also are the rapid changes in temperature during the transition seasons of spring and fall. (*After Troll.*)

Climatic data for icecap stations (*Fi*)

	J	F	M	A	M	J	J	A	S	O	N	D	Yr	Range
Eismitte (Wegner), interior Greenland (70°54′N, 40°42′W, 9,941 ft)														
Temp., °F	−42	−53	−40	−24	−4	+4	+12	+1	−8	−32	−46	−37	−22	65
Precip., in.	No data													
Little America, coastal Antarctica (79°S, 164°W)														
Temp., °F	22	(9)	(−7)	−24	−27	−29	−34	−34	−29	−14	9	24	−11.3	58
Precip., in.	No data													

and their cloud decks, whose long-wave radiation back to earth may produce increases of 20° or more. According to Sabbagh,[2] at South Pole Station on June 28 and 29, 1958, the passage of a depression with a warm and a cold front first raised surface temperatures by 43° and destroyed the surface inversion, and then caused the temperature to drop by 23°.

Precipitation

If little information is available concerning temperatures of icecap climates, still less is known about their precipitation. There is no doubt that it is meager, or that nearly all of it falls as snow, mostly in the form of dry, hard, sandlike particles which are readily driven before the wind. The origin of the precipitation over icecaps is not well understood, however. Although there are almost no streams in such regions, some water is lost by evaporation from the ice surface, as well as by glaciers moving out to the sea in the form of icebergs. There must be enough precipitation to more than offset these losses. Yet conditions on the icecaps are generally opposed to precipitation: abnormally low temperatures, low specific humidity, extreme stability of the air as indicated by the persistent strong surface temperature inversions, and the infrequent occurrence of upslope winds—all act to inhibit snowfall.

No doubt a portion of the inland precipitation is produced by the cyclonic storms that pass along the margins of the ice plateaus. Much more originates in the smaller number of cyclones that pass in over the icecaps (Fig. 11.14). A little precipitation may result from condensation in the form of fine ice particles, or of hoarfrost, within the generally subsiding air as it reaches the intensely cold ice surface. The rate of frost deposit is exceedingly slow, no doubt; but since it is fairly continuous, the total amount may be considerable.

At Eismitte in interior Greenland (9,941 ft) the

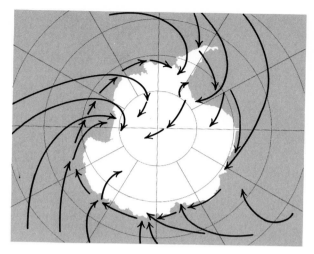

Figure 11.14 Winter storm tracks for the Antarctic region. They appear to swirl around the margins of the continent, moving inland mainly where continental elevation is lowest. (*After Rubin.*)

recorded average cloudiness in one study was between $5/10$ and $6/10$, with a maximum in the season of least cold. The cloud forms were predominantly stratus, cirrus, and cirrostratus. There were 204 days with precipitation, the larger part of which was cyclonic in origin. Considering the prevailing atmospheric conditions over icecaps, this number of days with precipitation seems unusually high. From a study of the snow layers marking the yearly accumulations of snow at Eismitte, it was established that the average annual snowfall over a period of 21 years was 12 in.—equivalent to 3 to 4 in. of water. Seasonal concentration was not conspicuous[3] (Fig. 11.15). It appears, then, that cyclones do cross the Greenland icecap, and they probably account for a large part of the snowfall. Certainly the weather element is a prominent feature of the climate, so that the concept of a permanent anticyclone is scarcely tenable.

[2] Michael E. Sabbagh, "A Preliminary Regional Dynamic Climatology of the Antarctic Continent," *Erdkunde,* Vol. 16, p. 99, 1962. See also Morton J. Rubin, "The Antarctic and the Weather," *Scientific American,* Vol. 207, pp. 84–94, September, 1962.

[3] Fritz Loewe, "Klima des Grönlandische Inlandeises," in W. Köppen and R. Geiger, *Handbuch der Klimatologie,* Vol. 2, part K, Gebrüder Borntraeger, Berlin, 1935. Also by the same author: "The Greenland Ice Cap as Seen by a Meteorologist," *Quart. J. Roy. Meteorol. Soc.,* Vol. 62, pp. 359–377, July, 1936.

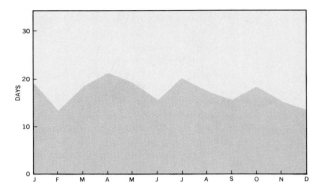

Figure 11.15 Days with precipitation at Eismitte on the Greenland icecap. (*After Loewe.*)

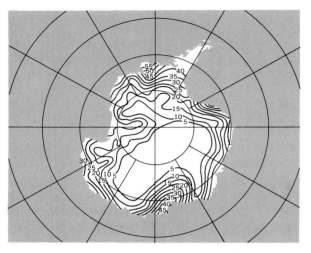

Figure 11.16 Annual accumulation of precipitation on Antarctica, expressed as water equivalent in centimeters. The mean value for the whole continent is about 14.5 cm. Most of the accumulation is along the marginal parts. (*After Rubin.*)

At South Pole Station in high interior Antarctica, there were only 17 days during 1958 with *measurable* precipitation, but 248 when some snow fell. A great majority of these had only a trace of snow. The total snowfall amounted to only 0.48 in. of water equivalent. There were 100 overcast days and 165 that were clear. It is believed that the whole Antarctic Continent receives an average of less than 4 in. of water equivalent in annual precipitation (Fig. 11.16). The least snow falls in the higher, colder, anticyclonic interior; the most occurs along the continent's seaward margins.

HIGHLAND CLIMATES (*H*)

Altitude and exposure as climatic controls. Next to the distribution of land and water, elevation above sea level is probably the most important cause of differences in climate in similar latitudes. The climatic effects of elevated land masses such as mountains and plateaus are expressed through the two factors altitude and exposure.

There is no such thing as a *highland type of climate,* however; not in the same sense that there is a tropical wet type or a boreal type. Almost endless varieties of local climates exist within a mountain mass: the atmospheric conditions change markedly with altitude and exposure, and of course with latitude as well. An enclosed valley or upland is climatically very different from an exposed peak.

Weather conditions on windward slopes contrast with those on leeward sides, while flanks inclined toward the sun are unlike those lying more in the shadow. Each of these in turn is different at various altitudes and latitudes. Representative temperature and rainfall curves for highland climates as a class do not exist, and only the most flexible generalizations are broadly applicable. One of the most valuable of these generalizations is that highland climates to a considerable degree are low-temperature variants of lowland climates in similar latitudes.

Methods of classification. It is impossible to indicate the world's almost limitless variety of highland climates on a small-scale map like the one inside the front cover. Not only are there very few weather stations in such regions, but the profusion of local climates makes it impossible for the weather data of most stations to be representative of anything but a very restricted area.

Because highland climates have a third, or vertical, dimension, it is difficult to fit this group into a classification intended for horizontal arrangement. It is true, of course, that temperature decreases both poleward and upward. As a consequence, the temperature of, say, latitude 40° should be duplicated at some altitude closer to the equator. Thus Quito, Ecuador, almost on the equator at 9,350 ft, has a mean annual temperature very similar to that of St. Louis, Missouri, in the middle latitudes. Actually, however, the two stations have totally different climates, for St. Louis has a range of 47° between its warmest and coldest months, while the range at Quito is less than 1°. Yet in some climatic classifications the two stations are included within the same type.

A unique feature of highland climates is their large difference in temperature between day and night—a difference caused by the intense solar radiation during the day and the equally rapid loss of heat by earth radiation at night. Throughout much of the tropics, the cool climates of the high altitudes show far greater temperature changes between day and night than they do between the extreme seasons. One is forced to conclude that a system of climatic classification designed for lowland climates is not especially applicable to highlands.

There appears to be no simple satisfactory way of representing the variety of highland climates on a small-scale map. Up to an altitude of about 4,000 ft, the climatic peculiarities due to altitude are not so noticeable. But above about 6,000 ft they become very prominent. So in the climate map inside the front cover, a compromise based on this fact has been worked out. Where there are extensive highlands of only moderate elevation above the surrounding lowlands, these have often been included within the general climatic type characteristic of the lowlands, even though they represent a modified form of the lowland climate. Highlands of greater elevation have been included within the general group of *H* climates. It is the distinctive characteristics of this group that are described here.

Atmospheric pressure in mountains

At low elevations the minor changes in air pressure from day to day, or from season to season, are not directly perceptible to the human body. Consequently it is as a climatic control rather than a climatic element that pressure has been discussed in the preceding chapters. In highland climates, however, the rapid decrease in the atmosphere's weight with increasing elevation, and the low pressures found in high mountains and plateaus, cause pressure as an element to be somewhat more important. At an elevation of about 17,500 ft above sea level, pressure is reduced to approximately one-half its sea-level value. The highest human habitations are found below this level, although there are said to be settlements in Tibet and the Bolivian Andes located fairly near it. Most people feel the physiological effects of decreased pressure (faintness, headache, nosebleed, sleeplessness, nausea, weakness) at altitudes above about 10,000 ft. Usually "mountain sickness" is a temporary inconvenience which passes away after a week or so of residence at high altitudes.

Solar radiation

Intensity of sunlight increases aloft in the cleaner, drier, thinner air of mountains. This is because dust, water vapor, cloud particles, and other principal reflecting, scattering, and absorbing elements of solar radiation in the atmosphere are much more abundant at lower elevations. On a clear day, probably three-fourths of the solar radiation received at the outer atmosphere penetrates to 6,000 ft, but only one-half to sea level. The great relative intensity of the sun's rays is usually noticeable by people arriving at a high-altitude location. On Mont Blanc (15,781 ft) the intensity of solar radiation is approximately 26 percent greater than at Paris (200 ft). At the time of the summer solstice, the amount of solar energy received by Tibet is nearly one and a half times as great as that falling upon the neighboring lowlands in India.

Solar radiation is not only more intense in the higher altitudes, but it is also proportionally richer in the shorter wavelengths of energy at the blue end of the spectrum. People therefore burn and tan quickly in mountain sunlight.

Soil temperature.   Associated with the greater intensity of solar radiation at higher elevations is the excess of surface or soil temperatures over air tem-

peratures. Although the thin, dry mountain air is incapable of readily absorbing and retaining either solar or terrestrial energy, the land surface easily absorbs the intense solar radiation. Thus it attains a relatively high daytime temperature compared with the cool air. The effects upon plants are striking.

Air temperature

Probably the most fundamentally important of the climatic changes resulting from increased elevation is the drop in air temperature (about 3.5° per 1,000

Figure 11.17   Thermoisopleths for a highland station near the equator. Temperature changes much more during the 24-hr diurnal period than it does during the 12 months of the year. (*After Troll.*)

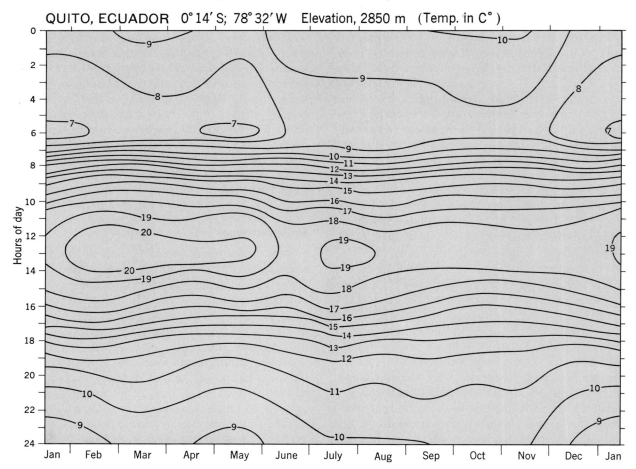

QUITO, ECUADOR   0° 14′ S; 78° 32′ W   Elevation, 2850 m   (Temp. in C°)

Climatic data for a high plateau station in the tropics

	J	F	M	A	M	J	J	A	S	O	N	D	Yr	Range
						Quito, Ecuador (9,350 ft)								
Temp., °F	54.5	55.0	54.5	54.5	54.7	55.0	54.9	54.9	55.0	54.7	54.3	54.7	54.7	0.7
Precip., in.	3.2	3.9	4.8	7.0	4.6	1.5	1.1	2.2	2.6	3.9	4.0	3.6	42.3	

ft)—this in spite of the increased intensity of solar radiation. Quito, Ecuador, on the equator at an elevation of 9,350 ft, has an average annual temperature of 55°, which is 25° lower than that of the adjacent Amazon lowlands (Fig. 11.17). But although the clear, rarified air at that elevation remains chilly, since it is incapable of absorbing and retaining much energy, the sunlight is intense. It is a climate of cool shade and hot sun. Viscount Bryce writes of the Bolivian plateau:

The keen air which this elevation gives has a fine bracing quality, yet there are disadvantages. One is never warm except when actually in the sunlight. . . . The inhabitants get accustomed to these conditions and shiver in their ponchos, but the traveler is rather wretched after sunset and feels how natural was Sun worship in such a country.

Vertical temperature gradients along mountain slopes are many times steeper than the most continental winter gradients in a horizontal direction on lowlands. In the low latitudes, a railroad trip into the mountains which takes only a few hours can transport the traveler from tropical to polar temperature zones. The steep vertical temperature gradients in highlands are particularly significant in the tropics, where people can escape the oppressive heat by going to the mountains. Partly because of their lower temperatures, tropical highlands commonly become centers of population concentration. In tropical Latin America, for instance, the capital cities of Venezuela, Colombia, Bolivia, Mexico, and five of the Central American republics are on highlands. In India, the so-called "hill stations" of the sub-Himalayas, at elevations of 6,500 to 7,500 ft, become havens for residents from the lowlands during the long hot season.

Vertical temperature zones in tropical highlands. Several altitude zones of climate, each with its characteristic wild vegetation and crops, are found in tropical highlands (Fig. 11.18). In mountainous parts of tropical Latin America, four such zones are commonly recognized: *tierra caliente* (hot lands), *tierra templada* (temperate lands), *tierra fría* (cool lands), and *tierra helada* (land of frost). Naturally these belts do not have identical altitudes throughout the tropics.

The lowest zone, or *caliente*, normally extends from sea level to 2,000 to 3,000 ft (the annual temperature is roughly 83 to 75°). Where precipitation is abundant, this belt is characterized by a luxuriant vegetation cover of trees or of trees and tall grass and by such crops as rubber, bananas, and cacao.

The *tierra templada* lies above the *caliente* and extends up to 6,000 to 6,500 ft (temperature is roughly 75 to 65°). A great variety of crops—among them coffee, maize, tea, cotton, and rice—is produced here.

Tierra fría, lying above the *templada*, reaches up to 10,000 to 11,500 ft (temperature is 65 to 54°). Middle-latitude crops, such as wheat, barley, apples, and potatoes, grow there, and the pastoral industries are frequently important.

Still higher, the *tierra helada* is above the tree line and beyond the agricultural zone. Here the alpine pastures lie, terminated along their upper margins by the permanent snowfields. Close to the equator, where a dry season is either absent or very short, the grass of the *tierra helada* is relatively thick and remains green throughout the year. This is known as the *paramo*. With increasing distance from the equator there is a longer dry season, so that the alpine grasses are dry for a part of the year and are burned over by native farmers. This is the

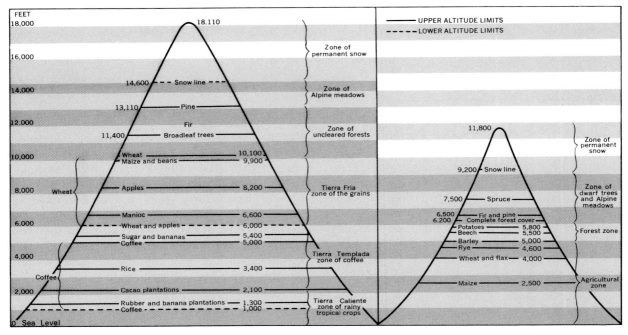

Figure 11.18 Vertical temperature zones and approximate altitudinal limits of selected crops and vegetation types on a tropical mountain (left) and a middle-latitude mountain (right). (*After Sapper.*)

Horizontal and vertical arrangement of the climatic vegetation belts in the tropical Andes

Number of humid months	Tierra caliente	Tierra templada	Tierra fría	Tierra helada
12 11 10	Tropical evergreen and semideciduous transition forest	Tropical mountain forest	Tropical altitude-and-fog forest	Paramo
9 8 7	Tropical moist savanna (forest and grassland)	Tropical moist valley vegetation (forest and grassland)	Tropical moist sierra vegetation (moist sierra brush)	Wet puna
6 5	Tropical dry savanna (forest and grassland)	Tropical dry valley vegetation (forest and grassland)	Tropical dry sierra vegetation (dry sierra brush)	Dry puna
4 3 2	Tropical thorn savanna (forest and grassland)	Tropical thorn valley vegetation (forest and grassland)	Tropical thorn sierra vegetation (thorn sierra brush)	Thorn puna
	Tropical semidesert	Tropical valley semidesert	Desert sierra	Desert puna
1 0	Tropical desert	Tropical valley desert		

Source: After Troll.

wet *puna*. As the dry season lengthens with increases in latitude, the grass cover becomes less dense, and the wet *puna* changes gradually to dry *puna*, then to thorn *puna*, and finally to desert *puna*. Thorn and desert *puna* are characterized by thorny shrubs and cacti. Grazing is the principal form of land use in the *paramo* and *puna*.

Above heights of 14,000 to 15,000 ft the zone of perpetual snow usually begins. On these tropical mountains local trade of considerable importance is fostered by the temperature zonation of products.

Middle-latitude highlands. Although mountains and plateaus within the tropics may be climatically desirable because of their lower temperatures, this same characteristic causes highlands in the middle latitudes to be climatically inferior to lowlands. After all, tropical lowlands have an excess of heat, so that any reduction of temperature with altitude usually is an advantage both for human comfort and for the greater variety of products that can be grown. In the middle latitudes, on the other hand, even the lowlands are commonly none too warm. Thus reduction of temperature with altitude, causing a cooler summer and shorter growing season, materially decreases the opportunities for agriculture. In other words, there are fewer *utilizable* temperature zones in middle-latitude highlands (Fig. 11.18).

Diurnal temperature changes. Mountains and high plateaus have been called the radiation windows of the earth, since the thin, dry air facilitates the entry of strong solar radiation by day and the loss of heat by terrestrial radiation at night. Rapid daytime heating and nighttime cooling are the result, so that large diurnal ranges of temperature are characteristics of highland climates (Fig. 11.19).

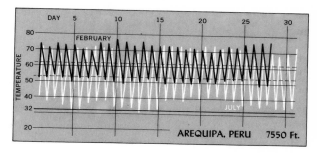

Figure 11.19 Daily maximum and minimum temperatures of the extreme months at a tropical mountain station of moderate altitude. Note the diurnal regularity of temperature change, indicating sun control. Diurnal range is greater in July, the drier season, when there is the least cloud.

In tropical highlands this marked temperature change between day and night stands in great contrast to the very slight temperature changes between the averages for the months and the seasons. At high altitudes in tropical highlands, especially in the *tierra helada*, the large diurnal range of temperature results in a very large number of days with night freezing and daytime thawing (Fig. 11.20). Such a frequent oscillation between freeze and thaw has major effects upon weathering processes, soil formation, and vegetation characteristics. In summary, a distinctive feature of the climate of high tropical mountains is the combination of large *diurnal* temperature variations and small *seasonal* ranges.

Seasonal temperatures and annual ranges. The lower temperatures in highlands have led to the statement that mountains in the tropics have perpetual spring. Quito's annual temperature of 55°, for instance, is not greatly unlike the May average at Madison, Wisconsin. However, the great variety

Climatic data for a representative high-altitude station in middle latitudes (*H*)

	J	F	M	A	M	J	J	A	S	O	N	D	Yr	Range
Longs Peak, Colorado (8,956 ft)														
Temp., °F	23	22	26	33	41	51	55	55	48	39	31	24	37	33
Precip., in.	0.7	1.2	2.0	2.7	2.4	1.6	3.6	2.2	1.7	1.7	0.9	0.9	21.6	

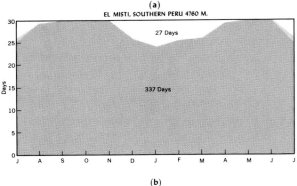

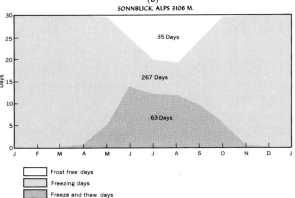

Frost free days

Freezing days

Freeze and thaw days

Figure 11.20 Annual number of freeze-and-thaw days on a tropical and on a middle-latitude mountain. A large number of days on which night freeze and day thaw occur is characteristic of high mountains in the tropics. This climatic feature expresses itself strongly in the nature of the wild vegetation, in weathering processes, and in soil characteristics. Middle-latitude highlands have many fewer freeze-and-thaw days and many more freezing days. (*After Troll.*)

of elevations within a tropical mountain mass obviously results in all gradations of temperature.

And although the temperature is lower on a tropical mountain than on an adjacent lowland, both locations have a similar uniformity in average monthly and daily means. Small seasonal ranges, and the same monotonous repetition of daily weather, belong alike to tropical highlands and tropical lowlands (Fig. 11.21). At Quito, for instance, the temperature difference between the warmest

and coolest months is under 1°, which is even smaller than that of the Amazon lowlands in the same latitude. Mexico City, at 7,474 ft and farther poleward than Quito, has an *average annual temperature* 17° below that for Veracruz on the coast; yet their *annual ranges* are almost identical—11.5 and 11°, respectively. One climatologist has aptly described both the difference and the similarity in the annual march of temperature between lowlands and highlands in the same latitude in the statement: "The pitch changes; the tune remains the same."

As a general rule, seasonal temperature contrasts, or mean annual range, increase with latitude in both highland and lowland stations. Also, the annual ranges for highland and lowland stations in similar latitudes are of the same general magnitude. Thus high-altitude regions in the tropics present the unique feature of a cold climate with a small mean annual range of temperature. This is quite in contrast to cold climates in middle latitudes. The small annual ranges in tropical highlands have some resemblance to those of marine locations farther poleward (Fig. 11.21).

Temperature inversions in highland climates. One of the characteristic phenomena of mountain regions, especially those of middle latitudes, is the larger diurnal and seasonal ranges of temperature in the valleys compared with the slopes. On summer

Figure 11.21 A comparison of the annual march of temperature at Iquitos, a tropical lowland station, and at Quito, a tropical highland station, both in equatorial latitudes. Note the generally lower temperature at Quito. On the other hand, a small annual range of temperature is characteristic of both stations.

ANNUAL MARCH OF TEMPERATURE

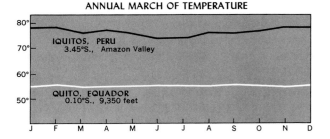

days the protected valley receives much reflected radiant energy from the surrounding slopes and may become warmer than the slopes above, where winds provide greater ventilation. But at night, especially in winter, the valleys become reservoirs of cold air draining from surrounding slopes. In mountainous regions, with their uneven surfaces and varied exposures, extensive air drainage causes local temperature inversions to be unusually well developed. This is further aided by the briefer daily period of sunshine, as well as the calmer, more stagnant air in the protected valley. The basins are not only colder but foggier. A pall of gray fog may shroud the Swiss foreland for a week at a time.

In descending into a mountain valley on a warm summer evening, one does not need an instrument to note the increasing dampness and coolness of the air. From the valleys of Carinthia in the Austrian Alps, where temperature inversions are well developed, comes the proverb, "If you climb in winter up a stair, one less coat you need to wear" (Hann). It often happens that on a winter day one can ascend from damp, cold, dark, sunless weather in the valley and come out suddenly on the slopes above, into a realm of sunshine and beauty where the air is clear, mild, and dry.

Highlands as temperature divides.   Since highlands function as barriers to the free movement of air masses, they commonly act as a shelter to areas on their leeward sides. Steep horizontal temperature gradients may be the result. Thus the cold continental air masses from central and northern Asia, which dominate the winters of eastern Asia, are prevented from invading Hindustan in winter by the great highland mass of the Himalayas and Tibet north of India and Pakistan. The protective effect of the mountains is made clear by a comparison of the January temperatures of stations in India-Pakistan and in China in similar latitudes:

Lahore, Pakistan 53°   Calcutta, India 65°
Shanghai, China 38°   Canton, China 50°

The northern highland rim of the Mediterranean Basin likewise blocks the free movement of cold

winter air into southern Europe. Consequently while Bucharest and Uskub in the Balkans have January temperatures of 26° and 29°, Salonika south of the highlands records 41°, Athens 46°, and Trieste 39°. See the accompanying table for the same kind of temperature effects caused by the Cascade Mountains.

The Cascade Mountains as a climatic divide

	Windward side (Tacoma)	Lee side (Yakima)
Temp., °F		
Mean winter	41	31
Mean summer	62	69
Mean daily range	15	29
Mean annual range	25	45
Absolute range	91	128
Rel. humidity, % mean, 1,630 hr	67	46
Sunshine duration, % possible, yr	43	65
No. of clear days, mean annual	73	113
Precip., in., mean annual	38.7	7.2

Source: After Landsberg.

Precipitation

Increased precipitation in highlands.   On a mean annual rainfall map of the earth, mountains ordinarily stand out as "islands" of increased precipitation. The additional precipitation in highlands has a multiple origin, a topic which was discussed in Chapter 4. The patterns of precipitation increase with altitude are somewhat different in low and in middle latitudes. In tropical mountains, total annual rainfall shows an initial increase at lower altitudes which continues up to about 5,000 ft (around 1,500 m). At higher elevations there is a continuous decline. Outside the tropics, however, total annual precipitation continues to increase, at least up to the levels of high peaks.

It is for dry climates, no matter in what latitude, that the heavier rainfall of highlands has critical importance. In regions of lowland drought, highlands are areas not only of heavier precipitation, but of heavier vegetation cover, and usually of more abundant agricultural production as well. In both arid and semiarid lands, highlands frequently bear a cover of forest in contrast to the meager grass-and-shrub vegetation of the adjacent drier lowlands. The Black Hills of western South Dakota are "black" because the color of their dark-green forests is such a contrast to the tawny-hued steppes surrounding them.

Not only are settlements in dry lands attracted to the humid slopes of mountains and to the well-watered mountain valleys, but streams descending from the rainier highlands carry the influence of highland climate far out on the surrounding arid lowlands. The Yemen Highlands in southern Arabia, and the adjacent lowlands watered by its rivers, are a garden spot as well as the principal center of population in that otherwise largely desert country. In eastern North Africa, the Ethiopian (Abyssinian) Highlands are a similar "culture island." The Nile floods have their origin in this same mountain knot. The waters of the Colorado River, whose principal sources are in the Rocky Mountains, make agriculture possible in the dry Imperial Valley of southern California over 700 miles away. From the Andes come the 50 or more small streams that, crossing the Peruvian coastal desert, nourish the irrigated strips of what is otherwise mainly wasteland.

The heavy precipitation associated with highlands is also significant in terms of power resources, for as the accumulation of water (including snow) at high elevations flows to lower levels, its energy may be harnessed to produce hydroelectric power. The amount of potential water power depends upon two factors: the volume of water (precipitation), and the extent of its fall, which in turn is directly related to the height of the mountain.

Windward and leeward slopes. As already discussed, highlands usually have wetter windward slopes, and their drier leeward side is described as the *rain shadow*. On the leeward side also, however, precipitation is likely to increase with elevation, just as it does on the windward side. This is because the ascending current on the windward side does

not begin to descend immediately upon reaching the crest of a mountain range, but continues rising for some distance beyond the crest. As a consequence the heavy windward precipitation spills over onto the higher leeward slopes.

Moreover, the lifting effect of a highland barrier upon approaching air masses, and hence the increase in precipitation, begins at some distance out in front of the marked change in slope. For example, there is a strong increase in rainfall within the flat lowlands of northern Java 25 miles or more away from the mountains. Lowlands just leeward of highlands are likely to be much drier, as shown in the table on the Cascade Mountains (page 365).

In summary, the windward sides of mountains are cloudier and rainier and have smaller temperature ranges, while the climate on leeward slopes is drier and sunnier, shows more variability in temperature, and has a tendency toward increased continentality.

Snowfall and the snow line. Because of lower temperatures in highlands, snow falls more frequently and abundantly, and remains on the ground for a longer period, than on the surrounding lowlands. It is the slow melting of deep accumulations of snow in highlands that gives many lowlands a large and continuous supply of water throughout the warmer seasons.

Considerable attention has been given to the permanent, or year-round, snowfields in highlands and to the changing elevation of the snow line, which marks the lower boundary of permanent snow. As a rule, the snow line becomes increasingly higher with decreasing latitude, a fact which reflects the decreasing temperatures away from the equator (Fig. 11.22). But as the table on page 367 shows, the highest snow line is not at the equator, where precipitation is heavy, but in two zones 15 to 25° to the north and south of it, where precipitation totals are smaller. Thus although latitude, or zonal temperatures, largely accounts for the general latitudinal variations in the altitude of the snow line, there are bound to be large departures from the zonal average caused by differences in amounts of precipi-

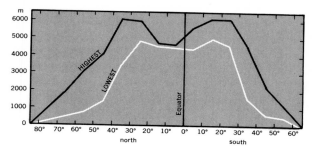

Figure 11.22 Highest and lowest limits of the snow line in the different latitudinal zones. In general, heights decline with latitude. But the greatest heights are in the subtropics, which have less cloud and precipitation than regions close to the equator. (*After V. Paschinger.*)

tation, distribution of precipitation throughout the year, and degree of exposure to the sun. The fact that the snow line on the warmer southern slopes of the Himalayas is 2,500 ft lower in elevation than on the cooler northern slopes reflects the much heavier precipitation on the former. As a rule, the snow line is lower on windward slopes, where precipitation is heavier, than on drier leeward slopes. It is also lower on shadier slopes inclined away from the sun. A winter maximum of precipitation tends to lower the snow line too. In some places these controls are cumulative in their effects; in others they counteract and oppose each other.

Winds

On exposed mountain slopes and summits, where ground friction is small, wind speed is high. Mountain valleys, on the other hand, are more protected against violent winds.

Owing to the great variety of relief and exposure

in highlands, there are a number of local winds characteristic of mountainous areas. The gravity wind, involving a downslope movement of cold, dense air, is one type. Another is the diurnal reversal of wind direction, upslope by day and downslope by night, which was discussed in the section on mountain and valley winds in Chapter 3. Most are unnamed.

Foehn or chinook. The foehn is a warm, dry, gusty downslope wind generated by a passing cyclone or anticyclone. In the western United States and Canada it goes by the name of chinook. Such local winds are fairly common in most highland regions which feel the effects of large-scale traveling disturbances. They are especially well known in the northern valleys of the Alps and along the eastern slopes of the Rocky Mountains from Colorado northward. When a well-developed cyclonic storm is positioned over the High Plains just east of the Rockies it acts to draw air down the lee slopes toward its low-pressure center. The warmth of the chinook may be due exclusively to heating by compression as the air descends the eastern slopes. But commonly this source of heat is supplemented by heat of condensation as air is drawn up the windward western slopes and over the crest of the range (Figs. 11.23 and 11.24).

Usually the temperature of the chinook in winter is not much over 40°, but 60 to 70° temperatures are not unknown. These appear very warm—by contrast, at least—after a spell of frigid anticyclonic weather. The warmth is made more noticeable by the accompanying desertlike relative humidities of 10 to 20 percent. If snow lies on the ground, it vanishes as if by magic before the warm, dry blast

Mean altitude of the snow line (Elevation in ft)

	Latitude							
	0–10°	10–20°	20–30°	30–40°	40–50°	50–60°	60–70°	70–80°
Northern Hemisphere	15,500	15,500	17,400	14,100	9,900	6,600	3,300	1,650
Southern Hemisphere	17,400	18,400	16,800	9,900	4,900	2,600	0	0

Source: After Landsberg.

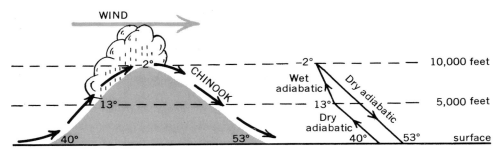

Figure 11.23 Diagrammatic representation of a chinook (*left*), and lapse rates associated with the chinook (*right*).

of the chinook. A rise in temperature of 40° within 24 hr is not unusual. Kipp, Montana, has the extraordinary record of a 34° rise within an interval of 7 min.

Foehns or chinooks have a number of conflicting economic effects. In the western parts of the North American High Plains, the chinook meliorates the winter cold and makes the snow cover less persistent, so that livestock can be grazed throughout the winter on the open range. An offsetting factor is the serious depletion of soil moisture, together with the evaporation of the snow mantle before its meltwater can penetrate the soil. Chinooks following spring planting may seriously damage the sprouted seeds.

In the northern valleys of the Swiss Alps, the average number of foehn days may be as high as

45 to 50 per year, with a minimum in summer and a maximum usually in spring. Some of the foehn valleys become climatic oases in winter. In spring the foehn valleys are freed early from their winter snow mantle, thus hastening the beginning of farming operations. Autumn foehns help advance the ripening of grapes and so lessen the danger from freeze. During a vigorous foehn, however, great precautions must be taken against the fire hazard associated with the excessive dryness and the strong winds.

Daily weather

In highlands the weather changes within the 24-hr period are likely to be greater than on adjacent lowlands. Striking contrasts between hot sun and cool

Figure 11.24 Barograph, thermograph, and humidity changes during a foehn at Altdorf in Switzerland. (*After Walter.*)

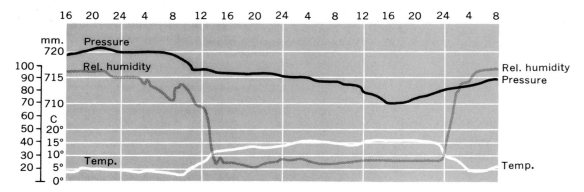

Average monthly frequency of the foehn at Sierre, Switzerland

	J	F	M	A	M	J	J	A	S	O	N	D	Yr
Average number of days	2	2	5	7	6	3	2	1	2	3	2	2	37

Source: After Landsberg.

shade, rapid shifts from chilly winds to calm, sudden gusts of rain or possibly snow, changing again to intense sunlight—such is the erratic nature of the daily weather. Even within tropical highlands, the complex sequence of daily weather contrasts with the uniformity of temperature conditions between the months.

At Quito, Ecuador, nights and early mornings are cold and raw, but the strong sun causes temperatures to mount rapidly. By noon it may be hot in the sun, though still cool in the shade. About midday or before, convectional clouds begin to form. In the afternoon there may be a violent thunderstorm, with heavy rain or snow showers, sometimes accompanied by hail. Storms and clouds die away as night approaches. The early mornings are fine, with clear air and good visibility. But in the afternoons clouds hang low over the gloomy landscape, so that the mountain peaks are ordinarily hidden and distant views are rare.

SELECTED REFERENCES FOR FURTHER STUDY OF TOPICS IN CHAPTER ELEVEN

American Geographical Society, The Antarctic atmosphere: climatology of the surface environment, Folio 8 of the Antarctic Map Folio Series, plates compiled by National Weather Records Center and W. S. Weyant, text by W. S. Weyant, 1968.

Antarctic Meteorology: Proceedings of the Symposium Held in Melbourne, Pergamon Press, New York, 1960.

Dorsey, Herbert G., Jr., "Arctic Meteorology," in Thomas F. Malone (ed.), *Compendium of Meteorology,* American Meteorological Society, Boston, 1951, pp. 942–951.

Hare, F. Kenneth, "Some Climatological Problems of the Arctic and Sub-Arctic," in Thomas F. Malone (ed.), *Compendium of Meteorology,* American Meteorological Society, Boston, 1951, pp. 952–964.

————, *Studies in Arctic Meteorology,* Publ. Meteorol. No. 10, Arctic Meteorology Research Group, McGill University, Montreal, 1958.

———— and Svenn Orvig, *The Arctic Circulation,* Publ. Meteorol., No. 12, Arctic Meteorology Research Group, McGill University, Montreal, 1958.

Rubin, M. J., "The Antarctic and the Weather," *Scientific American,* Vol. 207, pp. 84–94, 1962.

————, "Antarctic Weather and Climate," in *Research in Geophysics,* Vol. II, *Solid Earth and Interface Phenomena,* The M.I.T. Press, Cambridge, Mass., 1964, pp. 461–478.

Sabbagh, Michael E., "A Preliminary Regional Dynamic Climatology of the Antarctic Continent," *Erdkunde,* Vol. 16, pp. 94–111, 1962.

————, "Seasonal and Regional Variations of Temperature over the Antarctic Continent during 1958," *Geog. Ann.,* Vol. 45, pp. 52–75, 1963.

————, and Reid A. Bryson, *An Objective Precipitation Climatology of Canada,* Tech. Rept. No. 8, The University of Wisconsin Department of Meteorology, Madison, Wis., 1962.

Schneider-Carius, K., "Klimazonen und Vegetationsgürtel im tropischen und subtropischen Gebirge," *Erdkunde,* Vol. 2, pp. 303–313, 1943.

Shear, James A., "The Polar Marine Climate," *Ann. Assoc. Amer. Geographers,* Vol. 54, no. 3, pp. 310–317, 1964.

van Rooy, M. P. (ed.), *Meteorology of the Antarctic,* South African Weather Bureau, Pretoria, Norsk Polarinstitut, Oslo, 1957.

Wexler, H., "Seasonal and Other Temperature Changes in the Antarctic Atmosphere," *Quart. J. Roy. Meteorol. Soc.,* Vol. 85, pp. 196–208, 1959.

CHAPTER TWELVE

DRY CLIMATES (B)

The five groups of humid climates described in Chapters 8 through 11 are defined in terms of temperature. Only one of the great climatic groups, the dry type, has its boundaries fixed by rainfall values. Because it is the exception, this group is treated after the five humid groups, even though its letter, *B*, seems to place it earlier in the discussion. As mentioned in Chapter 7, *B* was Köppen's label for the dry climates, and his letter is retained here because the Köppen classification is widely known and used among American geographers.

Definition and boundaries. The essential feature of a dry climate is that the annual water loss through evapotranspiration at the earth's land-water surface exceeds the annual water gain from precipitation. (In a humid climate, of course, the reverse is true.) Thus dry climates do not have a constantly replenished groundwater supply, and permanent streams cannot originate within dry regions. However, permanent streams can cross areas with dry climates—as do the Nile, the Colorado, and the Indus, for instance—provided they have their sources in more humid regions.

Logically the boundary separating dry from humid climates is located where, on the average, annual precipitation equals annual evapotranspiration. Notice that since two variables are involved, no specific amount of annual rainfall can serve as the universal boundary of dry climates as a group. The amount of rainfall which is permissible will always vary with the amount of water lost by evapotranspiration.

But to compare water gains and water losses over the earth is not easy. The rain gauge measures precipitation with fair accuracy, so that rainfall distribution, at least over much of the land surface, is relatively well known. But no adequate instrument has been developed for measuring evapotranspiration. As a consequence much less is known about the distribution of water losses from the land-water surface to the atmosphere.

Because there are no actual measurements of evapotranspiration for large areas of the earth, the only alternative is to discover a relationship between potential evapotranspiration and other climatic elements for which more abundant world data do exist. Since the rate of evaporation depends upon the vapor-pressure difference between the surface and the air immediately above, and also upon turbulence which carries the moistened air upward, it has been determined that there is a fairly close relationship between air temperature and wind speed on the one hand and potential evapotranspiration on the other.

Thornthwaite and Holzman have developed the following formula for computing evaporation:

$$E = \frac{17.1\ (e_1 - e_2)\ (u_2 - u_1)}{t + 459.4}$$

where

E = evaporation in inches per day

e_1, e_2 = vapor pressure in inches of mercury at the two levels 1 and 2

u_1, u_2 = wind speed at the two levels in miles per hour

t = temperature in °F

Unfortunately, data on vertical gradients of vapor pressure and wind speed are rare, so that the formula is not very helpful for showing the distribution of the earth's dry climates. Other evaporation formulas based on more commonly measured quantities have been developed, but the usefulness of these too is impaired because of meager observational data.

There are cruder but more widely applicable evaporation formulas, often referred to as moisture indexes or indexes of aridity, that relate potential evaporation to air temperature alone, since temperature is the major control of evaporation potential. Lang's index of aridity, $I = P/T$, indicates that the effectiveness of precipitation varies directly with precipitation and inversely with temperature. De Martonne's moisture index, $I = P/(T + 10)$, and Köppens $I = P/(T + 7)$, are slight refinements of Lang's. All these use annual values of precipitation (in millimeters) and temperature given °C.

No such formulas seem to be entirely satisfactory or universally acceptable, however. They all do show that in the warmer lower latitudes, where the losses due to evapotranspiration are great, less of the total rainfall is effective for plant growth than in the cooler middle latitudes. It is quite possible that an amount of precipitation sufficient to produce a humid climate in regions with moderate temperatures might be classed as dry in a region of great heat. Paris, France, has an annual rainfall of only 23 in.; yet in that cool marine climate it produces a distinctly humid landscape. The same amount in tropical Africa could result in semiarid conditions. Moreover, if a large amount of the total annual rainfall comes in the warmer months, much more is lost through evapotranspiration than would be in a region where precipitation is evenly distributed over the 12 months or concentrated in the cool season.

Köppen's method for determining the amount of annual precipitation that defines the boundary between dry and humid climates, or the outer margins of the semiarid steppe, is expressed by three formulas:

$r = 2t + 14$ rainfall evenly distributed   (1)

$r = 2t$ rainfall concentrated in winter   (2)

$r = 2t + 28$ rainfall concentrated in summer   (3)

where

r = annual rainfall in centimeters

t = average annual temperature in °C

From the formulas it is obvious that the amount of rainfall r defining the humid boundary of dry climates will be greater where mean annual temperatures (and hence losses through evapotranspiration) are high, and less where the temperatures are low. Also, r will be highest where rainfall is concentrated in the warm season (formula 3), and least where the maximum is in winter, when evapotranspiration is least (formula 2). From the table below it can be seen that in a climate with a mean annual temperature of 50° (provided rainfall is well distributed seasonally), a yearly rainfall of only 13.5 in. or over is required to produce a humid climate; but where the annual temperature is 80°, the comparable figure is just about doubled, or 26.7 in. Of course, Köppen never meant his formulations to be definitive and exact boundary values, but only convenient approximations.

Köppen designed formulas 2 and 3 to be applied only in situations where there are marked seasonal contrasts both in rainfall and in temperature. In places without important seasonal temperature differences, winter rains obviously can be no more effective than summer rains. Köppen is never very definite concerning the exact meaning of "summer concentration," "winter concentration," or "even distribution" of the annual rainfall as used in his formulas. Some writers have used the figure of 70 percent to mean a concentration as applied to the

Boundary between humid and dry climates according to Köppen (no dry season)

Temp., °F	Precip., in.
50	13.5
60	17.9
70	22.3
80	26.7
90	31.1
100	35.5

summer (April through September) or winter (October through March) half-year. If less than 70 percent is concentrated in either 6-month period, the rainfall is said to be evenly distributed. Others choose the same definition of summer rain and winter rain that Köppen uses to identify w and s in his C climates, i.e., ten times as much rain in the rainiest summer month as in the driest winter month and three times as much rain in the wettest winter month as in the driest summer month.

When the three formulas given above are converted into their nearly exact equivalents in English units of measurement (R = annual rainfall in inches, T = average annual temperature in °F), they appear as follows:

$R = 0.44T - 8.5$
$\qquad$ rainfall evenly distributed   (1)

$R = 0.44T - 14$
$\qquad$ rainfall concentrated in winter   (2)

$R = 0.44T - 3$
$\qquad$ rainfall concentrated in summer   (3)

But in their transliterated form they tend to give a false impression of the degree of accuracy which Köppen intended for his formulas. Patton[1] suggests substituting three transliterated formulas with simpler constants which approximate their metric counterparts closely enough:

$R = \frac{1}{2}T - 12$ rainfall evenly distributed   (1)

$R = \frac{1}{2}T - 17$ rainfall concentrated in winter   (2)

$R = \frac{1}{2}T - 6$ rainfall concentrated in summer   (3)

Moreover, he continues the simplification process by transforming the three formulas into the single one

$$R = \frac{1}{2}T - \frac{1}{4}Pw$$

in which Pw is the observed winter percentage of rain at a particular place. From tests applied, this

[1] Clyde P. Patton, "A Note on the Classification of Dry Climates in the Köppen System," *Calif. Geographer,* Vol. 3, pp. 102–112, 1962.

single formula, simpler to use and to learn than three, gives results which do not depart significantly from those obtained by using the three standard Köppen formulas.

Dry climates are the most extensively developed over lands of any of the great climatic groups—occupying, according to Köppen, 26 percent of the continental area. Of this total, 14 percent is steppe, or semiarid, and 12 percent is desert, or arid. As might be expected, dry climates occupy a much smaller proportion of the earth's water surface—probably between 4 and 5 percent.

Temperature. Since dry climates are found in such a wide variety of latitudes and continental locations, few valid general comments can be made concerning their temperatures. On the whole, however, their characteristic interior and leeward locations on the continents, and their clear skies and dry atmosphere, tend to make them severe *for their latitude,* with relatively extreme seasonal temperatures and thus large annual ranges.

Their large daily ranges, however, are even more marked. Clear, cloudless skies and relatively low humidity permit an abundance of solar energy to reach the land surface by day, but likewise allow a rapid loss of earth energy at night. Large diurnal ranges in deserts also are associated with the meager vegetation cover, which permits the barren surface to become intensely heated by the sun. The surface of dry ground may reach a daytime temperature of 200°F. It is a physical law that the higher the temperature of a body, the more rapid is its loss of heat by radiation, and consequently the more rapid its reduction in temperature.[2] Deserts therefore not only acquire heat quickly, but also lose it very fast.

On the other hand, in humid regions with a relatively complete vegetation cover, more of the solar radiation reaching the earth's surface is utilized in evapotranspiration. A large part of the energy consumed in this process is taken from the moist soil surface or the transpiring plant cover. Thus evapotranspiration is a heat regulator, preventing excessively high surface temperatures. As a consequence, humid regions with a vegetation cover are unlikely to have such extreme ground and air temperatures as deserts.

Based upon temperature, two main subdivisions of dry climate may be recognized: that of the tropics-subtropics, or *hot* dry climate (*Bh*), and that of temperate-boreal latitudes, or *cold* dry climate (*Bk*)—"cold" meaning cold in winter.[3]

Winds. Dry regions are inclined to be windy places, since there is little frictional retardation of the moving air by the lowly and sparse vegetation cover. In this respect they are like the oceans. Moreover, the rapid daytime heating of the lower air over deserts leads to strong convectional overturning, and this interchange of lower and upper air tends to accelerate horizontal surface currents during warm hours when convection is at a maximum. "In the desert the wind is almost the only element of life and movement in the domain of death and immobility. A journey in the desert is a continuous strife against the wind charged with sand and, in moments of crisis, a painful physical struggle" (Gautier). Nights are likely to be much quieter, which is a partial explanation of the rapid nocturnal cooling of surface air in dry regions.

Because of the strong and persistent winds, desert air is often murky with fine dust which fills the eyes, nose, and throat, causing serious discomfort. Much of this dust is carried beyond the desert margins to form the loess deposits of bordering regions. The heavier, wind-driven rock particles, traveling close to the surface, are the principal tool of the wind in sculpturing desert landforms.

[2] The amount of heat radiated by a body is directly proportional to the fourth power of its absolute temperature.

[3] Köppen uses the mean annual isotherm of 18°C (64.4°F) as the boundary between the two main latitudinal subdivisions of dry climate. Many American geographers have substituted the January isotherm of 32° (0°C). In the classification employed here, the isoline of 8 months with a temperature of 50° (10°C) or above is accepted as the boundary between hot and cold dry climates. It is identical with the boundary separating subtropical and temperate climates.

Precipitation and humidity. Average annual rainfall in the dry climates is always deficient. In addition, it is extremely variable from year to year, so that the average is not to be depended upon (Fig. 4.34). One of the general principles of climatology, already discussed in Chapter 4's section on precipitation distribution, is that dependability of precipitation usually decreases with decreasing amount. Thus the handicaps of meagerness and unreliability of rainfall seem to go together. No part of the earth is known for certain to be absolutely rainless, although at Arica in northern Chile, the rainfall over a period of 17 years was only 0.02 in. During the whole 17 years there were only three showers heavy enough to be measured.

With a few exceptions, relative humidity is low in the dry climates, 12 to 30 percent being usual for midday hours. Conversely, potential evaporation is extremely high. Absolute humidity, on the other hand, is by no means always low, for hot desert air usually contains a considerable quantity of water vapor, even though it may be far from saturated. The amount of sunshine is great, and cloudiness small. Direct as well as reflected sunlight from the bare earth is blinding in its intensity.

Based on annual rainfall, two subdivisions of dry climate (which include both hot *Bh* and cold *Bk* types) are commonly recognized: the arid, or desert, type (*BW*), and the semiarid, or steppe, type (*BS*). In general, the steppe is a transitional belt surrounding the desert and separating it from the humid climates beyond. The boundary between arid and semiarid climates is an arbitrary one, which is defined by Köppen as one-half the amount separating steppe from humid climates. For example, if for a specified temperature, 19 in. of rainfall marks the outer, or humid, boundary of dry climates in general, then for a similar temperature, $9\frac{1}{2}$ in. may be taken as the boundary between steppe and desert. Thus in the three Köppen formulas given previously for determining the boundary of steppe and humid climates, the right-hand side of each equation must be divided by 2 in order to represent the boundary between desert and steppe. The same arbitrary treatment may be applied to the single Patton formula. A. Austin Miller, who has suggested the formula $r = t/3$ as representing the boundary between steppe and humid climates, proposes $r = t/5$ for defining the boundary between desert and steppe.

TROPICAL-SUBTROPICAL DRY CLIMATES (*BSh*, *BWh*)[4]

Type location. The heart of the tropical dry climates (see Figs. 7.2, 12.1, and the climate map) lies in the vicinity of latitudes 20 or 25°N and S and the average positions of their margins are at approximately 15 and 35° +. Thus they coincide fairly closely with the dry, subsiding air masses of the subtropical anticyclones. Subsidence and drought are not confined to the crests of the anticyclonic cells, for these characteristics extend out onto the equatorward slopes of the highs into the trade winds.

Ordinarily tropical dry climates do not reach to the eastern margins of the continents; humid climates characteristically take their places on the windward margins (Fig. 7.2). Along west coasts in these latitudes, by contrast, dry climates extend not only down to the sea margins, but even far beyond them over the eastern parts of the adjacent oceans. They also thrust abnormally far equatorward along some tropical western littorals (e.g., Peru-Ecuador and southwestern Africa).

As discussed in Chapter 7, this asymmetrical development skewed toward the west sides of continents is related to several factors: (1) the greater stability of air masses in the eastern as compared with the western parts of oceanic subtropical anticyclones, (2) the smaller number of atmospheric

[4] As shown in Chapter 7's list of definitions, the letter *h* indicates a hot, dry climate—8 months or more with an average temperature of 50° (10°C) or above.

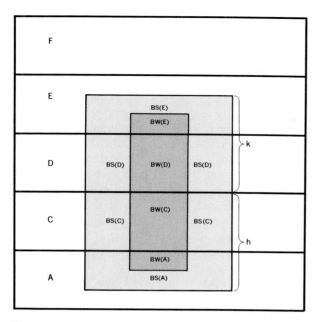

Figure 12.1 In this diagram, dry climates (*B*) are seen as intersecting four of the five great thermal zones (*A, C, D, E*), producing four subsets, *BA, BC, BD,* and *BE.* Two levels of aridity, steppe (*S*) and desert (*W*), are then added to produce the basic types shown in the diagram. *BA* and *BC* may be combined to form the hot (*h*), or tropical-subtropical, dry climates; and *BD* and *BE* to form the cold (*k*) dry climates. (*After James A. Shear.*)

Madagascar), while dry climates are carried equatorward beyond their normal latitudes along some western littorals. Low-latitude dry climates are characteristically bounded by *Aw* climates on their equatorward sides and by middle-latitude *C* (usually *Cs*) and *B* climates on their poleward and interior margins (compare Fig. 7.2 with the climate map inside the front cover).

The difference between type locations of desert and steppe is that desert climates are situated in the core area of subsidence, divergence, and temperature inversions within the subtropical anticyclone. These features are opposed to the development of fronts and atmospheric disturbances. Even when the intense surface heat of summer develops seasonal low-pressure centers over some of the low-latitude deserts, such convergent cyclonic circulations in the lower air are relatively shallow, and anticyclonic circulation still predominates aloft, damping any tendency toward convective overturning. Too far equatorward to be reached by the equatorward advance of middle-latitude fronts and cyclones on the poleward side, too far poleward to be affected by the ITC advancing from the low latitudes, and too far inland from eastern littorals to be affected by the humid onshore winds and the disturbances in the western oceanic trades, these low-latitude deserts lie outside the realms of the usual rainbringing winds and storms (Fig. 7.2).

Steppe typically surrounds the desert, except possibly on its western side (Fig. 7.2). It therefore constitutes a transition belt between desert and humid climates. Because steppe is less at the heart and more on the margins of dry, settling tropical air masses associated with the subtropical high-pressure cells, and is consequently one stage closer to the humid climates than the desert, steppe areas are encroached upon for a short period of the year by winds with rainbringing disturbances. It is this brief period of seasonal rains which causes them, although still having a dry climate, to be semiarid rather than arid.

In the case of North Africa, which may be taken as a fairly typical example, the climate map shows that there are two large and separate areas of tropi-

disturbances along the eastern parts of tropical oceans and the adjacent western coasts of tropical continents, and (3) the general humidity contrasts between windward and leeward coasts in the tropical easterlies. Along tropical west coasts, no doubt the principal drought maker is subsidence in the subtropical anticyclone. But the cool ocean currents, fed by upwelling, which usually parallel these west coasts may further intensify the aridity. Some of the more extensive deserts, particularly the Sahara and the Australian desert, are important source regions for tropical continental air masses.

It appears to be a general rule, then, that humid tropical climates extend unusually far poleward along the eastern (windward) sides of continents (eastern Brazil, eastern Central America, eastern cal steppe, each occupying a representative location. One borders the Sahara on the *north* and lies between it and the subtropical dry-summer (*Cs*) climate. This steppe region is encroached upon by the westerlies with their cyclones and fronts in winter, the time of their maximum equatorward migration. Consequently winter is the season of maximum rainfall. Other subtropical steppes with this same type location are found in southern Australia; in Mesopotamia, Arabia, and southern Iran, in western Asia; and in northwestern Mexico and adjacent parts of the southwestern United States.

The second steppe region of northern Africa lies along the *equatorward* margins of the Sahara, separating the desert from the *Aw* type. This steppe receives its meager rainfall at the time of high sun from the disturbances accompanying the poleward-displaced ITC. The semiarid regions in northern Australia, much of dry southwestern Africa, and northwestern India-Pakistan largely belong to this group. In Australia the two typical belts of steppe, one on the poleward and the other on the equatorward margins of the tropical desert, are connected by a crescentic belt of steppe along the eastern side of the arid area. It is only on the western sides of continents, therefore, that the tropical deserts usually extend down to the sea margins and have no semiarid transition belts.

Precipitation

Annual amount. Although no exact amount of rainfall can be accepted as defining the outer or humid margins of tropical-subtropical dry climates, the figure usually lies somewhere between 20 and 30 in. Half these amounts are more characteristic of the humid margins of deserts. Over much of the Sahara precipitation is under 5 in., as it is also in large parts of the other low-latitude deserts: the Kalahari in South Africa, the Atacama-Peruvian desert in western South America, the Australian desert, the Thar in northwestern India-Pakistan, and the Sonora in northwestern Mexico and the southwestern United States. At Cairo, Egypt, the annual rainfall averages 1.2 in.; at Lima, Peru, 2;

William Creek, Australia, 5.4; Yuma, Arizona, 3.3; and Port Nolloth in southwestern Africa, 2.3. In parts of northern Chile, rain may not fall for 5 to 10 years in succession. At Calama in northern Chile, located back of the Coastal Range, it is said that no rain has ever been recorded. Averages are of little value in giving a correct impression of desert rainfall, however, for not only is it small in amount, but it is also erratic. Over most low-latitude deserts, rainfall variability shows a 40+ percent departure from the normal (Fig. 4.35). It is therefore almost impossible to speak of a *typical* rainfall curve, or an *average* annual rainfall, for desert stations. At Iquique in northern Chile, for example, no rain fell during a period of 4 years. Then in the fifth year, one shower produced 0.6 in., which made the "average" annual rainfall for the 5-year period 0.12 in. On another occasion, 2.5 in. of rain fell in a single shower.

Rainfall in the steppes is also highly variable. Its undependability is probably more dangerous in semiarid than in arid lands, for in the desert there is not enough precipitation to tempt settlers to depend upon it. But the somewhat greater steppe rainfall, especially in more humid years, may lure the agriculturist into taking chances. For the most part it is only where irrigation supplements the variable rainfall that crop farming is reasonably safe. Animal grazing consequently takes on greater importance in steppe agriculture.

Seasonal distribution. Desert rainfall is so meager and so erratic in its time of fall that its distribution throughout the year is of almost no consequence. But the larger annual total characteristic of steppes makes its distribution vitally important to settlers. As noted in an earlier section, tropical-subtropical steppes with contrasting type locations have different rainfall regimes.

Steppes with low-sun rainfall (BShs).[5] These belts of steppe, which lie on the poleward sides of tropical deserts and close to *Cs* climate (see Fig. 7.2 and the climate map), have nearly all their rain

[5] The letter *s* indicates summer drought.

Climatic data for representative stations in tropical-subtropical deserts (*BWh*)

	J	F	M	A	M	J	J	A	S	O	N	D	Yr	Range
					·Jacobabad, Pakistan									
Temp., °F	57	62	75	86	92	98	95	92	89	79	68	59	79	41
Precip., in.	0.3	0.3	0.3	0.2	0.1	0.2	1.0	1.1	0.3	0.0	0.1	0.1	4.0	
					William Creek, Australia									
Temp., °F	83	83	76	67	59	54	52	56	62	70	77	81	68	30.5
Precip., in.	0.5	0.4	0.8	0.4	0.4	0.7	0.3	0.3	0.4	0.3	0.4	0.3	5.4	
					Phoenix, Arizona									
Temp., °F	51	54	60	69	77	86	91	89	84	72	59	53	70	40
Precip., in.	0.7	0.8	0.7	0.3	0.1	0.1	0.8	1.1	0.7	0.5	0.5	0.8	7.2	

in the cool seasons. Like the *Cs* climates they border, they receive their rain from fronts associated with cyclonic storms, which travel more equatorward routes in winter than in summer. During the warmer part of the year, however, *BShs* regions are dominated by the dry settling and diverging air masses of subtropical high-pressure cells. Because rainfall is concentrated in the cool season, evaporation is less, and consequently the modest amount that falls is relatively effective for plant growth.

At Benghazi in the steppe region north of the Sahara, 3 midsummer months—June, July, and August—are rainless, while December and January are the rainiest months (see the table given below). The mean annual precipitation is slightly less than 11 in., but this is derived from such totals for single years as 22 in., 24 in., 6.8 in., and 7.1 in.—all of which occurred within a span of 7 years. Variability is not so great, however, as in steppes which have a high-sun rainfall maximum.

In steppe lands with a low-sun rainfall, spells of cloudy, rainy weather associated with passing cyclonic storms are not unusual in the winter season.

Not only cloud and rainfall are involved, but changes in temperature as well: Air of tropical origin with higher temperatures precedes the low center, and higher-latitude, cooler air is advected following the storm. It is obvious from the climate map that these regions are far enough poleward to feel the effects of modified polar air masses. In spite of the fact that winter is the rainiest season, it is not a time of predominantly overcast skies, for a considerable part of the cool-season rain falls from cumulus clouds and has a short duration.

Steppes with high-sun rainfall (BShw).[6] Tropical steppe lands occupying the equatorward margins of deserts, and therefore lying between them and the *Aw*, are likely to have a very brief period of relatively heavy rains at the time of high sun, when the ITC is farthest poleward (see Fig. 7.2, the climate map, and the table of climatic data for Kayes, Mali). Seasonal rainfall periodicity is like that of the *Aw*, except that the dry season is longer and the total precipitation less.

[6] The letter *w* indicates winter drought.

Climatic data for a representative tropical-subtropical station with low-sun rainfall (*BShs*)

	J	F	M	A	M	J	J	A	S	O	N	D	Yr	Range
				Benghazi, Libya, 32°8′N (marginal)										
Temp., °F	56	57	60	65	71	75	77	78	76	73	67	59	68	22.5
Precip., in.	2.7	1.6	0.8	0.2	0.1	0.0	0.0	0.0	0.1	0.7	1.8	2.6	10.5	

Climatic data for a representative tropical-subtropical steppe station with high-sun rainfall (*BShw*)

	J	F	M	A	M	J	J	A	S	O	N	D	Yr	Range
					Kayes, Mali, 14° 20′N									
Temp., °F	77	81	89	94	96	91	84	82	82	85	83	77	85	19.2
Precip., in.	0.0	0.0	0.0	0.0	0.6	3.9	8.3	8.3	5.6	1.9	0.3	0.2	29.1	

At Dakar in Senegal, 6 months—December through May—are nearly rainless, while 86 percent of the year's total falls (principally in convectional showers) during July, August, and September. Since the rain arrives in the hot season when evaporation is at a maximum, less of it is effective for vegetation, and consequently these steppes bordering the savannas usually have more total rainfall than their poleward counterparts described above. This convectional rainfall is also more variable. Temperatures are not greatly different from those of the adjacent desert. As shown in the accompanying table, Kayes, a station with abnormally high temperatures, has its time of greatest heat preceding the rainy season. Since cloudiness is greatest during its rainy summer months, its highest temperatures do not coincide with the time of highest sun.

Intensity of rainfall. Tropical dry climates, like those of low latitudes in general, are characterized by convective rains that fall as heavy local showers from towering cumulonimbus clouds. Seven single convective storms brought nearly one-quarter of the total rain (30.7 in.) that fell at Helwan in the Egyptian Sahara in a 20-year period. In one of these storms 1.8 in. of rain fell, causing torrents to sweep down the wadies and do much damage to buildings and crops.

These sudden heavy downpours may cause much more harm than good. The wadies, which are entirely without water during most of the year, can become muddy torrents filled with debris after one of these flooding rains. In an earlier period, army regulations forbade French troops in the Sahara to pitch their overnight camps in the bottoms of dry wadies because of the danger of sudden floods—and this in spite of the fact that the high banks of the

wadies give protection against the disagreeable winds. After a serious desert flood, settlements suffer, roads, bridges, and railways may be damaged, and irrigation systems are often clogged with debris.

Because of the violence of tropical desert rains and the sparseness of the vegetation cover, temporary local runoff is excessive, and consequently less of the total fall becomes effective for vegetation or for the crops of the oasis farmer. This "dash" character of hot-desert showers, plus their local nature and their erratic seasonal distribution, makes them of little direct use for agriculture, so that no immediate dependence is placed upon them as a source of water. Much of the precipitation that reaches the earth is quickly evaporated by the hot, dry desert air, but some sinks in to replenish the underground water which appears at the surface in the form of springs or artesian flows.

Long-continued general rains, falling from an overcast cloud deck, are almost limited to steppes located on the poleward margins of tropical desert (*BShs*). In these, the winter rains are derived from cyclonic storms of the westerlies.

Cloudiness and sunshine. Skies are prevalently clear in the low-latitude deserts, so that sunshine is abundant. In the Sonoran Desert of the United States and Mexico, 75± percent of the possible sunshine is received in winter, and 90± percent in the other seasons. Yuma, Arizona, averages 88 percent of the possible sunshine for the year: 95 percent in May, 97 percent in June, and 78 percent in January. Over much of the Sahara, December and January have a cloudiness of only $\frac{1}{10}$, while from June to October it drops to about $\frac{1}{30}$. The pitiless glare of sunlight in the tropical deserts is such an essential characteristic of their landscapes that the occasional dark or rainy day, being so unusual, is said to be particularly depressing.

Strong surface heating, due to the intense solar radiation and the nearly bare ground, must give rise to vigorous surface convection currents. But because the stability associated with subsidence aloft is so great and the relative humidity so low, it is only infrequently that the heated surface air rises high enough to produce thunderheads. Dark cumulonimbus clouds do form occasionally, sometimes accompanied by thunder and lightning; but the streamers of rain that can be seen descending from them usually are evaporated in the arid atmosphere before they reach the earth. However, even though the air may be physiologically dry and may have intense evaporating power, there is usually a moderate amount of moisture in it. All that is lacking is a way to cause it to be condensed and precipitated. Thus the air at Yuma, Arizona, contains nearly as much moisture in July, and twice as much in January, as that at Madison, Wisconsin, in the same months. However, the relative humidity at Yuma (January, 47 percent; July, 34 percent—8 P.M. readings) is only one-half to two-thirds as great in either season. In spite of the low relative humidity, nocturnal surface cooling in the deserts frequently is sufficient to produce valley fogs and dew.

Owing to the high temperature and low relative humidity, potential evaporation is excessive, often twenty or more times the precipitation. At Yuma the average evaporation during the 5 hot months is 55 in., while the average rainfall during the same period is not quite 1 in. Relative humidities as low as 2 percent, with temperatures over 100°, have been recorded in the Egyptian Sahara. It was the extraordinarily dry air which allowed the Egyptians to mummify their dead.

Temperature

Annual and diurnal temperatures.   Annual ranges of temperature in low-latitude dry climates are larger than in any other type of climate within the tropics, for 20 to 30°+ is usual (Figs. 12.2 and 12.3). Aswan in the Sahara has mean temperatures

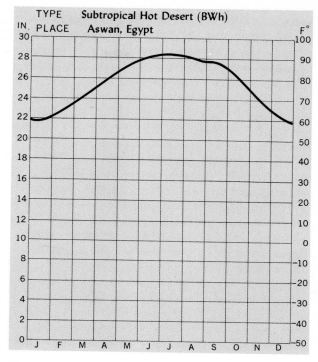

Figure 12.2 Average monthly temperatures for a hot-desert station in the Egyptian Sahara. Note the large annual range for a latitude of about 24°. Precipitation is not appreciable.

of 61° in January and 95° in July, resulting in an annual range of 34°. Such ranges, which even exceed those of some middle-latitude climates, reflect not only the clear skies, bare earth, and low humidity, but also the higher latitudes of the deserts and somewhat greater seasonal extremes of solar radiation compared with most of the *humid* tropics. Remember that in the tropical dry climates it is the excessive summer heat, rather than the winter cold, which leads to the marked differences between the seasons. This is the only tropical climate where annual range may occasionally equal daily range.

Daily ranges, however, are usually larger, averaging 25 to 45° and in rare instances even reaching 60°. The same conditions—dry air, cloudless skies, and bare, dry earth—that make for moderately large temperature differences between the extreme

TOMBOUCTOU, MALI (Temp. in C°) 16°49'N; 2°52'E

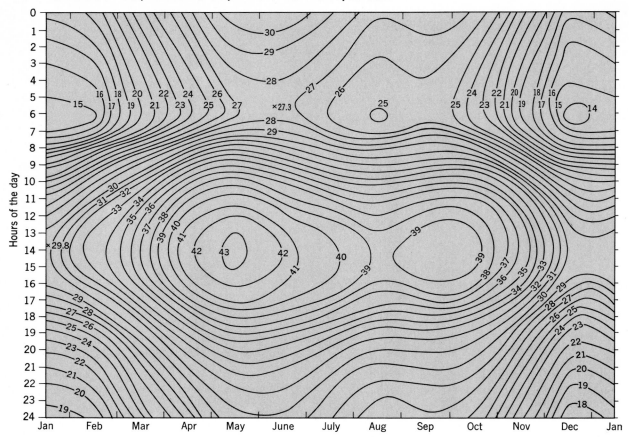

Figure 12.3 Thermoisopleths for a representative hot-desert station in the Mali Sahara of North Africa.

months are likewise conducive to wide differences within the diurnal period. At Bir Milrha in the Libyan Sahara, a minimum temperature of 31° and a maximum of 99° have been recorded on the same day.

The hot season. During the high-sun period, scorching, desiccating heat prevails. Hot-month temperatures average between 85 and 95° (Yuma, 91°; Tombouctou, 94.5°; Nullagine, Australia, 90°), and midday readings of 105 to 110° are common at this season (Fig. 12.4). In the western part of the Australian desert, temperatures have exceeded 100°

on 64 consecutive days and 90° on 150 consecutive days. At Greenland Ranch in Death Valley, California, the daily maxima in 1960 reached or rose above 100° on 136 consecutive days. At this time of year, although the lower night temperatures are a distinct relief by contrast with the days, they are by no means cool. At Phoenix, Arizona, the midsummer daily maxima usually exceed 100°, and the minima are close to 75 or 76°. In the northern Sahara, average daily maxima of 99° are followed by minima approximating 71°. At Azizia, 25 miles south of Tripoli, 136.4° has been recorded; this is the highest air temperature in the shade ever registered under

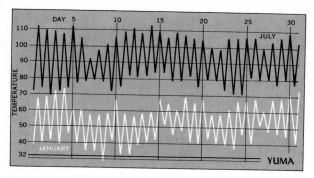

Figure 12.4 Daily maximum and minimum temperatures for the extreme months at a hot-desert station, located on the northern margins of the subtropics in Arizona. While solar control is dominant, modest nonperiodic air-mass effects associated with passing cyclones and anticyclones are present at this latitude, especially in winter.

standard conditions of thermometer exposure. The highest official air temperature ever recorded in the United States is 134°, in Death Valley in the California desert.

The low-sun period.   During the period of low sun the days still are warm, with the daily maxima usually averaging 60 to 70° and occasionally reaching 80° (Figs. 12.3 and 12.4). But nights are distinctly chilly at this season, with the average minima in the neighborhood of 50°±. The station In Salah in the Sahara during January has average daily maxima of about 70° and minima of around 40°. The average annual extremes there are 30° and 122°, and the temperature has been known to go as high as 124° and as low as 26°. There may even be occasional frosts in low spots in some subtropical deserts. A Saharan traveler gives the following description of the cool season in that arid region:

In truth the days are generally very trying, for there is either a blazing sun which burns through the thin cotton (clothing), or a bitter wind which pierces every bone in one's body. Sometimes there were both together, and then one side of one is frozen and the other baked! One's skin split and blistered under this treatment, but there was only one hour I shall never be able to forgive the desert. This was the moment when, at 5 A.M., one crept shivering out of one's warm flea-bag into pitch darkness, placed one's feet gingerly on icy cold sand, fumbled with numb fingers for a candle and matches, and proceeded to drag on cold, stiff garments from each of which fell a shower of sand.[7]

Daily weather

As is true of the rest of the tropics, the sun very much controls the weather in the low-latitude dry climates. Diurnal regularity is pronounced, so that there is a great deal of sameness between successive days. This is especially true of the summer season in the dry climates. But on their poleward margins, which lie close to the middle latitudes, they have occasional invasions of polar air with associated fronts and cyclonic storms, which now and then may cause spells of weather. This happens particularly during the winter season, when the westerlies and the cyclone belt are farthest equatorward.

The daily march of temperature at Yuma, Arizona, clearly indicates these nondiurnal oscillations of temperature in winter (Fig. 12.4). On the front of the cyclone, southerly winds of tropical origin advect higher temperatures. These are followed by several days of cooler weather as the winds swing around to the north on the rear of the low. Occasional gray days with some rain may result from a few of the cyclones, whose centers are usually well beyond the poleward margins of the dry climates.

On the equatorial margins of tropical dry climates, it is the high sun or summer period that has the most "weather." As the ITC shifts poleward with the sun, various kinds of tropical disturbances bring spells of showers, particularly to the steppes. The effects of these disturbances are chiefly confined to precipitation, although their clouds are likely to reduce temperatures somewhat during the brief rainy season.

[7] Rosita Forbes, *The Secret of the Sahara: Kufara*, Doubleday & Company, Inc., New York, 1921, pp. 168–169.

COOL WESTERN LITTORALS (Bn)[8]

Temperature

The usual characteristics of tropical deserts—high temperatures, large diurnal and annual temperature ranges, low relative humidity, and little cloud—are significantly modified along the littorals (usually western) of several of the low-latitude deserts, especially where cool ocean currents parallel the coasts (Fig. 12.5). The influence of cool water is especially marked along the desert coasts of Peru and northern Chile from about latitude 4 to 31°S, and the west coast of Africa from 8 to 32°S. Cool currents also have a noticeable effect on climate along the Atlantic coasts of the Moroccan Sahara from Casablanca to Senegal, Somali in northeastern Africa, and northwestern Mexico. (Some of these regions are steppes rather than true deserts.)

The low temperatures of the ocean waters are partly the result of advection from higher latitudes, for the currents are driven equatorward by the spiraling winds about the eastern end of a subtropical anticyclone. In part also, they are due to upwell-

[8] In the Köppen symbols, n signifies frequent fog (Nebel in German); n′ high relative humidity with haze and therefore low visibility, but not actual fog.

Figure 12.5 Distribution of cool coastal dry climates (Bn) in tropical and subtropical latitudes. Here fog and low stratus are common. Characteristically this subtype of tropical-subtropical dry climate is located along coasts which are paralleled by cool ocean currents fed by upwelling and which are affected by strong anticyclonic subsidence.

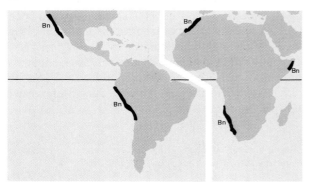

ings of colder subsurface water along these coasts. The land margins adjacent to the cool waters are unusually cool, with temperatures 10° or more lower than normal for the latitude (Fig. 12.6). Thus Callao on the Peruvian coast has an annual temperature of only 67°, and Mollendo 65°, while Bahía on the east coast of Brazil, in a similar latitude, averages 77°. The hottest month at Callao is only 71° (similar to July at Madison, Wisconsin), while the coldest is 62.5°. At Port Nolloth, a coastal desert station in South Africa which is paralleled by the cool Benguela Current, temperatures of the warmest and coldest months are 59 and 53°, while

Figure 12.6 Average monthly temperatures for a marine desert station located on a coast paralleled by a cool ocean current. Temperatures are abnormally low, and the annual range is very small. Compare with Fig. 12.2. Precipitation is not appreciable.

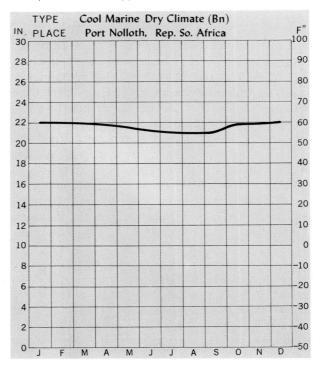

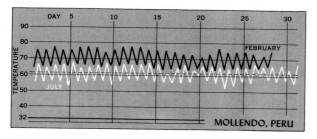

Figure 12.7 Daily maximum and minimum temperatures for the extreme months at a low-latitude desert station located on a coast paralleled by a cool ocean current. Note the abnormally low average temperatures and the small daily ranges. Compare with Fig. 12.4.

at Durban, across the continent on the east coast, comparable figures are 76 and 64°.

From the temperature data given above, it may be noted that the annual ranges of temperature are unusually small. The temperature difference between the extreme months is only 8.5° at Callao and 6° at Port Nolloth. These ranges should be compared with 34° at Aswan, Egypt, and 40° at Jacobabad, Pakistan. Diurnal ranges are also very small in these cool marine deserts of the low latitudes (Fig. 12.7). *BWn* regions are thus characterized by (1) lower-than-average annual temperatures, (2) distinctly cooler summers, and (3) smaller annual and diurnal ranges than are normal for most tropical dry climates.

Precipitation and fog

Annual rainfall along these cool coasts is extremely low (2.3 in. at Port Nolloth, 1.2 in. at Callao), and the drought conditions may extend to within a few degrees of the equator. Yet even so, these are regions of high relative humidity and abundant fog. Their intensified aridity is chiefly attributable to the very

stable air and low-level inversions characteristic of the eastern side of the subtropical anticyclone, which is strongly developed along these tropical and subtropical west coasts. Perhaps the cool coastal waters have a further stabilizing effect upon the *mT* air and hence may be an auxiliary cause for the decreased precipitation. In Peru, this cool, stable air mass must be forced upward along the flanks of the Andes to an altitude of about 5,000 ft before reaching the zone where there are dependable annual rains.

The fog and low stratus so frequent on these cool-water coasts are related both to advection from the cool water and to the low-level temperature inversion. Turbulence near the surface usually is sufficient, especially during the day, to prevent the fog from reaching to ground level, so that much of the time it hangs in the form of a low, uniform stratus deck. It is kept from rising and dissipating by the temperature inversion. In winter the fog and stratus lower and are more continuous, often reaching down to ground level at night. Occasionally the stratus produces mist and even a fine drizzle. At Swankopmund, South-West Africa, fog is recorded on 150 days during the year. In Peru the heavy fog, or "wet mist," is sufficient to create a meager showing of vegetation on the coastal hills. Darwin, in his book *The Voyage of the Beagle*, describes these Peruvian mists as follows:

A dull heavy bank of clouds constantly hung over the land, so that during the first sixteen days I had only one view of the Cordillera behind Lima. It is almost become a proverb that rain never falls in the lower part of Peru. Yet this can hardly be considered correct, for during almost every day of our visit there was a thick drizzling mist which was sufficient to make the streets muddy and one's clothes damp. This the people are pleased to call "Peruvian dew."

Climatic data for a representative desert station on a cool-water coast (*BWn*)

	J	F	M	A	M	J	J	A	S	O	N	D	Yr	Range
Lima, Peru														
Temp., °F	71	73	73	70	66	62	61	61	61	62	66	70	66	12.8
Precip., in.	0.0	0.0	0.0	0.0	0.0	0.2	0.3	0.5	0.5	0.1	0.0	0.0	1.8	

MIDDLE-LATITUDE DRY CLIMATES (*BSk, BWk*)[9]
(Temperate-Boreal)

Type location. Unlike the dry climates of low latitudes, middle-latitude steppes and deserts are not primarily a consequence of location within drought-making pressure and wind systems. Instead they are typically found in the deep interiors of the great continents, far from the oceans, which are the principal sources of the continental atmosphere's water vapor (Fig. 7.2 and the climate map). In both Eurasia and North America the aridity of the deep continental interiors is further intensified by highlands, which surround extensive regions in the interior and act to block the entrance of humid maritime air masses and rain-producing storms. Where high mountains closely parallel a coast, as they do in the western United States, arid climates may approach relatively close to the sea. Asia, the greatest land mass in middle latitudes, has the largest area of dry climates; North America is next in order.

Although tropical deserts characteristically extend down to the ocean margins on the leeward (western) sides of continents, the leeward (eastern) sides of land masses in the middle-latitude westerlies are far from dry. Eastern North America and Asia, for example, have mostly humid climates. This shifting of middle-latitude dry climates inland from the leeward coasts is associated with the presence of monsoons and cyclonic storms along the eastern sides of land masses in the westerlies. Owing to an unusual combination of circumstances, dry climates do reach the east coast in Patagonia (Argentina), but this is the exception. Outside of South America, no other Southern Hemisphere continent occupies sufficiently high latitudes to permit the development of extensive middle-latitude steppes and deserts. As in the tropics, steppe climates surround deserts, forming transition zones between them and humid climates.

Temperature

Although dry climates of middle latitudes duplicate the arid and semiarid climates of the tropics in their meager and undependable rainfall, they differ from them in having a season of severe cold. On the other hand, they resemble continental *humid* climate (*Dc*) of middle latitudes in their temperature and weather characteristics; they simply receive less rainfall.

The interior locations of most middle-latitude dry climates assure them of having relatively severe seasonal temperatures and thus large annual ranges (Fig. 12.8). Because their latitudinal spread is so wide (15 or 20° in both North America and Asia), it is difficult to speak of *typical* temperature conditions. For example, southernmost Colorado and

Figure 12.8

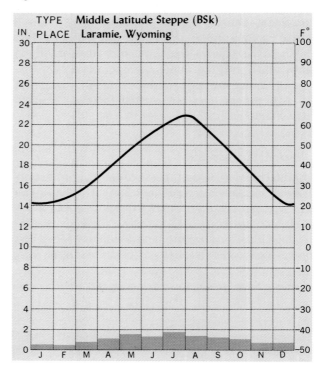

---

[9]The letter *k* indicates a cold (in winter) dry climate—less than 8 months with an average temperature of 50° (10°C) or above.

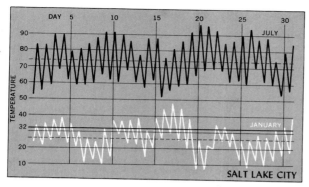

Figure 12.9 Daily maximum and minimum temperatures for a middle-latitude steppe station (*BS*). Note the strong nonperiodic air-mass control, especially in winter.

southern Alberta are both included within these climates; yet one is crossed by parallel 37°, while the other lies mostly poleward of 50°. Although naturally their temperature features are very different, each station has relatively severe temperatures for its latitude. Summers are likely to be warm or even hot, while winters are correspondingly cold. Tashkent in the U.S.S.R. at 41° N ranges from 32° in January to 81° in July, while for Ulan Bator, Mongolia, at 48°N, the comparable figures are −16 and 63°.

Because the ground is usually dry, the arrival of spring is relatively sudden, and the warm season advances rapidly. The quick rise in spring temperatures is much greater than in more humid continental climates, where a large part of the sun's energy is expended in melting the snow and evaporating the water rather than in heating the ground and the air. Diurnal ranges are apt to be large, for the same reasons given in the discussion of tropical steppes and deserts (Fig. 12.9).

The narrow Patagonian desert has seasonal temperatures quite unlike those of dry interior Eurasia and North America. Understandably, marine rather than continental conditions prevail, so that winter cold and summer heat are both largely absent and the annual range is small. No month's average temperature is as low as 32°, and none is as high as 60°.

Precipitation

Annual amount. Two factors chiefly account for the meager precipitation of middle-latitude dry climates: (1) They are either in the deep interiors of large continents or separated from the ocean by mountain barriers. (2) A well-developed seasonal anticyclone is characteristic of all but their subtropi-

Figure 12.10 Wide fluctuations in the location of the annual boundary separating dry from humid climates over a period of 5 successive years in the semiarid-subhumid region east of the Rocky Mountains. (*After Kendall.*)

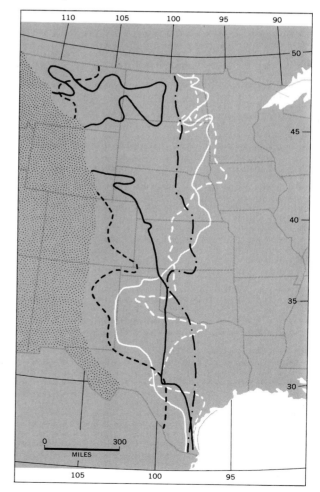

cal margins in winter. Moreover, the regions of most severe aridity are located in basins surrounded by highlands. Unlike the dry climates of the tropics, those of middle latitudes receive a portion of their precipitation in the form of snow, although the amount is characteristically small, and the winter snow cover is not deep.

Because of their generally lower temperatures, and therefore reduced evaporation potential, the humid boundaries of middle-latitude dry climates ordinarily have somewhat less rainfall than many tropical dry climates. This does not mean they are necessarily more arid. In fact the opposite is more often the case, for as a rule middle-latitude deserts are not so nearly rainless as some tropical ones.

Like rainfall in dry climates generally, that in middle-latitude *B* types varies greatly from year to year (Fig. 12.10). And in the middle as in the lower latitudes, this undependability has graver consequences in the steppe than in the desert. Parts, at least, of the steppe have enough rainfall to attract agricultural settlement. Moreover, a succession of more humid years may tempt settlers to push the agricultural frontier nearer to the desert. But drought years are sure to follow, bringing crop failures and economic disaster. In 30 to 40 percent of the years over a considerable area of the semiarid country in the United States, rainfall is less than 85 percent of the average. Over a period of 38 years at Minot, North Dakota, average annual rainfall was 15.7 in., but

the most humid year had 24.3 in. and the driest only 7.2.

Seasonal distribution.   It is not easy to generalize concerning seasonal distribution of precipitation in the dry climates of middle latitudes. In the more interior and continental locations poleward of about 40°, summer is usually the period of maximum precipitation (see the data for Ulan Bator and Williston, and also Fig. 12.9). This is related to the higher surface temperatures and associated convectional overturning, greater specific humidity, and tendency toward an inblowing system of monsoonal winds in summer. In winter the low temperatures and the predominantly anticyclonic circulation are opposed to abundant condensation. Most of interior Asia and the Great Plains region of North America are dry lands with a distinct summer maximum in their precipitation curves. Ulan Bator, Mongolia, for example, receives 84 percent of its average annual 7.6 in. of rainfall in the 3 summer months. Most middle-latitude dry climates do have a summer maximum of precipitation.

But those dry climates which lie close to subtropical latitudes, and especially to *Cs* climate, are likely to have somewhat more rain in winter than in summer (see data for Fallon, Nevada). This is caused by the same controls which produce subtropical dry-summer climate. Southern Turkestan in Soviet Russia and the southwestern part of the

Climatic data for representative stations in middle-latitude deserts (*BWk*)

	J	F	M	A	M	J	J	A	S	O	N	D	Yr	Range
Santa Cruz, Argentina														
Temp., °F	59	58	55	48	41	35	35	38	44	49	53	56	47.5	24
Precip., in.	0.6	0.4	0.3	0.6	0.6	0.5	0.7	0.4	0.2	0.4	0.5	0.9	6.1	
Kuche, Sinkiang, China (3,182 ft)														
Temp., °F	10	26	45	56	65	71	76	73	65	54	35	19	49	66
Precip., in.	0.1	0.1	0.2	0.1	0.1	1.3	0.7	0.3	0.2	0.0	0.1	0.3	3.4	
Fallon, Nevada (3,965 ft)														
Temp., °F	31	36	41	50	56	65	74	72	61	51	40	32	50.6	42.7
Precip., in.	0.6	0.5	0.5	0.4	0.6	0.3	0.1	0.2	0.3	0.4	0.3	0.6	4.7	

Climatic data for representative stations in middle-latitude steppes (*BSk*)

	J	F	M	A	M	J	J	A	S	O	N	D	Yr	Range
					Williston, North Dakota (marginal)									
Temp., °F	8	12	24	42	55	63	71	68	57	46	28	16	40.9	63
Precip., in.	0.6	0.5	0.7	0.9	1.4	3.3	1.9	1.5	1.1	0.7	0.6	0.5	13.7	
					Ulan Bator, Mongolia (3,800 ft)									
Temp., °F	−16	−4	13	34	48	58	63	59	48	30	8	−17	28	79
Precip., in.	0.0	0.1	0.0	0.0	0.3	1.7	2.6	2.1	0.5	0.1	0.1	0.1	7.6	
					Saratov, U.S.S.R.									
Temp., °F	11	13	21	43	60	69	74	69	56	42	28	17	43	63
Precip., in.	1.1	1.0	0.8	1.0	1.3	1.8	1.2	1.3	1.1	1.4	1.4	1.2	14.5	

Great Basin in the United States both have a winter-maximum regime. Such regions are simply poleward extensions of the tropical *BShs* steppes with their low-sun rainfall maximum.

Seasonal weather

As a rule, the nonperiodic weather element associated with passing cyclones and anticyclones is not so well developed in the continental dry climates as in the continental humid types. In *winter* the dominant influence in dry climates of the higher middle latitudes is the continental anticyclone, whose settling air and diverging winds produce

much clear sky and cold weather. Even so, the weather element is stronger in this season than in summer. On the whole, winds are not vigorous in winter, although violent winds sometimes do occur, chiefly with the passing of an occasional cyclone or with an anticyclonic surge from higher latitudes. The anticyclonic surge is known on the North American plains as a *blizzard,* already described in Chapter 10.

Summer weather is largely sun-controlled, and consequently the diurnal element is conspicuous in temperature, wind, cloud, and precipitation. Winds are often unpleasantly strong during the daylight hours, when convectional overturning is at a maximum.

REGIONAL FEATURES OF DRY CLIMATES

Probably in no other group of climates are there so many instances of local and regional departures from the world pattern.[10] A number of these consist of rainfall deficiencies in locations which, in terms of broad-scale circulation patterns, might be expected to be humid. Local and regional controls appear to be particularly effective in producing arid-

[10]For a more complete analysis of the dry anomalies than this section provides, see Glenn T. Trewartha, *The Earth's Problem Climates,* The University of Wisconsin Press, Madison, Wis., 1961.

ity. Many of the regional arid anomalies are not at present well understood. Significantly, a majority of them are located within tropical latitudes.

Africa.   Some of the earth's most striking drought anomalies are in Africa. One is the unusual extensiveness of the Sahara, both in a north-south and an east-west direction. Obviously the Sahara's size is made possible by the great breadth of the continent north of the equator. Still, this is not the only reason, for broad South America south of the equator in Saharan latitudes has only a minimal amount of dry climate. In northern Africa, with no terrain barrier such as the Andes to break the continuity of the subtropical anticyclone, a divergent, subsident circulation prevails across the entire land mass and on into Arabia, southwestern Asia, and northwestern India-Pakistan. Dry Saharan air pushes far south, and as a result humid climates are able to extend northward only to about 13 to 15°, even though the surface ITC reaches a mean July-August position of 20–21°N. Here, however, the vertical structure of the air is such that the surface ITC is situated well within the Sahara. Hot, dry, northerly Sahara air overruns a thin wedge of humid maritime southwesterly air, but the Saharan air is so dry that in spite of slow lifting it yields no rain. Only as the wedge of maritime air deepens toward the south are convective showers possible, and this condition is not reached for several hundred miles south of the surface ITC.

Probably the most impressive single climatic anomaly in all Africa is the widespread rainfall deficiency in tropical and equatorial East Africa (see the rainfall map inside the back cover). It seems as if this region's location on the eastern or windward side, together with its elevation, should cause it to be wet. Since North Africa has no oceanic east coast north of about latitude 13°N, it may not be so surprising that desert climate there reaches down to the Red Sea margins. The real puzzle is why Somali, fronting on the Indian Ocean (also southern and eastern Arabia, fronting on the Arabian Sea), should be so dry, for east-side deserts in the tropics are rare. Moreover, along this east side, dry climates extend right across the equator and for nearly 10° into the Southern Hemisphere. And much of what is designated on the climatic map inside the front cover as tropical wet-and-dry climate is so distinctly subhumid that it is actually close to being semiarid.

The explanations for such an extensive development of dry and subhumid climates in tropical East Africa are complicated. Contributing factors are:

1. Both the northerly (November-February) and southerly (June-September) circulations crossing the area are divergent and subsident over extensive areas. Evidence of this is the frequency of above-surface inversions of temperature.

2. Both of these so-called monsoons have a strongly meridional flow, so that they do not advect oceanic moisture deeply into the interior. Not infrequently they closely parallel the coastline, or in places may actually have an offshore component.

3. North of the equator, Somali (also Arabia) is dominated year-round by stable subsident air of anticyclonic origin. There is no Indian Ocean-Arabian Sea anticyclone whose western margins might provide unstable or neutral mT air.

4. The southwest winds paralleling the Somali coast in summer result in upwelled cool waters, which may act to intensify the coastal aridity and produce coastal fog. This is the only important east-side Bn climate.

A striking feature of the Kalahari dry region in southwestern Africa is its Namib coastal section. Within the Bn class of deserts, the Namib is distinctive in that next to Peru-Chile, it is latitudinally the most extensive and also the most arid (Fig. 12.11). An annual rainfall of under 4 to 5 in. is maintained along a narrow coastal strip for about 15° of latitude. Three stations along this section of coast record less than 1 in. Lowest precipitation amounts appear to coincide with the coastline, and there is a slow increase inland. Fog, low stratus, and mist are prevalent.

The controls operating to produce this climatic combination are associated with a locationally stable anticyclone with a maximum subsidence along the coast at about 25°S. There is a low and persistent inversion. Since the cell's eastern end terminates abruptly at the coastline, smoothly curved southerly atmospheric and oceanic circulations develop along the coast, with an accompanying upwelling of cold water. As long as the coast bends westward, so that it maintains contact with the atmospheric and cool oceanic circulations around the eastern end of the anticyclone, drought prevails. Where it bends eastward, so that contact with the air and water circulations is weakened, rainfall increases. Still, the effects of these circulations in terms of greatly re-

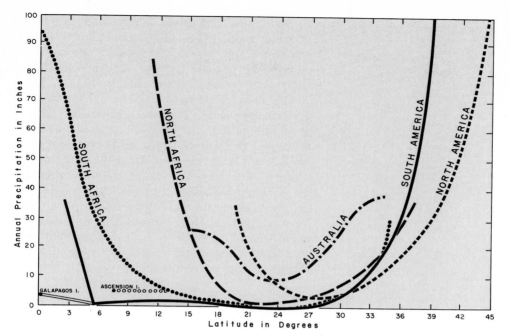

Figure 12.11   The annual precipitation curves plotted against latitude for five coastal deserts in tropical-subtropical latitudes. Note that the desert of Chile-Peru exhibits the highest intensity of aridity maintained throughout the greatest span of latitude. (*After Paul Lydolph.*)

duced rainfall are carried to within 5° or less of the equator along the coast.

The small enclave of semiarid, or subhumid *Aw*, climate in Ghana-Togo along the Guinea coast was referred to earlier in Chapter 8's discussion of tropical wet- and dry-climates.

Australia.   Here the chief anomaly within the extensive dry climates is the absence of any *Bn* involving a cool and intensely dry western littoral with a high frequency of fog and low stratus. Lowest rainfall along the west coast appears to be about 9 in., although it may decline to 6 in. in the deep interior. The reduced coastal aridity compared with similar locations on other continents probably is related to the absence of a strong, fixed anticyclonic cell whose smoothly curved easternmost circulation coincides with the Australian coastline. Consequently there is no cool current with upwelling of cold water. Instead of a positionally stable anti-

cyclone, western Australia is influenced by a succession of migratory high-pressure cells moving eastward, with occasional showers occurring in the troughs between individual cells.

South America.   Nowhere else on the earth is there a coastal desert of such intense aridity which spans so many degrees of latitude (25°±) as the Chilean-Peruvian desert (see Fig. 12.11 and the rainfall map inside the back cover). It is the *Bn* climate par excellence. From northern Peru, at about 5°S, to 30°S in northern Chile, the highest coastal rainfall is probably under 50 mm (2 in.), and a number of stations in northern Chile have recorded no rain over periods of a decade or more.[11] Even northward from Peru, dry climate in less intense form is con-

[11] Wolfgang Weischet, "Zur Klimatologie der Nordchilenischen Wüste," *Meteorol. Rundschau*, Vol. 19, pp. 1–7, 1966.

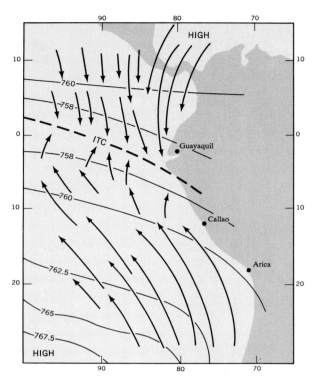

Figure 12.12 Pressure and wind patterns during one of the infrequent rainy episodes in the Peruvian desert. Note that the ITC, followed by northerly winds, had advanced to a position south of the equator along the coast of Peru. (*After Schott.*)

tinued along the Ecuador littoral to within a degree or less of the equator.

The combination of controls producing this phenomenal Peruvian-Chilean desert resembles that of southwestern Africa, except that in South America the drought makers, including anticyclonic circulation and cold current, maintain their effectiveness farther equatorward. This in turn is related to the concave coastline which continues to thrust westward to within a few degrees of the equator, so that the smoothly curved anticyclonic air and water circulations are held close to the coast. The high Andes also contribute by abruptly terminating the anticyclonic cell almost at the coastline. On rare occasions in northern Peru, when anticyclonic

circulation and cool current are temporarily displaced by an invasion of ITC and its accompanying northerly winds and warm ocean current, the northern part of the coastal desert is suddenly converted from aridity to tropical wet climate (Fig. 12.12).

The desert-steppe of Patagonia has the distinction of being the only extensive east-side dry climate in middle latitudes (refer to the climate and rainfall maps). What is more, the driest or desert parts are on the seaward side, where annual rainfall drops to 4 or 5 in. In view of the narrowness of the continent in these latitudes, with consequent marine control, the prevailing aridity is all the more remarkable. Its origin is related to the effects of the Andes upon the broad-scale westerly wind flow. The Andean terrain barrier sets up a system of long, meridional, upper-troposphere waves in the wester-

Figure 12.13 Annual march of precipitation at a subtropical desert station. Note the primary maximum in summer and a secondary maximum in winter.

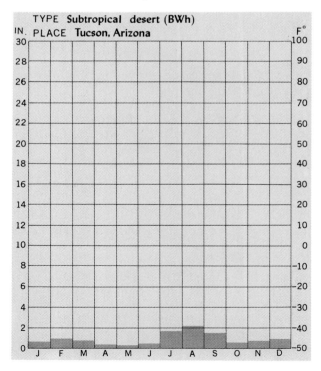

lies. These include a meridional anticyclonic ridge positioned over the narrow continent and a trough situated farther eastward over the ocean. Subsidence under the anticyclonic ridge acts to stifle rainmaking processes and damps the activity of cyclonic storms. If the continent were wider, so that its eastern parts were located underneath the trough, rainfall there would very likely be heavier.

Two other anomalous dry regions, one in easternmost Brazil and the other in coastal Colombia and Venezuela, already have been mentioned in the discussion of tropical wet-and-dry climates.

North America. The unexpected semiarid area in northwestern Yucatán (Mexico) may result from the fact that in summer, the rainy season, winds parallel the coast or blow slightly offshore. An offshore wind would result in coastline subsidence because of increased wind speed over the sea, where frictional drag is reduced. Paralleling winds would likewise result in divergence and subsidence, and also in the upwelling of cool water.

Along the west Gulf Coast both in northeastern Mexico and in adjacent parts of Texas, there is a semiarid region which appears out of place. Other parts of the Gulf Coast in both Mexico and the United States are wet. No adequate explanation for this dry region exists.

A distinctive feature of the dry intermontane region in the United States is its biannual rainfall maximum (Fig. 12.13). One maximum consistently occurs in winter. The second is more variable in its time of occurrence; in some parts it falls in spring or early summer, while in others it is a feature of mid- or late-summer. This dual maximum is in the nature of a transition type situated between the strong winter maximum of the Pacific Coast and the marked summer maximum of the Great Plains.

SELECTED REFERENCES FOR FURTHER STUDY OF TOPICS IN CHAPTER TWELVE

Bailey, Harry P., "A Simple Moisture Index Based upon a Primary Law of Evaporation," *Geog. Ann.*, Vol. 40, pp. 196–215, 1958.

Blumenstock, D. I., "The Humid-Arid Boundary Problem," *Geog. Rev.*, Vol. 29, pp. 681–682, 1939.

Bryson, Reid A., and Peter M. Kuhn, "Stress-differential Induced Divergence with Application to Littoral Precipitation," *Erdkunde*, Vol. 15, pp. 287–294, 1961.

————, and William P. Lowry, "Synoptic Climatology of the Arizona Summer Precipitation Singularity," *Bull. Amer. Meteorol. Soc.*, Vol. 36, pp. 329–339, 1955.

————, and David Baerreis, "Possibilities of Major Climatic Modification and Their Implications: Northwest India, a Case Study," *Bull. Amer. Meteorol. Soc.*, Vol. 48, pp. 136–142, 1967.

Lydolph, Paul E., "A Comparative Analysis of the Dry Western Littorals," *Ann. Assoc. Amer. Geographers*, Vol. 47, pp. 213–230, 1957.

————, "The Russian Sukhovey," *Ann. Assoc. Amer. Geographers*, Vol. 54, pp. 291–309, 1964.

Meigs, P., *World Distribution of Arid and Semiarid Homoclimates*. UNESCO Reviews of Research on Arid Zone Hydrology, Paris, 1953, pp. 203–209.

Patton, Clyde P., "A Note on the Classification of Dry Climates in the Köppen System," *Calif. Geographer*, Vol. 3, pp. 105–112, 1962.

Sawyer, J. S., "The Structure of the Intertropical Front over N.W. India during the S.W. Monsoon." *Quart. J. Roy. Meteorol. Soc.*, Vol. 73, pp. 346–369, 1947.

Sewell, W. R. D., Human Dimension of Weather Modification, Research Paper No. 105, Department of Geography, University of Chicago, Chicago, 1966.

Thornthwaite, C. W., and J. R. Mather, *The Water Balance*, Publ. Climatology, Drexel Institute of Technology, Philadelphia, Vol. 8, no. 1, 1955.

Trewartha, Glenn T., *The Earth's Problem Climates*, The University of Wisconsin Press, Madison, Wis., 1961, pp. 22–33, 35–39, 50–54, 59–64, 70–71, 80–82, 93–96, 106–110, 123–134, 144–148, 166–169, 260–270, 274–275.

White, Gilbert F. (ed.), *The Future of Arid Lands*, Publ. No. 43, American Association for the Advancement of Science, Washington, D.C., 1956.

KÖPPEN'S CLASSIFICATION
OF CLIMATES

Köppen recognizes five principal groups of world climate which are intended to correspond with five principal vegetation groups. The five climatic groups, each designated by a capital letter, are as follows: *A*, tropical rainy climates with no cool season; *B*, dry climates; *C*, middle-latitude rainy climates with mild winters; *D*, middle-latitude rainy climates with severe winters; and *E*, polar climates with no warm season. Each of these in turn is subdivided into climatic types based upon the seasonal distribution of rainfall or the degree of dryness or cold. The small letters *f*, *s*, and *w* indicate the seasonableness of precipitation; no dry season (*f*); dry season in summer (*s*); dry season in winter (*w*). The capital letters *S* and *W* are employed to designate the two subdivisions of dry climate; semiarid, or steppe (*S*), and arid, or desert (*W*). Capital letters *T* and *F* are similarly employed to designate the two subdivisions of polar climate; tundra (*T*), and icecap (*F*). The table on page 396 shows the Köppen scheme of five main climatic groups and eleven climatic types. The parentheses indicate that the combinations *As* and *Ds* rarely occur, and for this reason they are not recognized as among the principal climatic types.

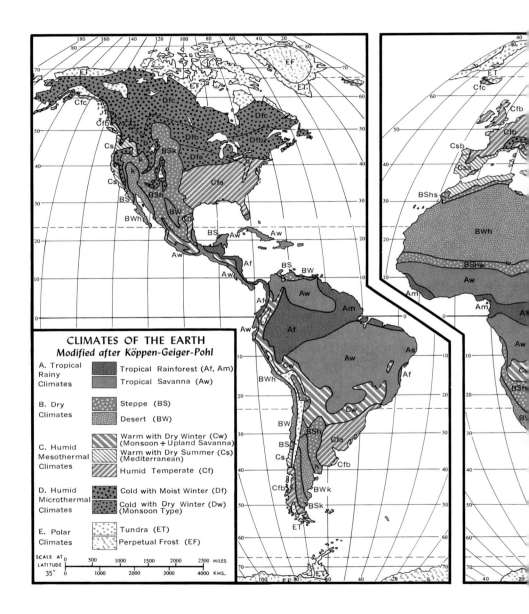

CLIMATES OF THE EARTH
Modified after Köppen-Geiger-Pohl

A. Tropical Rainy Climates
- Tropical Rainforest (Af, Am)
- Tropical Savanna (Aw)

B. Dry Climates
- Steppe (BS)
- Desert (BW)

C. Humid Mesothermal Climates
- Warm with Dry Winter (Cw) (Monsoon + Upland Savanna)
- Warm with Dry Summer (Cs) (Mediterranean)
- Humid Temperate (Cf)

D. Humid Microthermal Climates
- Cold with Moist Winter (Df)
- Cold with Dry Winter (Dw) (Monsoon Type)

E. Polar Climates
- Tundra (ET)
- Perpetual Frost (EF)

SCALE AT LATITUDE 35°
0   500   1000   1500   2000   2500 MILES
0   1000   2000   3000   4000 KMS.

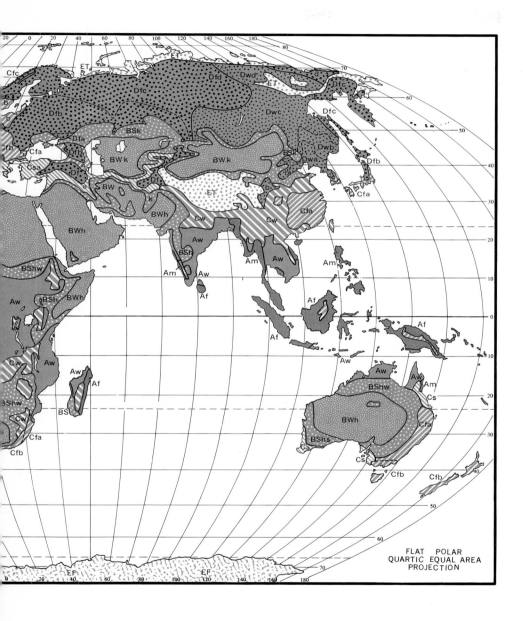

ET
Cfc
Dfd   Dwd'   Dfc
Dfc   ET   Dfc
Dfb   Dfc
Dfc   Dwc   Dfc
Dfa   BSk   Dwb
BWk   BWk   BSK   Dwa   Dfb
Cfa   Dfb
Csa   SBW   Df   Cfa
a   S BW   ET   b
k   Cw   Cw   Cfa
BWh   BWh   Aw
BWh   Am   BSh   Aw   Am
Aw   BShw   Am   Af
BShw   BWh   Af   Af
Aw   BSh   BWh   Af   Af
Cw   Aw   Af
Aw   Aw   Af
Aw   Aw   Am
BShw   BSt   Aw   Aw   Am
Cw   BShw   Cs
Cfa   BWh   Cfa
Cfb   BShs   Cs
Cfb   Cfb   Cfb

FLAT   POLAR
QUARTIC EQUAL AREA
PROJECTION

EF   EF

Köppen's climatic groups and types

Climatic group	Symbol	Dry period	Degrees of dryness or cold	
Tropical rainy climates	A	f (s) w		
Dry climates	B		S	W
Mild temperate rainy climates	C	f s w		
Cold snow-forest climates	D	f (s) w		
Polar climates	E		T	F

A CLIMATES

$A =$ Tropical rainy climates; temperature of the coolest month above 64.4° (18°C). With monthly temperatures lower than 64.4° certain sensitive tropical plants do not thrive. This is the realm of plants known as megatherms which need continuously high temperatures and relatively abundant precipitation. Within the A group of climates two main types are recognized: one has adequate precipitation throughout the year, while the other contains a distinctly dry season which affects vegetation adversely.

$Af =$ Tropical wet climate; f: rainfall of the driest month is at least 2.4 in. (6 cm). Within this climate there is a minimum of seasonal variation in temperature and precipitation, both remaining high throughout the year.

$Aw =$ Tropical wet-and-dry climate; w: distinct dry season in low-sun period or winter. A marked seasonal rhythm of rainfall characterizes Aw climates; at least 1 month must have less than 2.4 in. (6 cm). Temperature is similar to that in Af.

Other small letters used with A climates are as follows:

m (monsoon) $=$ short dry season, but with total rainfall so great that ground remains sufficiently wet throughout the year to support rainforest. Am is intermediate between Af and Aw, resembling Af in amount of precipitation and Aw in seasonal distribution. In both Aw and Am the rainfall of the driest month is below 2.4 in. (6 cm). Whether it is Aw or Am depends upon the total amount of rainfall and the amount occurring in the driest month. For example, when the total annual rainfall is 50 in. the boundary between Aw and Am is 1.94 in. for the driest month; less than this is Aw; more than this, but less than 2.4 in., is Am. At 60 in. annual total the Am/Aw boundary is 1.55: at 70 in., 1.15; at 90 in., 0.34. The boundary between Am and Aw is expressed by the formula $a = 3.94 - r/25$, where r is the annual rainfall in inches and a the rainfall during the driest month. If with a given annual rainfall the rainfall of the driest month is greater than the value of a as obtained by the preceding formula, the climate is Am; if smaller than a, it is Aw.

$w' =$ rainfall maximum in autumn.

$w'' =$ two distinct rainfall maxima separated by two dry seasons.

$s =$ dry season during high-sun period (rare).

$i =$ range of temperature between warmest and coldest months less than 9° (5°C).

$g =$ Ganges types of annual march of temperature; hottest month comes before the solstice and the summer rainy season.

B CLIMATES

$B =$ Dry climates, in which there is an excess of evaporation over precipitation. No surplus of water remains, therefore, to maintain a constant groundwater level so that permanent streams cannot *originate* within B climates. The amount of precipitation that falls is not sufficient data with which to determine the B climate boundaries, since the effectiveness of precipitation in providing moisture in the ground for plants is dependent upon the rate of evaporation, which in turn varies directly with temperature. Rain that falls during a hot summer obviously is less effective than the same amount falling in a cool winter. Köppen's formulas for identifying arid and semiarid climates, therefore, involve not only the annual temperature and the total annual rainfall, but likewise the season of maximum precipitation. There are two main subdivisions of B climates, the arid, or desert, type BW (W from the German word, *Wüste,* meaning "desert"), and the semiarid, or steppe, type BS (S from the word *steppe,* meaning dry grassland).

$BW =$ Arid climate or desert.

$BS =$ Semiarid climate or steppe; for precise definitions of steppe and desert boundaries see formulas at end of this section on B climates.

Other small letters used with B climates are as follows:

h (*heiss*) $=$ average annual temperature over 64.4° (18°C). BWh and BSh therefore are low-latitude, or tropical, deserts and steppes.

k (*kalt*) $=$ average annual temperature under 64.4° (18°C). BWk and BSk therefore are middle-latitude, or cold, deserts and steppes.

$k' =$ temperature of the warmest month under 64.4° (18°C).

$s =$ summer drought; at least three times as much rain in the wettest winter month as in the driest summer month (see footnote under Cs climate).

$w =$ winter drought; at least ten times as much rain in the wettest summer month as in the driest winter month (see footnote under Cw climate).

n (*Nebel*) $=$ frequent fog. BWn and BSn climates are usually found along littorals paralleled by cool ocean currents.

Formulas for identifying steppe (BS) **and desert** (BW) **margins** ($r =$ annual rainfall, in., $t =$ average annual temperature, °F)

	Boundary between BS and humid climates	Boundary between BW and BS
Rainfall evenly distributed	$r = 0.44t - 8.5$	$r = \dfrac{0.44t - 8.5}{2}$
Rainfall max. in summer; at least ten times as much rain in the wettest summer month as in driest winter month	$r = 0.44t - 3$	$r = \dfrac{0.44t - 3}{2}$
Rainfall max. in winter; at least three times as much rain in the wettest winter month as in driest summer month	$r = 0.44t - 14$	$r = \dfrac{0.44t - 14}{2}$

C CLIMATES (MESOTHERMAL)

$C =$ Mild temperate rainy climates; average temperature of coldest month below 64.4° (18°C) but above 26.6° (−3°C); average temperature of warmest month over 50° (10°C). The average temperature of 26.6° (−3°C) for the coldest month supposedly roughly coincides with the equatorward limit of frozen ground and a snow cover lasting for a month or more. Within the C group of climates three contrasting rainfall regimes are the basis for recognition of three principal climatic types: the f type with no dry season; the w type with a dry winter; and the s type with a dry summer.

$Cf =$ no distinct dry season; difference between the rainiest and driest months is less than for w and s, and the driest month of summer receives more than 1.2 in. (3 cm).

$Cw =$ Winter dry; at least ten times as much rain in the wettest month of summer as in the driest month of winter.[1] This type of climate has two characteristic locations: (1) elevated sites in the low latitudes where altitude reduces the temperature of the Aw climates which prevail in the adjacent lowlands, and (2) mild middle-latitude monsoon lands of southeastern Asia, particularly northern India and southern China.

$Cs =$ Summer dry; at least three times as much rain in the wettest month of winter as in the driest month of summer,[2] and the driest month of summer receives less than 1.2 in. (3 cm).

Other small letters used with C climates are as follows:

$a =$ hot summer, average temperature of warmest month over 71.6° (22°C).

$b =$ cool summer; average temperature of warmest month under 71.6° (22°C).

$c =$ cool short summer; less than 4 months over 50° (10°C).

$i =$ same as in A climates.

$g =$ same as in A climates.

$x =$ rainfall maximum in late spring or early summer; drier in late summer.

$n =$ same as in B climates.

D CLIMATES (MICROTHERMAL)

$D =$ Cold snow-forest climates; average temperature of coldest month below 26.6° (−3°C), average temperature of warmest month above 50° (10°C). The average temperature of 50° (10°C) for the warmest month approximately coincides with the poleward limits of forest. D climates are characterized by frozen ground and a snow cover of several months' duration. Two principal subdivisions of the D group are recognized: the one, Df, with no dry season, and the other, Dw, with dry season in winter.

$Df =$ Cold climate with humid winters.

$Dw =$ Cold climate with dry winters; characteristic of northeastern Asia, where the winter anticyclone is well developed.

Other small letters used with D climates are as follows:

$d =$ average temperature of coldest month below −36.4° (−38°C).

$f, s, w, a, b,$ and c are the same as in C climates.

---

[1]Alternative definition: 70 percent or more of the average annual rainfall is received in the warmer 6 months.

[2]Alternative definition: 70 percent or more of the average annual rainfall is received in the winter 6 months.

E CLIMATES

E = Polar climates; average temperature of the warmest month below 50° (10°C). In the higher latitudes, once the temperatures are well below freezing and the ground frozen it makes little difference to plant life how cold it gets. Rather, it is the intensity and duration of a season of warmth which is critical. For this reason a warm-month isotherm is employed as the poleward boundary of E climates. Two climatic subdivisions are recognized: one, ET, in which there is a brief growing season and a meager vegetation cover, and the other, EF, in which there is perpetual frost and no vegetation.

ET = Tundra climate; average temperature of warmest month below 50° (10°C) but above 32° (0°C).

EF = Perpetual frost; average temperature of all months below 32° (0°C). Such climates persist only over the permanent icecaps.

INDEX

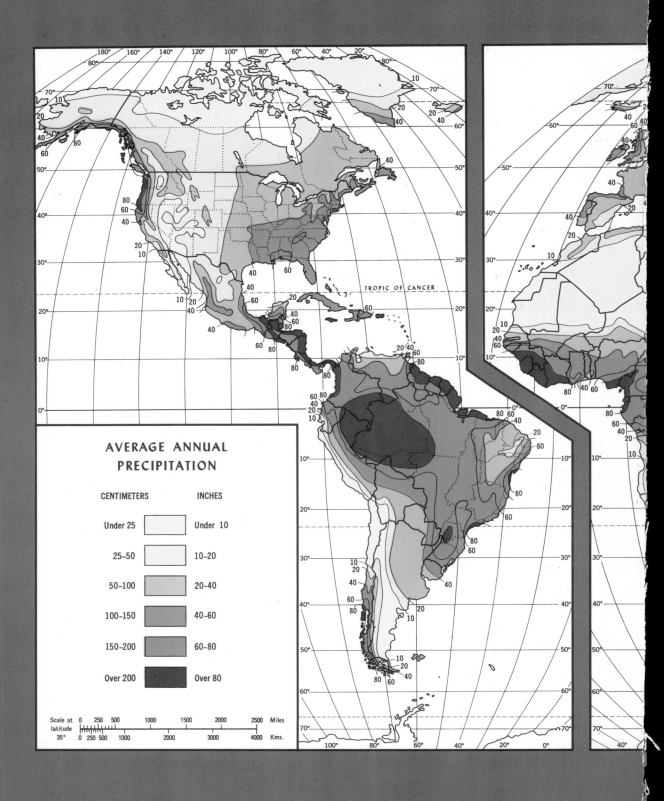

1350
11-?75

AVERAGE ANNUAL PRECIPITATION

CENTIMETERS		INCHES
Under 25		Under 10
25–50		10–20
50–100		20–40
100–150		40–60
150–200		60–80
Over 200		Over 80

Scale at latitude 35°

Miles: 0 250 500 1000 1500 2000 2500
Kms.: 0 250 500 1000 2000 3000 4000

TROPIC OF CANCER